# CHEMICAL BIOLOGY

# CHEMICAL BIOLOGY

## Approaches to Drug Discovery and Development to Targeting Disease

**Edited by**

**NATANYA CIVJAN**

A JOHN WILEY & SONS, INC., PUBLICATION

Published by John Wiley & Sons, Inc., Hoboken, New Jersey
Published simultaneously in Canada

For general information on our other products and services or for technical support, please contact our Customer Care Department within the United States at (800) 762-2974, outside the United States at (317) 572-3993 or fax (317) 572-4002.

Wiley also publishes its books in a variety of electronic formats. Some content that appears in print may not be available in electronic formats. For more information about Wiley products, visit our web site at *www.wiley.com*.

*Library of Congress Cataloging-in-Publication Data:*

Civjan, Natanya.
  Chemical biology : approaches to drug discovery and development to targeting disease / edited by Natanya Civjan.
     p. cm.
  Summary: "Based on the award winning Wiley Encyclopedia of Chemical Biology, this book provides a fairly general overview of the role of chemical biology in drug discovery and development and in understanding biological mechanisms of common diseases"–Provided by publisher.
  Includes bibliographical references and index.
  ISBN 978-1-118-10118-6 (hardback)
  1. Biochemistry. I. Title.
  QP514.2.C58 2012
  572–dc23

                                                                                    2012012322

Printed in the United States of America

10 9 8 7 6 5 4 3 2 1

# CONTENTS

# PREFACE

Chemical biology is a cross-disciplinary science at the interface of chemistry, biology, and medicine. Chemical biology aims at the elucidation of the chemical mechanisms of living systems and is important to drug discovery, playing a key role in the development of novel agents for the prevention, diagnosis, and treatment of diseases.

Containing carefully selected reprints from the award-winning *Wiley Encyclopedia of Chemical Biology*, which published in 2008, as well as new contributions, this book features a collection of chapters on topics of interest to scientists working in the areas of drug discovery and development. Chapters in Part I of this book cover a range of topics on the design, development, and optimization of drug candidates, the pharmacokinetics and properties of drugs, and drug transport and delivery. Part II covers the mechanisms of particular diseases and the strategies used to understand, target, and treat these diseases.

Written by prominent scholars, this book will be of interest to advanced students and researchers, both in academia and industry, in the field of chemical biology or related fields such as biochemistry, medicine, and pharmaceutical sciences, with particular interest in drug discovery and development.

NATANYA CIVJAN

# CONTRIBUTORS

**Cynthia M. Arbeeny,** Endocrine and Metabolic Diseases, Genzyme Corporation, Framingham, Massachusetts

**Glen B. Baker,** Neurochemical Research Unit, Department of Psychiatry, University of Alberta, Edmonton, Alberta, Canada

**Peter J. Barnes,** National Heart and Lung Institute, Imperial College, London, United Kingdom

**Francis Berenbaum,** Paris Universitas APHP Saint-Antoine Hospital, Paris, France

**Grant E. Boldt,** University of Oxford, Oxford, United Kingdom and The Scripps Research Institute, La Jolla, California

**Thomas M. Bridges,** Department of Pharmacology, Vanderbilt Institute of Chemical Biology, Vanderbilt University Medical Center, Nashville, Tennessee

**Doug A. Brooks,** Molecular Medicine Sector, Sansom Institute, University of South Australia, Adelaide, Australia

**Jianwei Che,** Genomics Institute of the Novartis Research Foundation, San Diego, California

**Li Di,** Wyeth Research, Princeton, New Jersey

**Yael Elbaz,** Department of Biologic Chemistry, Alexander Silberman Institute of Life Sciences, Hebrew University of Jerusalem, Jerusalem, Israel

**Maria Fuller,** Lysosomal Diseases Research Unit, Department of Genetic Medicine, Women's and Children's Hospital, Adelaide, Australia

**Marjolaine Gosset,** Paris Universitas, Paris, France

**Nathanael S. Gray,** Harvard Medical School, Dana Farber Cancer Institute, Biological Chemistry and Molecular Pharmacology, Boston

**Brian T. Hawkins,** Laboratory of Pharmacology, NIH/NIEHS, Research Triangle Park, North Carolina

**Claire Jacques,** Paris Universitas, Paris, France

**Hualiang Jiang,** Shanghai Institute of Materia Medica, Chinese Academy of Sciences, Shanghai, China

**J. Phillip Kennedy,** Department of Chemistry, Vanderbilt University, Nashville, Tennessee

**Edward H. Kerns,** Wyeth Research, Princeton, New Jersey

**Vincent H. L. Lee,** School of Pharmacy, The Chinese University of Hong Kong, Hong Kong SAR

**Honglin Li,** Shanghai Institute of Materia Medica, Chinese Academy of Sciences, Shanghai, China

**Craig W. Lindsley,** Departments of Pharmacology and Chemistry, Vanderbilt Institute of Chemical Biology, Vanderbilt University Medical Center, Nashville, Tennessee

**Yi Liu,** Genomics Institute of the Novartis Research Foundation, San Diego, California

**Xiaomin Luo,** Shanghai Institute of Materia Medica, Chinese Academy of Sciences, Shanghai, China

**Ferenc Martenyi,** Lilly Research Laboratories, Indianapolis, Indiana

**David S. Miller,** Laboratory of Pharmacology, NIH/NIEHS, Research Triangle Park, North Carolina

**Nicholas D. Mitchell,** Neurochemical Research Unit, Department of Psychiatry, University of Alberta, Edmonton, Alberta, Canada

**Andrew J. Pope,** Platform Technology & Science, GlaxoSmithKline, Collegeville, Pennsylvania, USA.

**Richard J.T. Rodenburg,** Nijmegen Center for Mitochondrial Disorders, Radboud University Nijmegen Medical Centre, Nijmegen, The Netherlands

**Heidi M. Sampson,** McGill University, Quebec, Canada

**Johanna C. Scheinost,** University of Oxford, Oxford, United Kingdom and The Scripps Research Institute, La Jolla, California

**Shimon Schuldiner,** Department of Biologic Chemistry, Alexander Silberman Institute of Life Sciences, Hebrew University of Jerusalem, Jerusalem, Israel

**Jérémie Sellam,** Paris Universitas, Paris, France

**David Selwood,** Wolfson Institute for Biomedical Research, University College London, London

**Sheo B. Singh,** Merck Research Laboratories, Rahway, New Jersey

**Jan A.M. Smeitink,** Nijmegen Center for Mitochondrial Disorders, Radboud University Nijmegen Medical Centre, Nijmegen

**David Y. Thomas,** McGill University, Quebec, Canada

**Miklos Toth,** Department of Pharmacology, Weill Medical College of Cornell University, New York, New York

**Joseph P. Vacca,** Merck Research Laboratories, West Point, Pensylvania

**Garry M. Walsh,** University of Aberdeen, School of Medicine, Aberdeen, Scotland, United Kingdom

**David Weaver,** Department of Pharmacology, Vanderbilt Institute of Chemical Biology, Vanderbilt University Medical Center, Nashville, Tennessee

**Paul Wentworth, Jr.,** University of Oxford, Oxford, United Kingdom and The Scripps Research Institute, La Jolla, California

**Ronald E. White,** Schering-Plough Research Institute, Kenilworth, New Jersey

**Mingyue Zheng,** Shanghai Institute of Materia Medica, Chinese Academy of Sciences, Shanghai, China

**Weiliang Zhu,** Shanghai Institute of Materia Medica, Chinese Academy of Sciences, Shanghai, China

**Zhong Zuo,** School of Pharmacy, The Chinese University of Hong Kong, Shatin, N.T., Hong Kong SAR

# PART I

# DRUG DISCOVERY
# AND DEVELOPMENT

# 1

# THE ROLE OF CHEMICAL BIOLOGY IN DRUG DISCOVERY

ANDREW J. POPE

*Platform Technology & Science, GlaxoSmithKline, Collegeville, Pennsylvania*

Chemical biology methods are central to the discovery of new medicines. The development of new approaches and techniques has accompanied, and been central to, major advances in understanding of disease biology and a move from pharmacology-driven to molecular-medicine-based drug discovery. This chapter will highlight some critical contributions from chemical biology methods in the context of a high-level overview of drug discovery as practiced currently.

## 1.1  INTRODUCTION

The discovery and development of new medicines is a highly complex, time-consuming, and expensive endeavor that has faced many changes and challenges over the last two decades (1). On the one hand, the pace of scientific discovery within the biomedical sciences has proceeded at unprecedented levels, fueled by dramatic advances in technology (e.g., Genomics (2) and High Throughput Screening (3, 4)). On the other hand, the development of new medicines has never been more challenging in terms of cost, time, and the regulatory environment (5–8). However, it is inevitable that societies will strive continually to improve their economic circumstances and improve human health and that discovery and access to medicines will play a critical role. The development and application of chemical biology methods are central to many aspects of the pursuit of medicines

---

*Chemical Biology: Approaches to Drug Discovery and Development to Targeting Disease*, First Edition.
Edited by Natanya Civjan.
© 2012 John Wiley & Sons, Inc. Published 2012 by John Wiley & Sons, Inc.

and are providing new approaches that may form the basis for a renaissance in the productivity of new medicine discovery in the future.

Drug discovery and development is a lengthy process, typically taking around 15 years from initiation of work on a target to final registration of a pharmaceutical product (9). As a result, there is a considerable lag between the development and application of new methods and their impact on clinical practice. Consequently, most of currently prescribed drugs were discovered during a completely different era of science and technology (10). This, coupled with the high level of attrition (i.e., failure at some point in the discovery/development process (5–9)), make it quite difficult to assess directly the impact of approaches applied in early drug discovery on the delivery of new medicines. In this chapter, the importance and potential impact of existing and emerging chemical biology methods is discussed, particularly in reference to their potential to increase the predictability and success rates of molecules in subsequent stages of the drug discovery/development process (e.g., to reduce attrition).

For the purposes of description, it is useful to think of drug discovery as consisting of a series of discrete steps. Figure 1.1 provides a schematic illustration that highlights some key activities underlying each phase. This chapter is organized to outline the drug discovery process from beginning to end, with an attempt to emphasize ways in which chemical biology may improve flow and success rates and, therefore, the overall delivery of new medicines. Obviously, the subject matter here is huge and this cannot provide a comprehensive overview. Many of the references provided are themselves detailed (and current) reviews of specific aspects, which should allow the reader to find more information readily. In addition, subsequent chapters cover specific aspects in much more detail.

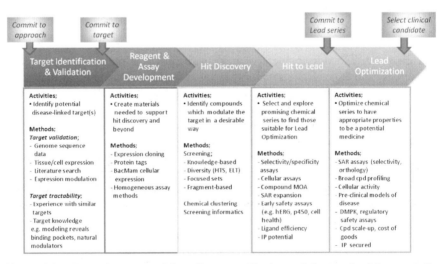

**Figure 1.1** The major stages of drug discovery. The boxes below each of the stages list some critical activities and methods used at each stage.

## 1.2 TARGET IDENTIFICATION AND VALIDATION

What is meant by a drug "target" and how are these identified? The generally accepted paradigm for safe and effective medicines is that they cause inhibition or activation of cellular function by binding specifically to one or more discrete protein targets and thus providing appropriate modulation of (patho)physiology. Indeed, the therapeutic targets for most currently marketed drugs, and the role of those targets in disease, is generally well established and involves specific interaction generally with a single protein target (10, 11).

Hence, most drug discovery efforts are initiated around one specific protein (or protein complex). Some exceptions to this approach are described as follows. Two major criteria need to be judged to assess the potential value of a drug discovery target:

1. Target Validation—i.e., evidence the modulation of the function or expression will have a disease-modifying effect in a safe and effective way).
2. Target Tractability—i.e., evidence to support the prospect for discovery and optimization of compounds that will perform the desired function in a safe and effective way.

If only the first criteria is achieved (i.e., modulation of the target by a small molecule seems unlikely to succeed), biopharmaceutical approaches may be pursued.

**Target Validation**. There are many ways in which evidence to link a specific gene product to disease may be obtained (for reviews, see References 12–14). Genetic studies, either from patient populations or in experimental systems such as transgenic animal models, often provide data about the pathophysiological consequences of specific proteins being modified (for example, via mutations), absent, or overexpressed (15). These approaches are often coupled with experiments in cellular systems in which the relevant proteins are modified in similar ways (16, 17). Often, multiple lines of evidence, each of which is relatively weak on its own, provide sufficient evidence to justify further exploration of the target. In many cases, questions of the role of the putative target can only be answered by the identification of potent and selective compounds that modulate its function (see the subsequent discussion).

Academic work reported in the scientific literature is often a source of targets for organizations that have drug discovery capabilities (i.e., pharmaceutical/biotechnology companies). Interestingly, recent studies have cast doubt on the reliability and reproducibility of this information in some cases, adding to the difficulty in selecting and validating drug targets (18).

**Target Tractability**. This aspect is concerned entirely with assessment of the likely success (and projected cost and time) of identifying compounds that modulate the target in the desired way and can then be further optimized, ultimately to produce a medicine. Computational methods have been successfully applied to this problem (see subsequent chapters 2 and 3). Tractability assessments are also often based on extrapolation from previous experience with related targets (e.g., from the same protein family). In the absence of any known pharmacophores against the target (or close homologs), judgments about tractability are largely subjective and based on the accrued experience and intuition of an experienced "drug hunter."

The decoding of the human genome has allowed the potential number of tractable drug targets (independent of their linkage to disease) to be estimated. Expansion of the protein classes known to be targets for existing drugs to include all related genes was used to produce an estimate of the "druggable genome" (19, 20). This study suggested that around 10% of all proteins encoded by the genome were potentially pharmacologically tractable, suggesting a rich potential source of novel tractable drug targets. For example, the human body contains >500 protein kinases (21), which highlights tremendous opportunities for exploitation of this class of proteins, but also caution in terms of the prospects for finding agents that only interact with a single kinase. In estimating the potential number of therapeutic targets, one must also consider the exploitation of additional new protein classes, for example, those linked to epigenetic processes (22–25). It seems reasonable to assume that there remain large numbers of biological targets that have not been exploited pharmacologically. Therefore, the major challenges facing biomedical discovery include improving the efficiency of exploration of targets and identifying better ways to triage them in terms of potential disease linkage.

Given the fact there seems to be no shortage of potentially tractable drug targets, but that evidence from genetic/biological/translational experiments is often not as clear as desired, several alternative approaches to the problem of selection of drug targets have been developed. Two of these will be highlighted here: phenotypic screening and chemical genomics. Figure 1.2 shows a schematic comparison of these different approaches to drug discovery targets.

Phenotypic approaches to drug discovery essentially move efforts away from a reductionist focus on a specific molecular target studied in isolation (for example, at the level of a single protein domain), to a more holistic approach in which an intact biological "system" is defined as the target. In most current approaches, the "system" is defined as a population of cells (26, 27). This strategy has some similarities to how drug discovery was performed before the development of molecular medicine approaches but uses high-tech, cell-based methods instead of primary tissue and animal pharmacology. Proponents of these methods are quick to point to studies that highlight the fact that many first-in-class medicines were discovered using approaches that may broadly be referred to a "phenotypic" (10).

Questions of chemical tractability apply as much to phenotypic approaches as to those directed toward specific protein targets, but they raise a different set of issues. As the precise mechanisms by which compounds will interact to produce the desired phenotype are unknown, at least initially, tractability can be difficult to assess (14). First, it may be difficult to determine whether compound activities in such assays are sufficiently specific to allow further progression because there are many ways in which they can act. A second, but related issue, is how to ultimately determine the Molecular Mechanism of Action (MMOA) of the compounds. Despite these drawbacks, this approach is attractive on the basis that the pharmacology observed may be more likely to reflect compound effects *in vivo* than, for example, simple measurements of binding to an isolated protein domain. As most proteins exist in complexes, are expressed in very specific compartments, and are heavily post-translationally modified, there is a strong argument that the more "natural" presentation of the target can result in finding better leads. A corollary is that phenotypic approaches can be prone to nonspecific effects as a result of, for example, low-grade compound toxicity (28, 29).

Chemical genomics takes an almost exactly opposite approach to phenotypic screening. In this case, targets are selected and prosecuted on their basis of putative tractability, with the idea of discovering chemical "probes" (Fig. 1.2; (18–20)). Provided these probes have the appropriate properties (potency, selectivity, and physicochemical properties), they can then be used to understand the role of targets in disease. In some cases, screening is applied to a target or pathway that already has some basis as a drug target in an attempt to find specific probes that will allow further development of that hypothesis. This is in large part the basis for academic screening efforts (30). Recently, attempts have been made to tackle whole classes of proteins that are currently unexploited pharmacologically, based on the ability to express and purify rapidly large numbers of proteins (*structural genomics* (31)). Chemical genomics is very much based around the identification of new chemical matter and therefore uses largely the same methods as those adopted for well-validated targets. These are discussed in a subsequent section (see Section 1.4).

## 1.3 REAGENT AND ASSAY DEVELOPMENT

Given the starting point of a defined biological "target(s)" (which could be a specific protein, pathway, or phenotypic cellular response, or even a panel of proteins), the next objective is to identify compounds that interact with the target in a desirable way. Initial efforts to achieve this are centered around the definition, creation, and procurement of the reagents and assays required to support lead discovery efforts. This is not just concerned with primary target and hit identification methods, but also with the critical path activities required to progress such compounds successfully (e.g., understand selectivity versus related targets and orthogonal approaches to confirm initial findings). The nature of the reagents and assays deployed is obviously somewhat dependent

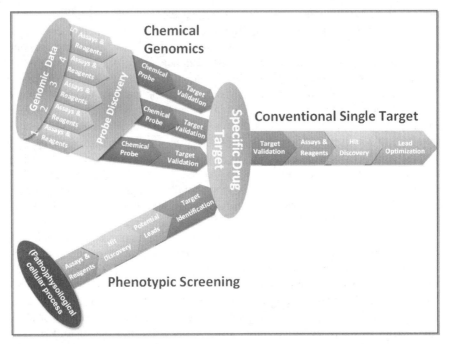

**Figure 1.2** Three approaches to the pursuit of new medicines. In *conventional single target* approaches, a specific protein is nominated as the target for therapeutic intervention at the outset and all activities are directed prosecution of that strategy. In *chemical genomics*, a range of targets is selected for discovery of chemical probes, which further enable target validation. In *phenotypic screening* approaches, an entire pathway of cellular process is screened, with a view to being able to tackle many potential targets simultaneously, with identification of the target mechanism of compounds occurring after they have been shown to produce a useful effect. Figure partly adapted from Reference 14.

on the target. For discrete protein targets (e.g., enzymes), an initial requirement is to clone, express, and purify the relevant protein in an appropriate form. Expression tags that aid in purification are almost universally used (for review, see Reference 32). For soluble protein targets, efforts may be focused even at an early stage on obtaining protein structural information. In some cases, virtually identical strategies may be deployed to produce protein for assays and structural work. Rapid methods for expression and purification of proteins is enabling for both single-target and multi-target "panning" approaches (Fig. 1.2 (33)).

For membrane targets, such as G-protein-coupled receptors (GPCRs) and ion channels, expression within cells is almost always used in conjunction with functional assays (for reviews, see References 34 and 35). Transient expression methods have greatly improved the speed and efficiency of generating appropriate recombinant cellular systems for assays (36). Phenotypic approaches emphasize the presentation of the target in the most natural setting, and so the cells used are often not recombinantly engineered (27).

In addition to the appropriate biological reagents, assays are required that allow the detection of test compound effects to be determined with appropriate robustness and sensitivity. The development of a range of technologies and accompanying chemical/biochemical reagents to allow screening has been a key enabler for hit identification and profiling in the last 15 years or so (for reviews, see References 37–39). The development of assay methods, linked to specialized detection, automation, and liquid handling instrumentation is a major commercial activity and commercial systems are used by most laboratories, both in industry and academia.

## 1.4   LEAD DISCOVERY

Armed with suitable reagents and assays, a research team is now in a position to test compounds in earnest. However, even the most sensitive and robust screen is only as good as the compounds against which it is tested! Hence, the selection of what to screen is absolutely crucial. Table 1.1 provides a high-level summary of hit identification approaches.

In some cases, sufficient knowledge already exists (for example from literature, knowledge of receptor ligands or substrate catalysis mechanism of an enzyme) to design compounds for testing. These types of approaches may by broadly described as *knowledge-based lead discovery*. It is worth noting that, prior to the advent of high-throughput methods, this (together with serendipitous observations of pharmacological effects) was the major way in which drug discovery efforts were started. Hence, together with phenotypic approaches, knowledge-based discovery is a major contributor to the origin of currently prescribed medicines (10). In retrospect, this was greatly assisted by the types of targets pursued. For example, GPCRs, which bind small aminergic ligands such as histamine and serotonin, provided fairly straightforward approaches based on modification of cognate receptor ligands. Knowledge-based approaches still contribute significantly to current drug discovery, but they tap into different data and resources. For example, in the 1990s, virtually all compounds of interest needed to be designed and synthesized, whereas very large numbers of structures (>20M) are now available from a network of chemical suppliers (40). Hence, electronic search of available chemical space tends to be a first line and economic way to tackle at least initial forays into knowledge-based design.

Given sufficient knowledge of the target (e.g., a high-resolution crystal structure), it is possible to approach the question of which compounds should be

**TABLE 1.1   High-level attributes of various *lead discovery* methods**

| | | Hit identification methods | | |
|---|---|---|---|---|
| Method | No. cpds tested | Collection emphasis | Initial screening condition | Hit detection mode |
| Knowledge based Discovery | 10–100's | detailed knowledge of specific target | ~10 μM or dose-response | funcional assays |
| Focused screening | $1 - 5 \times 10^4$ | chemical connectivity of similar target types | ~10 μM | functional assays |
| High through-put screening | $3 \times 10^5 - 2 \times 10^6$ | maximize diversity and best compound properties | ~10 μM | functional assays |
| Library Encoded Methods | $10^6 - >10^9$ | maximize diversity and best compound properties | <<1 μM | binding enrichment |
| Fragment Based Drug Discovery | $1 - 5 \times 10^3$ | maximize diversity with small mole "chunks" which bind efficently | ~1 mM | Binding measurement |

selected, designed, and screened entirely using computational methods. These approaches are the subject of subsequent chapters and so not covered in any detail here (see chapters 2 and 3). Such methods often play a crucial role once compounds have been identified from other methods (see the subsequent discussion). In most cases, chemical structures suggested by computational methods require confirmation of their effects in a biological assay. Hence, computational methods to understand and refine compound binding to a drug target tend to be used in conjunction with other approaches.

Another way in which knowledge is applied to lead discovery is what may be broadly referred to as *focused screening*. In this case, knowledge of the drug target is applied to select which compounds to test. Perhaps the most successful example of this approach is for protein kinases. This large and important class of proteins has a canonical role in cell signaling and thus provides potential targets for intervention in several diseases (41). Kinases share common structural/functional features (e.g., use of adenosine triphosphate (ATP) to transfer a phosphate group to a protein amino acid residue), which results in considerable "chemical connectivity" between kinase inhibitors and their targets. Therefore, a much higher level of hit enrichment is obtained when testing compounds designed against kinases than, say, a random set of compounds with similar chemical properties. For this reason, a set of ~10,000 carefully selected kinase compounds can provide a relatively good chance of obtaining potent hits (perhaps equivalent to ~$10^6$ randomly selected compounds; a 100-fold enrichment). This approach yields a rich source of starting points for this class of target, but it does have important limitations. As only a relatively narrow set of chemical space is interrogated, novelty (e.g., in compound structure, mechanism of binding, and scope for intellectual property creation) is more difficult to achieve than for other methods. Given that there are several hundred protein kinases in the human body, inhibitor selectivity may be a critical factor to achieve safe and efficacious medicines. It seems reasonable that the same chemical connectivity factors that make focused screening against kinases successful may make it more difficult to achieve high levels of target selectivity.

At least in principle, *focused screening* approaches can be applied to any target or attribute (e.g., predicted blood–brain penetration for compounds targeted toward neuroscience indications). However, even the best design principles do not guarantee successful hit enrichment, so focused sets should be continually refined based on their actual performance. Personal experience at GlaxoSmith-Kline suggests that many focused sets do not provide very large hit enrichment over "diverse" (see the subsequent discussion) compounds sets with similar properties (<10-fold).

The most widely adopted method for *lead discovery* during the last 10 years or so is high throughput screening (HTS). This approach grew out of the development of methods to test large numbers of natural product samples (the first form of "random" screening). Once capabilities were developed to test large numbers of natural product extracts, groups turned their attention to curating and testing internal collection of compounds synthesized in house. The development of this capability was contemporary with an explosion in combinatorial chemistry methods, which in turn drove the building and acquisition of large HTS libraries. Natural-product–based drug discovery is discussed in detail in chapter 4.

At the time of writing, HTS is currently the commonly practiced form of what can be broadly categorized as *diversity screening* methods. All of these types of methods operate on the same basic principle—which is to test a target against the broadest possible range of appropriate chemical space with the intent of finding novel pharmacophores. In some ways, this approach represents the antithesis of focused screening, in that selection of the set of targets does not take into account the nature of the target (i.e., the same set of compounds is applied to each target, irrespective of other information). However, it is important to emphasize that the diversity screening methods do not ever use "random" selection of compounds. HTS compound sets are often highly curated to maximize the quality of the hits discovered.

This needs to take several factors into account (42), including:

• Chemical properties (e.g., compounds must meet Lipinski, or more stringent, criteria (43))

- Chemical diversity coverage (e.g., compounds are spread in many diversity clusters (44))
- Compound quality (e.g., purity, stability, reactivity, and solubility (42))
- Known hit properties (e.g., promiscuity)
- Cost of acquisition, maintenance, and dispensing compounds (which, along with the cost of screening, places an upper limit on library size)

These factors mean that diversity screening strategies are often driven largely by pragmatic and economic factors and therefore vary considerably in scale. In large pharmaceutical companies, HTS collections typically currently contain 1–2M compounds (3). Smaller biotech and contract research organizations (CRO) typically use collections in the range of 100–500 K compounds. In most cases, all compounds in the library are tested initially at a single concentration (typically 10 μM), although some groups have proposed that there is additional value in testing the entire library at multiple doses (45). All other things being equal (although they never are!), the success of diversity screening should increase in direct proportion to the library size, but given the economic limits on the scale of HTS, the design of diversity collections and careful tracking of their performance are critical (44).

In recent years, the value of diversity screening to identify chemical probes has been recognized by academic institutions, with the development of an infrastructure similar to that of the pharmaceutical industry (46, 47). The most prominent example of this is the NIH-funded Molecular Libraries Screening Center (MLSCN) and Molecular Libraries Production Centers Network (MLPCN) (30). These centers were established with a mission of providing access to chemical probes to academics investigating putative disease mechanisms and putative intervention approaches, in some ways directly analogous to the early stage activities pursued by pharmaceutical companies. The MLPCN library consists, at the time of writing, of around 350,000 discrete compounds. All data produced from these screens are made publically available (48).

Given the massive numbers of possible chemical structures within "lead-like" chemical space that can be envisioned (49), it seems sensible to assume the success of HTS may be limited by its economic limits of scale. For this reason, methods allowing larger number of compounds to be tested provide the potential to complement, or even replace, HTS. Clearly, a different approach to the problem is required. This has been successfully achieved in some cases where large numbers of compounds are synthesized by using combinatorial chemistry methods employing compound tagging strategies that allow identification of which compounds bind to a target when many are tested simultaneously (50, 51). Recently, a method that uses DNA-barcode sequences to code for chemistry and binding affinity selection using DNA sequencing methodology was described (52). This allows the generation and testing of massive libraries ($>10^9$ compounds) and is now being exploited as a mainstream hit identification method at GlaxoSmithKline.

Another approach to provide coverage of chemical space is applied in *fragment-based drug discovery* (FBDD (53–55). In this method, small "chunks" of molecules (e.g., MW <200) are screened at a relatively high concentration (e.g., 1 mM) to identify compounds that bind weakly but with high ligand efficiency (52). By using this approach, quite a large coverage of chemical space can be achieved with relatively small numbers of compounds (<10,000). The number of possible drug "fragments" (i.e., <200 Da) is around 10 orders of magnitude smaller than the equivalent value for drug-sized molecules (i.e., <450 Da) (55).

Because fragments bind weakly, screening is usually used in conjunction with biophysical and protein structure methods to provide detailed characterization of binding to confirm that fragments are binding productively and to guide subsequent optimization (54). The ability to determine bound ligand structures is normally thought to be a prerequisite for effective use of these methods, so they are mainly applied to soluble targets such as enzymes. However, recent reports suggest that, given appropriate engineering of the target proteins, this technique has recently also been applied to GPCRs (56).

**Hit Confirmation**. Whatever hit identification method is used, a critical next step is to confirm that compounds identified as apparently active in screens are not a result of artifacts (57, 58). Often a first step is to eliminate false-positive signals by confirming that the initial result can be reproduced under the same conditions. This is then normally followed by dose response testing, to produce a candidate list of preliminary hits. At this point, the structures of hits are typically examined to determine whether preliminary patterns of structure activity relationships (SAR) are present (e.g., multiple hits occur within related chemical families).

Given a list of preliminary hits, a critical next step is to identify those that have promise for further exploration. Several factors are typically taken into account in this phase. First, it is important to determine which classes of compound bind to the target in a productive manner (e.g., with defined stoichiometry and mechanism). Several "nuisance" mechanisms exist that can result in apparent activity in assays but where the compounds are not progressable (e.g. 66, 67). For biochemical screening approaches, biophysical methods such as surface plasmon resonance and isothermal calorimetry of nuclear magnetic resonance may be useful in determining binding specificity, kinetics, and stoichiometry (59).

For cellular approaches, demonstration that positive compounds interact with the target is more difficult because direct methods to measure binding cannot generally be used. Hit confirmation for these targets is usually concerned with identifying those compounds that act in a specific manner without causing cellular toxicity. Specificity is usually tested by counter screening hits against a closely related system in which the signal is not dependent on the target. In the case of phenotypic approaches where, by definition, useful hits are not restricted to activity at a single specific target, the definition of specific hit activity may represent the starting point for experiments to elucidate compound MMOA. Proteomic methods, such as stable-isotope labeling by amino acids in cell culture (SILAC) have proved extremely powerful as a means to profile the interaction targets for compounds (60, 61).

Another important step in hit confirmation is to check that the observed activity is actually caused by the specific chemical structure of interest because activity in assays can occur as a result of compound breakdown products or minor contaminants. For this reason, putative hits are often confirmed by using a fresh sample of compound or, in some cases, via resynthesis of compounds of interest.

## 1.5  HIT TO LEAD/PROBE

Once activity in a screening assay has been confirmed and shown to be appropriately specific, potential compound series are selected for further characterization and expansion. The following types of characteristics may be used:

- The number of chemically related compounds identified, which can give initial evidence for the emergence of SAR
- Good affinity and/or ligand efficiency for the target (62, 63)
- Good selectivity (i.e., relative potency for the target versus related proteins)
- Productive binding mechanism (e.g., reversible, substrate competitive binders)
- Good chemical properties (e.g., MW <400, cLogP <4) and solubility
- Good chemical tractability (i.e., the ability to procure and/or synthesize analogs)
- Intellectual property (i.e., scope to generate intellectual property from subsequent optimization)
- Identification of liabilities that may affect subsequent progression (e.g., inhibiting P450 enzymes, drug transporters, and cardiac ion channels)

As a result of this triage, a small number (<10) of potential compound series is typically selected for further characterization and expansion. This normally starts with an exhaustive search of available analogs (40) coupled to some limited exploratory synthesis. The expansion of potential chemical lead series is also accompanied by further characterization of the mechanism and specificity of interaction with the target. For phenotypic approaches, this is the point at which experiments to determine the binding target often need to have produced clear results for the effort to be progressed further (14). In the case of chemical genomic approaches, compounds will have now met the criteria for use in further biological experiments to investigate the role of the target(s)—i.e., they are completed "chemical probes" (e.g., see References 64–67). For mainstream drug discovery programs, this phase of activity culminates in *lead series*, which meet all of the desired criteria for further optimization and thus allow the initiation of *lead optimization* and the investment of chemistry resources.

## 1.6   LEAD OPTIMIZATION

According to a recent SBS/IUPAC glossary of terms (68), a lead is defined as "compound (or compound series) that satisfies predefined minimum criteria for further structure and activity optimization." In the general case, this refers to a series of compounds for which there is already demonstrated SAR against the target of interest, scope for further modification, and at least the potential to create new intellectual property. These attributes must be combined with appropriate biological (i.e., potency, specificity, and selectivity) and physicochemical properties to provide a useful prospect for further optimization to a clinical candidate.

*Lead optimization* processes are discussed in detail in chapter 4. Fundamentally, this stage consists of iterative cycles of optimization of chemistry, each of which is associated with the measurement of particular aspects of the biological properties of compounds that aim to move continually closer to the selection of a single compound as a clinical candidate. *Lead optimization* is characterized by a complex balancing of a number of critical factors, some of which are listed in Table 1.2. The precise challenges presented in any particular lead optimization campaign are to some extent unique to that target; therefore, a highly flexible approach is required. Given the high levels of attrition that occur during drug clinical development (5–7), the identification of issues that may cause problems in development as early as possible is a key area of focus.

Optimization of a lead series to the point of nomination of a clinical candidate typically takes 1–3 years of work and involves around 1000 compounds made over perhaps 30–100 iterations of chemistry optimization. One critical factor is therefore supplying data to support the next round of chemical synthesis as rapidly as possible for the characterization of previous SAR to be built into subsequent chemistry plans (chapter 4). Subsequent chapters discuss other critical factors that need to be taken into account during *lead optimization*, such as ADME and pharmacokinetic properties (chapters 5 and 6), drug transport processes (chapter 7), and blood–brain barrier penetration (chapter 8). Several recent studies have pointed out the importance of focusing effort during this phase not only on biological data but also on compound physicochemical properties (for reviews, see References 69 and 70).

Naturally, the major focus in *lead optimization* is to find a molecule that has good potential to be useful therapeutically in terms of safety and efficacy. Loss of a drug candidate during clinical development can be highly costly and

TABLE 1.2    Some key properties of lead that are optimized on the pathway from during *lead optimization* to produce clinical candidates

| Optimization property | Rationale | Measured how? |
| --- | --- | --- |
| Potency and ligand efficiency against primary target(s) | Ensure optimal engagement of therapeutic target and balance of physical properties | Dose response profiling, biophysical assays, calculated properties |
| Selectivity for primary target(s) verus related biological systems | Minimize off-target effects, which could result in toxicity | As above |
| Activity in predictive models of disease | Increase confidence in therapeutic target approach and predict potential clinical dose | cellular and animal models |
| Potency in animal orthologues | Asses predictvity of pre-clinical studies for efficacy and safety | As above |
| Specificity relative to broad panels of pharmacologically relevant targets | Minimize off-target effects, which could result in toxicity | Broad profiling of compounds against panels of targets |
| Compound physico-chemical properties (solubility, lipophilicity etc.) | Good properties required for ADME | Some simple properties can be calculated, other are physically measured |
| ADME/Pharmacokinetic properties | Optimize bioavailability and subsequent pharmacokinetics | *In vitro* models, animal DMPK |
| Early safety assessment | Identify potential drug safety issues as early as possible | *In vitro* models (e.g. hERG channel assays, cell health measurements, AMES/genotox testing) |
| Intellectual Property generation/protection | Protect value of discovery | patent literature and authorship |

disruptive commercially (8), so an absolutely key activity during *lead optimization* is to attempt to minimize the risk of subsequent termination. Excluding commercial and strategic considerations, there are only two ways in which a drug candidate can fail: either inadequate effectiveness or safety concerns. Therefore, as compounds are optimized to improve their overall properties, more predictive information is sought around how target modulation by those molecules could translate into useful disease modification in the clinic (71). Inevitably, this is often linked to consideration of the potential commercial environment for such a medicine (e.g., what is the size of the unmet medical need? how does this medicine compare with others currently on the market or in development, and what end-points will the medicine need to achieve to add value and therefore maximize likelihood of reimbursement?).

The drug discovery process essentially ends at the point of identification of a *clinical candidate*. This is the nomination of a specific compound that will be the subject of *drug development* processes, which is outside of the scope of this chapter.

## 1.7 CONCLUSIONS

Chemical biology methods play a central role in drug discovery, particularly as it is practiced currently. In addition, new and established methods are allowing further exploration of interactions between compounds and biological processes at an unprecedented scale and quality. It is too soon to tell in most cases how current methods will contribute to the rate of successful development of new medicines. However, the combination of current and future (e.g., nano-engineering) technologies together with careful analysis of the lessons from the past seems poised to drive many new advances in the most noble of causes, that of improving our understanding, treatment, and prevention of disease.

## REFERENCES

1. Kola I, Landis J. Can the pharmaceutical industry reduce attrition rates? Nat. Rev. Drug Discov. 2004;3:711–715.
2. Lander ES. Initial impact of the sequencing of the human genome. Nature 2011;470:187–197.
3. Macarron R, Banks NM, Bojanic D, Burns DJ, Cirovic DA, Garyantes T, Green DVS, Hertzberg RP, Janzen WP, Paslay JW, Schopfer U, Sittampalam SG. Impact of high throughput screening on biomedical research. Nat. Rev. Drug Discov. 2009;10:188–195.
4. Mayr LM, Fuerst P (2008) The future of high-throughput screening. J. Biomol. Scr. 2008;13:443.
5. DiMasi JA, Feldman L, Seckler A, Wilson A. Trends in risks associated with new drug development: Success rates for investigational drugs. Clin. Pharmacol. Therapeut. 2010;87(3):272–277.
6. Arrowsmith J. Phase III and submission failures: 2007–2010. Nat. Rev. Drug Discov. 2011;10:1.
7. Paul SM, Mytelka, Dunwiddie CT, Persinger CC, Munos BH, Lindborg SR, Schacht AL. How to improve R&D productivity: The pharmaceutical industry's grand challenge. Nat. Rev. Drug Discov. 2010;9:203–214.
8. Morgan S, Grootendorst P, Lexchin J, Cunningham C, Greyson D. The cost of drug development: A systematic review. Health Policy 2011;100:4–17.

9. Dickson M, Gagnon JP. Key factors in the rising cost of new drug discovery and development. Nat. Rev. Drug Discov. 2004;3:417–429.

10. Swinney DC, Anthony J. How were new medicines discovered? Nat. Rev. Drug Discov. 2011;10(7):507–512.

11. Inning P, Sining C, Meyer A. Drugs, their targets and the nature and number of drugs. Nat. Rev. Drug Discov. 2006;5:821–834.

12. Chan JNY, Nislow C, Emili A. Recent advances and method development for drug target identification. Trends Pharmacol. Sci. 2010;31(2):82–88.

13. Rask-Anderson M, Almen MS, Schioth HB. Trends in the exploitation of novel drug targets. Nat. Rev. Drug Discov. 2011;10:579–590.

14. Tertappen GC, Schluipen C, Raggiaschi R, Gaviraghi G. Target deconvolution strategies in drug discovery. Nat. Rev. Drug Discov. 2007;6:891–903.

15. Hardy J, Singleton A. Genomewide association studies and human disease. N. Engl. J Med. 2009;360:1759–1768.

16. Prokop A, Michelson S. Springer Briefs in Pharmaceutical Science & Drug Development. 2m 69–76. 2012. Springer, New York.

17. Jackson AL, Linsley PS. Recognizing and avoiding siRNA off-target effects for target identification and therapeutic application. Nat. Rev. Drug Discov. 2010;9:57–67.

18. Prinz F, Schlange T, Asadullah. Believe it or not: How much can we rely on published data on potential drug targets? Nat. Rev. Drug Discov 2011;10:712.

19. Hopkins AL, Groom CR. The druggable genome. Nat. Rev. Drug Discov 2002;1:727–730.

20. Russ AP, Lampel S. The druggable genome: An update. Drug Discov. Today 2005;10(23–24):1607–1610.

21. Manning G, Whyte DB, Martinez R, Hunter T, Sudersanam S. The protein kinase complement of the human genome. Science 2002;298(5600):1912–1934.

22. Heightman T, Pope AJ. Epigenetic drug discovery. J. Biomol. Scr. 2011; 16:1135–1136.

23. Lu Q, Quinn AM, Patel MP, Semus SF, Graves AP, Bandyopadhyay D, Pope AJ, Thrall SH. Perspectives on the discovery of small-molecule modulators for epigenetic processes. J. Biomol. Scr. 2012;17(5):1–17.

24. Carey N. Epigenetics—an emerging and highly promising source of new drug targets. Med. Chem. Commun. 2012;3:162–166.

25. Sippl W, Jung Manfred M, Eds. Epigenetic Targets in Drug Discovery. Methods and Principles in Medicinal Chemistry. Vol. 42. Wiley-VCH, New York, 2009.

26. Frank A, Tolliday N. Cell-based assays for high-throughput screening. Mol. Biotechnol. 2010;45(2):180–186.

27. Lee JA, et al. Open innovation for phenotypic drug discovery. J. Biomol. Scr. 2011;16(6):588–602.

28. Shukla SJ, Huang R, Austin CP, Xia M. The future of toxicity testing: A focus on *in vitro* methods using a quantitative high-throughput screening platform. Drug Discov. Today 2010;15(23–24):997–1007.

29. Thorne N, Auld DS, Inglese J. Apparent activity in high-throughput screening: Origins of compound-dependent assay interference. Curr. Opin. Chem. Biol. 2010;14(3):315–324.

30. *http://mli.nih.gov/mli/*.

31. Weigelt J. Structural genomics—impact on biomedicine and drug discovery. Exp. Cell Res. 2010;316(8):1332–1338.

32. Koehn J, Hunt I. High throughput protein production (HTTP): A review of enabling technologies to expedite protein production. Methods Mol. Biol. 2009;498:1–18.

33. Hunt I. From gene to protein: A review of new an enabling technologies for multi-parallel protein expression. Protein Expr. Purif. 2005;40(1):1–22.

34. Wang D, Li Y, Zhang Y, Liu Y, Shi G. High throughput screening (HTS) in identification new ligands and drugable targets of G protein-coupled receptors (GPCRs) combinatorial chemistry & high throughput screening. Comb. Chem. High T. Scr. 2012;15(3):232–242.

35. Terstappen GC, Roncarati R, Dunlop, Peri R. Screening technologies for ion channel drug discovery. Future Med. Chem. 2010;2(5):715–730.

36. Kost A, Condrea P, Ames R. Baculovirus gene delivery: A flexible assay development tool. Curr. Gene Ther. 2010;10(3):168–173.

37. Inglese J. A Practical Guide to Assay Development and High Throughput Screening in Drug Discovery. 2010. CRC Press, Boca Raton, FL.

38. An WE, Tolliday N. Cell-based assays for high throughput screening. Mol. Biotechnol. 2010;45(2):180–186.

39. Pope AJ, Haupts UM, Moore KJ. Homogeneous fluorescence readouts for miniaturized high-throughput screening: Theory and practice. Drug Discov. Today. 1999;4(8):350–362.

40. *http://www.emolecules.com/*.

41. Cohen P. Targeting protein kinases for the development of anti-inflammatory drugs. Curr. Opin. Cell Biol. 2009;21(2):317–324.

42. Wigglesworth M, Wood T, eds. Management of Chemical and Biological Samples for Screening Applications. 2012. Wiley-VCH, New York.

43. Lipinski C. Lead and drug-like compounds: The rule of five revolution. Drug Discov. Today: Technologies 2004;1(4):337–341.

44. Harper G, Pickett SD, Green DVS. (2004) Design of a compound screening collection for use in high throughput screening. Comb. Chem. High T. Scr. 2004;7(1):63–70.

45. Inglese J, Auld DS, Jadhav A, Johnson RL, Simeonov A, Yasgar A, Zheng W, Austin CP. Quantitative high-throughput screening: A titration based approach that efficiently identifies biological activities in large chemical libraries. Proc. Natl. Acad. Sci. U.S.A. 2006;103:11473–11478.

46. Dove A. High throughput screening goes to school. Nat. Methods 2007; 4(6):523–532.

47. Frearson JA, Collie IT. HTS and hit finding in academia—from chemical genomics to drug discovery. Drug Discov. Today 2009;14:1150–1158.

48. *http://www.ncbi.nlm.nih.gov/pcassay*.

49. Medina-Franco JL, et al. Visualization of chemical space in drug discovery. Curr. Comput. Aided Drug Design 2008;4:322–333.

50. Baldwin, JJ, Horlbeck, EG. U.S. Patent 5 663 046, 1997.

51. Rokosz LL, Huang CY, Reader JC, Stauffer TM, Southwick EC, Li G, Chelsky D, Baldwin JJ. Exploring structure-activity relationships of tricyclic farnesyltransferase inhibitors using ECLiPS libraries. Comb. Chem. High T. Scr. 2006;9:545–558.

52. Clark MA, Acharya RA, Arico-Muendel CC, Belyanskaya SL, Benjamin DR, Carlson NR, Centrella PA, Chiu CH, Creaser SP, Cuozzo JW, Davie CP, Ding Y, Franklin GJ, Franzen KD, Gefter ML, Hale SP, Hansen NJV, Israel DI, Jiang J, Kavarana MJ, Kelley MS, Kollmann CS, Li F, Lind K, Mataruse S, Medeiros PF, Messer JA, Myers P, O'Keefe H, Oliff MC, Rise CE, Satz AL, Skinner SR, Svendsen JL, Tang L, Vloten KV, Wagner RW, Yao G, Zhao B, Morgan BA. Design, synthesis and selection of DNA-encoded small-molecule libraries. Nat. Chem. Biol. 2009;5(9):647–654.

53. Bembenek SD, Tounge BA, Reynolds CH. Ligand efficiency and fragment-based drug discovery. Drug Discov. Today 2009;14(5/6):278–283.

54. Edfeldt FNB, Folmer RHA, Breeze AL. Fragment screening to predict druggability (ligandability) and lead discovery success. Drug Discov. Today 2011;16(7/8):284–287.

55. Murray CW, Rees DC. The rise of fragment-based drug discovery. Nat. Chem. 2009;1:187–192.

56. Congreve M, Rich R, Myszka DG, Figaroa, Siegal G, Marshall F. Fragment screening of stabilized g-protein-coupled receptors using biophysical methods. Methods Enzymol. 2011;495:115–136.

57. Johnson P. Redox cycling compounds produce H2O2 in HTS buffers containing strong reducing agents—real hits of promiscuous artefacts. Curr. Opin. Chem. Biol. 2011;15(1):174–182.

58. Feng BY, Shelat A, Doman TN, Guy RK, Shoichet BK. High-throughput assays for promiscuous inhibitors. Nat. Chem. Biol. 2005;1(2):146–148.

59. Lundquvist T. The devil is still in the details—driving early drug discovery forward with biophysical experimental methods. Curr. Opin Drug Discov. Develop. 2005;8(4):513–519.

60. Drewes G. Chemical proteomics in drug discovery. Chemical proteomics: Methods and protocols. Methods Mol Biol 2012;803:15–21.

61. Ong SE, Mann M. Stable isotope labelling by amino acids in cell culture for quantitative proteomics. Methods Mol. Biol. 2007;359:37–52.

62. Perola E. 2010. An analysis of the binding efficiencies of drugs and their leads in successful drug discovery programs. J. Med. Chem. 2010;53:2986–2997.

63. Abad-Zapatero C, Metz JT. Ligand efficiency indices as guideposts for drug discovery. Drug Discov. Today 2005;10(7):464–469.

64. Cong F, Cheung AK, Huang SM. Chemical genetics-based target identification in drug discovery. Annu. Rev. Pharmcol. Toxicol. 2012;52:57–78.

65. Rix U, Superti-Furga. Target profiling of small molecules by chemical proteomics. Nat. Chem. Biol. 2009;5:616–624.

66. Edwards AE, Bountra C, Kerr DJ, Wilson TM. Open access chemical and clinical probes to support drug discovery. Nat. Chem. Biol. 2009;5:436–440.

67. Hauser AT, Jung M. Chemical probes: Sharpen your epigenetic tools. Nat. Chem. Biol. 2011;7:499–500.

68. *http://www.slas.org/education/glossary.cfm*.

69. Meanwell NA. Improving drug candidates by design: A focus on physicochemical properties as a means of improving compound disposition and safety. Chem. Res. Toxicol. 2011;24:1420–1456.

70. Waring MJ. Lipophilicity in drug discovery. Expert. Opin. Drug Discov. 2010;5(3):235–248.

71. Littman BH, Krishma R. Translational Medicine and Drug Discovery. 2011. Cambridge University Press, New York.

# 2

# COMPUTATIONAL APPROACHES TO DRUG DISCOVERY AND DEVELOPMENT

HONGLIN LI, MINGYUE ZHENG, XIAOMIN LUO, WEILIANG ZHU, AND HUALIANG JIANG

*Shanghai Institute of Materia Medica, Chinese Academy of Sciences, Shanghai, China*

Recently, computational approaches have been considerably appreciated in drug discovery and development. Their applications span almost all stages in the discovery and development pipeline, from target identification to lead discovery, from lead optimization to preclinical or clinical trials. In conjunction with medicinal chemistry, molecular and cell biology, and biophysical methods as well, computational approaches will continuously play important roles in drug discovery. Several new technologies and strategies of computational drug discovery associated with target identification, new chemical entity discovery, and lead optimization will be the focus of this review.

Drug research and development (R & D) is a comprehensive, expensive, and time-consuming enterprise, and it is full of risk throughout the process (1). Numerous new technologies have been developed and applied in drug R & D to shorten the research cycle and to reduce the expenses. Among them, computational approaches have revolutionized the pipeline of discovery and development (2). In the last 40 years, computational technologies for drug R & D have evolved very quickly, especially in recent decades with the unprecedented development of biology, biomedicine, and computer capabilities. In the postgenomic era, because of the dramatic increase of small-molecule and biomacromolecule

*Chemical Biology: Approaches to Drug Discovery and Development to Targeting Disease*, First Edition. Edited by Natanya Civjan.
© 2012 John Wiley & Sons, Inc. Published 2012 by John Wiley & Sons, Inc.

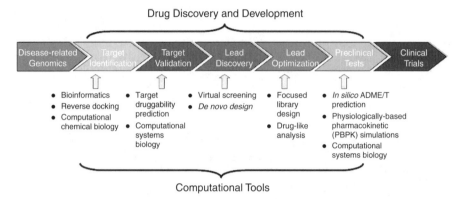

**Figure 2.1**   Computational approaches cover the drug-discovery pipeline, spanning from target identification to lead discovery to preclinical test.

information, computational tools have been applied in almost every stage of drug R & D, which has greatly changed the strategy and pipeline for drug discovery (2). Computational approaches span almost all stages in discovery and development pipeline, from target identification to lead discovery, from lead optimization to preclinical or clinical trials (see Fig. 2.1). In this review we highlight some recent advances of computational technologies for drug discovery and development; current to the time of writing this article in 2008; emphases are put on computational tools for target identification, lead discovery and ADME/T (absorption, distribution, metabolism, excretion, and toxicity) prediction.

## 2.1   TARGET IDENTIFICATION

Target identification and validation is the first key stage in the drug-discovery pipeline (see Fig. 2.1). By 2000, only about 500 drug targets had been reported (3, 4). The completion of human genome project and numerous pathogen genomes unveils that there are 30,000 to 40,000 genes and at least the same number of proteins; many of these proteins are potential targets for drug discovery. However, identification and validation of druggable targets from thousands of candidate macromolecules is still a challenging task (5).

   Numerous technologies for identifying targets have been developed recently. Experimental approaches such as genomic and proteomic techniques are the major tools for target identification (6, 7). However, these methods have been proved inefficient in target discovery because they are laborious and time consuming. In addition, it is extremely difficult to get relatively clear information related to drug targets from enormous information produced by genomics, expression profiling, and proteomics (5). As complementarities to the experimental methods, a series of computational (*in silico*) tools have also been developed for target identification in recent two decades (6, 8–12). In general, they can be categorized into sequence-based approach and structure-based approach. Sequence-based approach contributes to the processes of the target identification by providing functional information about target candidates and positioning information to biological networks. Such methods include sequence alignment for gene selection, prioritization of protein families, gene and protein annotation, and expression data analysis for microarray or gene chip. Here, we only introduce a special structure-based method for target identification—reverse docking.

Molecular docking has been a promising approach in lead discovery (2, 13) (also see discussion below). Reverse docking, which is opposite to the process of docking a set of ligands into a given target, is to dock a compound with certain biological activities into the binding sites of all the three dimensional (3-D) structures in a given protein database. The identified protein "hits" through this method then served as potential target candidates for further validation study. This computational approach in target identification has demonstrated to be efficient and cost effective (14–16). In general, identifying a protein target by using reverse docking includes four steps: inverse docking of a small molecule to a series of selected proteins; hit proteins postprocessing through bioinformatics analysis to select candidates; experimental validation by using biochemical and/or cellular assays; and finally, if it is possible and necessary, determination of the X-ray crystal (or nuclear magnetic resonance) structures of the small molecule-protein complexes to verify the target at the atomic level. To this end, it requires a sufficient number of known protein structures that cover a diverse range of drug targets (17). The protein structures are usually selected from the Protein Data Bank (PDB) or constructed with protein structure prediction method.

Based on DOCK4.0, we developed a reverse docking method, TarFisDock (Target Fishing Docking) (16), and we constructed a corresponding drug target database with 3-D structures, potential drug target database (PDTD) (17). Based on TarFisDock and PDTD, we also developed a reverse docking web server (*http://www.dddc.ac.cn/tarfisdock/*). TarFisDock has proved to be a tool of great potential value for identifying targets by more than 400 users from over 30 countries and regions. One typical example of TarFisDock application is that we found peptide deformylase (PDF) is a target for anti-*H. pylori* drugs.

Colonization of the human stomach by the bacterium *H. pylori* is a major causative factor for gastrointestinal illnesses and gastric cancer. However, discovering anti-*H. pylori* agents is a difficult task because of a lack of mature protein targets. Therefore, identifying new molecular targets for developing new drugs against *H. pylori* is obviously necessary. In a random screening of a diverse small-molecule and herbal extract library by using agar dilution method to identify active components against *H. pylori*, compound **1** (*N*-*trans*-caffeoyltyramine) isolated from *Ceratostigma willmottianum*, a folk medicine used to remedy rheumatism, traumatic injury and parotitis (18), showed inhibitory activity against *H. pylori* with MIC value of 180 μg/mL. Chemical modification on compound **1** afforded a number of analogs, among which, compound **2** ((*E*)-4-(3-oxo-3-(phenethylamino)-prop-1-enyl)-1,2-phenylene

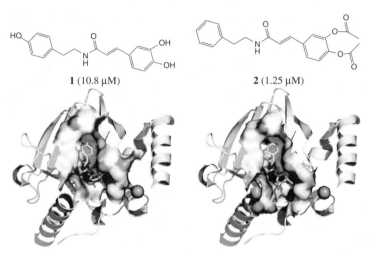

**Figure 2.2**   Chemical structures of compounds **1** and **2** and their X-ray crystal structures in complex with $Hp$PDF. Data inside the parentheses are inhibition activities ($IC_{50}$ values) of the compounds.

diacetate) is the most active one with improved MIC value of 100 μg/mL against *H. pylori* (see Fig. 2.2). Potential binding proteins of compound **1** were screened from PDTD using TarFisDock. Fifteen protein targets with interaction energies to compound **1** under $-35.0$ kJ/mol were identified (19). Homology search indicated that only two proteins of the fifteen candidates, diaminopimelate decarboxylase (DC) and PDF, have homologous proteins in the genome of *H. pylori*. Enzymatic assay demonstrated compounds **1** and **2** are the potent inhibitors against the *H. pylori* PDF ($Hp$PDF) with $IC_{50}$ values of 10.8 and 1.25 μM, respectively. X-ray crystal structures of $Hp$PDF and the complexes of $Hp$PDF with compounds **1** and **2** were determined, which reveal that these two inhibitors exactly locate at the $Hp$PDF binding pocket (see Fig. 2.2). All these indicate that $Hp$PDF is a potential target for screening new anti-*H. pylori* agents. In a similar approach, Paul et al. (15) successfully recovered the corresponding targets of four unrelated ligands with reverse docking method.

The advantage of reverse docking is obvious, in addition to identifying target candidates for active compounds, it is also possible to identify potential targets responsible for toxicity and/or side effects of a drug under the hypothesis that the target database contains all the possible targets (20). However, reverse docking still has certain limitations. The major one is that the protein entries in the proteins structure databases like PDB are not enough for covering all the protein information of disease related genomes. The second one is that this approach has not considered the flexibility of proteins during docking simulation. These two aspects will produce false negatives. Another limitation is that the scoring function for reverse docking is not accurate enough, which will produce false positives (16). One tendency to overcome these shortages is to develop new docking program including protein flexibility and accurate scoring function. Another tendency is to integrate sequence-based and structure-based approaches together.

## 2.2 LEAD DISCOVERY

Computational approaches for lead discovery are more mature than those for target identification. Some methods such as virtual screening and library design have become promising tools for lead discovery and optimization (2, 13). Here, we introduce two computational techniques, virtual screening, and focused combinatorial library design.

### 2.2.1 Virtual Screening

The drugs developed in the past 100 years are found to interact with approximately 500 targets; in the same period, about 20,000,000 organic compounds including natural products have been synthesized or isolated. Moreover, the genomic and functional genomic projects have produced additional 1500 druggable targets for drug intervention to control human diseases (21). Therefore, it is believable that a large number of new drugs, at least many leads or hits, hide in the existing chemical *mine*. However, how to dig out this source is a hard task. Collecting all the existing compounds and screening them randomly against all the potential targets one by one are extremely unpractical, because it is intolerably expensive and time consuming. However, virtual screening shows a dawning to satisfy this requirement. Indeed, virtual screening has been incorporated into the pipeline of drug discovery as a practical tool (22).

In essence, virtual screening is designed for searching large-scale hypothetical databases of chemical structures or virtual libraries by using computational analysis for selecting a limited number of candidate molecules likely to be active against a chosen biological receptor (23). Therefore, virtual screening is a logical extension of 3-D pharmacophore-based database searching (PBDS) (24) or molecular docking (25), which is capable of automatically evaluating very large databases of compounds. Two strategies have been used in virtual screening (see Fig. 2.3): 1) using PBDS to identify potential hits from the databases, mostly in the cases that 3-D structures of the targets are unknown and 2) using molecular docking approach to rank the databases if the 3-D structures of the targets are available. Normally, these two approaches are used concurrently or sequentially, because the former can filter out the compounds quickly and the latter can evaluate the ligand-receptor binding more accurately.

For the pharmacophore-based screening, a 3-D-pharmacophore feature is constructed by structure-activity relationship analysis on a series of active compounds (26) or is deduced from the X-ray crystal structure of a ligand-receptor complex (27). Taking this 3-D–pharmacophore feature as a query structure, 3-D database search can be performed to select the molecules from the available chemical databases, which contain the pharmacophore elements and may conform to the

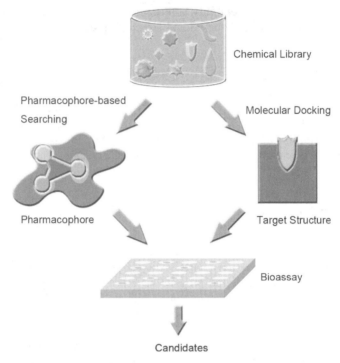

Chemical Library

Pharmacophore-based Searching

Molecular Docking

Pharmacophore

Target Structure

Bioassay

Candidates

**Figure 2.3**   The flowchart for virtual screening.

pharmacophore geometric constraints. Then the selected compounds are obtained either from commercial sources or from organic synthesis for the real pharmacologic assays (see Fig. 2.3).

Docking-based virtual screening (DBVS) requires the structural information of both receptors and compounds. The general procedure of DBVS includes five steps: receptor modeling (virtual screening mode construction), compound database generation, computer screening, hit molecules postprocessing, and experimental bioassay. The core step of virtual screening is docking and scoring. Docking is a process to place each molecule from a 3-D small-molecule database into the binding site of a receptor protein, optimize the relative orientation and conformation for a ligand interacting with a protein, and select molecules from the database that may bind to the protein tightly. Billions of possible conformations can be created for a flexible ligand even with a few freedom degrees in the rotation space alone, current optimization algorithms cannot sample exhaustively for all conformations and orientations, needless to say to account for the flexibility of protein with thousands of degrees of freedom. Therefore, instead of searching exhaustively in the search space, an optimization algorithm should sample the solutions effectively and rapidly close to the global optimum (28). Basically, the optimization algorithms employed widely by molecular docking program can be divided into three categories: numerical optimization methods, random or stochastic methods, and hybrid optimization methods. Essentially, there are three types of scoring functions (29), which include force-field based scoring functions, empirical scoring functions, and knowledge-based scoring functions. However, no single scoring function can perform satisfactorily for every system because many physical phenomena determining molecular recognition were not fully accounted for, such as entropic or solvent effects (13). Since Kuntz et al. (30) published the first docking algorithm DOCK in 1982, more than 20 docking programs have been developed in recent two decades. Of the existing docking program, DOCK, FlexX, AutoDock, and GOLD are most frequently used, other programs such as Glide, ICM, and Surflex have been applied successfully in virtual screening. However, many limitations and challenges still exist for molecular docking, such as accurate prediction of the binding conformation and affinity, protein flexibility, entropy, and solvent effects. Without doubt, the molecular docking process is still a complex and challenging project of computational chemistry and biology (25, 31).

Virtual screening has discovered numerous active compounds and leads, more than 50 compounds have entered into clinical trials, and some have been approved as drugs (2). Virtual screening enriched the hit rate (defined as the quotient in percentage of the number of active compounds at a particular concentration divided by the number of all compounds experimentally tested) by about 100-fold to 1000-fold over random screening (22, 32). Additionally, virtual screening provides an alternative way in fleetly finding new leads of some targets, whereas the techniques for high-throughput screening remain in development (e.g. potassium ion ($K^+$) channels). By using docking-based virtual screening in conjunction with electrophysiological assay, we discovered 10 new blockers of the eukaryotic *Shaker* $K^+$ channels, 4 natural product blockers (33), and 6 synthetic compounds (34).

### 2.2.2 Focused Combinatorial Library Design

What big pharmas and medicinal chemists are seeking is new chemical entities (NCEs), which can be strictly protected by compound patents. Nevertheless, all hits discovered by screening the existing compounds databases can only find the new medical usages for the existing compounds or old drugs. At least two kinds of computational approaches are available for NCE discovery: *de novo* drug design (35, 36) and combinatorial library design (37). The *de novo* drug design does not start from a database of complete molecules, but aims at building a complete molecule from molecular bricks ("building blocks") to fill the binding sites of a target molecule chemically. The complete chemical entries could be constructed through linking the "building blocks" together, or by growing from an "embryo" molecule with the guidance of evaluation of binding affinity. The "building blocks" could be either atoms or fragments (functional groups or small molecules). However, using atoms as "building blocks" is thought to be inefficient, which is seldom used currently. In the fragment-linking approach, the binding site is mapped to identify the possible anchor points for functional groups. Then, these groups are linked together to form a complete molecule. In the sequential-growing approach, the molecule grows in the binding site controlled by an appropriate search algorithm, which evaluates each growing possibility with a scoring function. Different from docking-based virtual screening, fragment based *de novo* design can perform sampling in whole compound space, which can obtain novel structures that are not limited in available databases. Nevertheless, the quality of a growing step strongly depends on previous steps. Any step chemically growing wrongly would lead to an unacceptable result. For the fragment linking approach, it remains a problem how to choose linkers to connect fragments together to form complete structures. The most remarkable drawback of this approach might be the synthetic accessibility of the designed structures.

To overcome the disadvantages of *de novo* design approaches, we developed a target-focused library design method to adopt the advantages of focused library and targeted library, as well as integrating technologies of docking-based virtual screening and drug-like (ADMET) analysis. A software package called LD1.0, was also developed (37) based on this technology. Starting with the structures of active hits and therapeutic target, the overall skeleton of potential ligands is split schematically into several fragments according to the interaction mechanism and the physicochemical properties of the binding site. Individual fragment library is constructed for each kind of fragment taking into account of the binding features of the fragments to the binding site. Finally, target-focused libraries regarding the studied target are constructed with the judgments from structural diversity, druglikeness (ADME/T) profiles, and binding affinities (37). During the target-focused library design, library-based genetic algorithm was applied to optimize our focused library, and our newly developed drug-likeness filter was used to predict the drug-like profile of the library. The molecular docking approach was employed to predict the binding affinities of the library molecules with the target.

One application example of target-focused library design is the discovery of new cyclophilin A (CypA) inhibitors (38, 39). CypA inhibitors are of therapeutic significance in organ transplantation. However, inhibitors of CypA are mainly derived from natural sources (such as cyclosporin A, FK506, rapamycin and san-glifehrin A) and peptide analogs, which are all structurally complex molecules, and little has been reported regarding the small-molecule CypA inhibitors. By using docking-based virtual screening, organic synthesis, and bioassay, we discovered dozens of binders of CypA with binding affinities under submicromolar or micromolar level, and 14 of them showed high CypA peptidyl-prolyl isomerase (PPIase) inhibition activities with $IC_{50}$s of 2.5–6.2 μM (38). To discover new chemical entities of CypA inhibitors with more potent activities, a focused library was designed based on the structures of the 14 inhibitors and their binding modes to CypA. Based on binding mechanism, each inhibitor was divided into three fragments, and three fragment sublibraries were constructed with sizes of 5, 3, and 17 fragments, respectively. Therefore, 255 molecules could be formed by conventional combinatorial library approach. With LD1.0, a CypA-focused

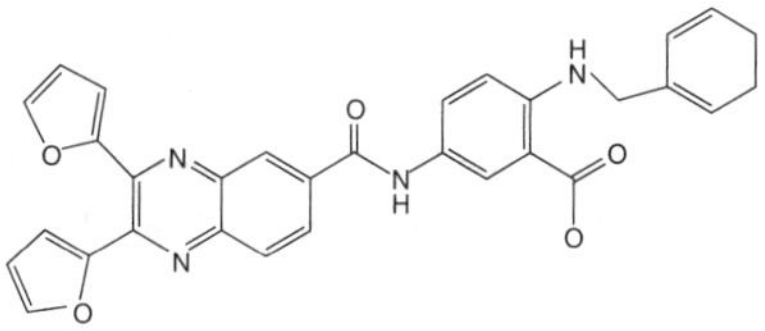

3a (0.25 μm)

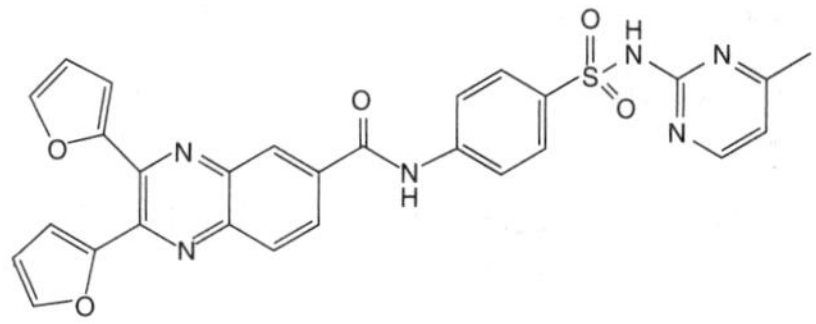

3b (6.43 μm)

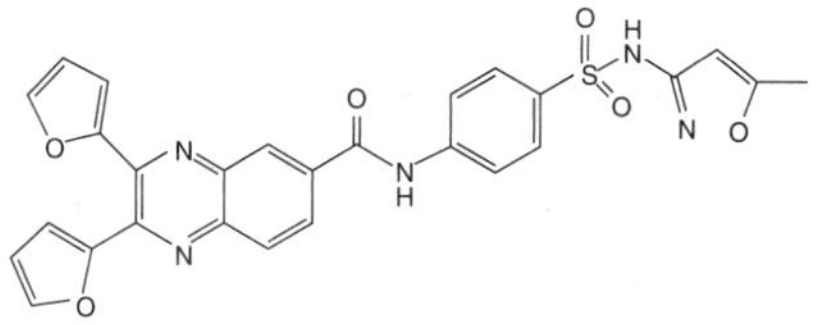

3c (2.86 μm)

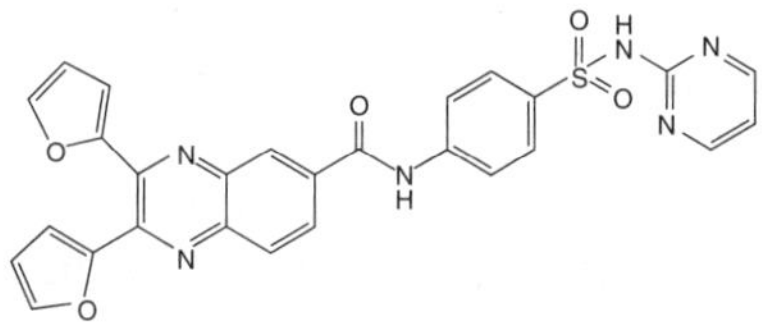

3d (0.96 μm)

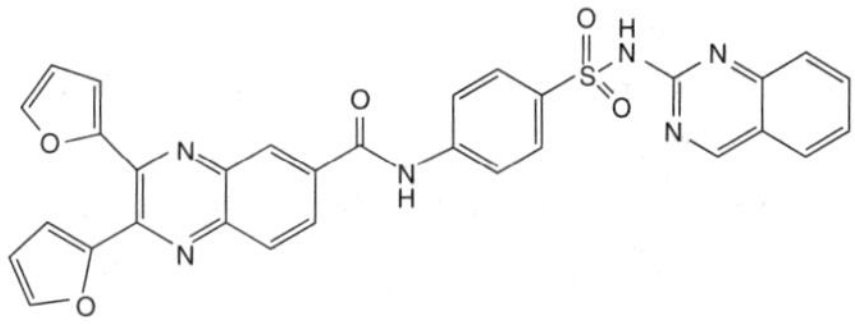

3e (0.78 μm)

**Figure 2.4** Chemical structures of the five CypA inhibitors discovered using focused library design. Left and right parts are the LD1.0 optimized fragments respectively binding to the small and large pockets of CypA, middle fragment is the LD1.0 optimized link binding to the saddle cleft between the two binding pockets of CypA (39). Data inside the parentheses are inhibition activities ($IC_{50}$ values) of the compounds.

library was optimized with a size of 2 × 2 × 4. Thus, the 16 compounds in the library were synthesized for bioassay. All these 16 molecules were determined to be CypA binders with binding affinities ($K_D$ values) that ranged from 0.076 to 41.0 μM, and five of them (**3a–3f**) are potent CypA inhibitors with PPIase inhibitory activities ($IC_{50}$ values) of 0.25–6.43 μM (see Fig. 2.4). The hit rates for binders and inhibitors are as high as 100% and 31.25%, respectively. Remarkably, both the binding affinity and inhibitory activity of the most potent compound increase ~10 times than that of the most active compound discovered previously. The high hit rate and the high potency of the new CypA inhibitors demonstrated the efficiency of the strategy for focused library design and screening. In addition, the novel chemical entities reported in this study could be leads for discovering new therapies against the CypA pathway.

## 2.3 ADME/T OPTIMIZATION

It has widely been appreciated that ADME/T should be involved in the early stage of drug discovery (lead discovery) in parallel with efficacy screening

(40, 41). Many high-throughput technologies for ADME/T evaluation have been developed, and the accumulated data make it possible to construct models for predicting the ADME/T properties of compounds before structural modifications. These computational models may reduce the redesign–synthesize–test cycles. Hereinafter, we will show how computational approaches predict and model the most relevant pharmacokinetic, metabolic and toxicity endpoints, which thereby accelerate the pace of new drug development.

### 2.3.1   Prediction of Intestinal Absorption

Absorption of drugs from the gastrointestinal tract is a complex process that can be influenced by not only physicochemical and physiological factors but also by formulation factors. Some published models and commercially available computer programs have been developed for predicting the intestinal absorption. Among them, physiologically based approaches are of particular interest because the human physiological environment and the processes involved in drug absorption are simulated.

Mixing tank models describe the intestine as one or more well-mixed tanks, in which dissolved and solid forms of drug have a uniform concentration and are transferred between the mixing tanks by first-order transit kinetics. This model has been implemented in commercial program Intellipharm PK (*http://www.intellipharm.com*) and OraSpotter (*http://www.zyxbio.com*). Tube models assume the intestine to be a cylindrical tube with or without changing radius, which includes the mass balance models and the models used in PK-Map and PK-Sim (*http://www.bayertechnology.com*). In the mass balance model, the difference between the mass flows in and out of the tube represents the mass absorbed per time unit, and it is proportional to the permeability and the concentration of the drug in the intestines. In models of PK-Map and PK-Sim, a transit function was introduced and the intestinal permeability coefficient can be calculated solely by the lipophilicity and molecular weight of the investigated compound.

Yu et al. (42) developed an absorption and transit (CAT) model, which formed a basis of many models and commercially available computer programs, such as GITA (GI-Transit Absorption), iDEA (*http://www.akosgmbh.de*) and GastroPlus (*http://www.simulations-plus.com*). iDEA uses an absorption model proposed by Grass et al. (43), and GastroPlus uses the ACAT (advanced compartmental absorption and transit) model developed by Agoram et al. (44). Parrott et al. (45) compared these two commercial programs both in their ability to predict fraction absorbed for a set of 28 drugs and in terms of the functionality offered. The results suggested that iDEA may perform better with measured input data, whereas GastroPlus presents a more sophisticated user interface, which shows strengths in its ability to integrate additional data.

### 2.3.2  Prediction of Blood–Brain Barrier (BBB) Permeability

The blood–brain barrier (BBB) is a physiological barrier in the circulatory system that restrains substances from entering into the central nervous system (CNS). The orally administrated drugs that target receptors and enzymes in the CNS must cross the BBB to produce desired therapeutic effects. At the same time, the peripherally acting agents should not cross it to avoid the side effects (46). To date, various methods that have been applied for model generation include the traditional statistical approaches and state-of-art machine-learning techniques.

Generally, current BBB models have shown acceptable capabilities depending on the approaches used and particularly the type and the size of the dataset under investigation. Some insights into the molecular properties that determine the brain permeation have also been gained. However, most current *in silico* modeling of BBB permeability start with the assumption that most drugs are transported across the BBB by passive diffusion. Absence of data regarding active transport or P-glycoprotein (P-gp) efflux limits the development of accurate and reliable BBB models that can more closely mimic the *in vivo* situation. Recently, Adeno et al. (47) reported their studies based on a large heterogeneous dataset (~1700 compounds) including 91 P-gp substrates. Likewise, Garg and Verma (48) took the active transport of the molecules into consideration in the form of their probabilities of becoming a substrate to P-gp, and their results indicated that the P-gp indeed plays a role in BBB permeability for some molecules.

### 2.3.3  Prediction of CYP-Mediated Metabolism

The human cytochromes P450 (CYP450) constitute a large family of heme enzymes. In drug development, the CYP-mediated metabolism is of particular interest because it may profoundly affect the initial bioavailability, desired activity, and safety profile of compounds. A great variety of *in silico* modeling approaches have been applied to CYP enzymes. Connections and integrations between the "ligand-based," "protein-based," and "ligand-protein interaction–based" groups of methods have been extensively described in the review of Graaf et al. (49). Apart from this methodological categorization, current *in silico* models mainly address the following three aspects of metabolism: CYP inhibition, isoform specificity, and site of CYP-mediated metabolism.

Inhibition of CYP enzymes is unwanted because of the risk of severe side effects caused by drug–drug interactions. A novel Line-Walking Recursive Partitioning method was presented that uses only nine chemical properties of the shape, polarizability, and charge of the molecule to classify a compound as a 3A4, 2C9, or 2D6 inhibitor (50). Recently, Jensen et al. (51) constructed Gaussian kernel weighted k-NN models to predict CYP2D6 and CYP3A4 inhibition, resulting in classification models with over 80% of overall accuracy.

CYP isoform specificity predicts the subtype responsible for the main route of metabolism. This type of study started from the work by Manga et al. (52), in which the isoform specificity for CYP3A4, CYP2D6, and CYP2C9 substrates was investigated. Terfloth et al. (53) reported a similar study on the same dataset using ligand-based methods, and the accuracy of prediction was improved substantially.

The site of metabolism (SOM) refers to the place in a molecule where the metabolic reaction occurs, which is usually a starting point in metabolic pathway investigation. Recently, Sheridan et al. (54) reported a study addressing the SOM mediated by CYP3A4, CYP2C9, and CYP2D6. In this study, pure QSAR-model was constructed with only substructure and physical property descriptors. The cross-validated accuracies range from 72% to 77% for the investigated datasets, which are significantly higher than that of earlier models and seem to be comparable with the results from MetaSite (*http://www.moldiscovery.com*), which is a more mechanism-based method of predicting SOM.

### 2.3.4   Prediction of Toxicity

As of 2008, a variety of toxicological tests needs to be conducted by the drug regulatory authorities for safety assessment. Computational techniques are fast and cheap alternatives to bioassays as they require neither experimental materials nor physically available compounds. Some examples of available computer programs that predict toxicity are Deductive Estimation on Risk from Existing Knowledge (*http://www.lhasalimited.org*), Multiple Computer Automated Structure Evaluation (*http://www.multicase.com*), and Toxicity Prediction by Komputer Assisted Technology (current DS TOPKAT, *http://accelrys.com/products/discovery-studio/toxicology*). Recently, developments in artificial intelligence research and the improvement of computational resources have led to efficient data-mining methods that can automatically extract structure toxicity relations (STRs) from toxicity databases with structurally diverse compounds. Helma et al. (55) described some general procedures used for generating (Q)STR models from experimental data. Here, we briefly present a study made to model the mutagenic probability (56).

Mutagenicity is the ability of a compound to cause mutations in DNA, which is one of the toxicological liabilities closely evaluated in drug discovery and their toxic mechanisms have been relatively well understood. In the development of (Q)STR models, it is essential to select the structural or chemical properties most relevant to the endpoint of interest. For mutagenicity, we designed a molecular electrophilicity vector (MEV) to depict the electrophilicity of chemicals, which is a major cause of chemical/DNA interaction. A model for the classification of a chemical compound into either mutagen or nonmutagen was then obtained by a support vector machine together with a $F$-score based feature selection. For model construction, the SVM+MEV method shows its superior efficiency in the data fitting, with a concordance rate of 91.86%. For validation, a prediction accuracy of 84.80% for the external test set was yielded, which is close to the experimental reproducibility of the Salmonella assay. The $F$-score based feature weighting analyses highlight the important role of atomic partial charges in describing the noncovalent intermolecular interactions with DNA, which is usually a hard part in the mutagenicity prediction.

## 2.4 THE FUTURE

The technological progress of computational chemistry and biology brought a paradigm change to both pharmas and research institutions. Computational tools have been considerably appreciated in drug discovery and development, which play increasingly important roles in target identification, lead discovery, and ADME/T prediction. In the future, in addition to improving individual existing computational techniques, such as perfection of the accuracy and effectiveness of virtual screening, one major tendency is to integrate computational chemistry and biology together with chemoinformatics and bioinformatics. This research is leading to a new topic known as pharmacoinformatics, which will impact the pharmaceutical development process and increase the success rate of development candidates (57). Another tendency is to use more accurate computational approaches, such as molecular dynamics simulation, to design specific inhibitors or activators against the pathway of target protein folding (58). By using this strategy, Broglia et al. (59) obtained inhibitors that may block HIV-1 protease monomer to fold into its native structure, which causes the inhibitors not to create resistance. Similarly, by targeting the $C$-terminal $\beta$-sheet region of an A$\beta$ intermediate structure extracted from molecular dynamics simulations of A$\beta$ conformational transition (60), we obtained a new inhibitor that abolishes A$\beta$ fibrillation using virtual screening in conjunction with thioflavin T fluorescence assay and atomic force microscopy determination (61).

After the completion of the human genome and numerous pathogen genomes, efforts are underway to understand the role of gene products in biological pathways and human diseases and to exploit their functions for the sake of discovering new drug targets (62). Small and cell-permeable chemical ligands are used increasingly in genomics approaches to understand the global functions of genome and proteome. This approach is referred to as chemical biology (or chemogenomics) (63). Another emerging field is computational systems biology to model or simulate intracellular and intercellular events using data gathered from genomic, proteomic, or metabolomic experiments (64). This field highly impacts drug discovery and development.

## REFERENCES

1. DiMasi JA, Hansen RW, Grabowski HG. The price of innovation: new estimates of drug development costs. J. Health. Econ. 2003;22:151–185.
2. Jorgensen WL. The many roles of computation in drug discovery. Science 2004;303:1813–1818.

3. Drews J. Drug discovery: a historical perspective. Science 2000;287:1960–1964.
4. Leader B, Baca QJ, Golan DE. Protein therapeutics: a summary and pharmacological classification. Nat. Rev. Drug Discov. 2008;7:21–39.
5. Huang CM, Elmets CA, Tang DC, Li F, Yusuf N. Proteomics reveals that proteins expressed during the early stage of Bacillus anthracis infection are potential targets for the development of vaccines and drugs. Genomics Proteomics Bioinformat. 2004;2:143–151.
6. Marton MJ, DeRisi JL, Bennett HA, Iyer VR, Meyer MR, Roberts CJ, Stoughton R, Burchard J, Slade D, Dai H, Bassett DE Jr., Hartwell LH, Brown PO, Friend SH. Drug target validation and identification of secondary drug target effects using DNA microarrays. Nat. Med. 1998;4:1293–1301.
7. Lindsay MA. Target discovery. Nat. Rev. Drug Discov. 2003;2:831–838.
8. Orth AP, Batalov S, Perrone M, Chanda SK. The promise of genomics to identify novel therapeutic targets. Expert Opin. Ther. Targets 2004;8:587–596.
9. Garcia-Lara J, Masalha M, Foster SJ. Staphylococcus aureus: the search for novel targets. Drug Discov. Today 2005;10:643–651.
10. Fliri AF, Loging WT, Thadeio PF, Volkmann RA. Biospectra analysis: model proteome characterizations for linking molecular structure and biological response. J. Med. Chem. 2005;48:6918–6925.
11. Lagunin A, Stepanchikova A, Filimonov D, Poroikov V. PASS: prediction of activity spectra for biologically active substances. Bioinformatics 2000;16:747–748.
12. Stepanchikova AV, Lagunin AA, Filimonov DA, Poroikov VV. Prediction of biological activity spectra for substances: evaluation on the diverse sets of drug-like structures. Curr. Med. Chem. 2003;10:225–233.
13. Kitchen DB, Decornez H, Furr JR, Bajorath J. Docking and scoring in virtual screening for drug discovery: methods and applications. Nat. Rev. Drug Discov. 2004;3:935–949.
14. Chen YZ, Zhi DG. Ligand-protein inverse docking and its potential use in the computer search of protein targets of a small molecule. Proteins 2001;43:217–226.
15. Paul N, Kellenberger E, Bret G, Muller P, Rognan D. Recovering the true targets of specific ligands by virtual screening of the protein data bank. Proteins 2004;54:671–680.
16. Li H, Gao Z, Kang L, Zhang H, Yang K, Yu K, Luo X, Zhu W, Chen K, Shen J, Wang X, Jiang H. TarFisDock: a web server for identifying drug targets with docking approach. Nucleic Acids Res. 2006;34:W219–224.
17. Gao Z, Li H, Zhang H, Liu X, Kang L, Luo X, Zhu W, Chen K, Wang X, Jiang H. PDTD: a web-accessible protein database for drug target identification. BMC Bioinf. 2008;9:104.
18. Yue JM, Xu J, Zhao Y, Sun HD, Lin ZW. Chemical components from Ceratostigma willmottianum. J. Nat. Prod. 1997;60:1031–1033.
19. Cai J, Han C, Hu T, Zhang J, Wu D, Wang F, Liu Y, Ding J, Chen K, Yue J, Shen X, Jiang H. Peptide deformylase is a potential target for anti-Helicobacter pylori drugs: reverse docking, enzymatic assay, and X-ray crystallography validation. Protein Sci. 2006;15:2071–2081.
20. Chen YZ, Ung CY. Prediction of potential toxicity and side effect protein targets of a small molecule by a ligand-protein inverse docking approach. J. Mol. Graph Model 2001;20:199–218.

21. Hopkins AL, Groom CR. The druggable genome. Nat. Rev. Drug Discov. 2002; 1:727–730.

22. Doman TN, McGovern SL, Witherbee BJ, Kasten TP, Kurumbail R, Stallings WC, Connolly DT, Shoichet BK. Molecular docking and high-throughput screening for novel inhibitors of protein tyrosine phosphatase-1B. J. Med. Chem. 2002;45:2213–2221.

23. McInnes C. Virtual screening strategies in drug discovery. Curr. Opin. Chem. Biol. 2007;11:494–502.

24. Ekins S, Mestres J, Testa B. In silico pharmacology for drug discovery: methods for virtual ligand screening and profiling. Br. J. Pharmacol. 2007;152:9–20.

25. Leach AR, Shoichet BK, Peishoff CE. Prediction of protein-ligand interactions. Docking and scoring: successes and gaps. J. Med. Chem. 2006;49:5851–5855.

26. Kurogi Y, Miyata K, Okamura T, Hashimoto K, Tsutsumi K, Nasu M, Moriyasu M. Discovery of novel mesangial cell proliferation inhibitors using a three-dimensional database searching method. J. Med. Chem. 2001;44:2304–2307.

27. Wolber G, Langer T. LigandScout: 3-D pharmacophores derived from protein-bound ligands and their use as virtual screening filters. J. Chem. Inf. Model 2005;45:160–169.

28. Li H, Li C, Gui C, Luo X, Chen K, Shen J, Wang X, Jiang H. GAsDock: a new approach for rapid flexible docking based on an improved multi-population genetic algorithm. Bioorg. Med. Chem. Lett. 2004;14:4671–4676.

29. Boehm HJ, Stahl M. The Use of Scoring Functions in Drug Discovery Applicationg, volume 18. 2002. Wiley, New York.

30. Kuntz ID, Blaney JM, Oatley SJ, Langridge R, Ferrin TE. A geometric approach to macromolecule-ligand interactions. J. Mol. Biol. 1982;161:269–288.

31. Moitessier N, Englebienne P, Lee D, Lawandi J, Corbeil CR. Towards the development of universal, fast and highly accurate docking/scoring methods: a long way to go. Br. J. Pharmacol. 2007;153:S7–26.

32. Shoichet BK. Virtual screening of chemical libraries. Nature 2004;432:862–865.

33. Liu H, Li Y, Song M, Tan X, Cheng F, Zheng S, Shen J, Luo X, Ji R, Yue J, Hu G, Jiang H, Chen K. Structure-based discovery of potassium channel blockers from natural products: virtual screening and electrophysiological assay testing. Chem. Biol. 2003;10:1103–1113.

34. Liu H, Gao ZB, Yao Z, Zheng SX, Li Y, Zhu WL, Tan XJ, Luo XM, Shen JH, Chen KX, Hu GY, Jiang HL. Discovering potassium channel blockers from synthetic compound database by using structure-based virtual screening in conjunction with electrophysiological assay. J. Med. Chem. 2007;50:83–93.

35. Honma T. Recent advances in de novo design strategy for practical lead identification. Med. Res. Rev. 2003;23:606–632.

36. Schneider G, Fechner U. Computer-based de novo design of drug-like molecules. Nat. Rev. Drug. Discov. 2005;4:649–663.

37. Chen G, Zheng S, Luo X, Shen J, Zhu W, Liu H, Gui C, Zhang J, Zheng M, Puah CM, Chen K, Jiang H. Focused combinatorial library design based on structural diversity, druglikeness and binding affinity score. J. Comb. Chem. 2005;7:398–406.

38. Li J, Chen J, Zhang L, Wang F, Gui C, Zhang L, Qin Y, Xu Q, Liu H, Nan F, Shen J, Bai D, Chen K, Shen X, Jiang H. One novel quinoxaline derivative as a potent human cyclophilin A inhibitor shows highly inhibitory activity against mouse spleen cell proliferation. Bioorg. Med. Chem. 2006;14:5527–5534.

39. Li J, Zhang J, Chen J, Luo X, Zhu W, Shen J, Liu H, Shen X, Jiang H. Strategy for discovering chemical inhibitors of human cyclophilin a: focused library design, virtual screening, chemical synthesis and bioassay. J. Comb. Chem. 2006;8:326–337.

40. van de Waterbeemd H, Gifford E. ADMET in silico modelling: towards prediction paradise? Nat. Rev. Drug Discov. 2003;2:192–204.

41. Selick HE, Beresford AP, Tarbit MH. The emerging importance of predictive ADME simulation in drug discovery. Drug Discov. Today 2002;7:109–116.

42. Yu LX, Lipka E, Crison JR, Amidon GL. Transport approaches to the biopharmaceutical design of oral drug delivery systems: prediction of intestinal absorption. Adv. Drug Deliv. Rev. 1996;19:359–376.

43. Grass GM. Simulation models to predict oral drug absorption from in vitro data. Adv. Drug Deliv. Rev. 1997;23:199–219.

44. Agoram B, Woltosz WS, Bolger MB. Predicting the impact of physiological and biochemical processes on oral drug bioavailability. Adv. Drug Deliv. Rev. 2001;50:S41–S67.

45. Parrott N, Lave T. Prediction of intestinal absorption: comparative assessment of GASTROPLUS (TM) and IDEA (TM). Eur. J. Pharm. Sci. 2002;17:51–61.

46. Cecchelli R, Berezowski V, Lundquist S, Culot M, Renftel M, Dehouck MP, Fenart L. Modelling of the blood-brain barrier in drug discovery and development. Nat. Rev. Drug Discov. 2007;6:650–661.

47. Adenot M, Lahana R. Blood-brain barrier permeation models: Discriminating between potential CNS and non-CNS drugs including P-glycoprotein substrates. J. Chem. Inf. Comput. Sci. 2004;44:239–248.

48. Garg P, Verma J. In silico prediction of blood brain barrier permeability: an artificial neural network model. J. Chem. Inf. Model. 2006;46:289–297.

49. de Graaf C, Vermeulen NPE, Feenstra KA. Cytochrome P450 in silico: an integrative modeling approach. J.Med. Chem. 2005;48:2725–2755.

50. Hudelson MG, Jones JP. Line-walking method for predicting the inhibition of P450 drug metabolism. J. Med. Chem. 2006;49:4367–4373.

51. Jensen BF, Vind C, Padkjaer SB, Brockhoff PB, Refsgaard HHF. In silico prediction of cytochrome P450 2D6 and 3A4 inhibition using Gaussian kernel weighted k-nearest neighbor and extended connectivity fingerprints, including structural fragment analysis of inhibitors versus noninhibitors. J. Med. Chem. 2007;50:501–511.

52. Manga N, Duffy JC, Rowe PH, Cronin MTD. Structure-based methods for the prediction of the dominant P450 enzyme in human drug biotransformation: consideration of CYP3A4, CYP2C9, CYP2D6. Sar Qsar Environ. Res. 2005;16:43–61.

53. Terfloth L, Bienfait B, Gasteiger J. Ligand-based models for the isoform specificity of cytochrome P450 3A4, 2D6, and 2C9 substrates. J. Chem. Inf. Model. 2007;47:1688–1701.

54. Sheridan RP, Korzekwa KR, Torres RA, Walker MJ. Empirical regioselectivity models for human cytochromes p450 3A4, 2D6, and 2C9. J. Med. Chem. 2007;50:3173–3184.

55. Helma C, Kazius J. Artificial intelligence and data mining for toxicity prediction. Curr. Comput.-Aid. Drug Des. 2006;2:1–19.
56. Zheng M, Liu Z, Xue C, Zhu W, Chen K, Luo X, Jiang H. Mutagenic probability estimation of chemical compounds by a novel molecular electrophilicity vector and support vector machine. Bioinformatics 2006;22:2099–2106.
57. Schuffenhauer A, Jacoby E. Annotating and mining the ligand-target chemogenomics knowledge space. Drug Discov. Today: BioSilico 2004;2:190–200.
58. Broglia R, Levy Y, Tiana G. HIV-1 protease folding and the design of drugs which do not create resistance. Curr. Opin. Struct. Biol. 2008;18:60–66.
59. Broglia RA, Tiana G, Sutto L, Provasi D, Simona F. Design of HIV-1-PR inhibitors that do not create resistance: blocking the folding of single monomers. Protein Sci. 2005;14:2668–2681.
60. Xu Y, Shen J, Luo X, Zhu W, Chen K, Ma J, Jiang H. Conformational transition of amyloid beta-peptide. Proc. Natl. Acad. Sci. USA 2005;102:5403–5407.
61. Liu D, Xu Y, Feng Y, Liu H, Shen X, Chen K, Ma J, Jiang H. Inhibitor discovery targeting the intermediate structure of beta-amyloid peptide on the conformational transition pathway: implications in the aggregation mechanism of beta-amyloid peptide. Biochemistry 2006;45:10963–10972.
62. Kopec KK, Bozyczko-Coyne D, Williams M. Target identification and validation in drug discovery: the role of proteomics. Biochem. Pharmacol. 2005;69:1133–1139.
63. Stockwell BR. Exploring biology with small organic molecules. Nature 2004;432:846–854.
64. Materi W, Wishart DS. Computational systems biology in drug discovery and development: methods and applications. Drug Discov. Today 2007;12:295–303.

## FURTHER READING

Afzelius L, Arnby CH, Broo A, Carlsson L, Isaksson C, Jurva U, Kjellander B, Kolmodin K, Nilsson K, Raubacher F, Weidolf L. State-of-the-art tools for computational site of metabolism predictions: comparative analysis, mechanistical insights, and future applications. Drug Metab. Rev. 2007;39:61–86.

Dutta A, Singh SK, Ghosh P, Mukherjee R, Mitter S, Bandyopadhyay D. In silico identification of potential therapeutic targets in the human pathogen Helicobacter pylori. In. Silico. Biol. 2006;6:43–47.

Ghosh S, Nie A, An J, Huang Z. Structure-based virtual screening of chemical libraries for drug discovery. Curr. Opin. Chem. Biol. 2006;10:194–202.

Helma C. In silico predictive toxicology: the state-of-the-art and strategies to predict human health effects. Curr. Opin. Drug Discov. Devel. 2005;8:27–31.

Klebe G. Virtual ligand screening: strategies, perspectives and limitations. Drug Discov. Today 2006;11:580–594.

Kumar N, Hendriks BS, Janes KA, de Graaf D, Lauffenburger DA. Applying computational modeling to drug discovery and development. Drug Discov. Today 2006;11:806–811.

León D, Markel S. In silico technologies. In: Drug Target Identification and Validation (Drug Discoveries Series). 2006, Boca Raton, FL, CRC Press.

Lindsay MA. Target discovery. Nat. Rev. Drug. Discov. 2003;2:831–838.

PDTD: *http://www.dddc.ac.cn/pdtd*

Sperandio O, Miteva MA, Delfaud F, Villoutreix BO. Receptor-based computational screening of compound databases: the main docking-scoring engines. Curr. Protein Pept. Sci. 2006;7:369–393.

TarFisDock: *http://www.dddc.ac.cn/tarfisdock*

Villoutreix BO, Renault N, Lagorce D, Sperandio O, Montes M, Miteva MA. Free resources to assist structure-based virtual ligand screening experiments. Curr. Protein Pept. Sci. 2007;8:381–411.

Wolber G, Seidel T, Bendix F, Langer T. Molecule-pharmacophore superpositioning and pattern matching in computational drug design. Drug Discov. Today 2008;13:23–29.

Zahler S, Tietze S, Totzke F, Kubbutat M, Meijer L, Vollmar AM, Apostolakis J. Inverse in silico screening for identification of kinase inhibitor targets. Chem. Biol. 2007;14:1207–1214.

## SEE ALSO

Chapter 6
Chapter 4

# 3

# DESIGN AND SELECTION OF SMALL MOLECULE INHIBITORS

JIANWEI CHE AND YI LIU

*Genomics Institute of the Novartis Research Foundation, San Diego, California*

NATHANAEL S. GRAY

*Harvard Medical School, Dana Farber Cancer Institute, Biological Chemistry and Molecular Pharmacology, Boston, Massachusetts*

The discovery and optimization of small-molecule inhibitors for the treatment of human disease has been the major focus of the pharmaceutical industry for more than a century. In the last decade, they have also attracted increased attention from academia because they provide a powerful means to interrogate biologic systems in an acute fashion. Although pharmaceutical companies have identified many small-molecule inhibitors for drug targets, to find a desirable small-molecule inhibitor for a specific target is still a major challenge for drug discovery and for chemical biology. Here, we present an overview of the computational approaches that can be exploited to discover small-molecule inhibitors for protein targets of interest.

Small molecule inhibitors are of special value to chemical biology and to drug discovery. Thousands of small molecules have been developed as drugs to treat various human diseases. Increasingly, small-molecule inhibitors are used to perturb the function of proteins to elucidate their biologic functions (1, 2). Complementary to the conventional genetics, small molecules allow rapid, conditional, and tunable inhibition. Furthermore, small molecules can often be identified that alter specifically one particular function of a multifunctional protein. The

*Chemical Biology: Approaches to Drug Discovery and Development to Targeting Disease*, First Edition.
Edited by Natanya Civjan.
© 2012 John Wiley & Sons, Inc. Published 2012 by John Wiley & Sons, Inc.

completion of the human genome sequencing project has resulted in the identification of a plethora of new protein targets for which small-molecule modulators
could be developed as potential therapeutics and as biologic "tools." Identifying a selective small-molecule inhibitor for each protein in human genome is a
formidable challenge for drug discovery and for chemical biology. Experimentally, high-throughput screening is the most widely used technology currently
to identify hits for a target. On the other hand, computational methods, such
as virtual screening and structure-based design, have become equally important
to identify and to optimize small-molecule inhibitors. In Fig. 3.1, we illustrate how computational and experimental approaches can complement each
other in the compound discovery and optimization process. The numbers of
compounds associated typically with each method is indicated. At early stages
of hit identification, high throughput methods such as virtual ligand screening
(VLS) and ultra high-throughput screening (uHTS) are used frequently. At later
stages, high accuracy algorithms combined with biologic assays can be used to
guide the selection and the synthesis of a limited set of analogs. As shown in
Fig. 3.1, computational and experimental approaches are not mutually exclusive processes. They are best exploited by using computation and experiment
in an iterative fashion to optimize compounds progressively for a particular target of interest. Without these interactions, none of the methods can achieve
its maximum potential. In this review, we focus on computational approaches
depicted in Fig. 3.1. As will become clear later in the chapter, most computational methods are not associated exclusively with predefined processes or
protocols. Significant redundancies exist in terms of when and what methods

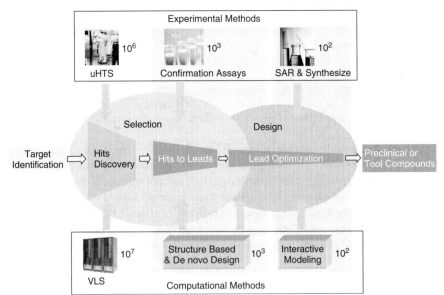

**Figure 3.1** Overview of computational methods used in the process of identifying small-
molecule inhibitors.

should be applied. The structure of the chapter follows the common flow of identifying small-molecule inhibitors. We will discuss the particular computational methodologies used at different stages of the process in subsequent sections.

## 3.1 VIRTUAL SCREENING IN SELECTING SMALL MOLECULE INHIBITORS

Virtual screening (VS) uses computational or theoretic methods to select molecules with desired biologic properties. It has broad applications in pharmaceutical industry. In this chapter, we limit ourselves to VS to identify biologically active small molecules. The estimated chemical space for small organic molecules is over $10^{60}$, and the registered organic and inorganic molecules in CAS are already over 30 million. Despite advances in experimental high throughput screening technologies, currently it is impossible to screen more than a few million compounds. With the rapid improvement of computational power and algorithms, VS methods have become an important complementary technique to experimental uHTS. It has many distinct advantages, such as investigating virtual molecules that can cover a broader area of chemical space. Compared with uHTS, VS is also a relatively inexpensive means to probe many molecular structures.

Generally, two types of virtual screening are used. One is receptor based, such as docking, and the other is ligand based, such as similarity search. In practice, both methods are often used synergistically. In Fig. 3.2, we illustrate how these two types of methods are used to select "hits" from a given small-molecule library.

## 3.2 RECEPTOR-BASED VLS

In receptor-based VLS, the atomic coordinates of receptor molecules are known experimentally or theoretically (e.g., homology models). The interactions between a small molecule and a protein are often characterized by a "lock-key" mechanism, in which high affinity small molecules often fit the protein-binding site well. The shape complementarities provide important enthalpy contributions from favorable van der Waals interactions. In addition, correctly positioned charges, hydrogen bonds, and hydrophobic contacts usually are required for higher affinity. Based on computational efficiency, receptor-based VLS can be divided into three tiers. The first tier is the popular high-throughput docking and scoring that can be computed for each ligand molecule within a few minutes on a typical desktop computer. In the rest of this chapter, we refer to high-throughput docking as docking and scoring for simplicity. The second tier includes the medium throughput methods that calculate the interactions more thoroughly and accurately. It often includes algorithms that account more accurately for desolvation energies, entropic penalties, and receptor flexibility. For example,

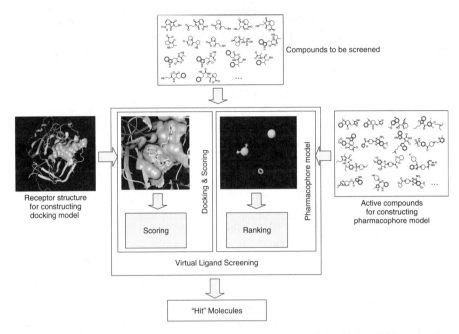

**Figure 3.2**  Workflow of computational algorithms used for identifying "hit" molecules from a compound library. The top panel indicates the compound collection from which the "hit" molecules are to be identified. The middle panel illustrates two virtual screening strategies. One is based on a known receptor 3-D structure (i.e., docking and scoring) and the other is based on active ligand molecules alone (i.e., pharmacophore searching and ranking). The docking/scoring uses existing 3-D structures of receptors to construct binding models, and the pharmacophore searching/ranking uses known active small molecules to construct pharmacophore models.

detailed implicit solvent models usually are applied in these calculations. It normally takes tens of minutes for each calculation. The third tier represents the slowest method but ultimately offers more accurate results. The typical examples are atomistic molecular dynamics (MD) or Monte Carlo simulations with explicit water molecules. The details of these calculations are out of the scope of this review. Interested readers can consulate reviews on these specific topics (3).

Among the three classes of methods, docking is the most widely used method in pharmaceutical industry. Many high throughput docking software packages are available publicly, such as Dock, Autodock, Gold, Glide, FlexX, QXP, and ICM. In addition, many companies have built their own proprietary docking technologies. A docking based VLS calculation usually consists of three stages: preparing, computing binding modes, and scoring. Currently, most docking programs treat receptors as rigid molecules. Under this approximation, a receptor field can be calculated for the binding site. A receptor field is the interaction potential created by receptor atoms. It consists of van der Waals, electrostatic, hydrogen bonding, and solvation contributions. Without atom thermal motions, the receptor field is also static, so it can be precalculated in the absence of ligand molecules. This feature provides enormous savings of computational time over explicit atom models. Different programs employ very different algorithms to sample ligand conformations, to orient small molecules in the binding site, to retain preferred binding modes, and to increase the computational efficiency (4–7). The predicted binding configuration is selected based on molecular interaction energies or an empirical scoring function. Finally, docking ranks all ligand molecules based on binding poses that possess the highest score, and virtual "hit" molecules are selected from the top of the list. This method allowed Hopkins et al. (8) to identify submillimolar Kinesin inhibitors from an 110,000 compound library. Novel casein kinase II inhibitors (9) and BCR-ABL tyrosine kinase inhibitors (10) were also discovered by docking. Docking can also be applied to discover small-molecule inhibitors for protein–protein interactions. Bonacci et al. (11) used docking (FlexX) to screen the chemical diversity set of 1990 compounds. Among 85 virtual hits, 9 of them inhibited SIGK binding with $G\beta_1\gamma_2$ with $IC_{50}$ that ranged from 100 nM to 60 $\mu$M. It was also used by Trosset et al. (12) to identify $\beta$-Catenin inhibitors for Wnt signal pathway from 177,000 compounds. Among 22 tested virtual hits, 3 hits were confirmed as binders to $\beta$-Catenin and Tcf4 competitors. Without experimental receptor structures, docking can be carried out using carefully constructed model structures. Becker et al. (13) and Salo et al. (14) reported the successful identification of small-molecule inhibitors of several G-protein coupled receptors (GPCR) systems based on docking.

One major challenge in docking-based VLS is the accuracy of scoring functions. Conceptually, binding free energy is a rigorous measure for correct binding conformation and compound affinity. However, the inherent errors in classic force fields and the extremely lengthy computation required preclude its immediate use for VLS. In practice, an empirical function (i.e., a scoring function) normally is used to estimate free energy of binding. A few examples of commonly used scoring functions are PMF (15), Chemscore (16), Drugscore (17), Goldscore (18), and Glidescore (19). The specific mathematic forms of Chemscore, Goldscore, and PMF are shown here to highlight their similarity and difference. Other scoring functions can be found in respective references.

$$\Delta G_{\text{Chemscore}} = \Delta G_0 + \Delta G_{\text{hbond}} + \Delta G_{\text{metal}} + \Delta G_{\text{lipo}} + \Delta G_{\text{rot}},$$

where $\Delta G_0$ is a fitting constant, $\Delta G_{\text{hbond}}$ is hydrogen bond energy between ligand and receptor, $\Delta G_{\text{metal}}$ is the energy between ligand atoms and metal ions in a receptor active site, $\Delta G_{\text{lipo}}$ is the ligand receptor lipophilic interaction, and $\Delta G_{\text{rot}}$ is for the rotatable bond penalty.

$$\Delta G_{\text{Goldscore}} = \Delta G_{\text{hbond}} + \Delta E_{\text{vdw}} + \Delta E_{\text{internal}},$$

where $\Delta E_{\text{hbond}}$ is the hydrogen bond energy between ligand and receptor, and $\Delta E_{\text{vdw}}$ is the van der Waals energy of ligand and receptor, and $\Delta E_{\text{internal}}$ is the van der Waals energy and torsional strain of the ligand molecule.

$$\Delta G_{\text{PMF\_score}} = \sum_{\substack{kl, \\ r < r_{cutoff}^{ij}}} A_{ij}(r) + \Delta E_{\text{vdw}},$$

$$A_{ij}(r) = -k_B T \ln \left[ f^j(r) \frac{\rho_{\text{seg}}^{ij}(r)}{\rho_{\text{bulk}}^{ij}} \right],$$

where $kl$ is a ligand receptor atom pair of type $ij$, $f^j(r)$ is ligand volume correction factor, $\rho_{\text{seg}}^{ij}(r)$ is the number density of a ligand receptor atom pair of type $ij$ at inter atomic distance $r$, $\rho_{\text{bulk}}^{ij}$ is the reference number density of ligand receptor pair of type $ij$, and $\Delta E_{\text{vdw}}$ is the van der Waals interaction between ligand and receptor. Although common features exist within different scoring functions, they often emphasize different aspects of the protein–ligand interaction. These empirical scoring functions bear significant errors because of crude approximations. Currently, as of 2008, no scoring functions exist that are able to rank diverse compound libraries consistently and accurately. The performance of docking based VLS varies considerably from system to system and from program to program (20). Although some programs tend to perform better than others do, it is difficult to predict the performance *a priori* on a specific system. In addition to ranking molecules based upon a single scoring function, combination of several scores (i.e., consensus score) or statistical methods (e.g., Bayesian statistics) are also used to select virtual hits.

Often, rigid receptor models provide a reasonable approximation and offer tremendous practical value. The majority of docking calculations used in industry adopt this approximation. In reality, however, receptor atoms undergo constant thermal motions, and some proteins have significant plasticity. One example is inactive and active conformations of kinases in which the activation loop moves in and out of active sites (21) with some atoms that move well over 10 Å. It is common to see a variety of conformational states for a given protein in PDB databank. It is also possible that such conformations observed crystallographically represent only a fraction of the transient dynamic states that may exist in solution. Although several methods (22–24) have been developed to account for protein flexibility, considerably more computational and experimental work will be required to create algorithms that can incorporate receptor flexibility accurately. In addition to receptor plasticity, another complicating factor is the difficulty to account accurately for the enthalpic and entropic contributions of discrete water molecules in the ligand bound and unbound states. Water molecules have been observed crystallographically to play important roles in mediating hydrogen bonds between protein atoms and between protein atoms and ligands. In most docking calculations, the water molecules are removed from binding sites, which assumes that to displace these water molecules by small molecules would give favorable entropic contribution. However, the entropic gain comes at the expense of enthalpy loss if the water molecules make strong hydrogen bonds to receptor atoms or mediate ligand–protein interactions. Several programs exist that attempt to include movable water molecules in docking calculations, however, often this results in a significant loss of computational efficiency.

Besides virtual screening of ligand libraries, docking is extremely useful for lead optimization. Successful docking can elucidate the important interactions between lead molecules and the target, explain important features of the structure activity relationships (SAR), and consequently help to design new molecules. Other than docking, useful receptor based VLS tools include MD simulation, MC simulation, and QM/MM calculation.

## 3.3 LIGAND-BASED VIRTUAL SCREENING (2-D SIMILARITY SEARCH, 3-D PHARMACOPHORE SEARCH)

Another category of VLS methods is to identify ligands for targets that have unknown 3-dimensional (3-D) structures. Generally, these methods are called ligand-based VLS. The most simple and probably the most used searches involve a physical, chemical, or topological similarity search based on one or several active compounds. The similarity between two molecules can be defined in various ways. Two-dimensional (2-D) methods refer to those that only consider atom connection tables to describe molecular structures, because atom connection tables are best represented by 2-D pictures such as the ones in the top panel of Fig. 3.2. These methods consider topological relationships between atoms in a molecule regardless their actual 3-Dl coordinates. The simplicity of these methods allows computationally efficient comparisons. It can screen millions of molecules in seconds. Here, we outline a few representative examples.

1. Molecular 2-D fingerprints are a class of methods that use chemical substructures to characterize similarities between molecules. A fingerprint is a signature that encodes the chemical topological information of a small molecule. It is usually a long binary bit string in which each bit represents the presence (1) or absence 0 of a particular chemical substructure. For example, an ATP molecule has substructures such as a pyrimidine and imidazole heterocycle but not benzene; therefore, the bits that represent pyrimidine and imidazole would be set to "1" and the bit that represents benzene would be set "0" in the fingerprint bit string of ATP. Many variations in detailed definitions of 2-D fingerprints exist. In one straightforward way, molecular similarities are measured by the number of bits two compounds have in common.

2. Pharmacophores are defined as essential structural features in a molecule that are responsible for its binding activity, and they can also be used to generate 2-D fingerprints. Examples of commonly used pharmacophores are specific molecular fragments (i.e., an isopropyl group), hydrogen bond donors and acceptors, and negatively and positively charged atoms or

groups. In two dimensions, the minimum number of bonds between two pharmacophores can be used as a distance metric. For example, the exocyclic amino group of ATP is located 3, 4, and 5 bonds away to the three acceptor nitrogens in the purine ring. This information is coded readily in a fingerprint bit string.

3. In addition to 2-D fingerprints, the topological relationships of atoms in a molecule can be explored in high feature dimensions and nonlinear fashion by other means, such as kernel methods (25), support vector machines (26), and self-organizing maps (27).

However, molecular recognition occurs in a 3-D world. It is well known that minor chemical modifications can alter the activity of a molecule drastically. On the other hand, very different chemical structures can have similar activity against the same target. The mere existence of particular pharmacophores and/or their topological relations are not sufficient for a molecule to be active against its target. To account for 3-D information, various VLS techniques based on activities of small molecules without receptor structures have been developed, such as 3-D quantitative structure activity relationship (QSAR) models. It is commonly accepted that a small molecule must exhibit a high degree of shape complementarity to bind to its protein target with high affinity. Finding the correct superposition of active ligand molecules is the key to a successful 3-D QSAR model. Of course, shape alone does not account for all contributions of small molecules. Other factors such as locations of hydrogen bond donors and acceptors, types of functional groups, electrostatic properties, and polarizabilities all play important roles for small-molecule activities. One attempt to capture this information is 3-D pharmacophore method. Like docking, many software programs exist to develop 3-D pharmacophore models, such as DISCO (28), Catalyst (28), GASP (28), PHASE (29), and many others (30, 31). Pharmacophore features are not mutually exclusive; for instance, a hydrogen bond acceptor (i.e., donating electron pair) can be negatively charged (e.g., side chain carboxylate from Asp or Glu). Most programs provide users with the flexibility to define their own pharmacophore features. The main merit of 3-D pharmacophore models is the ability to identify lead compounds that are diverse structurally.

Although software packages differ in how they construct pharmacophore models, they go through the following common steps: generate conformers, enumerate pharmacophore features, generate a hypothesis, and screen the ligand library. Usually, the active conformation of a small molecule is unknown without experimental evidence. A key component of a good pharmacophore method is to sample active conformations rapidly. Studies have shown that the active conformation of a small molecule may not necessarily be the same as the lowest energy one in the unbound state (32). Therefore, a set of possible conformers within a certain energy window (e.g., 15 kcal/mol above the lowest energy conformer) are collected and clustered. The feature-enumerating step is relatively straightforward. It identifies all the possible pharmacophore features in small molecules. Hypothesis generation is another challenging step in building pharmacophore models. In essence, this step involves finding a proper alignment of active molecules with a high degree of overlap in 3-D space in terms of their pharmacophore features. During this step, the spatial arrangements of pharmacophore features are identified so that they are common for all or a significant subset of active molecules. This step assumes that active molecules must satisfy a common pattern of pharmacophore features. However, it should not require a molecule to have all the features defined in a model to be active. The proper alignment of active molecules determines the quality of a 3-D pharmacophore model. Several algorithms have been developed to identify pharmacophore patterns (33–36) among the set of possible conformations for each active molecule. Pharmacophore hypothesis algorithms try to find and to rank the most probable pharmacophore models for known active molecules. It should also be noted that active molecules do not necessarily share a common pharmacophore pattern. Distinct binding modes have been observed in cocrystal structures (37). Moreover, different small molecules may bind to different binding sites of a receptor. Although a few algorithms exist to identify multiple pharmacophore patterns within a set of actives (29, 36), more robust methods are needed. The conformations of small molecules can be precomputed or sampled "on the fly." Once a valid pharmacophore model is constructed, it can be used to screen millions of small molecules. This strategy was used successfully by Singh et al. to identify some potent and novel VLA-4 antagonists (38). Three-dimensional pharmacophore models are used commonly for GPCR systems because generally, these membrane-localized target structures are unknown. A three-point pharmacophore model was constructed for Urotensin II receptor (39) and was used to search Aventis compound collection. Of 500 virtual "hits," 10 hits were active with the most potent compound exhibiting an $IC_{50}$ of 400 nM. Recently, the first GPR30 specific agonist was discovered by 2-D and shape similarity in combination with a pharmacophore search (40).

Not only can active molecules be used in pharmacophore modeling, structures of inactive molecules can also provide valuable information. In common pharmacophore models, information provided by inactive compounds is captured in the concept of "excluded volume" (29, 41). An "excluded volume" is a region that is so close to the receptor that no ligand atoms are tolerated. "Excluded volume" can be obtained automatically or defined manually. The method relies on the assumption that the binding modes of the compound series are similar to each other. Although many examples from cocrystal structures and SAR have confirmed this assumption, many exceptions to this empirical observation exist. Chemical similarity does not always predict similarity in biologic activities. When active compounds consist of multiple scaffolds because of different modes of action, it is very challenging to develop pharmacophore models and to classify the actives into proper categories automatically. In addition, a compound that satisfies a good pharmacophore model is not guaranteed to have activity. For example, entropic contributions, solubility, cell permeability, and stability can also have important implications on a compound's biologic activity.

Even though most pharmacophore models are built without receptor structures, binding site structures do help to construct proper pharmacophore models. The critical residues provide a complementary image of ligand pharmacophore field. This "negative image" can be converted to a ligand pharmacophore model. One advantage of receptor structure based pharmacophore models is the natural definition of "excluded volume." The pharmacophore model derived from a receptor structure can be used not only to search small-molecule libraries directly, but also to guide docking calculations (42). Similarly, docking can help to derive receptor based pharmacophore models as well. For instance, docking active compounds can help to identify the important pharmacophore features. Submicromolar tRNA-guanine transglycosylase inhibitors were discovered by combinations of pharmacophore searching and docking (43). Pharmacophore models can also be used for de novo design (44, 45) and building 3-D QSAR models (29, 46).

## 3.4  DESIGN OF SMALL MOLECULE INHIBITORS

VLS is a cost-effective and time-efficient approach to identify inhibitors for a given target from a small-molecule library. However, the hits from virtual screening usually are of low affinity and have properties that are not ideal for drug development or even for use as research tools (such as lacking cell permeability, potency, selectivity, and solubility). Furthermore, VLS is often performed using existing small-molecule libraries, which are biased heavily toward thoroughly explored regions of chemical space. An alternative and very important method to identify small-molecule inhibitors is through design. Two strategies are used in lead design. One strategy is based on known scaffolds or fragments from VLS, HTS, and fragment based screening to improve potency and desired pharmacological properties. The other strategy is de novo design, which generates small molecule leads from scratch. Figure 3.3 shows the relationships between these strategies and shows when they can be used appropriately.

## 3.5  SCAFFOLDS AND FRAGMENTS-BASED LEAD DESIGN

Computational methods offer ways to investigate chemical modifications automatically and rapidly that may be beneficial to the small-molecule binding. Based on a core fragment or scaffold of a hit, substituents are introduced to the core structure from a small fragment library to the core structure. The selection of substituents is biased toward favorable binding interactions. For each modification, a local energy optimization or full docking can be carried out. The new molecule is then evaluated based on a scoring function that measures interactions with the protein-binding site. The scoring function is similar to the one used in docking with necessary modifications to accommodate additional property requirements. The substituents can be added continuously until a given criteria, such as the

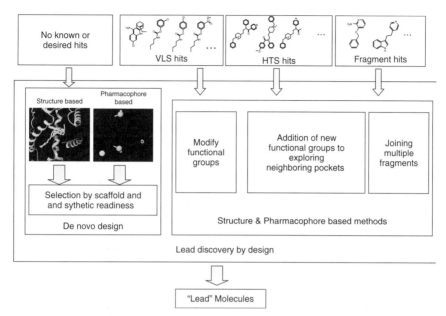

**Figure 3.3** Workflow of computational algorithms used to design lead molecules. The top panels indicate various sources of starting points on which "lead" design can be based. The middle panels illustrate the corresponding design strategies under the different starting conditions. The details of each method are described in the main text.

number of iterations, desired score, or maximum size of a molecule, is satisfied. LUDI (47) by Accelrys, Rachel and LeapFrog by Tripos, and molecular evolution developed internally at GNF are examples of the method. Programs differ in the details of how molecular structures are modified, optimized, and ranked. This strategy has been used to improve activities of HIV protease inhibitors (47) and other molecules (48).

Besides obtaining scaffolds from VLS, experimental fragment-based screening by NMR and crystallography offers a good starting point to design novel small-molecule inhibitors. The key to this method is cocrystal structures of the target protein and low-molecular-weight compounds, typically limited to molecular weights of between 120–250 Da to allow for subsequent additions. Usually, these compounds have weak affinities (Kds = 10 μM–10 mmol/L). To explore modifications that have the potential to increase binding affinities, the aforementioned computational methods can be applied readily. For structures with a single small molecule in a complex, the small compound can be extended automatically to maximize its interaction with target protein. During the progressive modifications, the binding mode for the initial fragment can be altered. However, certain constraints are applied to avoid drastic changes to the binding modes unless substantial modifications are introduced to the original structure. When evidence exists that multiple ligands bind to different pockets of an extended binding site in one or multiple crystal structures, computational methods can be used to propose

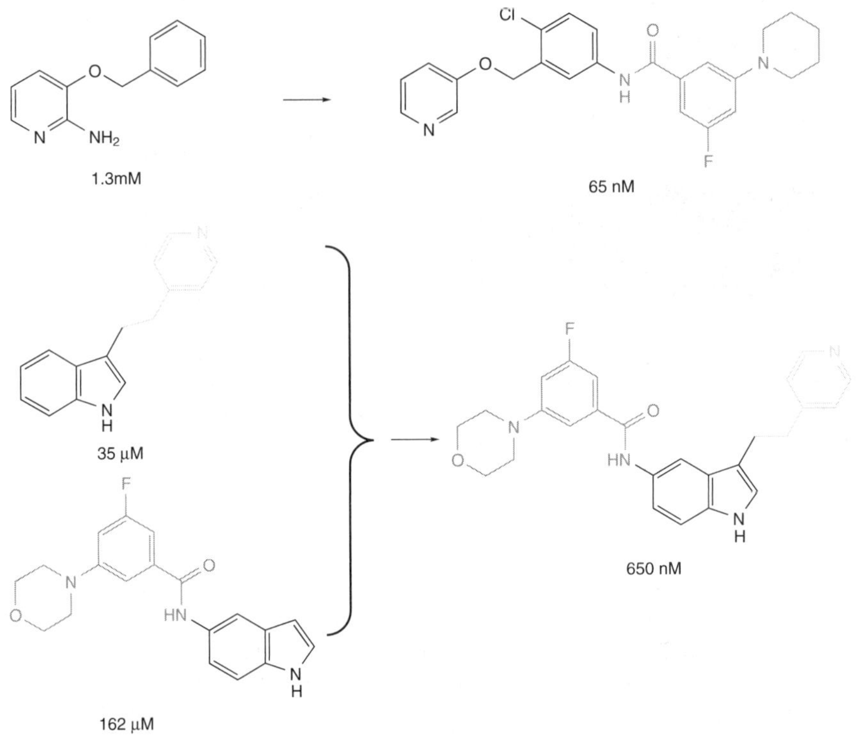

**Figure 3.4** Inhibitor design based on low-molecular-weight compounds. In the upper panel, the base fragment is shown in black and the added groups are shown in red. In the lower panel, a common core fragment for cojoining two fragments is shown in black and the different substituents are shown in red and light blue.

proper linkers or fuse-disjointed fragments to yield the synergistic effects. Gill et al. (49) have used this strategy to design P38α MAP kinase inhibitors based on hits with affinities in the millimolar range. For example, they identified a lead molecule with $IC_{50}$ = 65 nM by extending the base fragment to acquire additional interactions, and also to achieve a 100-fold increasing in potency rapidly by cojoining two overlapping fragment hits for an indole-derived compound. The structures are depicted in Fig. 3.4. It should be emphasized that crystal structures are extremely important in fragment-based design iterations. Because fragment hits are usually small and possess low affinity for the targets, to predict the correct binding modes by docking is difficult. In addition, the modifications to the base fragment may alter its binding modes fundamentally.

## 3.6 HYBRID DESIGN BASED ON KNOWN INHIBITORS

Through many years of drug discovery efforts, pharmaceutical companies have synthesized millions of small-molecule inhibitors. A large amount of information has been accumulated on how these inhibitors bind to their protein targets and to the QSAR of these inhibitors. Hybrid design is an emerging technique that attempts to take full advantage of this information to design new inhibitors for existing or new protein targets (50). The essence of the hybrid design is to recombine known inhibitors for a particular target in a rational way to create new inhibitors for the same target or for a new target. In a manner analogous to the evolutionary recycling of protein domains, the hope for chemical evolution through hybrid design is to create new inhibitors by recombining pharmacophores derived from different classes of compounds. Hybrid design starts by using inhibitor-receptor binding information obtained experimentally from crystallography or computationally by docking to overlay multiple known ligands, and then to predict how to recombine inhibitor substructures to create new compounds. This strategy is illustrated in Fig. 3.5, in which a potent Aurora A kinase inhibitor can be thought of as a hybrid of two inhibitors developed originally for two other kinases. This strategy has also been applied successfully to design novel kinase inhibitors that bind to a specific inactive kinase conformation (51, 52). Gleevec is a Bcr-Abl inhibitor that has been demonstrated crystallographically to bind to the nucleotide binding cleft of Abl, which uses both the adenine binding region and an adjacent hydrophobic pocket created by the activation loop being in a unique inactive conformation (53). A hybrid strategy involved appending a 3-trifluoromethylbenzamide group to a known ATP binding site inhibitor

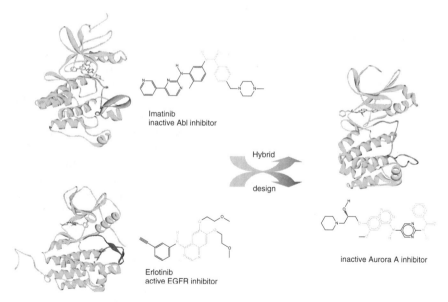

**Figure 3.5** Illustration of hybrid design strategy for small-molecule inhibitors. The activation loops of the proteins are colored differently to indicate active (blue) and inactive (orange) states. The important fragments of inhibitors are colored blue and orange to highlight the hybrid design concept.

(51). Of the hybrid compounds thus created, potent and selective inhibitors have been identified for Abl, c-Kit, p38, PDGFR, and Aurora kinases (52). Figure 3.5 shows how a benzamide group is attached to a quinazoline scaffold to create an inhibitor that binds to an inactive Aurora kinase conformation similar to that observed on Abl-Gleevec complex.

A computer program named BREED automates the process for hybrid design (54). The method imitates the common medicinal chemistry practice of joining fragments of two known ligands to generate a new inhibitor. The known active ligands are superimposed in their active conformation to identify all overlapping bonds, and the fragments on each side of each matching bond are swapped to generate a large set of novel inhibitors. This method has been demonstrated to have a high rate of success to identify novel inhibitors for HIV protease and protein kinases (54).

## 3.7   DE NOVO DESIGN

Small molecule inhibitors can also be generated by de novo design without the prior knowledge of other active ligands. In essence, the procedures are similar to those mentioned previously. For example, LeapFrog can start without a given small molecule. It will select a core randomly and will grow a ligand from its fragment library. Although it is still rare for a computer program to generate practical and active molecules from scratch, reports of novel molecules from "raw" output of automated computer program exist. SkelGen (55) was able to generate small-molecules inhibitors for estrogen receptor binding sites without hints from known actives. Among the 17 highest scoring structures, 5 structures were active with 4 structures that posses novel structures. We have also implemented a de novo ligand design algorithm in our in-house software GModE using evolutionary algorithm. A molecule is represented by a unique genetic code that is subject to normal genetic operations such as mutation, addition, deletion, and translocation. The population of ligand molecules is evolved toward increasing fitness. Our retrospective studies have shown that the method can produce lead molecules that are extremely similar to experimental nanomolar inhibitors.

If the target binding site structure is unavailable, de novo design can be carried out based on known active molecules. This circumstance may occur because none of the known actives that pose desired properties or that pose novel inhibitors are sought for patentability. However, the known inhibitors provide a basis to construct 3-D pharmacophore and other models such as Comparative Molecular Field Analysis (CoMFA) that can provide a template to guide the computational algorithm to maximize the fitness of the molecule "population" toward a given model. For example, LeapFrog can use CoMFA models in place of target binding site structures. It is also straightforward to apply pharmacophore models to novel molecular design with and without starting fragments. Novel structures are chosen based on its fitness to the given model. Unlike de novo design with known binding site structures, it is important to minimize the possibilities for resulting molecules to have steric clashes with receptor atoms. Even though these programs have produced promising results, the major drawback for such automated computer algorithms is that the "raw" computer derived molecules often are undesirable synthetically or implausible chemically. Some programs (55) have implemented chemistry rules to guide the generation of new molecules, but it is still far from satisfactory. Moreover, because a drug molecule interacts with its intended target in a complex biologic environment, a good lead molecule is selected by not only its binding affinity to its intended target, but also other properties such as crucial physical chemical properties, ability to expand SAR, selectivity relative to undesired off targets such as cytochrome P450 enzymes, and ion channels such as HERG. All are important considerations to advance a molecule toward clinical development. Various computational methods have been developed to predict these properties; however, normally they are not integrated into automated de novo design program.

## 3.8 HIT AND LEAD OPTIMIZATION

Computational algorithms and modeling have advanced significantly over the last two decades in terms of accuracy, efficiency, and ease of use. Many modeling procedures are automated; however, they are far from replacing the intuition and knowledge of an experienced computational chemist. Although automated programs are used widely in early stages of lead discovery with limited human intervention, lead optimization is a process in which a computational chemist makes most contributions out of his knowledge, insight, experience, and intuition. Before lead optimization, computational algorithms are designed to explore wide areas of chemical space and a wide range of activity data. During lead optimization, focus is placed on one or a few particular compound series. Small modifications are introduced to investigate local SAR and to improve other pharmacologic properties. Incremental improvements are often the objectives at this stage. Whereas the automated computational methods described previously are still important, they are not sufficiently precise to distinguish between modifications that result in 2 kcal/mol changes in binding affinities. Manual intervention, such as visual inspection and manual docking from an experienced chemist, are often of great importance for this phase.

## 3.9 STRUCTURE BASED

As we can observe from previous discussion, protein structures have played invaluable roles in small-molecule inhibitor selection and design. Currently, over 40,000 biomacromolecular structures are deposited in RCSB protein databank. Over 5000 new structures were deposited in 2006. The atomic level structural information provides an enormous wealth of mechanistic insights with regard to protein function. Approximately 6500 protein structures in the database are in complex with an inhibitor or substrate to provide chemists with important insights into the mechanism of ligand recognition. Because biologic systems are very diverse, each system requires specialized knowledge and treatment. Successful ligand design requires an intimate knowledge of all known inhibitors, which include how their activity and selectivity is changed as a function of chemical modification (SAR), rules for which interactions are required for high affinity binding, knowledge about the potential for conformational rearrangements, and insight into what is tractable synthetically. Success therefore requires the close collaboration of computational chemists, medicinal chemists, and structural biologists. Here, we will attempt only to describe a few common techniques that computational chemists use to optimize the activity of a small-molecule inhibitor.

Visual inspection is probably one of the oldest and most powerful methods in computer modeling. Large amounts of information can be acquired by looking at the 3-D structures of an inhibitor and a protein complex, such as important hydrogen bonds, hydrophobic contacts, and strong electrostatic interactions. It allows one to design a derivative to make stronger or additional hydrogen bonds, to explore unoccupied pockets, and even to alter the scaffold while maintaining important interactions. Of course, regular molecular modeling methods are used here to ensure the proposed molecules possess sensible conformations. In contrast to the automatic molecular generating programs mentioned previously, the modifications here are proposed directly by chemists; therefore, the molecules are tractable synthetically and often follow a particular experimentally derived SAR series. The iterative use of model building, synthesis, and biologic evaluation is the most powerful and efficient means to obtain small molecules with a desired biologic activity. When multiple cocrystal structures with different ligands exist, one can gain even more insights into conserved interactions. This insight can be very useful to identify key interactions and to help design new compounds to maximize possible key interactions. An algorithm developed by Deng et al. (56) uses a binary string to represent ligand protein interactions. By clustering the binary strings of different small-molecule binders, the authors were able to illustrate the similarity and diversity of small-molecule binding modes. This information can be used to select correct binding modes as well as to propose new molecules.

"Anchored docking" is another common technique used in lead optimization. Unlike automated docking, "anchor" points such as particular hydrogen bonding are specified by the user to limit the configuration space sampled by docking program. This limit is to ensure that the knowledge of a chemist about the system is enforced, because usually automated docking is sufficiently accurate to find correct binding modes consistently. The "anchor" points can be defined for a receptor, a small molecule, or both. The correct definition of "anchor" points largely relies on the users' knowledge of the system.

## 3.10   QSAR BASED

Even with amount of protein structures available, many interesting protein targets without 3-D structural information exist. GPCRs are typical examples. In this case, only SAR derived from the biologic assay can be used to guide optimization. Under these circumstances, QSAR has been an important tool. It is probably one of the earliest tools for computer aided drug design. Properly choosing descriptors for QSAR models is much like an art and requires a deep understanding of the system of interest. Most common descriptors are based on physical, chemical, geometric, and topological properties. Three-dimensional QSAR models may require active ligands to be aligned properly just like the pharmacophore model building. In fact, 3-D QSAR models can be generated directly by pharmacophore models, and many pharmacophore-generating programs provide this capability. One method developed early and widely used for 3-D QSAR is CoMFA.

## 3.11 DESIGN OF SELECTIVE SMALL MOLECULE INHIBITORS

For a small-molecule inhibitor to be useful as a therapeutic drug or as a tool for chemical biology study, it must have specificity for its intended target. A small-molecule drug often causes side effects or toxicity if it also modulates other proteins in the body (i.e., off-targets). Nonselective inhibitors will confound the analysis of resulting phenotypes. Therefore, the selectivity is a very important feature when considering design and selection of small-molecule inhibitors. Because ~30,000 proteins are encoded in the human genome, which can possess multiple ligand binding sites and can be present in vastly different concentrations, finding specific inhibitors can be an extremely difficult task. Analyzing the sequence and structural differences between a targeted protein and its closely related homologs is an effective way to design selectivity rationally.

## 3.12 TARGETING THE DIFFERENCES IN THE BINDING POCKET

The binding affinity of a ligand is determined by its complementarity to the size, shape, and physicochemical properties of the protein-binding site. To design selective small-molecule inhibitors rationally, the 3-D structure of the target-binding site is required. In principle, the structure of a protein-binding pocket is determined by its amino acid sequence, and it is straightforward to identify unique amino acids by sequence alignment. Once the unique residues are identified, they can be analyzed in the context of small-molecule binding to determine the residues that are important for protein ligand interaction. The inhibitors can then be designed to target them specifically for selectivity. Taunton et al. have used this approach to design selective inhibitors for p90 ribosomal protein S6 kinases (RSKs) (57). They identified two key selective determinants for RSKs relative to other kinases: One determinant is a small "gatekeeper" The amino acid, and the other is a Cys in the glycine-rich loop. Targeting these two residues, they were able to make very potent and selective inhibitors for RSKs.

Although successful for RSKs, it is often very difficult to identify unique residues suitable for selectivity for all proteins in a family. In many cases, a monoselective inhibitor among the protein family members is not feasible, so selectivity against a subset of the family members is desired. In this case, selectivity can be achieved by targeting the sequence differences in the binding sites between the target protein and the unwanted proteins. This strategy has been applied successfully to design selective inhibitors for targets within a closely related subfamily, such as p38 kinase $\alpha$, $\beta$, $\gamma$, and $\delta$ isoforms (58); protein kinase CDK2, and CDK4 (59); and COX-1 and COX-2 (60).

For protein targets that are closely related, the binding sites may look the same and no sequence difference is around the binding pocket. Therefore, a small molecule that binds to this primary site has very little selectivity. In these cases, a commonly used strategy is to search for a secondary site near the primary binding pocket. If it is less conserved, selective inhibitors can be designed by targeting the secondary site. For example, it is very difficult to develop a selective protein tyrosine phosphatase-1 B (PTP-1B) inhibitor, but a crystal structure of PTP-1B in complex with bis-(para-phosphophenyl) methane reveals a secondary aryl phosphate-binding pocket adjacent to the active site that can be targeted to design selective PTP-1B inhibitors (61).

Many protein structures are highly dynamic, and induced-fit effects are well known in small-molecule and protein binding (21). Although two proteins can have very similar binding pockets in static form, they can have different plasticity. Although protein flexibility is one major obstacle to predict the binding mode and the affinity of a ligand accurately, it provides a structural basis to design selective inhibitors. Usually, the conformations of the active enzymes within a family are more alike, but the inactive conformations are more different. For example, the ATP binding pockets of protein kinases in active states are highly similar, but considerable conformational variability exists among kinases in the inactive state that can be exploited for selective inhibitor design. The unique "DFG-out" inactive conformation has been applied already to the design of selective kinase inhibitors, which includes the already approved drug Gleevec and Nexavar. Lapatinib, one of the most selective kinase inhibitors, achieves its high selectivity by binding to an inactive conformation of EGFR with a methionine in the $\alpha$-C helix moved away from its normal position to accommodate the benzyloxy group of lapatinib (62). Exploiting differentially accessible conformational states among closely related targets has also been the structural basis for observed high selectivity of PI3$\delta$ kinase inhibits (63).

## 3.13 ENGINEERING SELECTIVITY USING CHEMICAL GENETICS

Identifying a selective small-molecule inhibitor can be time consuming. For a target in a large protein family, it is almost impossible to achieve monoselectivity. Inhibiting multiple targets by a small molecule leads to a significant complication in the analysis of target validation and studies of protein functions. To overcome the difficulty of identifying monoselective small-molecule inhibitors, a chemical genetic approach is applied to design as orthogonal receptor-ligand pairs (64, 65). In this approach, the targeted protein is mutated in the active site to create a structural distinction between the target and all other members of proteins in the same family. Then, a small molecule is designed to target the distinct structure so that it only binds the mutated protein. The first step is to study the binding mode of a small molecule to its target and identify an important interaction that is highly conserved for the ligand binding to all members in the protein family. Then, the ligand is modified in the conserved interaction region to disrupt the interaction so that it cannot bind effectively to any wild type protein, which includes the target. Two modifications are used commonly. One is to add a bulky substitute (a "bump") at the interacting position to create steric clashes, and the other is to disrupt the critical hydrogen bonding between the ligand and the protein. The next step is to mutate the residues around the bump to smaller ones (which creates a "hole") to accommodate the "bump" or to mutate residues to restore the critical hydrogen bonds. Although both approaches have introduced orthogonal protein–ligand pairs successfully, the "bump-hole" strategy is more suitable for design and is applicable to many different protein families. Shokat's lab has successfully applied the "bump-hole" strategy to generate monospecific

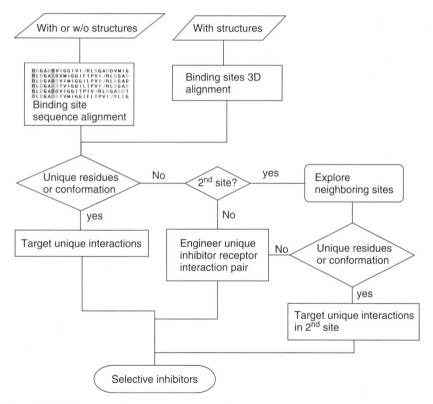

**Figure 3.6**   Flowchart of methods employed for designing selectivity of small-molecule inhibitors.

inhibitors for engineered kinases (64). The monospecific inhibitors have been used widely in kinase signaling pathway elucidation and target validation (66). Figure 3.6 outlines the strategies and workflows that we discussed for selective inhibitor design, in which the method details of each step in the flow chart have to be tailored toward specific systems.

## 3.14  SUMMARY

In this review, we described briefly some common computational techniques to select and to design small-molecule inhibitors. It is not a comprehensive list of all methods available. Instead, we focused on basic methods that are used commonly in drug discovery research. Because of the size limitation, it is also impossible to discuss all the details in each method. We hope that this short review can provide some fundamental concepts used in virtual screening and rational design, and can provide the interested reader with an entry point to explore the field of computational inhibitor design and selection. Although we described the computational methods in a linear flow, the methods used in one stage of small-molecule inhibitor discovery process can be applied at different stages whenever appropriate. To choose the proper methods for different purposes at different stages is crucial for computational modeling to be effective. In addition, computational modeling should always interact tightly with experimental investigation in order to make significant contributions. VS and rational design is also an ever-changing field. New algorithms and strategies are emerging continuously. We believe that computational methods will play increasingly important roles in discovering new small-molecule inhibitors.

## REFERENCES

1. Shogren-Knaak MA, Alaimo PJ, Shokat KM. Ann. Rev. Cell Dev. Biol. 2001; 17:405–433.
2. Schreiber SL. Nat. Chem. Biol. 2005;1:64–66.
3. Adcock SA, McCammon JA. Chem. Rev. 2006;106:1589–1615.
4. Kuntz ID, Blaney JM, Oatley S, Langridge R, Ferrin TJ. Mol. Biol. 1982; 161:269–288.
5. Morriss GP, Goodsell D, Halliday R, Huey R, Hart W, Belew R, Olson A. J. Comp. Chem. 1998;19:1639–1662.
6. Abagyan R, Totrov M, Kuznetsov DJ. Comp. Chem. 1994;15:488–506.
7. Che J. Chem. Theory Comput. 2005;1:634–642.
8. Hopkins SC, Vale RD, Kuntz ID. Biochemistry 2000;39:2805–2814.
9. Vangrevelinghe E, Zimmermann K, Schoepfer J, Portmann R, Fabbro D. Furet PJ. Med. Chem. 2003;46:2656–2662.
10. Peng H, Huang N, Qi J, Xie P, Xu C, Wang J, Yang C. Bioorg. Med. Chem. Lett. 2003;13:3693–3699.

11. Bonacci TM, Mathews JL, Yuan C, Lehmann DM, Malik S, Wu D, Font JL, Bidlack JM, Smrcka AV. Science 2006;312:443–446.

12. Trosset J, Dalvit C, Knapp S, Fasolini M, Veronesi M, Mantegani S, Gianellini LM, Catana C, Sundströ m M, Stouten PFW, Moll J. Proteins: Struct. Funct. Bioinfo. 2006;64:60–67.

13. Becker OM, Marantz Y, Shacham S, Inbal B, Heifetz A, Kalid O, Bar-Haim S, Warshaviak D, Fichman M, Noiman S. Proc. Nat. Sci.Acad. U.S.A. 2004; 101:11304–11309.

14. Salo OMH, Raitio KH, Savinainen JR, Nevalainen T, Lahtela-Kakkonen M, Laitinen JT, Jarvinen T, Poso AJ. Med. Chem. 2005;48:7166–7171.

15. Muegge IJ. Med. Chem. 2006;49:5895–5902.

16. Eldridge MD, Murray CW, Auton TR, Paolini GV, Mee RPJ. Comput. Aided Mol. Des. 1997;11.

17. Gohlke H, Hendlich M, Klebe GJ. Mol. Biol. 2000;295:337–356.

18. Verdonk ML, Cole JC, Hartshorn M, Murry C, Taylor R. Proteins 2003;52:609–623.

19. Friesner RA, Murphy RB, Repasky MP, Frye LL, Greenwood JR, Halgren TA, Sanschagrin PC, Mainz DTJ. Med. Chem. 2006;49:6177–6196.

20. Kontoyianni M, McClellan LM, Sokol GSJ. Med. Chem. 2004;47:558–565.

21. Huse M, Kuriyan J. Cell 2002;109:275–282.

22. Lorber D, Udo M, Shoichet BK. Protein Sci. 2002;11.

23. Kairys V, Gilson MKJ. Comp. Chem. 2002;23:1656.

24. Sherman W, Day T, Jacobson MP, Friesner RA, Farid RJ. Med. Chem. 2006;49:534–553.

25. Mahe P, Ueda N, Akutsu T, Perret J, Vert JJ. Chem. Inf. Model. 2005;45:939–951.

26. Jorissen RN, Gilson MKJ. Chem. Inf. Model. 2005;45:549–561.

27. Schneider G, Nettekoven MJ. Comb. Chem. 2003;5:233–237.

28. Patel Y, Gillet VJ, Bravi G, Leach AR. J. Comp. Aided Molec. Design 2002;16:653–681.

29. Dixon SL, Smondyrev AM, Rao SN. Chem. Biol. Drug Des. 2006;67:370–372.

30. Handschun S, Wagener M, Gasteiger JJ. Chem. Inf. Comput. Sci. 1997;38:220–232.

31. Holliday J, Willett PJ. Mol. Graph. Model. 1997;15:203–253.

32. Nicklaus M, Wang S, Driscoll J, Milne G. Bioorg. Med. Chem. 1995;3:337–469.

33. Jones G, Willett P, Glen RC. In: Pharmacophore Perception, Development, and Use in Drug Design. 2000. International University Line, La Jolla, CA. pp. 85–106.

34. Campos-Martinex J, Waldeck JR, Coalson RDJ. Chem. Phys. 1992;96:3613.

35. Martin YC, Bures MG, Danaher EA, DeLazzer J, Lico I, Pavlik PA. J. Comp. Aided Molec. Design 1993;7:83–102.

36. Feng J, Sanil A, Young SJ. Chem. Inf. Comput. Sci. 2006;46:1352–1359.

37. Muchmore SW, Smith RA, Stewart AO, Cowart MD, Gomtsyan A, Matulenko MA, Yu H, Severin JM, Bhagwat SS, Lee CH, Kowaluk EA, Jarvis MFLJC. J. Med. Chem. 2006;49:6726–6731.

38. Singh J, van Vlijmen H, Liao Y, Lee W, Cornebise M, Harris M, Shu I, Gill A, Cuervo JH, Abraham WM, Adams SP. J. Med. Chem. 2002;45:2988–2993.

39. Flohr S, Kurz M, Kostenis E, Brkovich A, Fournier A, Klabunde TJ. Med. Chem. 2002;45:1799–1805.
40. Bologa C, Revankar C, Young SM, Edwards BS, Arterburn JB, Kiselyov AS, Parker MA, Tkachenko SE, Savchuck NP, Sklar LA, Oprea TI, Prossnitz ER. Nat. Chem. Biol. 2006;2:207–212.
41. Kurogi Y, Guner O. Curr. Med. Chem. 2001;8:1035–1055.
42. Goto J, Kataoka R, Hirayama N. J. Med. Chem. 2004;47:6804–6811.
43. Brenk R, Naerun L, Grädler U, Gerber H, Garcia G, Reuter K, Stubbs MT, Klebe G. J. Med. Chem. 2003;46:1133–1143.
44. Tschinke V, Cohen N. J. Med. Chem. 1993;36:3863–3870.
45. Bohm HJ. J. Comp. Aided Molec. Design 1994;8:243–356.
46. McGregor MJ, Muskal SM. J. Med. Chem. 1999;39:569–574.
47. Böhm H-J. J. Comp. Aided Mol. Design 1992;6.
48. Pisabarro MT, Ortiz AZ, Palomer A, Cabre F, Garcia L, Wade RC, Gago F, Mauleon D, Carganico G. J. Med. Chem. 1994;37:337–341.
49. Gill AL, Frederickson M, Cleasby A, Woodhead SJ, Carr MG, Woodhead AJ, Walker MT, Congreve MS, Devine LA, Tisi D, et al. J. Med. Chem. 2005;48:414–426.
50. Lazar C, Kluczyk A, Kiyota T, Konishi Y. J. Med. Chem. 2004;47:6973–6982.
51. Okram B, Nagle A, Adrian FJ, Lee C, Ren P, Wang X, Sim T, Xie Y, Xia G, Spraggon G, Warmuth M, Liu Y, Gray NS. Chem. Biol. 2006;13:779–786.
52. Liu Y, Gray NS. Nat. Chem. Biol. 2006;2:358–364.
53. Schindler T, Bornmann W, Pellicena P, Miller WT, Clarkson B, Kuriyan J. Science 2000;289:1938–1942.
54. Pierce AC, Rao G, Bemis GW. J Med. Chem. 2004;47:2768–2775.
55. Firth-Clark S, Willems HMG, Williams A, Harris W. J. Chem. Inf. Comput. Sci. 2006;46:642–647.
56. Deng Z, Chuaqui C, Singh J. J. Med. Chem. 2004;47:337–344.
57. Cohen MS, Zhang C, Shokat KM, Taunton J. Science 2005;308:1318–1321.
58. Wilson KP, McCaffrey PG, Hsiao K, Pazhanisamy S, Galullo V, Bemis GW, Fitzgibbon MJ, Caron PR, Murcko MA, Su MS. Chem. Biol. 1997;4:423–431.
59. Honma T, Yoshizumi T, Hashimoto N, Hayashi K, Kawanishi N, Fukasawa K, Takaki T, Ikeura C, Ikuta M, Suzuki-Takahashi I, Hayama T, Nishimura S, Morishima H. J Med. Chem. 2001;44:4628–4640.
60. Gierse JK, McDonald JJ, Hauser SD, Rangwala SH, Koboldt CM, Seibert K. J. Biol. Chem. 1996;271:15810–15814.
61. Puius YA, Zhao Y, Sullivan M, Lawrence DS, Almo SC, Zhang ZY. Proc. Nat. Acad. Sci. U.S.A. 1997;94:13420–13425.
62. Wood ER, Truesdale AT, McDonald OB, Yuan D, Hassell A, Dickerson SH, Ellis B, Pennisi C, Horne E, Lackey K, et al. Cancer Res. 2004;64:6652–6659.
63. Knight ZA, Gonzalez B, Feldman ME, Zunder ER, Goldenberg DD, Williams O, Loewith R, Stokoe D, Balla A, Toth B, et al. Cell 2006;125:733–747.
64. Bishop A, Buzko O, Heyeck-Dumas S, Jung I, Kraybill B, Liu Y, Shah K, Ulrich S, Witucki L, et al. Annu. Rev. Biophys. Biomol. Struct. 2000;29:577–606.

65. Clemons PA, Gladstone BG, Seth A, Chao ED, Foley MA, Schreiber SL. Chem. Biol. 2002;9:49–61.
66. Shokat K, Velleca M. Drug Discov. Today 2002;7:872–879.

## FURTHER READING

Brooijmans N, Kuntz I. Molecular recognition and docking algorithms. Annu. Rev. Biophys. Biomol. Struct. 2003;32:335–373.
Tollenaera J, De Winter H, Langenaeker W, Bultinck P, eds. Computational Medicinal Chemistry and Drug Discovery. 2004. Marcel Dekker, New York.

# 4

# LEAD OPTIMIZATION IN DRUG DISCOVERY

CRAIG W. LINDSLEY

*Departments of Pharmacology and Chemistry, Vanderbilt Institute of Chemical Biology, Vanderbilt University Medical Center, Nashville, Tennessee*

DAVID WEAVER AND THOMAS M. BRIDGES

*Department of Pharmacology, Vanderbilt Institute of Chemical Biology, Vanderbilt University Medical Center, Nashville, Tennessee*

J. PHILLIP KENNEDY

*Department of Chemistry, Vanderbilt University, Nashville, Tennessee*

Lead optimization in drug discovery has changed significantly over the years and no longer is fragmented into separate hit-to-lead and lead optimization phases. Chemical lead optimization from high-throughput screening (HTS) to clinical candidate identification is now one seamless process that draws on new technologies for accelerated synthesis, purification, and screening of directed, iterative compound libraries. Advances In high-throughput screening technologies allow detection of new allosteric modes of target modulation, which provides new chemotypes and target opportunities. With the incorporation of drug metabolism and pharmacokinetics (DMPK) inputs early in the lead optimization work flow, molecules are not optimized solely for target potency and selectivity. Moreover, at the time of writing, "closed-loop" work flows are in place such that synthesis and primary screening operate on a 1-week turnaround for up to 48 compounds/week with DMPK data cycling every other week to guide compound design, which provides expedited timelines for the development of proof-of-concept compounds and clinical candidates with limited human resources.

*Chemical Biology: Approaches to Drug Discovery and Development to Targeting Disease*, First Edition. Edited by Natanya Civjan.
© 2012 John Wiley & Sons, Inc. Published 2012 by John Wiley & Sons, Inc.

The competitive drug discovery environment, whether in industry or academia, requires constant innovation and refinement as a prerequisite for success. A combination of market, patient, and regulatory concerns requires that new chemical entities act on truly novel, and therefore not clinically validated, molecular targets. With the high attrition rates and limited human resources, drug discovery efforts must focus on a large and diverse collection of molecular targets and judiciously employ enabling technologies and new paradigms to develop simultaneously multiple early stage programs. Importantly, the goal at the outset of a nascent program is to provide rapid target validation *in vivo* with a novel small molecule or to deliver a quick kill for the program so that resources can be reassigned. Coupled with these concerns is the need to establish intellectual property to support broad generic patent claims early in the development process, as chemical space is shrinking at an alarming rate (1).

Historically, the scope, mission, and technology platforms of lead optimization groups varied considerably across the drug discovery industry, which led to highly variable success rates (2–5). Some organizations had defined "handoff" criteria and fragmented lead optimization into a hit-to-lead phase and a chemical lead optimization phase. Hit-to-lead focused on optimizing screening hits, usually by library synthesis (solution phase and/or solid phase), for target potency with minimal concern for selectivity, ancillary pharmacology, and pharmacokinetics (PK). Leads that met certain potency criteria and displayed robust structure–activity relationships (SAR) then would be "handed-off" to a second group for the lead optimization phase, wherein more classic medicinal chemistry [single compound synthesis and intense drug metabolism and pharmacokinetics (DMPK) profiling] would occur (2–5).

In the past five years, lead optimization in drug discovery has changed significantly and no longer needs to be fragmented into separate hit-to-lead and lead optimization phases (2–5). Major advances have been made in HTS technologies, which have enabled detection of novel modes of target modulation. Once limited to detection of classic agonists and antagonists by HTS, kinetic imaging plate readers, such as Fluorescence Detection Screening System (FDSS) and Fluorescence Imaging Plate Reader (FLIPR), allow for the HTS identification of positive and negative allosteric modulators of both known and novel targets, which offers new chemotypes as well as improved selectivity and safety profiles (6, 7). Chemical lead optimization, from evaluation of screening hits to clinical candidate identification, now can be a seamless process that draws on new technologies for accelerated synthesis, purification, and screening. Directed, iterative compound libraries now are employed throughout the lead optimization continuum with single compound synthesis restricted to an "as needed" basis. With the incorporation of DMPK inputs at the initiation of a lead optimization program, molecules are not optimized solely for target potency and selectivity but also for the optimization of protein binding and pharmacokinetics and for diminishing CYP inhibition (1–11). Moreover, "closed-loop" work flows are in place such that chemical synthesis and primary screening data operate on a 1-week turnaround for hundreds of compounds/week, with DMPK data cycling every other week to guide compound design and to provide expedited timelines for the development of proof-of-concept compounds to validate/kill novel molecular targets and to deliver clinical candidates with limited human resources. To avoid the negative stigma of combinatorial chemistry, both industrial and academic laboratories, this new paradigm for lead optimization is coined "technology-enabled synthesis" or "TES"; however, a more accurate moniker would be "technology-enhanced medicinal chemistry" (12–21).

## 4.1   ADVANCES IN HIGH-THROUGHPUT SCREENING TECHNOLOGIES

The first step toward a successful lead optimization campaign begins in a state-of-the-art screening facility, which is ideally based on a philosophy that values the ability to automate complex biologic assays to allow screening of difficult-to-screen targets and to detect novel mechanisms of target modulation. Historical HTS paradigms valued the use of automation only to increase throughput; however, the focus now is to execute faithfully complex tasks with high precision. Modern HTS facilities employ automated screening systems composed of state-of-the-art liquid handling, plate readers, incubators, and other instruments to support a wide variety of cell-free and cell-based assays that range from enzyme assays on purified proteins to phenotypic screens on model organisms like *C. elegans* and zebrafish embryos (6, 7, 21–24). Advances in analysis software allow for information-rich assay forms, primarily in cell-based or organism-based environments, with read modes based on either parallel acquisition of kinetic data that use instruments like the Hamamatsu FDSS kinetic imaging plate readers (21, 25) or on object-based screening that uses high spatial resolution devices like automated microscopes or the BlueShift Isocyte (21, 26). Both of these read modes yield complex, information-rich data sets. The analysis and storage of such data can be challenging; however, the success of a lead optimization campaign is linked directly to the ability to acquire, synthesize, store, and present compounds as well as to the ability to collect/analyze data from the biologic systems for which we hope to discover proof-of-concept compounds and clinical candidates (6, 7, 21, 26).

### 4.1.1   A Triplicate Screen to Identify Classic and Allosteric Modes of Target Modulation

Miniaturization of assays that employ kinetic imaging plate readers allow for the development of robust high-throughput calcium mobilization-based assays that detect the activation/inhibition of molecular targets through both classic and allosteric modes of target modulation. For instance, we can measure receptor-induced intracellular release of calcium by using an imaging-based plate reader that makes simultaneous measurements of calcium levels in each well of a 384-well plate. In a novel triplicate-screening paradigm (Fig. 4.1), either vehicle or

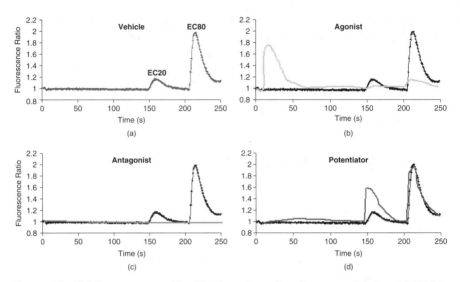

**Figure 4.1** Triplicate screen to identify allosteric modes of target modulation. (a) Vehicle with an $EC_{20}$ and $EC_{80}$ of agonist, (b) waveform profile of an agonist; (c) waveform profile of an antagonist (flowup necessary to distinguish orthosteric versus allosteric antagonist), and (d) waveform profile of a potentiator, aka, positive allosteric modulator. A single screen generates an entire spectrum of hits.

a test compound was added to cells expressing a G Protein-coupled receptor (GPCR) that has been loaded with fluorescent dye, Fluo-4. After a 2.5-minute incubation period, a submaximally effective ($EC_{20}$) concentration of orthosteric agonist was added, followed by a nearly maximal ($EC_{80}$) concentration added 1 minute later. In this manner, we can screen for and identify classic agonists/antagonists, allosteric potentiators, and antagonists simultaneously, which maximizes the efficiency of each screen and delivers a diverse collection of hits for chemists to optimize. This paradigm affords the medicinal chemists with options, both in terms of a modulatory mechanism for their therapeutic target and in terms of a chemotype, for the lead optimization campaign in a manner previously unavailable (27).

Of course, technology has not advanced only for the screening of GPCRs but also for kinases and ion channels. Kinase screens now employ both low and high concentrations of ATP to identify both ATP-competitive and allosteric inhibitors. Numerous technology platforms have appeared for ion channels targets, such as highly automated Ion-Works and Q-patch, which avoid the need for burdensome and slow single patch-clamp experiments (21–27).

## 4.2 SOLUTION-PHASE PARALLEL SYNTHESIS FOR LEAD OPTIMIZATION

The chemical technologies and platforms for chemical lead optimization have undergone a major paradigm shift in the past 10 years. In the 1990s, hit-to-lead efforts were driven by combinatorial chemistry and characterized by large (1000–10,000 member) solid-phase libraries that required months to synthesize and characterize (28). Often, by the time the library was ready for screening, the SAR of the program and/or lead series had moved on, and the value of the library was minimal (28, 29). As a result, most pharmaceutical companies disbanded their combinatorial chemistry groups, and lead optimization relied primarily on single compound synthesis or small collections (less than 12) of compounds. Driven to make the lead optimization process more efficient, the concept of solution-phase parallel synthesis began to gain favor, and technologies rapidly began to develop to create this new approach (30–34). In the last five years, major advances were made in the availability of polymer-supported reagents and scavengers and in the advent of precision-controlled, single-mode microwave synthesizers for organic synthesis. Along with the development of robust mass-directed, preparative HPLC purification, platforms have revolutionized and accelerated lead optimization (20, 30–36).

### 4.2.1 Solution-Phase Parallel Synthesis (SPPS)

Key to the success of SPPS was the development of "scavenging reagents." Scavenging (quenching) reagents are highly effective tools for the rapid purification and isolation of the desired product(s) from a solution-phase reaction by forming either covalent or ionic bonds with excess reactants and/or reaction by-products. In general terms, scavenging can be considered a "phase switching" technique wherein a chemo-selective reaction is employed to switch the phase of one product relative to another by virtue of a "tag" attached to the scavenging reagent (30–34). Three major classes of scavenging reagents are categorized by the nature of the phase tag: solid-phase polymers, ionizable functional groups, and fluoroalkyl chains (30–34, 37). In a typical scenario, an excess of reactant **B** is combined with **A** to provide product **P** along with **B** and other reaction by-products **X** in a homogeneous solution-phase reaction. Then, **B** and **X** are chemo-selectively removed from solution in a subsequent "scavenging" step with a scavenging reagent **1** linked to a phase tag. After separation of the resulting phases, the product, **P**, is obtained in high purity by simply evaporating the solvent (Fig. 4.2).

The most commonly used tags are solid-phase polymers, and hence, a wealth of literature centers on the applications of polymer-supported scavenger reagents

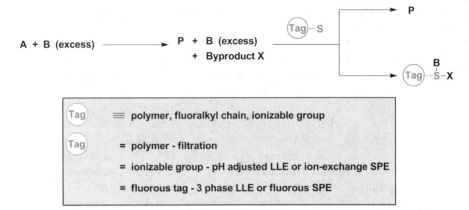

**Figure 4.2**   SPPS and "phase switching".

to transfer a captive species from the organic liquid phase to the solid phase for removal by filtration. Indeed, this approach has gained widespread acceptance because of the commercial availability of a diverse array of electrophilic and nucleophilic polymer-supported scavenging reagents along with an abundance of polymer-supported reagents. Moreover, because of site isolation, "cocktails" of polymer-supported reagents and scavengers can be used simultaneously (30–34).

Another commonly used tagging strategy involves linking a scavenger to an ionizable functional group, such as a COOH (pKa < 5) or an $NR_2$ (pKa > 10). In this instance, the captured species can be phase transferred selectively by either pH-adjusted liquid/liquid extraction or by solid-phase extraction (SPE) on an ion-exchange cartridge that leaves the desired product either in the organic liquid phase or in the SPE cartridge eluent (34). SPE is a very attractive method for purification because a crude reaction simply is applied to a disposable silica plug and grafted with either a sulfonic acid (SCX—strong cation exchange) or a tertiary amine (SAX—strong anion exchange), and neutral molecules are eluted off with methanol, whereas ionizable functional groups are retained on the SPE cartridge. Unfortunately, this strategy impacts the diversity of a library by limiting the presence of ionizable groups to either neutral or orthogonally charged library members (34).

Relying on the affinity that fluoroalkyl chains have for each other and the phobia that they exhibit toward both organic molecules/solvents and aqueous solvents, researchers began examining fluorous tags as a means of phase switching (35). Initially, efforts centered on "heavy" fluorous tags (60% or more fluorine content by molecular weight; for example, 18 or more difluoromethylene, $CF_2$, groups) that used liquid/liquid phase separation to isolate fluorous-tagged molecules from untagged organics. Typically, a three-phase liquid/liquid extraction, which requires an organic layer, an aqueous layer, and a fluorous layer (a perfluorohexane such as FC-72), delivers pure material. More recently, fluorous solid-phase extraction (FSPE) that employs fluorous silica gel (reverse-phase silica gel with a fluorocarbon bonded phase) has been developed to separate effectively both "heavy" fluorous-tagged molecules as well as "light" fluorous-tagged molecules (4 to 10 $CF_2$ groups) from untagged organics. The FSPE columns, referred to as Fluoro*Flash* (Fluorous Technologies, Pittsburgh, PA) columns, retain the fluorous-tagged material when eluted with a fluorophobic solvent, such as 80/20 MeOH/$H_2O$, which allows the untagged organic molecule to elute rapidly from the column. Homogeneous reaction kinetics, generally with respect to charged and neutral functional groups and a variety of efficient phase-separation options, have spurred a dramatic increase in the development of fluorous scavenging reagents and protocols (37).

### 4.2.2 Microwave-Assisted Organic Synthesis (MAOS)

MAOS, fueled by the development of precision-controlled, single-mode microwave reactors, has a profound impact on organic and parallel synthesis. Reaction times typically are cut by orders of magnitude, and it is usual to observe a diminution in side product formation. Moreover, MAOS reactions tend to be general in scope and lend themselves to the synthesis of libraries to develop SAR rapidly. These advantages, easily appreciated when considering established routes with successful reactions, are even more valuable when working out robust conditions for a synthesis. Exploratory reactions can be conducted in minutes to hours instead of days, and speculative, higher-risk ideas can be pursued with minimal time investment. Indeed, MAOS allows any chemistry to be pursued in parallel and allows chemistries that historically were avoided for library synthesis (multicomponent reactions, organometallics, transition-metal catalyzed couplings, etc.) to be completed successfully in minutes (20, 38, 39).

Beyond the speed advantage, two additional merits of MAOS and modern reactors should be highlighted: precision and reaction scope. As has been noted in these pages and elsewhere, the benefits of MAOS have been studied in multimode "kitchen microwaves" for decades; what prevented acceptance in the wider community was irreproducibility because of a lack of pressure and temperature control (Fig. 4.3). In addition, kitchen microwaves employ multimode resonators that lead to a heterogeneous field and local "hot" spots (Fig. 4.3a); despite this disadvantage, early work demonstrated the use of MAOS. Modern systems (Fig. 4.3b) provide a homogeneous field and precise control of temperature and pressure, and they bare little resemblance to kitchen microwaves. Importantly, MAOS relies on dipolar oscillations and ionic conduction, for example, molecular friction, to generate heat and to afford uniform heating of the sample. In contrast, conventional thermal heating relies on heat transfer from the walls of a reaction vessel and affords nonuniform heating of the sample (Fig. 4.3c). This uniform heating and rapid time to set temperature delivers reproducible results with fewer side products and, as a result, higher chemical yields (20).

Also, MAOS technology significantly has impacted library design and synthesis. For instance, when presented with a small heterocycle as a hit from an HTS,

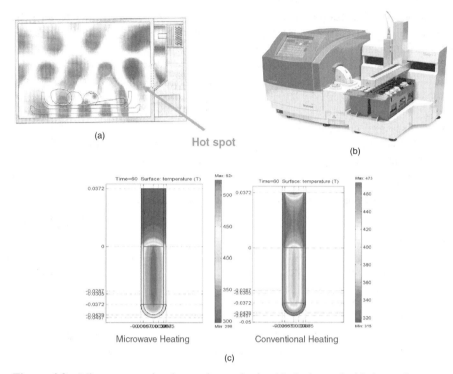

**Figure 4.3** Microwave-assisted organic synthesis. (a) A domestic kitchen microwave with local "hot spots." (b) A single-mode microwave reactor for organic synthesis. (c) Comparison of surface temperature between microwave and conventional heating.

MAOS technology allows one not only to synthesize and evaluate substitutions on the parent heterocyclic scaffold rapidly but also to synthesize and evaluate multiple heterocyclic templates with diverse substituents in parallel (Fig. 4.4). Therefore, a single library will contain multiple heterocyclic cores with varying degrees of basicity and topology, while broadening the generic scope for a composition of matter patent (38).

### 4.2.3  Mass-Directed Preparative HPLC

Despite the purity obtainable by SPPS scavenging (typically >90%) and the high purities obtained by MAOS (also typically >90%), modern lead optimization programs require >95% purity of all compounds that contribute to the development of SAR and that advance into DMPK assays. For years, many labs employed UV-directed preparative HPLC and often multichannel units to increase throughput. Although this approach worked, purification of a single sample might lead to 30–40 fractions per sample, which then required analysis by analytical Liquid Chromatography/Mass Spectrometer (LCMS) to identify which fractions

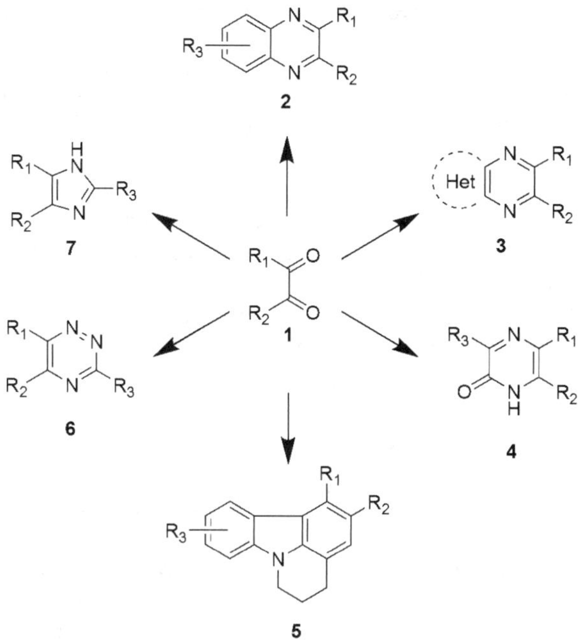

**Figure 4.4**  MAOS to access rapidly diverse heterocyclic scaffolds form a common intermediate.

contained the desired product (34, 40). In 2000, several vendors launched preparative LCMS units that offered mass-directed fractionation. Now, purification of a single crude sample afforded only one or two pure fractions—a significant advance. Additional modifications for library purification included DMSO slugs to bracket sample injections or "at-column dilution" to provide robust chromatography and prevent in-line sample precipitation before the column. These modified systems were capable of purifying, in a single pass, 60 to 80 compounds per day with purity levels exceeding 98%; however, the systems required an expert chromatographer to develop custom gradients for each sample in a library (34, 40). Recently, several vendors launched an analytical-to-preparative LCMS software package that addressed the need for a dedicated, expert chromatographer to operate each prep LCMS instrument. With analytical-to-preparative software, a file containing the compound ID and exact mass for each sample in a library to be purified is uploaded into the preparative LCMS system, which then electronically accesses the analytical LCMS data and extracts the retention time of the mass of interest from the crude sample chromatograms. The preparative LCMS system analytical-to-preparative software then calculates a customized gradient for each sample in a library and therefore reduces the need for an experienced chromatographer to achieve excellent first-pass purification results. This feature also allows the instrument to run overnight unattended and additionally increases operational efficiency.

### 4.2.4 Postpurification Sample Handling and Compound Characterization

Modern parallel synthesis laboratories, for example, high-throughput medicinal chemistry laboratories, have borrowed a page from the automotive industry and have developed highly efficient assembly lines for postpurification sample handling and compound characterization. Automated weighing systems with bar code readers scan and record weights on unique bar-coded vials into which pure compounds from the preparative LCMS systems are transferred for concentration in a sample evaporator. After the dry-down step, the bar-coded vials with pure, solid sample are transferred to a liquid handling robot. This instrument scans each bar code, weighs the vial, and determines the net weight of the pure product. This data file is merged with a registration file that contains the molecular weight of the compound, and the system software then calculates the volume of DMSO required to dilute the samples to a preset concentration for screening. The system then dilutes the samples, transfers the DMSO stock solution to a 96-well plate, and generates an electronic plate map file for submission to the primary screen, *in vitro* drug metabolism assays, pharmacokinetic cassettes, and for flow cell NMR (*vide infra*). With this highly automated workflow, a single scientist can oversee the postpurification sample handling of thousands of samples per week (1–5).

Lack of complete compound characterization has been a major shortcoming of combinatorial chemistry and early high-throughput medicinal chemistry laboratories that led to poor adoption by traditional medicinal chemists. Once again, technology has advanced such that every member of a compound library is fully characterized to the same standard as a single compound prepared by a traditional medicinal chemist. After purification by preparative LCMS, a final analytical LCMS is generated for each sample at two wavelengths (214 nM and 254 nM) and evaporative light scattering detection (ELSD). The LCMS vials then are delivered to a quadrupole time of flight (QTOF) mass spectrometer system with a 100-position autosampler for accurate mass measurement (high-resolution mass specification) determinations. For each sample, $^1$H NMR spectra are obtained. Initially, NMR tubes were prepared for each sample, and chemists took advantage of NMRs with autosamplers. More recently, flow-cell NMR and solvent suppression software allow for quality NMR spectra to be obtained from DMSO stock solutions in 96-well plate (41). Importantly, this ensures high-purity samples, prepared in directed libraries, to drive lead optimization programs and generate quality SAR at the same level as compounds prepared by singleton synthesis.

### 4.3 EXPEDITED DRUG METABOLISM AND PHARMACOKINETICS

Lead optimization involves more than just optimizing for target potency. New technology allows early lead optimization campaigns to address and consider multiple parameters and inputs for each round of iterative library synthesis. These inputs allow for the rapid development of potent compounds with drug-like profiles, as opposed to just potent compounds. These data also provide "quick kills" to individual leads or series and allow the lead optimization effort to re-direct resources toward more productive leads (1–5) and (8–20).

### 4.3.1 High-Throughput *in vitro* Drug Metabolism Assays

Significant effort has been applied to the miniaturization and DMSO compatibility of *in vitro* drug metabolism assays. Now, a 48-member library in a 96-well plate of DMSO stock solution can be evaluated rapidly in cytochrome P450 inhibition assays (CYP3A4, 2D6, 2C9), protein binding assays (rat, dog, and human), logP, hERG binding, and other standard assays (8–11, 42, 43). Previously, chemists were forced to choose which compounds to evaluate in these assays, and often only the most potent analogs would be selected; the most potent analogs were not necessarily the ones with the most promise as clinical candidates. Being able to acquire these data for an entire library provides opportunities to pursue leads within a series with the most balanced potency and DMPK profiles. This ability is of critical importance in lead optimization to ensure that drug-like leads are being pursued and additionally refined. Similarly, cassette dosing of compounds in liver microsomes and hepatocytes enables evaluation of an entire library in short order (8–11, 42, 43). This timely evaluation is especially valuable in the lead optimization of a backup clinical candidate program, wherein the clinical candidate is the positive control and new compounds (typically five to six per cassette) are viewed qualitatively as more or less stable than the first clinical candidate. Once an *in vitro/in vivo* correlation can be established for a given series, these rapid cassette experiments can drive a lead optimization program and require only intermittent *in vivo* experiments (8–20, 42, 43).

### 4.3.2 Pharmacokinetic Cassettes

Resource and practical constraints prohibit acquiring single rat and dog pharmacokinetics (PK) (i.v. and p.o.) for every member of a library; however, the compound with the best PK may not be the most potent analog in a library, and knowing this data is crucial for lead optimization. In fact, chemical lead optimization programs have been guided solely by optimization of PK. With the success of *in vitro* cassette paradigms for microsomal and hepatocyte stability, the concept was extended to *in vivo* PK in rats and dogs (Fig. 4.5) so that an entire library could be evaluated *in vivo* employing a limited number of animals (8–11). However, some caveats exist. Combinatorial oral dosing to determine oral bioavailability (%F) in cassette format generally proved not very reproducible nor in agreement with single PK experiments. In contrast, *intravenous* cassette dosing in both rats and dogs proved highly reproducible and within the error of a single PK experiment and is a valuable tool to determine qualitative rates of clearance between five to six new compounds and an internal control of known clearance (8–11). PK cassettes employ an overall low dose of test compounds to minimize potential drug–drug interactions. These rapid cassette experiments prioritize which compounds from a library then should be studied in

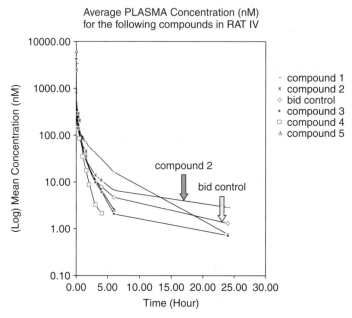

**Figure 4.5**  Pharmacokinetic evaluation of libraries: cassette dosing to evaluate clearance rates relative to a bid control compound.

single i.v./p.o. single animal PK studies. As shown in Fig. 4.5, compound 2 has qualitatively lower intrinsic plasma clearance rate than an internal control compound with known bid (twice daily predicted dosing) PK (Cl = 12 mL/min/kg), whereas the other four compounds in the cassette have higher clearance and need not be studied more.

## 4.4   EXPEDITED, "CLOSED-LOOP" WORK FLOW FOR LEAD OPTIMIZATION

Combining all above-mentioned technologies and paradigms for synthesis, screening and DMPK evaluation affords an aggressive, expedited process for chemical lead optimization (1, 12–21). This protocol allows one to two synthetic chemists to support a chemical lead optimization effort with accelerated timelines that deliver proof-of-concept compounds within 6 months and clinical candidates within 12 months of the initiation of a lead optimization campaign.

### 4.4.1   Library Design

Independently, the technologies and strategies described herein provide improvements for chemical lead optimization; however, when they become

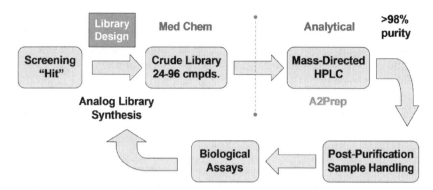

**Figure 4.6** Expedited "closed-loop" lead optimization paradigm.

closely aligned with screening and DMPK resources in a "closed-loop" paradigm, the impact on drug discovery is exponential (Fig. 4.6). Starting from an HTS hit, considerable attention is devoted first to library design, without question the most important component of a successful lead optimization effort. Library design changes over the course of a lead optimization campaign. The initial design strategy is to explode SAR around a screening hit and to be as diverse as possible with respect to monomer input and analog synthesis to rapidly identify productive changes for additional optimization. In addition, this component of lead optimization is conducted often in parallel, wherein a single chemist simultaneously will synthesize diversity libraries around four to six hits to identify expediently the best leads for additional optimization. After this initial diversity-oriented explosion, library design must become more focused to impact drug discovery goals: *Random libraries do not accelerate programs*. It is important to approach directed library design from a medicinal chemistry perspective and to assemble the library as a collection of single compounds designed to address a particular issue. For example, the design of a 24-member library should involve careful thought regarding what four single compounds would be synthesized first to increase potency, improve PK, and so forth. Then, for each of the first four analogs synthesized, designers should consider what the next four analogs should be if the first changes were productive or nonproductive. This exercise in library design generates quality data that drive a lead optimization program toward proof-of-concept compounds and clinical candidates very quickly (1, 12–21).

### 4.4.2 Division of Labor

Another key feature of the "closed-loop" approach to lead optimization involves division of labor and the transfer of samples from medicinal chemists to the analytical chemists. In this paradigm, the medicinal chemists design and synthesize the compound libraries (24 to 96 compounds) and obtain analytical LCMS reports for each member of the library. At this point, the medicinal chemists transfer the crude samples to the analytical chemists who purify the libraries by mass-directed preparative HPLC to >98% by using analytical-to-preparative software and perform all postpurification sample handling and coordinate submission of samples, in a 96-well plate format, to the biologists and DMPK personnel for screening (*vide supra*) (12–20). If resources allow, this division of labor affords opportunities for the medicinal chemists to focus on library design, develop and optimize new chemistries, and pursue multiple lead series in parallel (1, 12–21).

### 4.4.3  Data Turnaround

The success of this paradigm hinges on rapid screening and dissemination of data to the medicinal chemists so that the next iteration of library synthesis can be initiated. To facilitate this, the delivery of compounds is coordinated with the biologists and assays are run the same day the compound libraries are delivered. Biologic data then is returned within 24 hours of receipt of the libraries. This allows lead optimization to operate on a 1 week turnaround between the initiation of chemical synthesis and the generation of primary assay data. Secondary and/or selectivity data typically trail primary data by 1 to 2 days. As these data trigger the need for DMPK information, DMPK data typically follow 1 week after the initial assay data is obtained. Overall, this expedited process parallels traditional singleton medicinal chemistry work flows but generates data on hundreds of compounds in the time it used to take to evaluate just a few compounds. Moreover, this protocol allows one to two synthetic chemists to support a chemical lead optimization effort with accelerated timelines delivering proof-of-concept compounds within 6 months and clinical candidates within 12 months of the initiation of a lead optimization campaign. It is important to note that this lead optimization paradigm requires collaboration, close and frequent communication with biology and DMPK colleagues, sophisticated databases to store the volumes of data generated, and a major investment in technology (1, 12–20).

### REFERENCES

1. Glaser V. Optimizing strategies for hit-to-lead process. Gen. Engin. Biotech. News. 2005;25:31–37.
2. Gillespie P, Goodnow RA Jr. The hit -to- lead process in drug discovery. Ann. Rep. Med. Chem. 2004;39:293–304.
3. Keseru GM, Makara GM. Hit discovery and hit -to- lead approaches. Drug Discov. Today 2006;11:741–748.
4. Bleicher KH, Boehm H-J, Mueller K, Alanine AI. A guide to drug discovery: hit and lead generation: beyond high-throughput screening. Nat. Rev. Drug Disc. 2003;2:369–378.
5. Meador V, Jordan W, Zimmermann J. Increasing throughput in lead optimization in vivo toxicity screens. Curr. Opin. Drug Discov. Devel. 2002;5:72–78.

6. Wood M, Smart D. Real time receptor function *in vitro*: microphysiometry and the fluorometric imaging plate reader (FLIPR). In: Receptors: Structure and Function. 2nd edition. Stanford C, Horton R, eds. 2001. Oxford University Press, Oxford. pp 175–191.

7. Wood MD, Jerman J, Smart D. The FLIPR: a major advance in the study of neurotransmitter receptors. Rec. Res. Dev. Neurochem. 2000;3:135–142.

8. Huang R, Qian M, Chen S, Lodenquai P, Zeng H, Wu J-T. Effective strategies for the development of specific, sensitive and rapid multiple-component assays for cassette dosing pharmacokinetic screening. Int. J. Mass Spect. 2004;238:131–137.

9. Halladay JS, Wong S, Jaffer SM, Sinhababu AK, Khojasteh-Bakht SC. Metabolic stability screen for drug discovery using cassette analysis and column switching. Drug Metabol. Lett. 2007;1:67–72.

10. Bu H-Z, Magis L, Knuth K, Teitelbaum P. High-throughput cytochrome P450 (CYP) inhibition screening via a cassette probe-dosing strategy. VI. Simultaneous evaluation of inhibition potential of drugs on human hepatic isozymes CYP2A6, 3A4, 2C9, 2D6 and 2E1. Rapid Commun. Mass Spect. 2001;15:741–748.

11. Cai Z, Sinhababu AK, Harrelson S. Simultaneous quantitative cassette analysis of drugs and detection of their metabolites by high performance liquid chromatography/ion trap mass spectrometry. Rapid Commun. Mass Spect. 2000;14:1637–1643.

12. Lindsley CW, Wisnoski DD, Leister WH, O'Brien JA, Lemiare W, Williams DL Jr, Burno M, Sur C, Kinney GG, Pettibone DJ, Miller PR, Smith S, Duggan ME, Hartman GD, Conn PJ, Huff JR. Discovery of positive allosteric modulators for the metabotropic glutamate receptor subtype 5 from a series of *N*-(1,3-Diphenyl-1H-pyrazol-5-yl) benzamides that potentiate receptor function *in vivo*' J. Med. Chem. 2004;47:5825.

13. Lindsley CW, Zhao Z, Leister WH, Robinson RG, Barnett SF, Defeo-Jones D, Jones RE, Hartman GD, Huff JR, Huber HE, Duggan ME. Allosteric Akt (PKB) kinase inhibitors. Discovery and SAR of isozyme selective inhibitors. Bioorg. Med. Chem Lett. 2005;15:761–765.

14. Wolkenberg SW, Lindsley CW. MAOS: a 'diversity engine' for parallel synthesis. Discov Devel 2004;4:1–5.

15. Zhao Z, O'Brien JA, Lemiare W, Williams DL Jr, Jacobson MA, Sur C, Pettibone DJ, Tiller PR, Smith S, Hartman GD, Lindsley CW. Synthesis and SAR of GlyT1 inhibitors derived from a series of *N*-((4-(morpholine-4-carbonyl)-1-(propylsulfonyl)piperidin-yl) methyl) benzamides' Bioorg. Med. Chem.Lett. 2006;16:5968–5972.

16. Nanda KK, Nolt MB, Cato MJ, Kane SA, Kiss L, Spencer RH, Jixin Wang J, Lynch JL, Regan CP, Stump GL, Li B, White R, Yeh S, Bogusky MJ, Bilodeau MT, Dinsmore CJ, Lindsley CW, Hartman GD, Wolkenberg SE, Trotter BW. Potent antagonists of the atrial $I_{Kur}$/Kv1.5 potassium channel: synthesis and evaluation of analogous *N,N*-diisopropyl-2-(pyridine-3-yl)acetamides. Bioorg. Med. Chem. Lett. 2006;16:5897–5901.

17. Wolkenberg SE, Zhao Z, Kapitskaya M, Webber AL, Pertukhin K, Tang YS, Dean DC, Hartman GD, Lindsley CW. Identification of potent agonists of Photoreceptor-Specific Nuclear Receptor (NR2E3) and preparation of a radioligand. Bioorg. Med. Chem. Lett. 2006;16:5001–5004.

18. Lindsley CW, Zhao Z, Leister WH, O'Brien JA, Lemiare W, Williams DL Jr, Chen T-B, Chang RSL, Burno M, Jacobson MA, Sur C, Kinney GG, Pettibone DJ, Tiller PR, Smith S, Tsou NN, Duggan ME, Conn PJ, Hartman GD. Design, synthesis and *in vivo* efficacy of novel glycine transporter-1 (GlyT1) inhibitors derived from a series of [4-Phenyl-1-(propylsulfonyl)piperidin-4-yl] methyl benzamides. ChemMedChem. 2006;1:807–811.

19. Zhao Z, Wisnoski DD, O'Brien JA, Lemiare W, Williams DLJr, Jacobson MA, Wittman M, Ha S, Schaffhauser H, Sur C, Pettibone DJ, Duggan ME, Conn PJ, Hartman GD, Lindsley CW. Challenges in the development of mGluR5 positive allosteric modulators: the discovery of CPPHA. Bioorg. Med. Chem. Lett. 2007;17:1386–1391.

20. Lindsley CW, Wolkenberg SE, Shipe W. Accelerating lead development by microwave-enhanced medicinal chemistry. Drug Discov. Today: Technol. 2005; 2:155–161.

21. Lindsley CW, Weaver D, Conn PJ, Marnett L. Preclinical drug discovery research and training at Vanderbilt. ACS Chem. Biol. 2007;2:17–20.

22. Ueki T. High throughput screening in the process of drug discovery. Nippon Yakuri-gaku Zasshi 2007;129:276–280.

23. Macarron R. Critical review of the role of HTS in drug discovery. Drug Discov. Today 2006;11:277–279.

24. DeSimone RW. The use of combi chem, high-speed analog chemistry and HTS in drug discovery. Drug Discov. Today 2003;8:156.

25. For information on Hamamatsu FDSS. *www.hamamtsu.com*.

26. For information on BlueShift Isocyte. *www.blueshiftbiotech.com*.

27. Rodriguez A, Williams R, Jones C, Niswender C, Meng X, Lindsley CW, Conn PJ. Novel positive allosteric modulators of mGluR5 discovered in HTS have in vivo activity in rat behavioral models of antipsychotic activity. Nat. Chem. Bio. In press.

28. Dorwald FZ. Organic Synthesis on Solid Phase. 2000. Wiley-VCH, Weinheim.

29. Ellingboe JW Solid-phase synthesis in lead optimization and drug discovery. Curr. Opin. Drug Disc. Devel. 1999;2:350–357.

30. Kuroda N, Hird N, Cork DG. Further development of a robust workup process for Solution-Phase High-Throughput Library Synthesis to address environmental and sample tracking issues. J. Comb. Chem. 2006;8:505–512.

31. Altorfer M, Ermert P, Faessler J, Farooq S, Hillesheim E, Jeanguenat A, Klumpp K, Maienfisch P, Martin JA, Merrett JH, Parkes KEB, Obrecht J-P, Pitterna T, Obrecht D. Applications of parallel synthesis to lead optimization. Chimia. 2003;57:262–269.

32. Booth RJ, Hodges JC. Solid-supported reagent strategies for rapid purification of combinatorial synthesis products. Acc. Chem. Res. 1999;32:18–26.

33. Kaldor SW, Siegel MG. Combinatorial chemistry using polymer-supported reagents. Curr. Opin. Chem. Bio. 1997;1:101–106.

34. Siegel MG, Hahn PJ, Dressman BA, Fritz JE, Grunwell JR, Kaldor StW. Rapid purification of small molecule libraries by ion exchange chromatography. Tetrahedron Lett. 1997;38:3357–3360.

35. Leister WH, Strauss KA, Wisnoski DD, Zhao Z, Lindsley CW. Development of a custom high-throughput preparative liquid chromatography/mass spectrometer platform for the preparative purification and analytical analysis of compound libraries. J. Comb. Chem. 2003;5:322–329.

36. Kyranos JN, Lee H, Goetzinger WK, Li LYT. One-minute full-gradient HPLC/UV/ELSD/MS analysis to support high-throughput parallel synthesis. J. Comb. Chem. 2004;6:796–804.

37. Gladysz JA, Curran DP, Horvath IT, Eds. Handbook of Fluorous Chemistry. 2004. Wiley-VCH, New York.

38. Shipe WD, Yang F, Zhao Z, Wolkenberg SE, Nolt MB, Lindsley CW. Convenient and general microwave-assisted protocols for the expedient synthesis of heterocycles. Heterocycles. 2006;70:665–689.

39. Kappe CO, Dallinger D. The impact of microwave synthesis on drug discovery. Nat. Rev. Drug Disc. 2006;5:51–63.

40. Blom KF. Two-pump at-column-dilution configuration for preparative liquid chromatography-mass spectrometry. J. Comb. Chem. 2002;4:295–301.

41. Curran SA, Williams DE. Design and optimization of an NMR flow cell for a commercial NMR spectrometer Appl. Spectros. 1987;41:1450–1454.

42. Kariv I, Rourick RA, Kassel DB, Chung TDY. Improvement of "hit-to- lead" optimization by integration of in vitro HTS experimental models for early determination of pharmacokinetic properties. Comb. Chem. High Throughput Screen. 2002;5:459–472.

43. Kenakin T. A guide to drug discovery: predicting therapeutic value in the lead optimization phase of drug discovery. Nat. Rev. Drug Discov. 2003;2:429–438.

## FURTHER READING

Ardrey A. Liquid Chromatography-Mass Spectrometry: An Introduction. 2005. John Wiley & Sons, New York.

Burgess KD, Ed. Solid Phase Organic Synthesis. 2000. John Wiley & Sons, New York.

Sucholeiki I, Ed. High-Throughput Synthesis. 2001. Marcel Dekker, Inc., New York.

Tierney JP, Lidstrom P Microwave Assisted Organic Synthesis. 2005. CRC Press, Boca Raton, Florida.

## SEE ALSO

Chapter 6

# 5

# PHARMACOKINETICS OF DRUG CANDIDATES

RONALD E. WHITE

*Schering-Plough Research Institute, Kenilworth, New Jersey*

Pharmacokinetics is a mathematically based discipline that describes the time course of uptake, distribution, and elimination of a drug in an organism. This chapter explains the physiologic basis and clinical interpretation of the various pharmacokinetic parameters, such as half-life, clearance, and oral bioavailability. Because of the central importance of pharmacokinetics to the clinical use of drugs, the determination of pharmacokinetics is required at several points in the research and development cycle of new drug candidates. The rationale and methods of determination are discussed.

The therapeutic effectiveness of drugs generally can be related to the concentrations achieved at the physiologic site of action and to the length of time that effective concentrations are maintained (1). For instance, molecules of an antibiotic must reach the site of infection (e.g., inner ear, lung, or bladder) to interact with the molecular target (typically an enzyme) in the infecting bacteria. Similarly, cough suppression or relief of clinical depression may require a medicine to penetrate the blood–brain barrier to access molecular targets (typically pharmacologic receptors) in the central nervous system. Therefore, an assessment of the effectiveness of the therapy partly depends on our ability to monitor concentrations of the drug in these organs. However, because often it is not feasible to measure intraorgan concentrations of drugs directly, we usually measure blood concentrations and then infer the corresponding tissue concentrations by using predetermined relationships for the particular drug and organ in question. In

*Chemical Biology: Approaches to Drug Discovery and Development to Targeting Disease*, First Edition.
Edited by Natanya Civjan.
© 2012 John Wiley & Sons, Inc. Published 2012 by John Wiley & Sons, Inc.

practice, we establish the blood concentration that corresponds to the actual tissue concentration necessary for therapeutic effectiveness for each clinical drug, which allows us to discuss therapy in terms of blood concentrations only without reference to tissue levels. A useful clinical drug must exhibit several practical characteristics (2): *efficacy* (the intrinsic ability of the compound to produce a desired pharmacologic effect), *availability* (the ability of the compound to reach the target organ), *safety* (the sufficient selectivity at the therapeutic dose so that undesirable pharmacologic actions are acceptably mild), and *persistence* (the sufficient residence time in the body to allow a clinically useful duration of action, usually expressed as the plasma elimination half-life).

*Pharmacokinetics* (PK) is a mathematically based scientific discipline that describes the time course of uptake, distribution, and elimination of a drug in an organism. A central goal of PK is to determine the length of time that target pharmaceutical receptors are exposed to pharmacologically effective concentrations of the drug molecules. Therefore, PK is of the utmost importance in understanding drug action, especially the duration of action, optimal dose size, individual variation in response, interactions between drugs, and developed resistance or tolerance to drugs.

## 5.1  CLINICAL PHARMACOKINETICS

The following discussion addresses only small-molecule drugs, chemical compounds with molecular weights less than 1,000 Daltons. Large-molecule drugs (biologics) such as vaccines, cytokines, antibodies, and genes also can be studied by the methods of pharmacokinetics, but they are distributed and cleared in fundamentally different ways than small molecules. (3, 4).

### 5.1.1  Dose and Administration

*Dose* refers to the amount of the drug to be administered to the patient and may be expressed in several ways. The most common form is the amount (expressed in milligrams) taken in an oral formulation such as a tablet, capsule, or syrup. Often, the formulation is provided by the manufacturer in several dose strengths, such as 10, 25, and 50 mg. Some medicines are delivered by a parenteral route such as inhalation, topical, subcutaneous, or intravenous to deliver the medicine more directly to the target organ, to control the rate of delivery, or to substitute for when oral dosing is impractical. Depending on the rate of elimination of the drug from the body, additional doses are administered at intervals to maintain adequate therapeutic blood levels. Because the presence of food in the gastrointestinal tract may affect the absorption of some drugs, the patient may be instructed to take the drug either with or between meals. Thus, a dosing regimen refers to a particular dose strength in a particular formulation given at a particular interval or time–of day in relation to meals, such as a 10-mg tablet taken once a day after dinner.

### 5.1.2 Meaning and Interpretation of Pharmacokinetic Parameters

Plasma, the fluid component of blood left after the separation of cells, commonly is used for pharmacokinetic determinations. Figure 5.1 illustrates the time course of drug concentrations in plasma following either an intravenous (iv) or an oral (po) dose of a hypothetical drug. In the case of the iv bolus dose, the drug is delivered completely to the bloodstream essentially instantaneously and then concentrations immediately begin to decline as the drug distributes within and is eliminated from the body (see the dashed curve in Fig. 5.1). The solid curve shows the drug concentration profile for the same dose given orally. In this case, because the drug first must be absorbed from the gastrointestinal tract to reach the blood, concentrations initially rise (*absorption phase*); however, when the rate of absorption equals the rate of elimination, the input and output processes reach a balance and the peak concentration ($C_{max}$) is seen. Thereafter, when absorption is complete and only the elimination process is left, the *elimination phase* is reached and concentrations decline as the body acts to remove the drug.

During the absorption phase, the plasma concentration attains the *minimum therapeutic threshold* accompanied by onset of the efficacious effect. A desirable medicine has a large window between the efficacious concentration and a concentration that produces undesirable side effects (*adverse event threshold*). As concentrations fall after $C_{max}$, the duration of effect is determined by the time above the therapeutic threshold. Thus, the illustrated drug loses effect after about 12 hours and requires a second dose by that time.

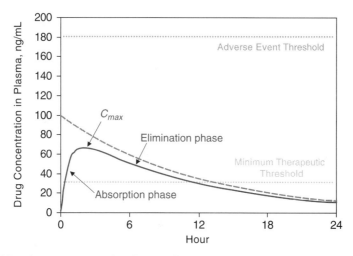

**Figure 5.1** Plasma concentration time profiles for a single 42-mg dose of a hypothetical drug with a half-life of 8 hours, $V_d$ 42 liters, and oral bioavailability of 80. Dashed curve: intravenous bolus dose. Solid curve: oral dose. Dotted horizontal lines represent plasma concentrations required for efficacy and for the onset of adverse events.

PK usually is expressed in several quantitative, clinically meaningful parameters, including half-life ($T_{1/2}$), area under the plasma concentration *versus* time curve ($AUC$), oral bioavailability ($F$), volume of distribution ($V_d$), clearance ($CL$), maximum observed plasma concentration ($C_{max}$), time after dose administration that $C_{max}$ occurs ($T_{max}$), and minimum concentration between successive doses ($C_{min}$).

$T_{1/2}$ describes the persistence of the drug in the body. An ideal drug follows *first-order kinetics* for elimination from the body so that a constant fraction is eliminated in fixed time intervals. It is convenient to select 50% as the constant fraction eliminated and to express the time required as $T_{1/2}$. Thus, at one half-life after dosing, half of the drug has been eliminated; at two half-lives, 75% has been eliminated; at three half-lives, 87.5% has been eliminated, and so forth. A property of first-order declines is that a plot of the natural logarithm of the amount remaining versus time is linear with the slope equal to $-0.693/T_{1/2}$, providing a straightforward means to determine half-life (Fig. 5.2). For many therapeutic indications, a half-life long enough (12–24 hours) to allow once-a-day dosing is desirable for the obvious reasons of patient convenience and compliance. Because many drugs do not exhibit ideal kinetics and do not follow a simple log-linear decline, we may use an operational parameter called *effective half-life*, which usefully approximates the decline in blood concentrations over most of the elimination curve (5).

Most drugs are given multiple times in a course of therapy so that when the second dose is taken a portion of the first dose still exists in the body

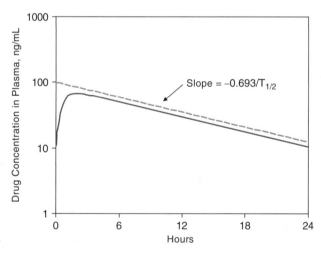

**Figure 5.2** Plasma concentration time profiles for the drug in Fig. 5.1 displayed with a logarithmic scale for plasma concentration. Dashed curve: intravenous dose. Solid curve: oral dose. The half-life may be calculated from the slope. Note that the concentration values first must be converted to the natural logarithms. $T_{1/2} = 0.693 / [(\ln 100 - \ln 12) / 24 \text{ hour}] = 8$ hours.

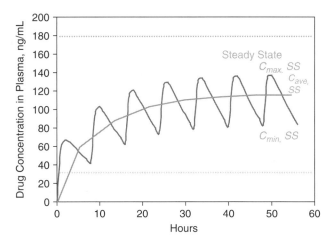

**Figure 5.3** Plasma concentration time profiles for repeated oral doses of the drug given every 8 hours, which illustrates the approach of the plasma concentrations to steady state.

and the $C_{max}$ achieved by the second dose will be that much higher than the first. A similar carryover will occur between the second and third doses, which makes the third $C_{max}$ even higher. However, as blood concentrations rise, the rate of elimination rises proportionally (a direct consequence of first-order kinetics). Thus, the tendency of the drug level to rise because of the incomplete elimination from sequential doses is compensated for by a rising elimination rate, which results after several doses in a balanced situation called *steady state* in which the same $C_{max}$ and $C_{min}$ values are observed after each successive dose (Fig. 5.3). The time-weighted average of plasma concentrations over a single dosing interval at steady state ($C_{ave, SS}$) is higher than that on the first dose, and the ratio of these two is called the *Accumulation Index* (*R*). A successful dosing regimen will keep $C_{min, SS}$ well above the therapeutic threshold at all times while never allowing $C_{max, SS}$ to approach the adverse event threshold.

$AUC$ is a measure of the total exposure of the body to the drug and can be estimated as the sum of the areas of a series of trapezoids formed between successive measured concentration time points (Fig. 5.4 and Equation 5.1). Because a drug may not necessarily be completely absorbed or may undergo presystemic metabolism or other elimination during the absorption process, the related parameter $F$ indicates the fraction of an orally administered dose that actually reaches the systemic circulation. As a practical matter, $F$, which usually is expressed as a percentage, is calculated as the ratio of the AUC measured from an oral dose to the AUC from an equivalent intravenous dose that is delivered completely to the bloodstream (Equation 5.2). Because two different drugs generally have different $C_{max}$ values, bioavailabilities, and half-lives, the observed exposures (i.e., $AUCs$) will be different even though equal doses of the drugs may have

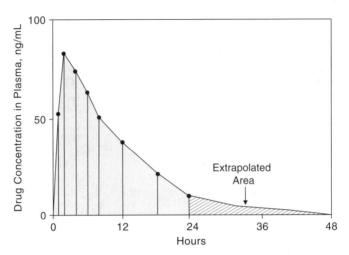

**Figure 5.4** Estimation of the total AUC for an oral dose of the drug by use of the Trapezoidal Rule. Vertical lines from measured concentration time points define a family of trapezoids; the sum of the areas of these trapezoids approximates the area under the curve. The area beyond the last time point (24 hour) is estimated by multiplying the last measured concentration by the factor $T_{1/2}$ / 0.693; *Area* (24-infinity) $= C_{final\ time} \times T_{1/2}$ / 0.693.

been given.

$$AUC = \sum_{i=2}^{n} \frac{1}{2}(C_i + C_{i-1}) \times (t_i - t_{i-1}) \tag{5.1}$$

$$F = \frac{AUC_{oral}}{AUC_{iv}} \times \frac{Dose_{iv}}{Dose_{oral}} \tag{5.2}$$

$V_d$ measures the volume of plasma in which the drug appears to be dissolved and is calculated by dividing the amount of drug in the body by the observed concentration in plasma at that time (Equation 5.3). For a drug that is confined to the plasma compartment, $V_d$ is the volume of plasma in the body [ca. 3 liters (L) in humans] (6). A drug that freely passes into and out of the cells of tissues will have a $V_d$ value of about 42 L (i.e., nominal total body water for an adult human) (6). However, many drugs show $V_d$ values that are much greater than 42 L, which cannot correspond to any physiologic compartment. In these cases, the drug must be sequestered significantly outside of the plasma compartment into organs and tissues. For this reason, $V_d$ sometimes is called the *apparent* volume of distribution. Because it is difficult to know at any moment the exact amount of drug in the body because of continuous elimination, we usually cannot apply

Equation 5.3 directly to determine $V_d$ and we must use an indirect method.

$$V_d = \frac{amount}{C_{plasma}} \tag{5.3}$$

Finally, $CL$ describes how quickly the drug is eliminated completely from plasma or blood and has the units of volume per unit time. $CL$ is calculated easily from the plasma $AUC$ observed for a given dose (Equation 5.4). For an intravenous dose $F = 1$, Equation 5.4 simplifies to $CL = Dose/AUC$. For an oral dose, if the bioavailability is not known, then the $Dose/AUC$ calculation yields $CL/F$ rather than the true clearance.

$$CL = \frac{Dose_{iv}}{AUC_{iv}} = \frac{F \times Dose_{oral}}{AUC_{oral}} \tag{5.4}$$

$CL$ is a useful working concept because it is related to the passage of drug-containing blood through a clearance organ such as liver or kidney. A drug that is cleared only by the liver (by liver metabolism, for instance) can be cleared only as fast as blood flows to the liver. This mass transport limitation is expressed in Equation 5.5, where $Q$ is the total blood flow to the liver [about 1.4 L per minute (min)], $CL_{int}$ is the intrinsic ability of the liver to clear the drug if blood flow were not a limitation, and $f_u$ is the fraction of the drug not bound to proteins in plasma (7). We can see from Equation 5.5 that in the two extreme cases, CL is equal to $CL_{int}$ when $CL_{int}$ is small compared with $Q$, and CL is equal to $Q$ when $CL_{int}$ is large compared with $Q$.

$$CL = \frac{Q \times f_u \times CL_{int}}{Q + f_u \times CL_{int}} \tag{5.5}$$

Similarly, renal clearance is limited ultimately by kidney blood flow (ca. 1,100 mL/min). Clearance by kidneys comprises three processes: *glomerular filtration*, the rate at which the kidneys filter plasma (ca. 120 mL/min); direct *secretion* of drugs into urine; and *reabsorption* of drugs from urine back into blood. A final useful characteristic of clearance is that it is additive. In other words, the total systemic clearance is the sum of the individual organ clearances. For instance, for a drug that has both hepatic and renal routes of elimination, $CL_{tot} = CL_h + CL_r$. Such a drug still can be cleared, albeit more slowly, in patients with either liver or kidney failure.

For an ideal, so-called *one-compartment* drug (i.e., one that is in rapid equilibrium with all tissues), the parameters half-life, clearance, and volume of distribution are interrelated by Equation 5.6. Because half-life and clearance can be determined independently, it is possible to calculate $V_d$. Equation 5.6 shows that two drugs that have the same volume of distribution but different clearances will have different half-lives, which means that one drug will maintain plasma concentrations longer than the other and may need less frequent dosing.

$$T_{1/2} = \frac{0.693 \times V_d}{CL} \tag{5.6}$$

As mentioned above, many drugs do not conform to the simple one-compartment model. These cases may require a two- or three-compartment model characterized by a bi- or tri-exponential decline (8). Alternatively, a simpler, commonly used approach is *noncompartmental analysis*, in which the concentration time profile is treated descriptively by the method of statistical moments (9). Whereas *CL* has exactly the same definition in noncompartmental analysis, persistence in the body is described by a new parameter, *mean residence time* (*MRT*, the average time that an drug molecule resides in the body) rather than terminal elimination half-life. For many drugs, *MRT* is a better indicator of the clinically effective duration. The difference between *MRT* values for oral and intravenous doses gives a descriptor of the rate of intestinal absorption called the *mean absorption time* (*MAT*). Noncompartmental analysis also defines a new volume parameter, $V_{dss}$, the volume of distribution at steady state, which is useful for calculations of plasma levels after multiple dosing or during intravenous infusion. A second descriptor of volume, $V_z$, represents the volume of distribution during the terminal elimination phase and is calculated exactly the same as $V_d$ in the one-compartment model (see Equation 5.6). $V_z$ usually is different from $V_{dss}$ and seldom is used in clinical dosing regimen considerations.

## 5.2   ROLE OF PK IN THE DRUG DISCOVERY AND DEVELOPMENT PROCESS

### 5.2.1   Discovery

Basic research elucidates the complex biochemical events that comprise a biologic process, such as the regulation of blood glucose, and identifies key control points that are mediated by enzymes or receptors. Most drugs exert their pharmacologic action by modulating the activity of one of these enzymes or receptors within cells of the abnormally functioning organ. So, the next step in the drug discovery process is *lead generation*, the design or discovery of a small molecule that will bind to the molecular target to modulate its activity. Once a lead has been generated, medicinal chemists typically synthesize thousands of analogs by systematically varying the structure to create a compound that has been optimized with respect to potency and selectivity toward the molecular target, PK characteristics, and safety. This compound then is designated as a clinical candidate, and it enters the development process (10).

Given the importance of PK in the clinical use of drugs, a suitable human PK profile is an important criterion in the selection from many discovery compounds of a single candidate for advancement to clinical development. Therefore, contemporary drug discovery screening of compounds includes pharmacologic potency, selectivity, and PK properties in the acceptable range for the medical indication (11). Because drug discovery normally takes place before the clinical phase, the human PK characteristics of a drug candidate must be inferred from *in vivo* animal models or from various human-derived *in vitro* systems (12).

### 5.2.2  PK Screening

***5.2.2.1  In Vitro* Studies**  Because clearance at the whole-body level often is determined by metabolism at the cellular level, it is possible to use a variety of human-derived *in vitro* systems to determine rates of metabolism. These systems include pure human enzymes (such as cytochrome P450 enzymes) (13) and human liver subcellular fractions (*S*9 and microsomes) (14). However, with enzymes and subcellular fractions, some information is lost because the whole-cell integration of subcellular processes has been disrupted. The use of cultured human hepatocytes retains the whole-cell integration at the expense of greater experimental complexity (15). Each system provides a different window on the metabolic processes, is relatively easy to use, and can be obtained from commercial sources. Rates and pathways of metabolism may be compared with a series of discovery compounds to identify those with the greatest relative metabolic stability or with a benchmark compound of known human PK characteristics to provide a more absolute estimate of hepatic metabolic clearance.

***5.2.2.2  In Vivo* Studies**  PK studies in typical laboratory animals (typically mice, rats, dogs, or monkeys) are useful because they directly determine the various PK parameters discussed above, which affords an understanding of the whole-body characteristics of absorption, distribution, metabolism, and elimination. Because these translate with some fidelity to humans, animal PK commonly is used to assess the PK acceptability of discovery compounds. Unfortunately, these studies tend to be too slow to permit rapid evaluation of dozens or hundreds of discovery compounds. Therefore, two modifications have been introduced that provide whole-body relevance with rates of screening that are sufficient to keep pace with the production of favorable discovery drug candidates (16).

The first modification, cassette dosing, is based on the simultaneous dosing of several compounds to the animals, which thereby provides *in vivo* PK information of the several candidates in parallel (17). Unfortunately, cassette dosing has a limitation of potential drug–drug interactions, which possibly distorts the PK parameters (18). The second modification, Rapid Rat Assay, uses a highly regimented process of single-compound dosing of several discovery compounds to several rats in parallel (19). Because each animal receives only one compound, drug–drug interactions are avoided. The Rapid PK procedure can be applied to dogs and monkeys as well but with lower throughput than with rats. The combination of human *in vitro* and animal *in vivo* assessment of PK works well to screen out compounds with unsuitable PK properties.

### 5.2.3 Clinical Candidate Nomination

Once a discovery compound has been identified with overall "drug-like" characteristics (good potency, selectivity, safety, and PK), that compound will be selected for advancement into clinical development. The next important PK activity is the quantitative projection of the human PK. Such information is useful for the planning of the preclinical and clinical development programs. Equation 5.6 can be used to estimate the clinical half-life if we can estimate $CL$ and $V_d$. Two approaches can be used to estimate $CL$.

First, if the compound was cleared mainly by hepatic metabolism in the animal species tested and if human hepatocytes *in vitro* suggest the same will be true in humans, then the measured hepatocyte clearance may be used in a process called *in vitro/in vivo scaling* to provide an estimate of the human intrinsic clearance. The application of Equation 5.5 then gives an estimate of the human systemic clearance. Second, the animal PK parameters of $CL$ and $V_d$ can be subjected to *allometric scaling* whereby the PK parameter is related to a measurable allometric variable such as body mass, body surface area, heart rate, and so forth. (21) by fitting these parameter–variable pairs for several species to an empirical power equation of the form

$$PK \text{ parameter} = \alpha \times (\text{allometric variable})^\beta \qquad (5.7)$$

where $\alpha$ and $\beta$ are adjustable constants. A common application is to scale $CL$ according to body mass.

$$CL = \alpha \times (\text{Body Mass})^\beta \qquad (5.8)$$

Both $CL$ and $V_d$ from the animal PK studies may be scaled allometrically to provide the estimates of human $CL$ and $V_d$ needed to project the human half-life. In addition to calculating the half-life, one also can estimate the human therapeutic dose if the blood levels that correspond to efficacy in a relevant animal pharmacologic model have been measured. The $AUC$ measured in the animal model can be equated with the $AUC$ needed for efficacy in humans, and then the application of Equation 5.4 with the estimated human $CL$ gives an estimated human dose. For drugs intended for oral administration, the bioavailability $F$ often is assumed to lie in the range between the lowest and highest values observed in the animal species tested. These projected human PK parameters need not be extremely accurate to be useful for planning.

### 5.2.4 Preclinical Development

Before a new chemical entity can be tested in humans, its safety in various *in vitro* and *in vivo* pharmacologic and toxicity tests must be assessed (22). This assessment includes an administration of high doses to animals (typically mice or rats and larger animals such as dogs or monkeys) and *in vitro* tests of genotoxic potential (such as the Ames test). PK normally will be determined in the animal species to be used for safety testing to assess the relationship between the oral dose and the exposure for dose selection and to select time points for the monitoring of exposure during the safety studies. In parallel with the safety testing, a large-scale chemical synthetic process must be developed to produce tens of kilograms of the drug candidate to enable clinical trials. Finally, a clinical formulation must be devised to allow adequate and reproducible exposure in the clinical trials, and animal PK studies are helpful in the selection process. These studies usually examine $C_{max}$, $T_{max}$, and $AUC$ for several possible formulations to allow matching to the desired clinical delivery profile. Data from the discovery and preclinical activities form the basis of a petition to government health authorities for permission to begin testing in humans. Once this dossier, called the Investigational New Drug Application (IND) in the United States, is approved, clinical trials may proceed (23). Information about the IND and other regulatory guidelines and documentation is available through the web link to the United States Food and Drug Administration at the end of this chapter.

### 5.2.5  Clinical Development

Clinical trials are divided formally into four distinct but temporally overlapping phases, defined by the objectives of each phase (24). *Phase I* tests the safety and tolerability of the drug candidate in normal healthy volunteer subjects. An important goal is to determine the maximum dose that may be safely given to people. An additional goal is the determination of the pharmacokinetic parameters, which are needed to plan the next phases. *Phase II* investigates whether the candidate modulates the pharmacologic target in ten to twenty patients (*proof of activity*) and whether this activity actually results in the desired therapeutic benefit (*proof of concept*). Also, the optimum dose is determined based on balancing benefit and safety. With an optimized dose identified, *Phase III* tests the drug in much larger groups of patients to establish the comprehensive clinical profile of the candidate, including confirming the therapeutic benefit with adequate statistics and investigating the variation of therapeutic response and the safety in various severities of the disease and in special populations such as ethnic groups or comorbidities. *Phase IV* usually occurs after the drug has been approved and is on the market and involves testing even larger populations for longer periods of time and testing against additional disease indications.

*5.2.5.1  Phase I*   Human PK is determined for the first time in Phase I, initially as a series of single doses escalated through several dose levels to establish the dose–exposure relationship and also the half-life. Because an oral dose may have an extended period of absorption, it is not possible to get the true clearance or volume of distribution from an oral dose; a confounding with bioavailability may occur. Furthermore, sometimes absorption is slower than elimination and a phenomenon known as "flip-flop kinetics" occurs in which the observed half-life really is a reflection of absorption and not elimination. Thus, when feasible, an intravenous dose is given to allow absolute oral bioavailability and the true values of total clearance, renal clearance, volume of distribution, and half-life to be determined. Administration of an intravenous dose requires the development of a sterile solution formulation and acute intravenous animal safety studies and may not be possible always with poorly soluble drugs. The next step is multiple dosing at several dose levels within the range determined to be safe from the single-dose study. In addition to more investigation of clinical safety, the main PK purpose of multiple dosing is to establish how many doses are required to reach steady state and what steady state $C_{max}$ and $C_{min}$ values are reached. A well-behaved drug will reach steady state in about five doses (see Fig. 5.3), and a general rule is that if doses are given at intervals of one half-life, then the steady state $C_{max}$ will be about twice that following a single dose. Dosing more frequently than the half-life results in a more than twofold accumulation, and dosing less frequently than the half-life results in a less than twofold accumulation. The average plasma level to be expected at steady state can be calculated by Equation 5.9 where $\tau$ is the dosing interval,

$$C_{ave,\ SS} = \frac{F \times Dose}{CL \times \tau} \tag{5.9}$$

or can be determined from a numerical simulation using the *Principle of Superposition*. Importantly, a significant deviation of the accumulation index $R$-value from that predicted by the single-dose PK indicates a time-dependent alteration of clearance. For example, continuing exposure of the body to some drugs causes an adaptive response called *autoinduction* in which drug-metabolizing enzymes are upregulated to allow the body to eliminate the drug more rapidly (26). Such a drug would then show an $R$-value of less than one, which means that subsequent doses would lead to lower $C_{max}$ and $AUC$ values. It is possible for autoinduction to be so severe that the drug becomes ineffective after a few doses. The opposite situation, although it is rare, also can occur; that is, $R$ values can be much larger than expected, which means that when multiple dosing, the body becomes less able to clear the drug and plasma levels continue to rise. Greater–than–expected accumulation usually indicates some form of toxicity, such as organ toxicity or irreversible inhibition of clearance enzymes. The cytochrome P450 enzyme designated CYP3A4 is especially susceptible to irreversible inhibition (27), so the discovery process usually includes a screen for this type of behavior to prevent such compounds from reaching the clinic. Again, this type of nonideal PK behavior could be enough to cause the drug candidate to be clinically unusable.

Additional Phase I PK studies include the determination of the effect of food on orally administered drugs (28), formulation testing (29), detection of circulating metabolites of the drug (30), and checking for PK interactions between coadministered drugs (31). The most common form of drug–drug PK interaction is the competition of the two drugs for the same clearance mechanism, especially the CYP-family of drug-metabolizing enzymes, which results in a limitation on which drugs ultimately can be used clinically with the new agent (32). All clinical PK studies rely on some method of quantitation of drug levels in blood or plasma, and the major bioanalytic technique in use today for this purpose is liquid chromatography coupled to tandem mass spectrometry (LC–MS/MS) (33, 34). Current technology permits the determination of picogram-per-mL levels of a drug in the presence of many interfering substances, with sample-to-sample cycle times less than 2 minutes. Because of individual biologic variability and inevitable experimental error, the measured drug concentrations at each time point will show variation among a group of subjects, which results in the need to express derived pharmacokinetic parameters as a mean value with an associated coefficient of variation.

In addition to the conventional first-in-human approach described above, another available approach allows rapid access to the clinic to answer specific questions addressable by a single dose in humans. The U.S. Food and Drug Administration allows a sponsor to proceed to limited clinical dosing with an Exploratory IND (35), which has more moderate requirements than the standard IND. Determination of human PK is the most common use of the new approach, which can be thought of as an extension of the discovery process to check human PK before the decision is made to commit full development resources to pursue the conventional IND. One option within the Exploratory IND is *microdosing*, which has even lower requirements (36, 37). A microdose is defined as no more than 1/100 of the expected therapeutic dose but not exceeding 100 micrograms. The advantages of a microdose are that minimal safety testing is required because only minute quantities of the drug candidate are administered and that large-scale chemical synthesis can be postponed until acceptable clinical PK is confirmed.

***5.2.5.2 Phase II*** Although the main goal of Phase II clinical trials is the determination of efficacy toward the disease indication, PK is a fundamental part of this determination because to understand the clinical use of the new drug candidate, it is necessary to determine the temporal relationship between the plasma levels and the pharmacodynamics of the beneficial effect (PK/PD). A secondary PK goal in Phase II is to determine if the disease state affects the PK of the candidate drug compared with the PK observed in healthy subjects.

During the dose range-finding portion of Phase II, several doses are assessed for efficacy in patients with the disease to be treated. The PK sampling at these doses is critical because it allows the construction of a pharmacokinetic/pharmacodynamic (PK/PD) model whereby the observed drug concentrations are used to predict the effect of the drug. The simplest PK/PD model is the *direct* or $E_{max}$ model in which a direct relationship exists between the concentration and the effect; changes in plasma levels are reflected immediately in the pharmacologic response. Often, however, as concentrations increase, eventually the pharmacologic responses reach a peak or nadir and greater increases in concentration result in only a small change in effect. Ideally, the drug candidate is safe enough to dose to the maximal effect. However, it is more common to strike a balance between the magnitude of response and the incidence of adverse side effects. Some drugs follow a more complex PK/PD relationship called the *indirect* model in which the pharmacologic response temporally lags behind changes in plasma levels (*hysteresis*). In this model, the response seen from a particular plasma concentration in the declining phase of the PK curve may be much greater than it was for the same plasma concentration during the rising phase of the curve. Hysteresis occurs when the initial stimulus of the drug binding to the receptor is uncoupled in time to the observed clinical response because a biologic cascade of several biochemical or physiologic events must occur for the clinical response to become manifest. After it is determined which type of relationship applies, a theoretical PK/PD model can be fit to the concentration–response data, which thereby generates a quantitative understanding of the action of the drug and the ability to predict the response to be expected in any situation.

With the availability of a PK/PD model, clinical trial simulation (39, 40) can be performed to explore alternative dosing regimens and to help achieve the optimal design of the Phase III program. The inputs for simulation include the number of patients to be tested, the duration of the testing, the potential clinical end points, and most importantly, the dose(s) to be tested. With Monte Carlo methods to simulate the variability in PK and response observed in Phases I and II, hundreds of computer simulations are conducted for each design and dose level to estimate the statistical probability of a successful outcome (i.e., one that meets predetermined end points of efficacy and safety). The goal is to find a design and dose that balances a high likelihood of a successful trial with a low incidence of adverse events. Ideally, this simulation activity is done in collaboration with government health authorities to minimize the approval time for the Phase III design and the chance of a failed Phase III study (41).

*5.2.5.3 Phase III*   In Phase III, large-scale trials are mounted with the goal of showing statistically significant benefit against either placebo or, more usually, standard-of-care therapy. The use of many patients affords the opportunity to create a *Population PK model* for the drug, which is a statistical means to account for individual variation in PK by means of clinically measurable *covariates* such as age, gender, race, ethnicity, body weight, kidney function, and so forth (42). In this way, the PK characteristics of a particular patient can be estimated before dosing, which allows a rational means to select the dose most likely to provide good benefit while minimizing risk. After Phase III is complete, the sponsor of the drug candidate may submit an application (called the New Drug Application in the United States) for approval to government health authorities of all countries in which marketing is desired (43).

## 5.3  SUMMARY

Pharmacokinetics studies the relationships for a given drug between blood concentrations, uptake, distribution, elimination, pharmacologic effect, and time. Because drugs that require large doses must be dosed frequently; cannot be dosed orally; must be timed in relation to food intake or have strong interactions with other drugs; may have issues of patient compliance, variable response, or toxicity, the pharmacokinetic aspects of the behavior of a drug are critical to its successful clinical use. Accordingly, the scientists who design drugs must use preclinical pharmacokinetic screening methods to maximize the likelihood that a drug candidate is acceptable pharmacokinetically. Correspondingly, the researchers who test the drug in clinical trials must ensure that all aspects of pharmacokinetic behavior have been characterized thoroughly before submitting the drug for approval to regulatory authorities.

## REFERENCES

1. Rowland M, Tozer TN. Clinical Pharmacokinetics. Concepts and Applications. 3rd edition. 1995. Lippincott Williams & Wilkins, Philadelphia, PA.

2. White RE. High-throughput screening in drug metabolism and pharmacokinetic support of drug discovery. Ann. Rev. Pharmacol. Toxicol. 2000;40:133–157.

3. Lobo ED, Hansen RJ, Balthasar JP. Antibody pharmacokinetics and pharmacodynamics. J. Pharm. Sci. 2004;93:2645–2668.

4. Ho RJY, Gibaldi M. Biotechnology and Biopharmaceuticals. Transforming Proteins and Genes into Drugs. 2003. John Wiley & Sons, Hoboken, NY.

5. Boxenbaum H, Battle M. Effective half-life in clinical pharmacology. J. Clin. Pharmacol. 1995;35:763–766.

6. Wilkinson GR. Pharmacokinetics. The dynamics of drug absorption, distribution, and elimination. In: The Pharmacological Basis of Therapeutics. 10 edition. Hardman JG, Limbird LE, Gilman AG, eds. 2001. McGraw-Hill, New York. pp. 3–29.

7. Benet LZ, Hoener B. Changes in Plasma Protein Binding Have Little Clinical Relevance. J. Clin. Pharmacol. Ther. 2002;71:115–121.

8. Bonate PL. Compartmental models. In: Pharmacokinetics in Drug Development, Volume 1: Clinical Study Design and Analysis. Bonate PL, Howard DR, eds. 2004. AAPS Press, Arlington, VA. pp. 291–320.

9. Muir KT, Gomeni RO. Non-compartmental analysis. In: Pharmacokinetics in Drug Development. Volume 1: Clinical Study Design and Analysis. Bonate PL, Howard DR, eds. 2004. AAPS Press, Arlington, VA. pp. 235–265.

10. Natarajan C, Rohatagi S. Role of preclinical pharmacokinetics in drug development. In: Pharmacokinetics in Drug Development. Volume 2: Regulatory and Development Paradigms. Bonate PL, Howard DR, eds. 2004. AAPS Press, Arlington, VA. pp. 127–161.

11. White RE, A Comprehensive Strategy for ADME Screening in Drug Discovery. pp. 431–450 in Borchardt, Ronald T. Edward H. Kerns Christopher A. Lipinski Dhiren R. Thakker, and Binghe Wang, eds. Pharmaceutical Profiling in Drug Discovery for Lead Selection. 2004. AAPS Press, Arlington, VA, USA.

12. Lin JH. Challenges in drug discovery: lead optimization and prediction of human pharmacokinetics. In: Pharmaceutical Profiling in Drug Discovery for Lead Selection. Borchardt RT, Kerns EH, Lipinski CA, Thakker DR, Wang B, eds. 2004. AAPS Press, Arlington, VA. pp. 293–325.

13. Rodrigues AD. Applications of heterologous expressed and purified human drug-metabolizing enzymes: an industrial perspective. In: Handbook of Drug Metabolism. Woolf TF, ed. 1999. Marcel Dekker, Inc., New York. pp. 279–320.

14. Ekins S, Mäenpää J, Wrighton SA. In Vitro metabolism: subcellular fractions. In: Handbook of Drug Metabolism. Woolf TF, ed. 1999. Marcel Dekker, Inc., New York. pp. 363–399.

15. Sinz MW. In vitro metabolism: hepatocytes. In: Handbook of Drug Metabolism. Woolf TF, ed. 1999. Marcel Dekker, Inc., New York. pp. 401–424.

16. Cox KA, White RE, Korfmacher WA. Rapid determination of pharmacokinetic properties of new chemical entities: in vivo approaches. Comb. Chem. High Throughput Screen. 2002;5:29–37.

17. Manitpisitkul P, White RE. Whatever happened to cassette-dosing pharmacokinetics? Drug Discov.Today. 2004;9:652–658.

18. White RE, Manitpisitkul P. Pharmacokinetic theory of cassette dosing in drug discovery screening. Drug Metab. Dispos. 2001;29:957–966.

19. Korfmacher WA, Cox KA, Ng KJ, Veals J, Hsieh Y, Wainhaus S, Broske L, Prelusky D, Nomeir A, White RE. Cassette-accelerated rapid rat assay: a systematic procedure for the dosing and liquid chromatography/atmospheric pressure ionization tandem mass spectrometric analysis of new chemical entities as part of new drug discovery. Rapid Communicat. Mass Spectrom. 2001;15:335–340.

20. Kwon Y. Handbook of Essential Pharmacokinetics, Pharmacodynamics and Drug Metabolism for Industrial Scientists. 2001. Kluwer Academic/Plenum Publishers, New York.

21. Davies B, Morris T. Physiological parameters in laboratory animals and humans. Pharmaceut. Res. 1993;10:1093–1095.

22. Lathia CD, Shah A. First-time-in-man studies. In: Pharmacokinetics in Drug Development. Volume 1: Clinical Study Design and Analysis. Bonate PL, Howard DR, eds. 2004. AAPS Press, Arlington, VA. pp. 3–30.

23. Troetel WM. The investigational new drug application and the investigator's brochure. In: New Drug Approval Process. 4th edition. Accelerating Global Registrations. Guarino RA, ed. 2004. Marcel Dekker, Inc., New York. pp. 63–100.

24. Coburn WA, Heath G. Efficient and effective drug development. In: Applications of Pharmacokinetic Principles in Drug Development. Krishna R, ed. 2004. Kluwer Academic/Plenum Publishers, New York. pp. 1–20.

25. Gibaldi M, Perrier D. Pharmacokinetics. 2nd edition. 1982. Marcel Dekker, Inc., New York.

26. Greenblatt DJ, von Moltke LL. Drug-drug interactions: clinical perspective. pp. 565–584 in Rodrigues A. David, ed. Drug-Drug Interactions. 2002. Marcel Dekker, Inc. New York.

27. Jones DR, Hall SD. Mechanism-based inhibition of human cytochromes P450: in vitro kinetics and in vitro-in vivo correlations. In: Drug-Drug Interactions. Rodrigues AD, ed. 2002. Marcel Dekker, Inc. New York. pp. 387–413.

28. Fleisher D, Reynolds L. Food-drug interactions: drug development considerations. In: Applications of Pharmacokinetic Principles in Drug Development. Krishna R, ed. 2004. Kluwer Academic/Plenum Publishers, New York. pp. 195–223.

29. Barrett JS. Bioavailability and bioequivalence studies. In: Pharmacokinetics in Drug Development. Volume 1: Clinical Study Design and Analysis. Bonate PL, Howard DR, eds. 2004. AAPS Press, Arlington, VA. pp. 91–119.

30. Baillie TA, Cayen MN, Fouda H, Gerson RJ, Green JD, Grossman SJ, Klunk LJ, LeBlanc B, Perkins DG, Shipley LA. Drug metabolites in safety testing. Toxicol. Appl. Pharmacol. 2002;182:188–196.

31. Bjornsson TD, Callaghan JT, Einolf HJ, Fischer V, Gan L, Grimm S, Kao J, King SP, Miwa G, Ni L, Kumar G, McLeod J, Obach SR, Roberts S, Roe A, Shah A, Snikeris F, Sullivan JT, Tweedie D, Vega JM, Walsh J, Wrighton SA. The conduct of in vitro and in vivo drug-drug interaction studies: a PhRMA perspective. J. Clin. Pharmacol. 2003;43:443–469.

32. Huang S-M. Drug-drug interactions. In: Applications of Pharmacokinetic Principles in Drug Development. Krishna R, ed. 2004. Kluwer Academic/Plenum Publishers, New York. pp. 307–331.

33. Hopfgartner G, Husser C, Zell M. High-throughput quantification of drugs and their metabolites in biosamples by LC-MS/MS and CE-MS/MS: possibilities and limitations. Ther. Drug Monit. 2002;24:134–143.

34. Powell ML, Unger SE. Bioanalytical methods: challenges and opportunities in drug development. In: Applications of Pharmacokinetic Principles in Drug Development. Krishna R, ed. 2004. Kluwer Academic/Plenum Publishers, New York. pp. 21–52.

35. United States Food and Drug Administration. Exploratory IND Studies. Guidance for Industry, Investigators, and Reviewers. January 2006. *http://www.fda.gov/cder/guidance/7086fnl.htm*.

36. Lappin G, Kuhnz W, Jochemsen R, Kneer J, Chaudhary A, Oosterhuis B, Drijfhout WJ, Rowland M, Garner RC. Use of microdosing to predict pharmacokinetics at the therapeutic dose: experience with 5 drugs. Clin. Pharmacol. Ther. 2006;80:203–215.

37. Sandhu P, Vogel JS, Rose MJ, Ubick EA, Brunner JE, Wallace MA, Adelsberger JK, Baker MP, Henderson PT, Pearson PG, Baillie TA. Evaluation of microdosing strategies for studies in preclinical drug development: demonstration of linear pharmacokinetics in dogs of a nucleoside analog over a 50-fold dose range. Drug Metab. Dispos. 2004;32:1254–1259.

38. Rohatagi S, Martin NE, Barrett JS. Pharmacokinetic/Pharmaco dynamic Modeling in Drug Development. pp. 333–372 in Krishna, Rajesh, ed. Applications of Pharmacokinetic Principles in Drug Development. 2004. Kluwer Academic/Plenum Publishers, New York.

39. Sale M. Clinical trial simulation. In: Pharmacokinetics in Drug Development. Volume 1: Clinical Study Design and Analysis. Bonate PL, Howard DR, eds. 2004. AAPS Press, Arlington, VA. pp. 531–550.

40. Bonate PL. Clinical trial simulation in drug development. Pharmaceut. Res. 2000;17:252–256.

41. Lesko LJ. Paving the critical path: how can clinical pharmacology help achieve the vision? Clin. Pharmacol. Ther. 2007;81:170–177.

42. Mould DR. Population pharmacokinetics: applications in industry. In: Pharmacokinetics in Drug Development. Volume 1: Clinical Study Design and Analysis. Bonate PL, Howard DR, eds. 2004. AAPS Press, Arlington, VA. pp. 501–529.

43. Guarino RA. The new drug application, content and format. In: New Drug Approval Process. 4th edition. Accelerating Global Registrations. Guarino RA, ed. 2004. Marcel Dekker, Inc., New York. pp. 113–172.

**FURTHER READING**

Cytochrome P450 Homepage. *http://drnelson.utmem.edu/cytochromeP450.html*.

Derendorf H, Lesko LJ, Chaikin P, Colburn WA, Lee P, Miller R, Powell R, Rhodes G, Stanski D, Venitz J. Pharmacokinetic/Pharmaco dynamic Modeling in Drug Research and Development. J. Clin. Pharmacol. 2000;40:1399–1418.

Lee JS, Obach RS, Fisher MB, eds. Drug Metabolizing Enzymes. Cytochrome P450 and Other Enzymes in Drug Discovery and Development. 2003. Marcel Dekker' Fontis Media, The Netherlands.

Holford NHG, Kimko HC, Monteleone JPR, Peck CC. Simulation of clinical trials. Annu. Rev. Pharmacol. Toxicol. 2000;40:209–234.

International Society for the Study of Xenobiotics. *http://www.issx. org/*.

Mahmood I, ed. Clinical Pharmacology of Therapeutic Proteins. 2006. Pine House Publishers, Rockville, MD.

Pharmacokinetic and Pharmacodynamic Resources. *http://www.boomer. org/pkin/*.

U.S. Food and Drug Administration, Center for Drug Evaluation and Research. *http://www.fda.gov/cder/*.

Wilkinson GR. Pharmacokinetics. The dynamics of drug absorption, distribution, and elimination. In: The Pharmacological Basis of Therapeutics. 10th edition. Hardman JG, Limbird LE, Gilman AG, eds. 2001. McGraw-Hill, New York. pp. 3–29.

# 6

# ADME PROPERTIES OF DRUGS

Li Di and Edward H. Kerns

*Wyeth Research, Princeton, New Jersey*

ADME properties have tremendous impact on the success of drug candidates. They are a set of fundamental physico-chemical and biochemical properties of drug molecules that can be affected by physiologic conditions *in vivo*: solubility, permeability, stability, metabolism, and others. ADME properties can be improved through structural modification by medicinal chemists. They are important both *in vitro* and *in vivo*. A successful drug must possess a balance of good potency and drug-like properties. High-throughput ADME assays have been developed and implemented widely in the pharmaceutical industry to reduce the attrition of drug candidates. The data are used to identify potential liabilities, guide structural modification, prioritize lead series, select compounds for *in vivo* studies, and diagnose *in vivo* assay results. Studies have demonstrated that early screening of ADME properties is a successful approach to enhance the quality of drug candidates.

Drug discovery typically starts from an unmet medical need (1), followed by target identification–validation, HTS (high-throughput screening) and hit finding, as well as hit-to-lead and lead optimization (Fig. 6.1). Once an active compound is identified, preclinical development starts. If the compound is efficacious and safe in animals, it will be evaluated in human clinical trials after an IND (investigational new drug) is filed. Phase I clinical trials involve safety and pharmacokinetic studies in a small group of healthy volunteers. Phase II involves efficacy studies in a small group of patients. Phase III continues safety and efficacy studies in a large group of patients. If a compound is safe and efficacious in patients, an NDA (new drug application) is filed for FDA review and approval. Once an NDA is

*Chemical Biology: Approaches to Drug Discovery and Development to Targeting Disease*, First Edition.
Edited by Natanya Civjan.
© 2012 John Wiley & Sons, Inc. Published 2012 by John Wiley & Sons, Inc.

**Figure 6.1**   Stages of drug discovery.

**TABLE 6.1   Reasons for drugs to fail development**

| Reason for failure | Percentage |
| --- | --- |
| Toxicity | 22% |
| Lack of efficacy | 31% |
| Market reasons | 6% |
| Poor biopharmaceutical property | 41% |

approved, the drug can be marketed. The safety of the drug continues to be monitored in postmarketing studies. Fifty discovery programs, each with hundreds to thousands of compounds prepared, will lead to 10 development candidates and result in one marketed drug (2). The process takes an average of 12 years, costs more than $800 million, and has a success rate of only 10% in the clinic (2, 3). Drug discovery and development is costly and high risk; it takes a long time. A study in the early 1990s showed that many reasons explain why drugs fail in development, such as toxicity, lack of efficacy, or market reasons (Table 6.1). It was striking to see that, at that time, more than 40% of the drugs failed development because of poor biopharmaceutical properties (3, 4). As a result, a major initiative in the pharmaceutical industry developed to address ADME/TOX (absorption, distribution, metabolism, excretion, and toxicity) issues during drug discovery so that property information can be generated in parallel with activity data and so that successful drugs can be developed with good potency and drug-like properties (Fig. 6.2) (5). In the same time period, other new technologies also were implemented in drug discovery, such as HTS, combinatorial chemistry, molecular biology, genomics, and proteomics (6, 7). They all have a tremendous impact on the way drug discovery is conducted today. The term ADME originally referred to *in vivo* processes that govern pharmacokinetics (PK). It has been expanded to include *in vitro* physico-chemical and biochemical properties that affect *in vivo* PK. This chapter discusses various *in vitro* ADME properties that influence the overall success of a drug candidate *in vivo*. For detailed discussion of PK and metabolism, please refer to Chapter 5 and resources in the further reading list at the end of this chapter. It is important to point out that suitable ADME properties not only are essential for the success of *in vivo* studies but also critical for the success of *in vitro* assays (5).

Traditionally, ADME/TOX assays were not high throughput (8). Each project team only had a few slots per month for ADME/TOX profiling. If a team were making a lot of compounds, they would have to rely solely on potency and

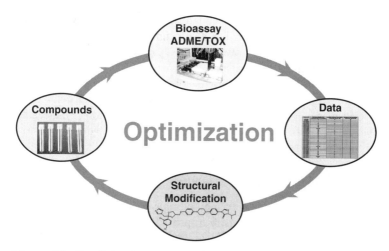

**Figure 6.2** Parallel optimization of potency and ADME/TOX properties.

selectivity to triage compounds for *in vivo* studies, and the compound selected potentially could have poor properties and not be worthwhile to pursue. Nowadays, ADME/TOX assays are high throughput. Depending on company philosophy and available resources, all newly synthesized compounds can be profiled for ADME/TOX properties (5, 8, 9). Teams can triage compounds for *in vivo* studies not only based on potency and selectivity but also on ADME/TOX properties so that informed decisions can be made.

## 6.1 SOLUBILITY

Solubility impacts every aspect of drug discovery and development. *In vivo*, an orally administered solid drug has to disintegrate, dissolve, pass through the gastrointestinal membrane into the portal vein, go through the liver, and finally reach systemic circulation. Solubility, along with permeability and metabolism, affects oral absorption and oral bioavailability. For IV dosing, a compound needs to be completely soluble in a dosing vehicle. Thus, good solubility is essential for the success of an IV formulation. Low solubility can cause abnormal PK profiles, such as second peak phenomena, which is caused by precipitation and redissolution. Formulation development for insoluble compounds can be costly and time-consuming, and it can lead to delayed introduction of lifesaving drugs to the market.

Solubility not only affects *in vivo* assay results, but it also has tremendous impact on *in vitro* assays (10, 11). Compounds that have poor aqueous solubility can produce erratic assay results, erroneous SAR (Structure-Activity Relationship), discrepancy between assays, and artificially low potency because of the right shift of $IC_{50}$ curves. They can cause low HTS hit rates, low confirmation of hits from bench-top assays, and underestimation of toxicity. For some compounds, the potency and properties just cannot be measured because of the extremely low concentration. Low solubility can cause a lot of frustration and loss of productivity in drug discovery.

Structural modifications to improve solubility include reducing Log P by introducing ionizable groups, increasing hydrogen bonds and polar groups, and reducing crystal packing by adding out-of-plane substitution (8). Because of the importance of solubility, many methodologies have been developed to address specific issues related to solubility (5, 12).

### 6.1.1  High-Throughput Kinetic Solubility

Several high-throughput solubility methods have been developed and are used widely in the pharmaceutical industry to provide solubility information for thousands of discovery compounds per year. The data are used to provide an early alert for potential solubility issues so that strategies can be developed to address them. High-throughput methods often are used to measure kinetic solubility. In kinetic solubility methods, solid material is not necessarily in equilibrium with the solution. Compounds are dissolved in DMSO, which is added to aqueous buffers. The concentration of the compound in solution is determined by direct UV measurement (13), turbidity (10), nephelometric (14), or flow cytometry (15). Kinetic solubility usually is higher than equilibrium solubility for three reasons: 1) a small amount of DMSO is present in the buffer ($\sim 1\%$), 2) precipitates tend to be amorphous materials, which have a higher solubility than crystalline solid, and 3) equilibrium is not fully established, and the solution is supersaturated. Because compounds are dissolved in DMSO first, the samples do not have memory of the crystal form, so kinetic solubility is not sensitive to crystal form. Kinetic solubility provides an optimistic estimate of the equilibrium solubility and insights for discovery experiments in which compound is first dissolved in DMSO.

### 6.1.2  Equilibrium Solubility

Equilibrium solubility typically is used in late-stage drug discovery and development, when crystal forms of the material are well characterized. However, a recent trend exists to move the assay much earlier in drug discovery to screen compounds in dosing vehicles and to select compounds for *in vivo* studies. Equilibrium solubility typically is measured by adding a solvent to solid material and shaking for 24 hours. When the solution reaches equilibrium, it is filtered; the filtrate is diluted accordingly and assayed by LC-UV (liquid chromatography-ultra violet) or LC-MS (liquid chromatography-mass spectrometry). Equilibrium solubility also is used for formulation development and regulatory filing.

### 6.1.3  Solubility in Bioassay Medium

Solubility in bioassay medium usually is initiated when a team suspects that bioassay results might be incorrect because of low solubility. The assay is performed by using the exact protocol used in the bioassay, including buffer composition, dilution protocol, incubation time, and temperature. The solution is filtered, and the filtrate is diluted accordingly and quantified using LC-MS. The solubility data are used to guide assay optimization and address questions related to bioassays, such as erratic assay results, erroneous SAR, irreproducible data, and discrepancy between assays (11).

## 6.2  PERMEABILITY

Drugs have to go through many biologic barriers before they can reach the therapeutic targets. The barriers can be the gastrointestinal membrane for oral absorption, the blood–brain barrier for brain penetration, and the cell membrane for cell-based assays. Many transport mechanisms exist for absorption (16). Most drugs are absorbed by passive diffusion across the lipid membrane. Certain compounds can be absorbed by active transport. This transport requires the compounds to have certain structural motifs that can be recognized by transporters, such as amino acids, peptides, nucleosides, and so forth. Polar compounds with a small molecular weight (<200 Dalton) can be absorbed by the paracellular route through the junctions between the cells. Endocytosis can enhance uptake of highly lipophilic compounds by forming vesicles that engulf the molecules. Efflux occurs when efflux transporters, such as Pgp, pump the molecules out of cells. Pgp efflux can impair brain penetration. It also affects oral absorption, especially for highly potent compounds, when the low concentration in the gastrointestinal tract is not likely to saturate the efflux transporters.

Strategies to improve permeability include increasing lipophilicity with an optimal Log $D_{7.4}$ between 1 and 3, reducing hydrogen-bonding capacity, and decreasing polarity and rotatable bonds (5, 8). Prodrugs also are used to enhance the permeability of compounds.

Several software programs are available to predict permeability and absorption. The simple descriptors of Log P and PSA (polar surface area) gave good predictability for oral absorption (17). Several assays described below commonly are used in the pharmaceutical industry to determine the permeability of drug candidates.

### 6.2.1  PAMPA

PAMPA stands for parallel artificial membrane permeability assay. The assay was developed in 1998 (18). It has been used widely in the industry to measure permeability (13). The assay uses a 96-well filter plate coated with lipids (e.g., 2% phosphatidyl choline in dodecane) as a permeability barrier between two aqueous wells. The donor well contains the test compound in solution, and the acceptor well contains only buffer at the beginning of the assay. Molecules cross the lipid barrier from the donor to the acceptor well, driven by the concentration gradient. After incubation for a period of time, the concentrations in the donor and acceptor wells are determined and the effective permeability is calculated. PAMPA is a single-mechanism assay. It accounts for passive diffusion only. Different variations of the original PAMPA assay have been developed, including a different lipid composition, different filter material and thickness, additives in the donor and acceptor to create "sink" conditions (19), and addition of cosolvent to solubilize compounds that are otherwise insoluble. Specific PAMPA models have been optimized to predict oral absorption [Double Sink PAMPA (19), BAMPA (20)], brain penetration [PAMPA-BBB (21)], and skin permeability [PAMPA-Skin (22)]. PAMPA has good correlation with Caco-2 (human colonic adenocarcinoma cells) and Multi-drug Resistant-Madin Darby Canine Kidney for compounds that predominantly permeate by passive diffusion (23). PAMPA has much higher throughput and lower cost than Caco-2 and MDCK assay. PAMPA is applied widely in drug discovery to predict permeability. Computational Quantitative Structure-Activity Relationship models also have been developed for variations of PAMPA (24).

### 6.2.2 Caco-2

Caco-2 is a monolayer cell permeability assay. It was designed originally as an oral absorption model, and now it also is used commonly to investigate transporters. Bidirectional transport can be determined from apical to basolateral direction (A–B) and from basolateral to apical direction (B–A). Efflux ratio is the permeability from B to A direction divided by permeability from A to B direction. A high efflux ratio (e.g., >2.5) is indicative of an efflux substrate. Inhibitors (e.g., verapamil for Pgp) can be added to confirm whether a compound is a substrate for a specific efflux transporter. The system commonly is used to diagnose efflux. Caco-2 not only expresses efflux transporters (Pgp, MRP) but also uptake transporters, such as PEPT1. It is used to screen compounds for substrates of PEPT1 to increase absorption via transporters. Caco-2 is a multiple mechanism assay including passive diffusion, paracellular route, active transport, as well as efflux and metabolism. To increase throughput, a 96-well Caco-2 assay has been developed (25). The Caco-2 assay involves a 21-day cell culture, which takes significant resources. An automated cell seeding and feeding workstation can be used to increase efficiency.

### 6.2.3 MDCK

MDCK is a monolayer cell transport assay. Similar to Caco-2, bidirectional transport can be determined. MDRI-MDCKII cells are transfected with the human MDRI gene to express human Pgp. The assay is a very sensitive Pgp assay and is used widely in the industry to identify Pgp substrates, especially for CNS (central nervous system) targets (26). MDCK only requires 3 days of cell culture, which significantly reduces resources needed for the assay.

## 6.3 STABILITY

Drugs encounter many stability challenges after oral dosing. In the gastrointestinal tract, compounds can degrade in different pH environments: acidic pH in the stomach and neutral to basic pHs in the small and large intestine. They can be metabolized in the intestine and liver by various metabolic enzymes. In blood, compounds can be hydrolyzed by hydrolytic enzymes. Physico-chemical and metabolic stability is essential for drug candidates to reach the therapeutic target with an efficacious concentration.

### 6.3.1 Metabolic Stability

Metabolism occurs in two phases. Phase I metabolism oxidizes or reduces the drug molecule or exposes the functional groups to increase water solubility. It involves oxidation, reduction, hydrolysis, hydration, isomerization, and others. CYP450 is a major metabolic enzyme family. Figure 6.3 shows an example of Phase I metabolism. Phase II metabolism involves synthesis and conjugation reactions, such as glucuronidation, sulfation, methylation, and others. Phase II metabolism increases the water solubility of the compound or detoxifies a reactive species. Examples of Phase II metabolism are shown in Fig. 6.4.

Metabolism affects clearance and, therefore, half-life and oral bioavailability. It dictates how often and how much (dose regimen) a drug should be dosed (5, 8, 27). *In vitro* metabolic stability assays can be used to predict *in vivo* clearance and oral bioavailability, guide structural modification to enhance stability, develop structure–metabolism relationships, and triage compounds for

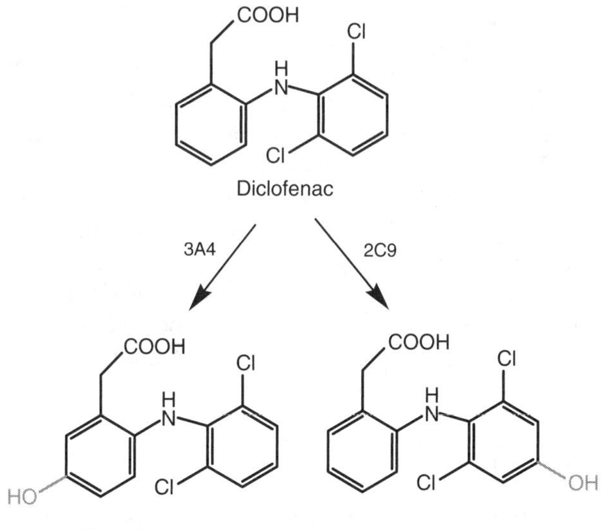

**Figure 6.3** Example of Phase I metabolism.

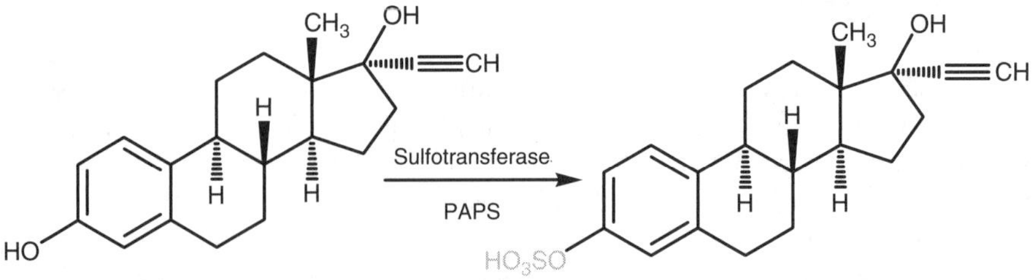

**Figure 6.4**  Examples of Phase II metabolism: glucuronidation and sulfation.

*in vivo* studies. Strategies to improve Phase I metabolic stability are as follows: 1) Block the labile sites, 2) remove the labile groups, 3) reduce lipophilicity, and 4) introduce polar functional groups. For Phase II metabolism, several approaches can be applied to reduce metabolism: 1) Introduce electron-withdrawing groups and steric hindrance, 2) employ isosteric replacement of hydroxyl and carboxylic acid groups, and 3) use prodrugs.

The microsomal stability assay is the most common platform for assessing metabolic liability of a compound. High-throughput assays have been developed to screen large numbers of compounds. Several strategies have been developed to increase throughput and shorten data turnaround time, including the single-time point approach, sample pooling, online extraction, parallel analysis of multiple samples, and the use of a combined ESI–APCI ion source (ESCi) to increase sample coverage. Assay conditions are critical for the success of a microsomal stability assay (28). Substrate concentration cannot be too high ($\leq 1$ μM), otherwise it can saturate the enzymes. The concentration of DMSO should be kept less than or equal to 0.2%. High DMSO concentration can inhibit the CYP450 enzyme. Microsomal protein concentration needs to be low (0.5 mg/mL) because nonspecific binding can affect assay results. Microsomal batches can vary greatly from batch to batch and from vendor to vendor; therefore, quality control is essential to generate consistent data. Insoluble compounds can produce artificially high stability estimates (29) because of precipitation and nonspecific binding to plasticware. Cosolvent can be used to overcome solubility effects.

Hepatocytes also are used for metabolic studies (30). Other metabolic enzymes are present in hepatocytes, which are not found in microsomes. It was found that hepatocytes give better predictions for Phase II glucuronidation than microsomes, even though both systems underestimated the extent of glucuronidation *in vivo* (31). Molecules have to permeate hepatocyte cell membrane to be metabolized, which is a closer mimic of *in vivo*. Heptaocytes are used commonly for studies of metabolic enzyme induction.

### 6.3.2  Plasma Stability

Plasma stability is important for compounds that contain hydrolyzable functional groups, such as esters, amides, carbamates, sulfonamates, lactams, and lactones.

It also is used commonly to screen prodrugs and to select antedrugs. Plasma contains many hydrolytic enzymes, such as esterase, lipase, peptidase, phosphatase, sulfatase, and others. They show species and intersubject differences in activity. A high-throughput plasma stability methodology has been developed (32). Plasma stability is less sensitive to assay conditions than microsomal stability. Enzyme activity varies greatly from batch to batch. Quality control is important for generation of consistent data. Hydrolytic products can be identified to guide structural modification.

### 6.3.3 Solution Stability

Solution stability is important for *in vitro* assays as well as for *in vivo* studies (33). Compounds need to have good stability in bioassay buffers to give reliable assay data. For oral dosing, stability in acidic and basic conditions, as well as stability in gastrointestinal fluids (SGF Simulated gastro Fluid and SIF Simulated Intestinal Fluid), is essential to reach the target with an efficacious concentration and to achieve high oral bioavailability. Solution stability is important, especially at the hit-to-lead stage, to identify potential liabilities of chemotypes. The decisions made at this stage can have far-reaching impact on the success of a discovery program.

### 6.4 CYTOCHROME P450 INHIBITION

Several drugs have been withdrawn from the market because of toxicity caused by drug–drug interactions; they include Posicor, Seldane, Hismanal, and Baycol (34). If a compound shows CYP450 inhibition preclinically, clinical trials must be conducted to evaluate drug–drug interaction potential, which can be costly (35). If a compound might cause drug–drug interaction, the information must be included in the product label, which can be a marketing disadvantage. In drug discovery, project teams make every effort to reduce CYP450 inhibition.

When a single drug is administrated, it will be metabolized mainly by CYP450 enzymes and eliminated from the body. However, if the drug is coadministrated with a second drug, the drugs might compete for the same metabolic enzymes, which leads to a higher blood concentration for the first drug and causes toxicity. The first drug commonly is referred to as the "victim" drug and the second drug as the "perpetrator". For example, Seldane is an antihistamine, and it is metabolized mostly by CYP3A4. When it is coadministrated with Erythromycin (an antibiotic), the blood level of Seldane increases because Erythromycin inhibits CYP3A4, which leads to toxicity of Seldane. Because of this occurrence, Seldane was withdrawn from the market. Its active metabolite currently is marketed as Allegra, which has a metabolism that is not dependent solely on CYP3A4.

CYP3A is the most abundant enzyme in human liver microsomes (28%), followed by CYP2 C (18%) and CYP1A2 (13%) (Table 6.2). Even though only 2% of CYP2D6 exists in human liver microsomes, 30% of commercial drugs are

**TABLE 6.2   Distribution and metabolism of CYP450 isozymes**

| CYP450 isozymes | Abundance in HLM (%) | Metabolism of drugs (%) |
| --- | --- | --- |
| CYP3A[*] | 28 | 50 |
| CYP2D6 | 2 | 30 |
| CYP2C[*] | 18 | 10 |
| CYP1A2 | 13 | 4 |

*3A and 2C family.

metabolized primarily by this isozyme. Substrates of CYP2D6 tend to contain basic amines and they are mostly CNS drugs. About 50% of drugs are metabolized by CYP3A, 10% by CYP2 C, and 4% by CYP1A2. For the reasons above, the four most important isozymes for drug–drug interaction studies are 3A4, 2D6, 2 C9, and 1A2.

PhRMA (Pharmaceutical Research and Manufacturers of America) recently provided general guidelines on *in vitro* and *in vivo* drug–drug interaction studies, and leaders in the field are working toward standardization of the assays (36). The current, as of 2008, acceptance criteria for CYP450 inhibition is $C_{max}$ / $K_i$ < 0.1 or $IC_{50}$ > 10 $\mu$M for early drug discovery.

Several methodologies have been developed to determine CYP450 inhibition. Some of these methodologies are implemented widely in the pharmaceutical industry (37).

### 6.4.1   Fluorescent-Based Assay

The fluorescent assay uses singly expressed recombinant human CYP450 isozymes. Substrates are coumarin-based or fluorescein-based compounds that can be metabolized by CYP450 enzymes to form fluorescent metabolites (38). If a compound is an inhibitor of a CYP enzyme, the metabolite formation will decrease and lead to the reduction of the fluorescent signal, from which percent inhibition is calculated. The assay is amendable to high throughput. Both the 96- and 384-well format are used commonly in the industry for screening of drug discovery compounds. The assay is used widely in the pharmaceutical industry to assess drug–drug interaction potential. MetaChip (Solidus Biosciences, Inc., Troy, NY) microarray technology has been developed with 11,200 reactions on a single chip and >1000-fold reduction in assay volume (39). The limitation of fluorescent-based assays is that if a test compound is autofluorescent or forms fluorescent metabolites, it can cause false-negative results. As such, fluorescent-based assays are not recommended for FDA submission. If a compound is a fluorescent quencher, false-positive results can be generated.

### 6.4.2   LC–MS-Based Assay

LC–MS-based assays use "drug-probes" that are specific substrates for certain isozymes. The formation of metabolites is monitored by using LC–MS–MS.

Human liver microsomes are used commonly in the assay. It is important to keep the protein concentration low so that substrate conversion is less than 10% to minimize metabolism and generate relevant kinetic information (40). To increase throughput, a cocktail approach, which uses a mixture of probe substrates for the different isozymes and human liver microsomes (41), or the double cocktail assay, which uses a mixture of rh-CYP isozymes and a mixture of drug probes, may be used (42, 43). LC–MS-based assay is used typically for regulatory submission or *in vivo* drug–drug interaction studies (36).

### 6.4.3 Luminescence-Based Assay

In the luminescence-based assay, luciferin derivatives are used as substrates and incubated with recombinant human CYP450 isozymes (44). Luciferin is released as a metabolite, which is converted to des-carboxyluciferin with light emission in the presence of luciferase and ATP. The assay is amendable to high-throughput screening of many compounds in 96-, 384-, or 1536-well format. It has the advantages of no fluorescent interference, a large dynamic range, and high sensitivity. Compounds that enhance or inhibit light emission directly can cause false negatives or false positives.

## 6.5 OTHER ADME/TOX PROPERTIES

Several other ADME/TOX properties are profiled commonly in drug discovery, such as uptake and efflux transporters, blood–brain barrier, plasma protein binding, lipophilicity, pKa, metabolite identification, hERG, and pharmacokinetics. These ADME/TOX properties are discussed in other chapters and elsewhere (5).

## REFERENCES

1. Dickson M, Gagnon JP. Key factors in the rising cost of new drug discovery and development. Nature Rev. Drug Disc. 2004;3:417–429.
2. Hann MM, Oprea TI. Pursuing the leadlikeness concept in pharmaceutical research. Curr. Opin. Chem. Biol. 2004;8:255–263.
3. Kola I, Landis J. Can the pharmaceutical industry reduce attrition rate? Nature Rev. Drug Disc. 2004;3:711–716.
4. Prentis RA, Lis Y, Walker SR. Pharmaceutical innovation by the seven UK-owned pharmaceutical companies (1964–1985). Br. J. Clin. Pharmacol. 1988;25:387–396.
5. Kerns EH, Di L. ADME/TOX Optimization. Drug-Like Properties: Concepts, Structure Design and Methodology. 2008. Elsevier, New York.
6. van de Waterbeemd H, Smith DA, Beaumont K, Walker DK. Property-based design: optimization of drug absorption and pharmacokinetics. J. Med. Chem. 2001;44:1313–1333.

7. Hann MM. Considerations for the use of computational chemistry techniques by medicinal chemists. In: Medicinal Chemistry. Principles and Practice. King FD, ed. 1994. Royal Society of Chemistry, Cambridge, United Kingdom, pp. 130–142.

8. Di L, Kerns EH. Application of pharmaceutical profiling assays for optimization of drug-like properties. Curr. Opin. Drug Disc. Devel. 2005;8:495–504.

9. Kerns EH, Di L. Pharmaceutical profiling in drug discovery. Drug Disc. Today 2003;8:316–323.

10. Lipinski CA, Lombardo F, Dominy BW, Feeney PJ. Experimental and computational approaches to estimate solubility and permeability in drug discovery and development settings. Adv. Drug Del. Rev. 1997;23:3–25.

11. Di L, Kerns EH. Biological assay challenges from compound solubility: strategies for bioassay optimization. Drug Disc. Today 2006;11:446–451.

12. Kerns EH. High throughput physicochemical profiling for drug discovery. J. Pharmaceut. Sci. 2001;90:1838–1858.

13. Avdeef A. Physicochemical profiling (solubility, permeability and charge state). Curr. Topics Med. Chem. 2001;1:277–351.

14. Bevan CD, Lloyd RS. A high-throughput screening method for the determination of aqueous drug solubility using laser nephelometry in microtiter plates. Anal. Chem. 2000;72:1781–1787.

15. Goodwin JJ. Rationale and benefit of using high throughput solubility screens in drug discovery. Drug Disc. Today: Technol. 2006;3:67–71.

16. Li AP. Screening for human ADME/Tox drug properties in drug discovery. Drug Disc. Today 2001;6:357–366.

17. Egan WJ, Merz KM, Baldwin JJ. Prediction of drug absorption using multivariate statistics. J. Med. Chem. 2000;43:3867–3877.

18. Kansy M, Senner F, Gubernator K. Physicochemical high throughput screening: parallel artificial membrane permeation assay in the description of passive absorption processes. J. Med. Chem. 1998;41:1007–1010.

19. Avdeef A. The rise of PAMPA. Expert Opin. Drug Metabol. Toxicol. 2005; 1:325–342.

20. Sugano K, Takata N, Machida M, Saitoh K, Terada K. Prediction of passive intestinal absorption using bio-mimetic artificial membrane permeation assay and the paracellular pathway model. Intl. J. Pharmaceut. 2002;241:241–251.

21. Di L, Kerns EH, Fan K, McConnell OJ, Carter GT. High throughput artificial membrane permeability assay for blood-brain barrier. Euro. J. Med. Chem. 2003;38:223–232.

22. Ottaviani G, Martel S, Carrupt P-A. Parallel artificial membrane permeability assay: a new membrane for the fast prediction of passive human skin permeability. J. Med. Chem. 2006;49:3948–3954.

23. Kerns EH, Di L, Petusky S, Farris M, Ley R, Jupp P. Combined application of parallel artificial membrane permeability assay and Caco-2 permeability assays in drug discovery. J. Pharmaceut. Sci. 2004;93:1440–1453.

24. Verma RP, Hansch C, Selassie CD. Comparative QSAR studies on PAMPA/modified PAMPA for high throughput profiling of drug absorption potential with respect to Caco-2 cells and human intestinal absorption. J. Computer-Aided Molec. Design 2007;21:3–22.

25. Balimane PV, Patel K, Marino A, Chong S. Utility of 96 well Caco-2 cell system for increased throughput of P-gp screening in drug discovery. Euro. J. Pharmaceut. Biopharmaceut. 2004;58:99–105.

26. Doan KMM, Humphreys JE, Webster LO, Wring SA, Shampine LJ, Serabjit-Singh CJ, Adkison KK, Polli JW. Passive permeability and P-glycoprotein-mediated efflux differentiate central nervous system (CNS) and non-CNS marketed drugs. J. Pharmacol. Experim. Therapeut. 2002;303:1029–1037.

27. van de Waterbeemd H, Gifford E. ADMET in silico modelling: towards prediction paradise? Nature Rev. Drug Disc. 2003;2:192–204.

28. Di L, Kerns EH, Hong Y, Kleintop TA, McConnell OJ, Huryn DM. Optimization of a higher throughput microsomal stability screening assay for profiling drug discovery candidates. J. Biomol. Screen. 2003;8:453–462.

29. Di L, Kerns EH, Li SQ, Petusky SL. High throughput microsomal stability assay for insoluble compounds. Intl. J. Pharmaceut. 2006;317:54–60.

30. Li AP. Human hepatocytes: isolation, cryopreservation and applications in drug development. Chemico-Biological Interact. Hepatocytes Drug Devel. 2007;168:16–29.

31. Miners JO, Knights KM, Houston JB, Mackenzie PI. In vitro-in vivo correlation for drugs and other compounds eliminated by glucuronidation in humans: pitfalls and promises. Biochem. Pharmacol. 2006;71:1531–1539.

32. Di L, Kerns EH, Hong Y, Chen H. Development and application of high throughput plasma stability assay for drug discovery. Internat. J. Pharmaceut. 2005;297:110–119.

33. Di L, Kerns EH, Chen H, Petusky SL. Development and application of an automated solution stability assay for drug discovery. J. Biomolec. Screen. 2006;11:40–47.

34. Shah RR. Can pharmacogenetics help rescue drugs withdrawn from the market? Pharmacogenomics 2006;7:889–908.

35. Rodrigues ADE. Drug-Drug Interactions. 2002. Marcel Dekker, New York.

36. Bjornsson TD, Callaghan JT, Einolf HJ, Fischer V, Gan L, Grimm S, Kao J, King SP, Miwa G, Ni L, Kumar G, McLeod J, Obach RS, Roberts S, Roe A, Shah A, Snikeris F, Sullivan JT, Tweedie D, Vega JM, Walsh J, Wrighton SA. The conduct of in vitro and in vivo drug-drug interaction studies: a pharmaceutical and manufacturers of America (PhRMA) perspective. Drug Metab. Dispos. 2003;31:815–832.

37. Zlokarnik G, Grootenhuis PDJ, Watson JB. High throughput P450 inhibition screens in early drug discovery. Drug Disc. Today 2005;10:1443–1450.

38. Crespi CL, Miller VP, Penman BW. Microtiter plate assays for inhibition of human, drug-metabolizing cytochromes P450. Anal. Biochem. 1997;248:188–190.

39. Lee M-Y, Park CB, Dordick JS, Clark DS. Metabolizing enzyme toxicology assay chip (MetaChip) for high-throughput microscale toxicity analyses. Proc. Natl. Acad. Sci. U.S.A. 2005;102:983–987.

40. Walsky RL, Obach RS. Validated assays for human cytochrome p450 activities. Drug Metab. Dispos. 2004;32:647–660.

41. Dierks EA, Stams KR, Lim H-K, Cornelius G, Zhang H, Ball SE. A method for the simultaneous evaluation of the activities of seven major human drug-metabolizing cytochrome P450s using an in vitro cocktail of probe substrates and fast gradient liquid chromatography tandem mass spectrometry. Drug Metab. Dispos. 2001;29:23–29.

42. Di L, Kerns EH, Li SQ, Carter GT. Comparison of cytochrome P450 inhibition assays for drug discovery using human liver microsomes with LC-MS, rhCYP450 isozymes with fluorescence, and double cocktail with LC-MS. Internat. J. Pharmaceut. 2007;335:1–11.

43. Weaver R, Graham KS, Beattie IG, Riley RJ. Cytochrome P450 Inhibition using recombinant proteins and mass spectrometry/multiple reaction monitoring technology in a cassette incubation. Drug Metab. Dispos. 2003;31:955–966.

44. Cali JJ, Ma D, Sobol M, Simpson DJ, Frackman S, Good TD, Daily WJ, Liu D. Luminogenic cytochrome P450 assays. Expert Opin. Drug Metab. Toxicol. 2006; 2:629–645.

## FURTHER READING

Borchardt RT, Kerns EH, Hageman MJ, Thakker DR, Stevens JL, eds. Optimizing the "Drug-Like" Properties of Leads in Drug Discovery. 2006. Springer, New York.

Borchardt RT, Kerns EH, Lipinski CA, Thakker DR, Wang B, eds. Pharmaceutical Profiling in Drug Discovery for Lead Selection. Proceedings of the Workshop held May 19–21, 2003 in Whippany, New Jersey. 2004. AAPS Press, Arlington, VA.

Dressman JB, Lennernäs H. Oral Drug Absorption: Prediction and Assessment. 2000. Marcel Dekker, New York.

Kerns EH, Di L. ADME/TOX Optimization. Drug-Like Properties: Concepts, Structure Design and Methodology. 2008. Elsevier, New York.

Lee JS, Obach RS, Fisher MB. Drug Metabolizing Enzymes: Cytochrome P450 and Other Enzymes in Drug Discovery and Development. 2003. Marcel Dekker, New York.

Pardridge WM. Introduction to the Blood-Brain Barrier. 2006. Cambridge University Press, Cambridge, United Kingdom.

Skett P. Introduction to Drug Metabolism. 2001. Nelson Thornes, Glos, United Kingdom.

Smith DA, van de Waterbeemd H, Walker DK. Pharmacokinetics and Metabolism in Drug Design. 2001. Wiley-VCH, Weinheim, Germany.

Testa B, van de Waterbeemd H, Folkers G, Guy R, eds. Pharmacokinetic Optimization in Drug Research: Biological, Physiological, and Computational Strategies. 2001. Verlag Helvetica Chimica Acta, Zurich, Switzerland.

van de Waterbeemd H, Lennernäs H, Artursson P. Drug Bioavailability: Estimation of Solubility, Permeability, Absorption and Bioavailability. 2003. Wiley-VCH, Weinheim, Germany.

Yan Z, Caldwell GW, eds. Optimization in Drug Discovery: In Vitro Methods. 2004. Humana Press, Totowa, NJ.

# 7

# DRUG TRANSPORT IN LIVING SYSTEMS

Yael Elbaz and Shimon Schuldiner

*Department of Biologic Chemistry, Alexander Silberman Institute of Life Sciences, Hebrew University of Jerusalem, Jerusalem, Israel*

Living organisms are constantly assailed by a host of harmful chemicals from the environment. Because of the diversity of these "xenobiotics", cellular survival mechanisms must deal with an immense variety of molecules. Polyspecific drug transporters supply one such strategy. They recognize a wide range of dissimilar substrates that may differ in structure, size, or electrical charge, and they actively remove them from the cytoplasm. As such, these transporters provide an essential survival strategy for the organism. However, given that the substrates of these multispecific transporters include many antibiotics as well as antifungal and anticancer drugs, they are associated with the phenomenon of multidrug resistance (MDR) that poses serious problems in the treatment of cancers and infectious diseases, and some of them are coined multidrug transporters (MDTs). As expected from their central role in survival, these transporters are ubiquitous, and in many genomes, several genes coding for putative MDTs have been identified. MDTs are found in evolutionary unrelated membrane transport protein families, which suggests that they might have developed independently several times during the course of evolution. In this review, we discuss some basic concepts regarding drug transport from bacteria to humans.

Drug elimination in living systems is a multistep process that may involve metabolism, binding to proteins in the circulatory system and binding to specific receptors and excretion processes. A common way by which living organisms, from bacteria to humans, protect themselves against the harmful effect of a toxic

*Chemical Biology: Approaches to Drug Discovery and Development to Targeting Disease*, First Edition.
Edited by Natanya Civjan.
© 2012 John Wiley & Sons, Inc. Published 2012 by John Wiley & Sons, Inc.

compound is the removal of these molecules from the cell, which thus reduces the amount of drug that reaches its target site of action. Membrane transport proteins recognize these drug molecules and in an energy-dependent manner transport them across the plasma membrane and away from the target, either out of the cell or into subcellular organelles. The scope of this chapter encompasses numerous biologic networks and a variety of transport systems operating in concert. We will only discuss some major systems involved in transport of so-called drugs across biologic membranes in organisms from bacteria to humans. We chose to focus on polyspecific drug transporters, and by examination of some of the best characterized, we extract some general concepts.

## 7.1   MULTIDRUG TRANSPORTERS (MDTS)

The discovery of P-glycoprotein (P-gp/MDR1) in the mid 1970s introduced the concept of a single protein that can confer resistance to a relatively large number of structurally diverse drugs (1). In the following decades it has become clear that all living cells, be it a bacterial cell avoiding the deleterious effect of an antibiotic, a kidney cell eliminating an environmental toxin such as nicotine from the body, or a cancerous cell evading chemotherapeutic agents, express multidrug efflux transporters. Multidrug transporters (MDTs) recognize a wide range of dissimilar substrates that may differ in structure, size, or electrical charge, and they actively remove them from the cytoplasm. As such, these transporters provide an essential survival strategy for the organism. However, given that the substrates of these multispecific transporters include many antibiotic, antifungal, and anticancer drugs, they are associated with the phenomenon of multidrug resistance (MDR) that poses serious problems in the treatment of cancers and infectious diseases (2–4).

The phenomenon of multidrug transport is an intriguing one because it seems to challenge the basic model of an enzyme binding specifically to a single substrate in an optimized set of interactions as a prerequisite to efficient catalysis. MDTs, on the other hand, are polyspecific proteins that recognize a remarkably broad array of substrates. The occurrence of MDTs in evolutionary unrelated membrane transport proteins families indicates that they have originated independently several times during the course of evolution (5, 6). MDTs have been identified in several families, based on their amino-acid sequence similarities [References (7) and (8); Table 7.1 (9–16)].

### 7.1.1   ATP Binding Cassette (ABC) Superfamily

This superfamily contains primary active transporters that couple ATP binding and hydrolysis to substrate translocation across the lipid bilayer.

**TABLE 7.1  MDTs: families, examples, and selected references**

| Superfamily | Examples | Recent reviews and additional reading |
|---|---|---|
| ABC | P-Glycoprotein | (9) |
| MFS | MdfA OATs (SCL22 family) OCTs (SCL22 family) | (10) (11) (12) |
| MATE | NorM | (13) |
| RND | AcrB | (14) |
| DMT | EmrE (SMR family) | (15, 16) |

### 7.1.2  Multidrug and Toxic Compound Extrusion (MATE) Family

A fairly newly identified family of MDTs called MATE includes representatives from all domains of life (17). These transporters contain 12 putative transmembrane helices. This family of MDTs was first identified with the characterization of NorM, which is a $Na^+$/drug antiporter from *Vibrio parahaemolyticus* (18, 19).

### 7.1.3  Major Facilitator Superfamily (MFS)

The MFS is one of the largest and most widespread families of transporters and is composed of numerous subfamilies of diverse functions. This superfamily consists of single-polypeptide transporters from bacteria to higher eukaryotes involved in the symport, uniport, or antiport of a variety of small solutes. Among the various substrates of MFS transporters are sugars, drugs, neurotransmitters, Krebs cycle metabolites, amino acids, and many more (20, 21). Most MFS transporters are about 400–600 amino acids in length and possess either 12 or 14 transmembrane helices. Most proteins in this superfamily are secondary-active transporters that couple ion gradients to the translocation of a substrate. Evidence demonstrating that proteins within several MFS families consist of duplicated six transmembrane units supports the possibility that MFS permeases developed by an internal gene duplication event (20).

One of the many families included in the MFS is the SLC (solute carrier) superfamily. The SLC superfamily represents approximately 300 genes in the human genome that encode for either facilitated transporters or secondary active symporters or antiporters. Members of the SLC superfamily transport various ionic and nonionic endogenous compounds and xenobiotics. The SLC22 family includes anion and cation transporters (Organic Anion Transporters, OATs; Organic Cation Transporters, OCTs).

The organic anion and cation transporters of the SLC22 gene family share a common theme with the MDTs discussed above. These polyspecific transporters handle an impressive broad substrate range. Many substrates are noxious to the organism, and these transporters actively remove them. The reasons they are not considered MDTs are therefore probably just historic.

### 7.1.4   Resistance–Nodulation–Cell Division (RND) Superfamily

RND is a large ubiquitous superfamily of transporters with representations in all domains of life. Composed typically of about 1000 amino-acid residues, they are arranged as 12 transmembrane helices proteins with two large hydrophilic extracytoplasmic loops between helices 1 and 2 and helices 7 and 8. It has been postulated that these proteins developed from an internal gene duplication event. The members of the RND family are also secondary active transporters that catalyze the proton–motive–force driven transport of a range of substrates, including hydrophobic drugs, bile salts, fatty acids, heavy metals, and more (22).

### 7.1.5   Small Multidrug Resistance (SMR) Family

The smallest known secondary active transporters belong to the SMR family within the drug metabolite transporter (DMT) superfamily (23).

This family of small and tightly packed proteins of about 100 amino acids in length is arranged in 4 membrane spanning helices. Members of the SMR family are antiporters that catalyze pH gradient-driven multidrug efflux. SMR transporters have been found only in prokaryotic organisms. The best characterized SMR member is EmrE from *Escherichia coli* that serves as a model for SMR function. It is intriguing to understand how the SMR proteins can accomplish the task that in other families of MDTs needs much larger complexes. Studies of EmrE have shown that the functional unit is a homodimer, and this seems to be the case for several other homologs (24), whereas others function as heterooligomers, probably heterodimers. Transport systems that function as heterooligomers are encoded by a pair of genes that belong both to the SMR family and are located in close proximity or even in an overlapping region in the genome. The only heterodimer characterized thus far is the EbrAB transport system from *Bacillus subtilis* (25).

Members of several RND, MFS, and ABC subfamilies have been proposed to function in combination with a periplasmic membrane fusion protein (MFP) and an outer membrane protein (OMP) to generate a single energy-coupled tripartite efflux system across both membranes of the gram-negative bacteria. The best characterized tripartite systems are MexAB-OprM from *Pseudomonas aeruginosa* and, by far, AcrAB-TolC from *Escherichia coli*.

## 7.2   INSIGHTS INTO STRUCTURES OF MDTS

Much research is directed toward deciphering the mechanisms by which MDTs confer resistance against cytotoxic drugs. Such understandings hold promise for tremendous clinical implications. Our knowledge of the multidrug transport mechanism is based on a wealth of data obtained from extensive research using biochemical and genetic approaches and on recently achieved structural data. In our attempts to understand how these transporters confer multidrug resistance, several questions should be addressed. The first and most obvious question regarding this class of transporters is the question of substrate recognition: What is the architecture of a binding site that enables the binding of numerous dissimilar drugs? Next to be answered are questions regarding the drug translocation pathway across the membrane, the coupling of ion/drug fluxes and conformational changes that must occur throughout the transport cycle. Acquisition of high-resolution structures of membrane proteins using X-ray crystallography is not an easy task, and so, at the time of writing in 2008, reliable high-resolution structural models were obtained only for a few ion-coupled transporters from the MFS and RND families. As of 2008, no structural information exists to date for transporters from the MATE family. A low-resolution (7Å) projection map of 2-D crystals of EmrE is the closest to atomic resolution data that exists for the SMR class (26). Based on the combined evidence gathered up to date, several general perceptions can be drawn as to the function mechanism of MDTs.

The first structural evidence that approaches atomic resolution of an MFS transporter was supplied by the interpretation of the three-dimensional structure of the oxalate transporter, OxlT, obtained at the low resolution of 6.5 Å (27, 28). Later on, structures of two proteins from this superfamily were achieved at higher resolution: LacY, the lactose permease (29), and GlpT, the glycerol-3-phosphate transporter (30). While LacY and GlpT are both MFS members, they share a mere 21% sequence identity and differ in their mechanism of function. Although LacY acts as a symporter of protons and β-galactosides, GlpT is an organic phosphate/inorganic phosphate ($P_i$) antiporter. Surprisingly, despite these differences, both structures present a highly similar overall fold. In addition, also the general overall structure of OxlT seems to be very similar to the LacY and GlpT structures. The finding that the fold might be better conserved than the sequence suggests the possibility that a general architecture exists for the 12 transmembrane MFS proteins and that the specific function of each transporter is achieved by subtle changes in the substrate binding site and translocation pathway. Following this line of thought, various structural representations of MFS transporters have been generated based on structural homology to LacY and GlpT, among which we can find models of VMAT (31), TetAB (31), MdfA (32), and rOCT1 (33).

Thus, although neither one of the solved MFS structures is a multidrug transporter, it is reasonable to assume that general insights can still be extracted and applied.

Another structure of an MFS multidrug transporter, the structure of EmrD from *Escherichia coli* solved to a 3.5 Å resolution, was published (34). Regrettably, the structure was solved without any substrate bound, and, at the time of writing, no biochemical data exist to support the mechanism speculated by the authors.

On the other hand, two structures of AcrB, which is a well-characterized RND multidrug transporter from *Escherichia coli*, with and without bound substrates, provide valuable information regarding transport mechanisms of tripartite multidrug efflux systems, alongside newly arising questions (35, 36). Functional implications rising from these publications will be discussed in more detail.

## 7.3 SUBSTRATE RECOGNITION: MULTIPLE DRUG BINDING PROTEINS

Substrate recognition by MDTs is the first and most obvious puzzle that arises when considering the remarkable broad-ranged substrate specificity of these proteins. How can a binding pocket accommodate so many structurally dissimilar substrates? This characteristic of MDTs has long been considered paradoxical because of its ostensible contradiction to basic dogmas of enzyme biochemistry (37, 38).

### 7.3.1  Soluble Regulatory Proteins

The first structural information to shed light on the nature of the multi-specific binding pocket came from the solved 3-D structures of soluble multidrug-recognizing proteins. As membrane proteins are not easy to crystallize, the focus of structural analysis attempts has been directed also to the soluble drug-binding proteins that regulate the expression of MDR proteins. The underlying rationale is that the regulatory proteins and the transporters share some substrates, and so the information obtained from these structures would be applicable to the binding sites of the MDTs. BmrR, which is a transcription regulator from *Bacillus subtilis*, promotes the expression of Bmr, which is an MDR transporter of cationic lipophilic drugs. The structure of BmrR bound to $TPSb^+$, which is an analog of the coactivator $TPP^+$, reveals the drug-binding domain. The drug-binding pocket contains hydrophobic and aromatic residues that interact with the drug molecule. The phenyl rings of the substrate molecule interact via van der Waals and stacking interactions with aromatic residues, and an essential electrostatic interaction takes place between a negatively charged residue and the positively charged substrate. Complementary electrostatic interactions are found between $TPSb^+$ and carboxylate groups of several acidic residues positioned in close vicinity (39). Valuable information was also achieved from the solved structures of QacR, which is a *Staphylococcus aureus* multidrug binding protein that represses the transcription of the QacA multidrug transporter gene. Structures of QacR were obtained in complex with six different lipophilic drugs, both monovalent and bivalent cations and many of them substrates of QacA (40). The structure of QacR exhibits a large drug-binding site with several aromatic residues forming stacking interactions with the substrate's rings, and several polar residues that mediate drug-specific interaction through hydrogen bonds. Four glutamate residues within the binding pocket maintain electrostatic interactions with the substrate's positively charged moiety. The different structures of QacR–drug complexes reveal a large, multifaceted drug-binding pocket in which different drugs can bind to different and partially overlapping sets of residues.

### 7.3.2  Transport Proteins

With MDTs the only detailed structural evidence is provided by AcrB. The structure of AcrB in the presence of substrates reveals a substrate binding pocket in each of the three protomers (35). Bound antibiotic-drug molecules, minocycline and doxorubicin, are observed only in one of the three protomers. The substrate binding pocket is rich in aromatic amino-acid residues, mostly Phe residues. Such aromatic side chains may interact with the drug molecules through hydrophobic or aromatic interactions. Several polar residues located in close vicinity may form hydrogen bonds with the drug molecules. The structure implies a voluminous binding pocket in which different sets of residues are used for binding of the different kinds of substrates. Such a strategy is similar to that previously observed in the regulatory protein QacR.

The critical role of aromatic side chains in the binding of lipophilic substrates is also demonstrated by site-directed mutagenesis studies of other MDTs. Three aromatic residues (Tyr 40, Tyr60, and Trp63) in EmrE, which is an extensively characterized SMR transporter from *Escherichia coli*, play a role in substrate recognition (41, 42). The importance of aromatic side chains in the equivalent positions has been demonstrated for other SMR proteins as well (43).

A distinguishing feature of MDTs that recognize cationic drugs seems to be the presence of a membrane-embedded negatively charged residue, namely glutamate or aspartate, that neutralizes the positively charged moiety of the drug. EmrE, which is a proton-coupled SMR transporter, has only a single membrane embedded charged residue, Glu14. Through site-directed mutagenesis studies, Glu14 has been shown to be an essential part of the binding site of both substrate and protons (44, 45). In LmrP, which is an MFS multidrug transporter from *Lactococcus lactis*, two acidic residues have been shown to be essential for the recognition of the bivalent cation Hoechst 33342, whereas a single negatively charged residue in the binding pocket is sufficient for recognition of the monovalent cation ethidium (38). Also in the soluble regulatory proteins the equivalent of such a buried residue has been visualized. Glu253 of the transcription factor BmrR contributes to the protein–drug interaction through its carboxylate group (39, 47). In QacR, which is another regulatory protein, four partially buried glutamates in the drug-binding site exist, at positions 57, 58, 90, and 120, that surround the drug-binding pocket and interact with the positively charged moiety of the substrates (40).

Another structure of the soluble regulatory protein QacR, this time in complex with bivalent diamidine substrates, reveals a novel manner of drug binding through electrostatic interactions. In these structures the electrostatic neutralization of the positively charged substrate was achieved through drug interactions with the negative dipoles of several oxygen atoms from nearby side chains and the peptide backbone, and not through interaction with the carboxylate of an acidic residue (48). QacA and QacB are both MFS transporters from *Staphylococcus aureus* that are closely related. Although QacA confers resistance to both monovalent and bivalent cations, QacB confers resistance to monovalent cations only. Site-directed mutagenesis studies provided evidence that the substrate specificity differences are caused by the presence of an acidic residue at position 323 in QacA that plays a critical role in conveying resistance to bivalent cations (8).

In MdfA, however, an MFS multidrug transporter, it has been demonstrated that no single acidic residue plays an irreplaceable role, and although efficient transport of positively charged substrates does require a negative charge at position 26 (either Glu or Asp), neutralization of this position does not necessarily abolish the interaction of MdfA with cationic drugs (49, 50). In the recently solved structure of AcrB in complex with minocycline and doxorubicin, no acidic residues seem to participate in cationic-substrate binding, and this finding may be a reflection of the fact that substrates of AcrB can be also uncharged and negatively charged (35).

The organic anion (OATs) and cation (OCTs) and multidrug transporters share a fundamental trait of polyspecificity. It is interesting to compare, from the knowledge that has accumulated so far, the strategy adopted by each such group of multispecific transporters that allows for broad-range substrate recognition.

The OATs transport various small amphiphilic organic anions, uncharged molecules, and even some organic cations that are usually around the molecular weight of 400–500 Da and include clinically relevant drugs such as anti-HIV therapeutics, anti-tumor drugs, antibiotics, anti-hypertensives, and anti-inflammatories (11). OCTs substrates are organic cations and weak bases that are positively charged at physiologic pH, as well as noncharged compounds, with molecular mass of <500 Da. Among transported substrates of OCTs are endogenous compounds, drugs, and xenobiotics (12).

Many SLC22 family members are expressed in the boundary epithelia of the kidney, liver, and intestines and play a major role in drug absorption and excretion. Substrate translocation by most OATs is energized by coupling the uptake of an organic anion into the cell to the extrusion of another organic anion from the cell; thus, OATs use the existing intracellular versus extracellular gradients of anions, such as $\alpha$-ketoglutarate, as a driving force (11). Similarly, OCT1-3 from the SLC22 family are exchangers of organic cations (12). Several OAT members have been identified and cloned from various eukaryotic origins. Human OAT1-4 and Urate1 were more thoroughly studied than others, and so knowledge regarding substrate specificity, drug transport, regulation, and overall characteristics has accumulated. In contrast, a description regarding the functional characteristics of OAT5-9 is still lacking.

In the organic anion transporter rOAT3, site-directed mutagenesis studies lead to the proposal that rOAT3 contains a large binding pocket with several inter-action domains responsible for the high-affinity binding of structurally diverse substrates. Two essential membrane-embedded basic residues (Arg 454 and Lys 370) attract negatively charged substrates through electrostatic interactions, but their role in determining specificity is not identical. Interestingly, replacement of the basic amino acids at positions 370 and 454 with the corresponding residues of OCTs, thus generating the mutant R454DK370A, changed the charge selectivity of the protein that could now transport the cation MPP$^+$ (41). Several conserved aromatic residues that have been shown to be important for the transport of dif-ferent substrates by rOAT3 are assumed to mediate substrate recognition through aromatic interactions that could include $\pi$-$\pi$ interactions with ring containing substrates (42). A conservative replacement of Asp475 in rOCT1 to glutamate yielded a protein that exhibited higher affinities for some of its cationic sub-strates but also impaired transport rates. The authors suggested that the mutation at position 475 from Asp to Glu alters the structure of the cation binding site and harms the translocation step, thus altering the selectivity of rOCT1 (53). Additional studies of the binding site of rOCT1 reveal six more amino acids critical for substrate affinity, among which are aromatic and polar amino acids (33).

The general feature emerging is that of a multifaceted binding pocket, in which different drugs bind to separate, yet overlapping, sets of residues, many of them aromatic.

## 7.4    TRANSPORT MECHANISMS OF MDTS

### 7.4.1    AcrB

Structures of the RND transporter AcrB (35, 36) suggest a possible mechanism for this class of transporters. AcrB functions together with the auxiliary MFP AcrA and the OMP TolC as a tripartite transport system (Fig. 7.1). The AcrB structures reveal an asymmetric trimer, in which each protomer adopts a differ-ent conformation according to its proposed role in the catalysis of the transport reaction (Fig. 7.2) (54). The "binding" protomer is occupied by bound substrate, and the "extrusion" protomer is outwardly open in a way that suggests it is the form present just after the extrusion of the substrate through the TolC funnel; the "access" protomer seems to be in the state just before substrate binding. The three different protomer conformations are suggested to represent consecutive states of the transport cycle that result in the guiding of the substrate through AcrB toward TolC. How is the proton gradient coupled to the drug efflux and to the confor-mational changes that shift the protomers between states? In AcrB it seems that the proton translocation pathway is separate from that of the substrate transloca-tion pathway. Site-directed mutagenesis studies in AcrB and MexB, which is an AcrB homolog from *Pseudomonas aeruginosa*, show that any replacements of

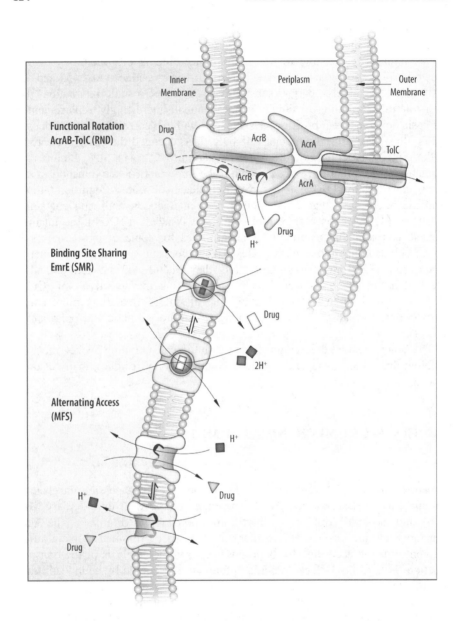

**Figure 7.1**  Hypothetical mechanisms of transport by MDTs. Three mechanisms have been suggested for the function of MDTs from different families, and they are discussed in the text.

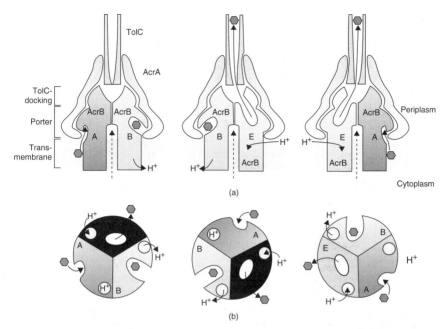

**Figure 7.2** The structure of the AcrB–drug complex and proposed mechanism of drug transport (54). (a) The complex is observed from the side, with the drug shown as a red hexagon. The dotted line indicates a possible pathway for substrates moving from the cytoplasm. The complex contains three molecules of AcrB, AcrA accessory proteins, and the TolC channel to the exterior. The drug is proposed to enter AcrB when it is in the access (A) conformation before binding more closely to the porter domain of AcrB in the binding (B) conformation. It is then transported to the opposite face and is released from the extrusion (E) conformation of AcrB. Transport of the xenobiotic is powered by the proton ($H^+$) gradient across the membrane. (b) The proposed ordered multidrug binding change mechanism of the three-unit AcrB complex.

either of the three transmembrane charged residues Asp407, Asp408, and completely abolishes the ability of the protein to confer drug resistance (55, 56). No other charged residues reside within the transmembrane domains making this triplet the most likely proton translocation pathway. The structure reveals that in the "access" and "binding" protomers Lys940 is coordinated by salt bridges with Asp407 and Asp408, whereas in the "extrusion" protomer, Lys940 is turned toward Thr978 and the salt bridges are eliminated. Additional studies identify also Thr978, which is close to the essential Asp–Lys–Asp network, as important for AcrB's function and as a putative part of the proton relay network (57). Some unanswered questions remain on the way to grasp fully the AcrB transport mechanism, among which is how this movement caused by protonation and deprotonation is transmitted to the large conformational changes that must occur throughout the transport cycle. Also, the essential role of AcrA and the interaction with TolC have yet to be decoded.

### 7.4.2  MFS Transporters

In the case of MFS transporters (Fig. 7.1) it is tempting to speculate that they may be acting by an alternative access mechanism similar to that suggested originally by Jardetzky (58) and recently supported for the extensively character-ized MFS transporter LacY, which is a lactose permease that cotransports protons and sugar (29). The substrate and proton binding sites and translocation pathways are proposed to be separate, and the coupling is transmitted by a series of confor-mational changes induced on substrate binding. The sugar-binding site located in the approximate middle of the molecule at the apex of a deep hydrophilic cavity has alternating access to either side of the membrane as a result of reciprocal opening and closing of hydrophilic cavities on either side of the membrane.

### 7.4.3  EmrE

The simplest coupling mechanism between drug and proton transport has been described for the MDTs from the SMR family (Fig. 7.1). A single membrane-embedded charged residue, Glu14 in the case of EmrE, an *E. coli* SMR, is evolutionary conserved throughout the SMR family and is essential for function. This residue provides the core of the coupling mechanism because its deproto-nation is essential for substrate binding (59, 60). Conversely, substrate induces proton release, and both reactions (substrate binding and proton release) have been observed directly in the detergent solubilized preparation of EmrE (61). The estimated $pK_a$ for Glu14 is unusually increased (about 8.3–8.5 compared with 4.25 for the same carboxyl in aqueous environment). The fact that the bind-ing sites for both substrates and protons overlap and that the occupancy of these sites is mutually exclusive provides the basis of the coupling mechanism (45, 59). The fine-tuning of the $pK_a$ is essential because replacement of Glu14 with Asp results in a decrease in the $pK_a$ of the carboxyl and generates a protein that at physiologic pH has already released the previously bound protons, can still bind substrate but cannot longer couple the substrate flux to the proton gradient (44, 61).

SMR proteins function through a binding site shared by protons and substrates, which can be occupied in a mutually exclusive manner and provide the basis for the simplest coupling of two fluxes (15, 59).

### 7.5  PHYSIOLOGIC ROLES AND NATURAL SUBSTRATES

MDTs provide a survival strategy for living organisms that are constantly assailed by a host of harmful chemicals from the environment. Because of the diversity of these "xenobiotics", cellular survival mechanisms must deal with an immense variety of molecules. MDTs supply one such strategy. The patterns of abundance of drug efflux systems in different organisms do not correlate directly with the above-suggested role. Although MDTs are highly abundant in the soil and envi-ronmental bacteria, they exist in relatively large numbers in other organisms as well, such as intracellular bacteria (5). It is not clear, however, whether this is the primary function of all of those identified as MDTs. It is also not clear why so many different ones are required and why is there such a high redundancy in substrate specificities among different MDTs within the same genome. This may be from the necessity to cover a range as wide as possible of substrates, and overlapping provides a backup strategy in case of failure of one of the systems.

In some cases, a direct correlation is found between MDTs and adaptation to specific environments. The natural environment of an enteric bacterium such as *E. coli* is enriched in bile salts and fatty acids. Bile acids are amphipatic molecules that act as emulsifying agents, thus possessing antimicrobial activity. Deletion of the *acrAB* operon leads to increased susceptibility to bile acids and fatty acids, such as decanoate, which suggests that efflux of these compounds may be one of the physiologic roles of the AcrAB efflux system in *E. coli*. Consistent with that, the expression of AcrAB was increased by growth in the presence of 5 mM decanoate (62). In another study, both bile and mammalian steroid hormones (estradiol and progesterone) have been shown to be substrates of the two major efflux systems of *E. coli*, AcrAB-TolC and the MFS EmrAB-TolC (63). CmeABC, which is a tripartite efflux pump from *Campylobacter jejuni*, was also shown to contribute to the bile resistance of the bacterium (64) and so have similar efflux systems from other bacteria, including *Salmonella typhimurium* (65) and *Vibrio cholerae* (66). In some cases, deletion of MDTs decreases the pathogenicity of given microorganisms, most likely because of their impaired adaptability (67). It has been suggested that some bacterial MDTs might be involved in export of signals for cell–cell communication (68). For some MDTs, a specific function has been well documented. BLT, which is an MFS multidrug transporter of *Bacillus subtilis*, has been shown to function as a substrate-specific transporter of the polyamine spermidine, which is a natural cellular constituent. BLT is encoded in an operon with BltD, a spermidine/spermine acetyltransferase, which emphasizes its possible physiologic role in downregulating cellular spermidine levels through efflux, in concert with an enzyme that chemically modifies spermidine (69).

The vesicular monoamine transporters (VMATs) from the SLC18 family provide another example of polyspecific proteins with a very defined physiologic role. These essential proteins are involved in the storage of monoamines in the central nervous system and in endocrine cells in a process that involves exchange of $2H^+$ with one substrate molecule. The VMATs interact with various native substrates and clinically relevant drugs, and they display the pharmacologic profile of multidrug transporters (70).

More recently possible roles in supporting pH homeostasis and alkali tolerance have been suggested for TetL, which is a tetracycline efflux protein from *Bacillus subtilis*, and the multidrug transporter MdfA (71).

In higher multicellular organisms, the strategy for drug elimination is a multistep process that may involve metabolism, binding to proteins in the circulatory system and binding to specific receptors and excretion processes. A variety of polyspecific transporters (from the ABC, MATE, RND, and MFS superfamilies) are expressed throughout the organism. Some of these transporters provide mechanisms that protect the brain and other sanctuaries from exposure to toxins, both endogenous and exogenous. Others are found in the boundary epithelia of kidney, liver, and intestine and play a major role in drug absorption and excretion.

The discussion of the role of MDTs is tightly connected with the question of the evolution of polyspecificity. One can suggest two possible paradigms of multidrug transporter evolution: The first one states that because MDTs confer broad-range drug resistance, they evolved to protect cells from environmental "xenobiotics". A second possible scenario is that each MDT has evolved as a substrate-specific transporter, and its ability to transport toxins is only an opportunistic side effect (72).

These issues and the others discussed above provide us with a glimpse of this fascinating and complex world of polyspecific proteins. These ubiquitous proteins in many cases play central roles in survival of organisms by providing the means to remove "xenobiotics" away from their targets. On the other hand, in some cases, they present a serious threat for treatment of drug-resistant cancers and infectious diseases. Understanding of the molecular mechanisms that underly the mechanism of action of these proteins is therefore highly relevant.

## ACKNOWLEDGMENT

YE is an Adams Fellow from the Israel Academy of Sciences and Humanities. SS is Mathilda Marks-Kennedy Professor of Biochemistry at the Hebrew University of Jerusalem. Work in the author's laboratory is supported by Grant NS16708 from the National Institutes of Health, Grant 2003-309 from the United States Israel Binational Science Foundation, Grant 119/04 from the Israel Science Foundation, and Grant P50 GM73210 from The Center for Innovation in Membrane Protein Production.

## REFERENCES

1. Juliano RL, Ling V. A surface glycoprotein modulating drug permeability in Chinese hamster ovary cell mutants. Biochim. Biophys. Acta 1976;455:152–162.

2. Gottesman MM, Ling V. The molecular basis of multidrug resistance in cancer: the early years of P-glycoprotein research. FEBS Lett. 2006;580:998–1009.

3. Nikaido H. Multiple antibiotic resistance and efflux. Curr. Opin. Microbiol. 1998;1:516–523.

4. Gottesman MM, Fojo T, Bates SE. Multidrug resistance in cancer: role of ATP-dependent transporters. Nat. Rev. Cancer 2002;2:48–58.

5. Paulsen IT, Chen J, Nelson KE, Saier MH Jr. Comparative genomics of microbial drug efflux systems. J. Mol. Microbiol. Biotechnol. 2001;3:145–150.

6. Paulsen IT. Multidrug efflux pumps and resistance: regulation and evolution. Curr. Opin. Microbiol. 2003;6:446–451.

7. Saier MH Jr. and Paulsen IT. Phylogeny of multidrug transporters. Semin. Cell Dev. Biol. 2001;12:205–213.

8. Paulsen IT, Brown MH, Littlejohn TG, Mitchell BA, Skurray RA. Multidrug resistance proteins QacA and QacB from Staphylococcus aureus: membrane topology and identification of residues involved in substrate specificity. Proc. Natl. Acad. Sci. U.S.A. 1996;93:3630–3635.

9. Deeley RG, Westlake C, Cole SP. Transmembrane transport of endo- and xenobiotics by mammalian ATP-binding cassette multidrug resistance proteins. Physiol. Rev. 2006;86:849–899.

10. Lewinson O, Adler J, Sigal N, Bibi E. Promiscuity in multidrug recognition and transport: the bacterial MFS Mdr transporters. Mol. Microbiol. 2006;61:277–284.

11. Rizwan AN, Burckhardt G. Organic anion transporters of the SLC22 family: biopharmaceutical, physiological, and pathological roles. Pharm. Res. 2007;24:450–470.

12. Koepsell H, Lips K, Volk C. Polyspecific organic cation transporters: structure, function, physiological roles, and biopharmaceutical implications. Pharm. Res. 2007;24:1227–1251.

13. Omote H, Hiasa M, Matsumoto T, Otsuka M, Moriyama Y. The MATE proteins as fundamental transporters of metabolic and xenobiotic organic cations. Trends Pharmacol. Sci. 2006;27:587–593.

14. Lomovskaya O, Zgurskaya HI, Totrov M, Watkins WJ. Waltzing transporters and 'the dance macabre' between humans and bacteria. Nat. Rev Drug Discov. 2007;6:56–65.

15. Schuldiner S. When biochemistry meets structural biology: the cautionary tale of EmrE. Trends Biochem. Sci. 2007;32:252–258.

16. Schuldiner S, Granot D, Steiner S, Ninio S, Rotem D, Soskin M, Yerushalmi H. Precious things come in little packages. J. Mol. Microbiol. Biotechnol. 2001;3:155–162.

17. Brown MH, Paulsen IT, Skurray RA. The multidrug efflux protein NorM is a prototype of a new family of transporters. Mol. Microbiol. 1999;31:394–395.

18. Morita Y, Kodama K, Shiota S, Mine T, Kataoka A, Mizushima T, Tsuchiya T. NorM, a putative multidrug efflux protein, of Vibrio parahaemolyticus and its homolog in Escherichia coli. Antimicrob. Agents Chemother. 1998;42:1778–1782.

19. Morita Y, Kataoka A, Shiota S, Mizushima T, Tsuchiya T. NorM of vibrio parahaemolyticus is an Na(+)-driven multidrug efflux pump. J. Bacteriol. 2000;182:6694–6697.

20. Pao SS, Paulsen IT, Saier MH Jr. Major facilitator superfamily. Microbiol. Mol. Biol. Rev. 1998;62:1–34.

21. Saier MH Jr, Beatty JT, Goffeau A, Harley KT, Heijne WH, Huang SC, Jack DL, Jahn PS, Lew K, Liu J, et al. The major facilitator superfamily. J. Mol. Microbiol. Biotechnol. 1999;1:257–279.

22. Tseng TT, Gratwick KS, Kollman J, Park D, Nies DH, Goffeau A, Saier MH Jr. The RND permease superfamily: an ancient, ubiquitous and diverse family that includes human disease and development proteins. J. Mol. Microbiol. Biotechnol. 1999;1:107–125.

23. Jack DL, Yang NM, Saier MH Jr. The drug/metabolite transporter superfamily. Eur. J. Biochem. 2001;268:3620–3639.

24. Ninio S, Rotem D, Schuldiner S. Functional analysis of novel multidrug transporters from human pathogens. J. Biol. Chem. 2001;276:48250–48256.

25. Masaoka Y, Ueno Y, Morita Y, Kuroda T, Mizushima T, Tsuchiya T. A two-component multidrug efflux pump, EbrAB, in Bacillus subtilis. J. Bacteriol. 2000;182:2307–2310.

26. Ubarretxena-Belandia I, Baldwin JM, Schuldiner S, Tate CG. Three-dimensional structure of the bacterial multidrug transporter EmrE shows it is an asymmetric homodimer. EMBO J. 2003;22:6175–6181.

27. Hirai T, Heymann JA, Shi D, Sarker R, Maloney PC, Subramaniam S. Three-dimensional structure of a bacterial oxalate transporter. Nat. Struct. Biol. 2002;9:597–600.

28. Hirai T, Heymann JA, Maloney PC, Subramaniam S. Structural model for 12-helix transporters belonging to the major facilitator superfamily. J. Bacteriol. 2003;185:1712–1718.

29. Abramson J, Smirnova I, Kasho V, Verner G, Kaback HR, Iwata S. Structure and mechanism of the lactose permease of Escherichia coli. Science 2003;301:610–615.

30. Huang Y, Lemieux MJ, Song J, Auer M, Wang DN. Structure and mechanism of the glycerol-3-phosphate transporter from Escherichia coli. Science 2003;301:616–620.

31. Vardy E, Arkin IT, Gottschalk KE, Kaback HR, Schuldiner S. Structural conservation in the major facilitator superfamily as revealed by comparative modeling. Protein Sci. 2004;13:1832–1840.

32. Sigal N, Vardy E, Molshanski-Mor S, Eitan A, Pilpel Y, Schuldiner S, Bibi E. 3D Model of the Escherichia coli Multidrug Transporter MdfA Reveals an Essential Membrane-Embedded Positive Charge. Biochemistry, 2005;44:14870–14880.

33. Popp C, Gorboulev V, Muller TD, Gorbunov D, Shatskaya N, Koepsell H. Amino acids critical for substrate affinity of rat organic cation transporter 1 line the substrate binding region in a model derived from the tertiary structure of lactose permease. Mol. Pharmacol. 2005;67:1600–1611.

34. Yin Y, He X, Szewczyk P, Nguyen T, Chang G. Structure of the multidrug transporter EmrD from Escherichia coli. Science 2006;312:741–744.

35. Murakami S, Nakashima R, Yamashita E, Matsumoto T, Yamaguchi A. Crystal structures of a multidrug transporter reveal a functionally rotating mechanism. Nature 2006;443:173–179.

36. Seeger MA, Schiefner A, Eicher T, Verrey F, Diederichs K, Pos KM. Structural asymmetry of AcrB trimer suggests a peristaltic pump mechanism. Science 2006;313:1295–1298.

37. Neyfakh AA. The ostensible paradox of multidrug recognition. J. Mol. Microbiol. Biotechnol. 2001;3:151–154.

38. Neyfakh AA. Mystery of multidrug transporters: the answer can be simple. Mol. Microbiol. 2002;44:1123–1130.

39. Heldwein EE, Brennan RG. Crystal structure of the transcription activator BmrR bound to DNA and a drug. Nature 2001;409:378–382.

40. Schumacher MA, Miller MC, Grkovic S, Brown MH, Skurray RA, Brennan RG. Structural mechanisms of QacR induction and multidrug recognition. Science 2001;294:2158–2163.

41. Elbaz Y, Tayer N, Steinfels E, Steiner-Mordoch S, Schuldiner S. Substrate-induced tryptophan fluorescence changes in EmrE, the smallest ion-coupled multidrug transporter. Biochemistry 2005;44:7369–7377.

42. Rotem D, Steiner-Mordoch S, Schuldiner S. Identification of tyrosine residues critical for the function of an ion-coupled multidrug transporter. J. Biol. Chem. 2006;281:18715–18722.

43. Paulsen I, Brown M, Dunstan S, Skurray R. Molecular characterization of the staphylococcal multidrug resistance export protein QacC. J. Bacteriol. 1995;177:2827–2833.

44. Yerushalmi H, Schuldiner S. An essential glutamyl residue in EmrE, a multidrug antiporter from Escherichia coli. J. Biol. Chem. 2000;275:5264–5269.

45. Yerushalmi H, Schuldiner S. A common binding site for substrates and protons in EmrE, an ion-coupled multidrug transporter. FEBS Lett. 2000;476:93–97.

46. Mazurkiewicz P, Konings WN, Poelarends GJ. Acidic residues in the lactococcal multidrug efflux pump LmrP play critical roles in transport of lipophilic cationic compounds. J. Biol. Chem. 2002;277:26081–26088.

47. Zheleznova EE, Markham PN, Neyfakh AA, Brennan RG. Structural basis of multidrug recognition by BmrR, a transcription activator of a multidrug transporter. Cell 1999;96:353–362.

48. Murray DS, Schumacher MA, Brennan RG. Crystal structures of QacR-diamidine complexes reveal additional multidrug-binding modes and a novel mechanism of drug charge neutralization. J. Biol. Chem. 2004;279:14365–14371.

49. Sigal N, Molshanski-Mor S, Bibi E. No single irreplaceable acidic residues in the Escherichia coli secondary multidrug transporter MdfA. J. Bacteriol. 2006;188:5635–5639.

50. Adler J, Lewinson O, Bibi E. Role of a conserved membrane-embedded acidic residue in the multidrug transporter MdfA. Biochemistry 2004;43:518–525.

51. Feng B, Dresser MJ, Shu Y, Johns SJ, Giacomini KM. Arginine 454 and lysine 370 are essential for the anion specificity of the organic anion transporter, rOAT3. Biochemistry 2001;40:5511–5520.

52. Feng B, Shu Y, Giacomini KM. Role of aromatic transmembrane residues of the organic anion transporter, rOAT3, in substrate recognition. Biochemistry 2002;41:8941–8947.

53. Gorboulev V, Volk C, Arndt P, Akhoundova A, Koepsell H. Selectivity of the polyspecific cation transporter rOCT1 is changed by mutation of aspartate 475 to glutamate. Mol. Pharmacol. 1999;56:1254–1261.

54. Schuldiner S. Structural biology: the ins and outs of drug transport. Nature 2006;443:156–157.

55. Guan L, Nakae T. Identification of essential charged residues in transmembrane segments of the multidrug transporter MexB of Pseudomonas aeruginosa. J. Bacteriol. 2001;183:1734–1739.

56. Murakami S, Nakashima R, Yamashita E, Yamaguchi A. Crystal structure of bacterial multidrug efflux transporter AcrB. Nature 2002;419:587–593.

57. Takatsuka Y, Nikaido H. Threonine-978 in the transmembrane segment of the multidrug efflux pump AcrB of Escherichia coli is crucial for drug transport as a probable component of the proton relay network. J. Bacteriol. 2006;188:7284–7289.

58. Jardetzky O. Simple allosteric model for membrane pumps. Nature 1966;211:969–970.

59. Yerushalmi H, Schuldiner S. A model for coupling of H( + ) and substrate fluxes based on "time-sharing" of a common binding site. Biochemistry 2000;39:14711–14719.

60. Muth TR, Schuldiner S. A membrane-embedded glutamate is required for ligand binding to the multidrug transporter EmrE. EMBO J. 2000;19:234–240.

61. Soskine M, Adam Y, Schuldiner S. Direct evidence for substrate induced proton release in detergent solubilized EmrE, a multidrug transporter. J. Biol. Chem. 2004;279:9951–9955.

62. Ma D, Cook DN, Alberti M, Pon NG, Nikaido H, Hearst JE. Genes acrA and acrB encode a stress-induced efflux system of Escherichia coli. Mol. Microbiol. 1995;16:45–55.

63. Elkins CA, Mullis LB. Mammalian steroid hormones are substrates for the major RND- and MFS-type tripartite multidrug efflux pumps of Escherichia coli. J. Bacteriol. 2006;188:1191–1195.

64. Lin J, Michel LO, Zhang Q. CmeABC functions as a multidrug efflux system in Campylobacter jejuni. Antimicrob. Agents Chemother. 2002;46:2124–2131.

65. Lacroix FJ, Cloeckaert A, Grepinet O, Pinault C, Popoff MY, Waxin H, Pardon P. Salmonella typhimurium acrB-like gene: identification and role in resistance to biliary salts and detergents and in murine infection. FEMS Microbiol. Lett. 1996;135:161–167.

66. Colmer JA, Fralick JA, Hamood AN. Isolation and characterization of a putative multidrug resistance pump from Vibrio cholerae. Mol. Microbiol. 1998;27:63–72.

67. Piddock LJ. Multidrug-resistance efflux pumps - not just for resistance. Nat. Rev. Microbiol. 2006;4:629–636.

68. Yang S, Lopez CR, Zechiedrich EL. Quorum sensing and multidrug transporters in Escherichia coli. Proc. Natl. Acad. Sci. U.S.A. 2006;103:2386–2391.

69. Woolridge DP, Vazquez-Laslop N, Markham PN, Chevalier MS, Gerner EW, Neyfakh AA. Efflux of the natural polyamine spermidine facilitated by the Bacillus subtilis multidrug transporter Blt. J. Biol. Chem. 1997;272:8864–8866.

70. Schuldiner S, Shirvan A, Linial M. Vesicular neurotransmitter transporters: from bacteria to human. Physiol. Rev. 1995;75:369–392.

71. Krulwich TA, Lewinson O, Padan E, Bibi E. Do physiological roles foster persistence of drug/multidrug-efflux transporters? A case study. Nat. Rev. Microbiol. 2005;3:566–572.

72. Neyfakh A. Natural functions of bacterial multidrug transporters. Trends Microbiol. 1997;5:309–313.

## FURTHER READING

Phylogeny–transport classification database. *www.tcdb.org*.

Membrane Proteins of Known Structure. *http://blanco.biomol.uci.edu/Membrane_ Proteins_xtal.html*.

*http://www.mpibp-frankfurt.mpg.de/michel/public/memprotstruct.html*.

# 8

# BLOOD-BRAIN BARRIER: CONSIDERATIONS IN DRUG DEVELOPMENT AND DELIVERY

David S. Miller and Brian T. Hawkins

*Laboratory of Pharmacology, NIH/NIEHS, Research Triangle Park, North Carolina*

The brain capillary endothelium, which comprises the blood-brain barrier, provides an exceptionally effective barrier to the entry of foreign chemicals into the central nervous system (CNS). As such, it limits pharmacotherapy of major CNS diseases, which include neurodegenerative diseases, epilepsy, brain cancer, and neuro-AIDS. Two elements of the endothelium are largely responsible for its barrier properties. First, tight junctions between endothelial cells present a physical barrier. Second, endothelial cells themselves express high levels of adenosine triphosphate (ATP)-driven drug efflux transporters, which present a selective/enzymatic barrier. Together, the physical and enzymatic barriers effectively limit blood to CNS delivery of both small-molecule and macromolecule drugs. Developing strategies to circumvent these barriers is a major challenge for both medicinal chemists and pharmacologists.

For a drug to be effective, it must access its site of action and be present for a long enough time and at a high enough concentration to have a therapeutic effect. In this regard, drug delivery to the brain is a serious and long-standing problem in pharmacotherapy of central nervous system (CNS) disorders. On the one hand, invasive approaches that involve the direct introduction of drugs into the brain (e.g., intracranially or intraventricularly) often present serious problems. On the other hand, peripheral administration and vascular delivery is often ineffective. Indeed, more than 98% of drug candidates for CNS disorders fail to reach the

*Chemical Biology: Approaches to Drug Discovery and Development to Targeting Disease*, First Edition.
Edited by Natanya Civjan.
© 2012 John Wiley & Sons, Inc. Published 2012 by John Wiley & Sons, Inc.

clinic (1). For most of these drugs, the major confounding issue is their inability to cross the blood-brain barrier at a sufficient level to have a therapeutic effect. This limiting barrier resides within the brain's capillary endothelium and has been an object of study for more than 100 years. Research on the blood-brain barrier has occurred in distinct stages. Initial work focused on the barrier's physiological properties (i.e., the ability to prevent movement of macromolecules between blood and the CNS). The ultrastructural basis of the diffusive barrier was determined to reside primarily with the tight junctions that connect the endothelial cells. The molecular components of the junctional complexes were identified as were the specific transport proteins that importantly determined whether and to what extent specific solutes can permeate the cells. Over the past several years, research on all of these aspects has continued within the context of three important themes: the barrier as a dynamic tissue that can respond to its environment; the endothelium as part of an integrated, neurovascular unit; and the barrier as a target for pharmacotherapy.

It is within this perspective that the current chapter was written. Our aims are to 1) describe the unique biological and physicochemical characteristics of the blood-brain barrier, 2) explain how these characteristics generate a formidable but selective barrier to both small-molecule and macromolecular drugs, and 3) discuss strategies, current to the time of writing this chapter, to move therapeutic drugs across the barrier without compromising its important protective properties.

## 8.1  CELLULAR PHYSIOLOGY OF THE BLOOD-BRAIN BARRIER

The brain capillary endothelium forms a dynamic interface between the blood and the brain parenchyma that carries out multiple complex tasks, which include the control of solute and water exchange between blood and the CNS (2). In doing this, the endothelium provides ions and nutrients for a metabolically active CNS, removes potentially toxic waste products of metabolism, protects against neurotoxicants, and thus maintains brain homeostasis. Not surprisingly, the blood-brain barrier phenotype is unique among the capillary beds of the body. Compared with their counterparts in the periphery, brain capillary endothelial cells are marked by a lack of fenestrations, low pinocytotic activity, the presence of high-resistance (low permeability) tight junctions, polarized expression of membrane-bound transport proteins, and enrichment of metabolic enzymes, some of which can metabolize drugs (Phase I and Phase II). Brain capillary endothelial cells also exhibit a high mitochondrial content, which is thought to be necessary for their enhanced active transport and metabolic activity.

### 8.1.1  The Blood-Brain Barrier Phenotype

The blood-brain barrier phenotype is not an intrinsic property of the endothelium itself, but rather it is an emergent property of the endothelium and its immediate microenvironment. The grafting experiments of Stewart and Wiley (3) demonstrated this in two ways: 1) Capillaries invading embryonic neural tissue grafted into a non-neural cavity developed a blood-brain barrier-like phenotype and 2) somatic tissue grafted into the cerebral ventricles was invaded by capillaries that lacked tight junctions and were leaky. These results suggest that differentiation of an endothelium into the blood-brain barrier phenotype is driven by the tissue into which capillaries grow. Moreover, a genomic comparison of freshly isolated brain capillaries to primary brain endothelial cell cultures indicates that the expression of many genes—which includes those involved in regulation of angiogenesis and ion transport—is dramatically decreased after only a few days' separation from the brain (4). Thus, it seems that the unique environment of the brain is necessary not only for the development of the blood-brain barrier phenotype but also for its maintenance.

The microenvironment that surrounds brain capillaries is strongly influenced by other cell types, and the emerging functional relationships among these cells suggest that they act in concert as a neurovascular unit. At the microscopic level, a cerebral capillary observed in cross-section consists of a lumen enclosed by a single endothelial cell (Fig. 8.1). Pericytes, which contain nonmuscle actin

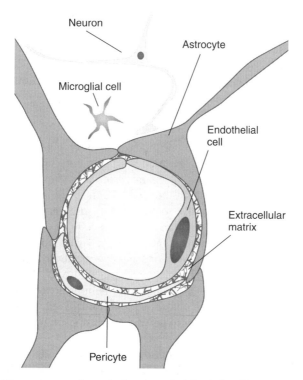

**Figure 8.1**   The neurovascular unit showing spatial relationships among constituents.

and thus can contract, are attached firmly to the abluminal (brain-side) membrane at irregular intervals. A basement membrane (basal lamina) that consists of matrix proteins such as collagen, laminin, and fibronectin surrounds the endothelial cells and pericytes and is itself contiguous with astrocytes foot processes which surround cerebral capillaries (Fig. 8.1). Astrocytes, pericytes, and matrix proteins have been shown to promote the expression of tight junction proteins and/or transport proteins in brain endothelial cells in culture, implying that all are involved in the formation and maintenance of the blood-brain barrier (2). We are just beginning to learn the details of how astrocytes and pericytes communicate with capillary endothelial cells to influence barrier properties. Finally, cerebral microvessels can be innervated by several neuronal cell types. These neuronal projections are thought to be involved in functional hyperemia (metabolic coupling) via both direct synapses on the vasculature (5) and juxtovascular astroglia (6). Little is known about what direct influence, if any, neurons have on the barrier functions of the microvasculature.

### 8.1.2 Altered Blood-Brain Barrier in Disease

It is important to note that "disruption" of the blood-brain barrier—that is to say, increased permeability—can signify one of at least three distinct classes of events, as follows: 1) penetration of peripheral immune cells into the parenchyma; 2) disruption of the paracellular diffusion barrier to large molecules and polar solutes, which ranges from extravasation of plasma proteins or hemorrhage to "size-selective" opening of the tight junctions to low-molecular weight molecules; and 3) diminished activity of efflux transporters, which leads to greater penetration of (generally more lipophilic) substrates. In addition, recent studies show that the expression of drug efflux transporters can be increased in disease, with the result being a selectively tightened barrier to those drugs that are substrates for transport.

The "immune-privileged" status of the brain is often conflated with the diffusion barrier to molecules; in fact, regulation of immune cell entry into the brain is both spatially and mechanistically separate from that of molecules (7). Recruitment of peripheral immune cells to the CNS occurs at the postcapillary venules, most likely through a transcellular route rather than a paracellular route. Thus, leukocytes can cross the endothelium while paracellular tight junctions remain intact; indeed, studies in experimental autoimmune encephalitis (a model for multiple sclerosis) have shown that areas of leukocyte infiltration do not necessarily correlate with areas of increased molecular permeability of the blood-brain barrier. Moreover, even cells that cross the endothelium do not necessarily reach the neuropil, as the glia limitans constitutes a formidable physical and functional barrier to the additional migration of leukocytes; it traps them in the perivascular space and initiates apoptosis via CD95/CD95L interactions.

Severe disruption of the blood-brain barrier to molecules occurs in response to acute insults such as traumatic brain injury and ischemia, and it can result in extravasation of plasma proteins and even hemorrhage. Indeed, evidence suggests that enhanced penetration of MRI contrast agents into the parenchyma is predictive of hemorrhagic transformation in stroke patients who receive thrombolytic therapy (8). Smaller, "size-selective" openings of the paracellular pathway have been identified in numerous animal models of disease (9), and these openings precede larger openings in experimental ischemia (10). Such changes in permeability are often associated with decreased expression and altered localization of tight junction proteins (2). In addition, changes in the expression of efflux transporters, especially p-glycoprotein (see below), at the blood-brain barrier have been associated with the development of Alzheimer's and Parkinson's diseases and of drug-resistant epilepsy. At the time of writing, it was not clear whether such changes in transporter expression play a role in susceptibility to or development of these and other neurodegenerative diseases (11). It is, however, critical to understand how the barrier function of the endothelium changes in disease, because such changes can have a profound impact on both the efficacy of CNS-acting drugs and the neurotoxicity of peripherally acting drugs that do not normally enter the CNS.

## 8.2  DRUG DELIVERY TO THE BRAIN

Multiple components of the brain capillary endothelium combine to present a formidable barrier to the entry of therapeutic drugs into the brain from the vasculature (Fig. 8.2). First, very tight junctional complexes effectively prevent the extensive diffusion of solutes between endothelial cells usually observed in peripheral capillaries. Indeed, measurements of transendothelial electrical resistance indicate that these tight junctions are among the least permeable found in the body (2). Second, as in other cells, the lipid cores of endothelial cell plasma membranes are a diffusion barrier to hydrophilic molecules. Third, low basal rates of transcytosis limit the capacity of vesicle-based mechanisms to move solutes across the endothelium through nonspecific (e.g., bulk flow) and receptor-mediated processes (9). Finally, high expression of adenosine triphosphate (ATP)-driven, drug efflux transporters on the luminal (blood-facing) plasma membrane of the endothelial cells restricts uptake of many lipophilic drugs that, based on structure, should readily diffuse across membranes. Of these transporters, we have the most complete picture of blood-brain barrier function for p-glycoprotein. Because of its high level of expression in the luminal plasma membrane, its potency as an efflux pump and its ability to handle an extremely wide range therapeutic drugs, p-glycoprotein (MDR1) is considered to be the major selective barrier to CNS pharmacotherapy with small-molecule drugs (11). Multispecificity also makes p-glycoprotein a potential source for interactions between drugs from different classes and with very different structures. Other efflux pumps expressed at the luminal membrane [e.g., multidrug resistance-associated proteins (MRP1,2,4) and breast cancer-related protein (BCRP and ABCG2)] are likely relevant for a smaller number of drugs (11, 12).

Together, these four components form an overlapping network of functional barriers to drug entry. For small, hydrophilic drugs and macromolecules (e.g.,

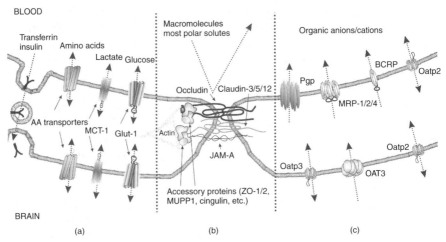

**Figure 8.2** Molecular constituents of the blood-brain barrier. (a) Carrier-mediated transport mechanisms, such as glucose transporter (GLUT)-1, monocarboxylate transporter (MCT)-1, and several amino acid transporters, facilitate brain entry of essential nutrients. Large signaling molecules, such as transferrin and insulin, can cross the blood-brain barrier by receptor-mediated endocytosis (24). (b) A hallmark of the BBB endothelium is the presence of epithelial-like tight junctions. Tight junctions are morphologically characterized by the extremely close apposition of adjacent cell membranes, which results in the effective fusion of these membranes into a continuous, selectively permeable surface. This task is accomplished via the interaction of membrane-spanning (or "integral") proteins that oligomerize with binding partners on adjacent cells, which include occludin, several subtypes of the claudin family, and junctional adhesion molecule (JAM)-A. These membrane-spanning proteins are linked via a complex of cytoplasmic (or "accessory") proteins to the actin cytoskeleton. Among these proteins are members of the membrane-associated guanylate kinase-like (MAGUK) homolog family, such as zonula occludens (ZO)-1 and -2, which are characterized by multiple domains specialized for facilitating protein-protein interactions (2). (c) Several members of the ATP-binding cassette (ABC) transporter family are expressed on the luminal membrane of the BBB, which include p-glycoprotein, multidrug resistance-related proteins (MRP)-1, -2, and -4, and breast cancer resistance protein (BCRP). These proteins act in concert with organic anion transporters (OATs) and organic anion transporting polypeptides (Oatps) to mediate active efflux of a wide array of substrates, which include many therapeutic drugs.

peptides, antibodies and RNA), the primary obstacles to CNS entry are the tight junctions and the plasma membranes. For lipophilic drugs that should readily permeate cell membranes, the primary obstacles are the drug efflux transporters. In reality, these barriers are not absolute. All solutes do exhibit finite barrier permeabilities, but for many drugs, actual permeabilities can be low enough to prohibit effectively the delivery of a sufficient drug to provide a therapeutic effect to the CNS.

## 8.2.1  CNS Complexity and Drug Distribution

The brain is a large heterogeneous organ with an amazing degree of spatial organization (e.g., extended neural circuits that are essential for normal brain function). Disease and injury can affect small regions of the brain that control limited, specific functions or larger regions that control multiple functions. However, with a few exceptions, currently it is impossible to use the vasculature to deliver an agent solely to a specific region of the brain. Moreover, all brain regions are exposed to the drug, although not necessarily uniformly, because blood flow can vary with the region of the brain. This consideration is important when trying to limit drug neurotoxicity. However, focal disruptions of the brain's normal physiology can lead to profound changes in the local vasculature, which can be transient or long lasting. Consider the following three examples: cancer, epilepsy, and stroke. First, capillaries around and in brain tumors can exhibit increased or decreased permeability to chemotherapeutics depending on tumor type (brain-derived or metastatic) and whether the vessels are newly formed under the influence of the tumor (13). Second, drug resistance in epilepsy has been linked to seizure-induced, focal overexpression of drug efflux transporters at the blood-brain barrier. This resistance could lead to reduced net blood to brain antiepileptic drug transport (transporter-based drug resistance) in just the region of the brain to which drug needs to be delivered (14). Finally, focal ischemic stroke causes complex, time-dependent changes in barrier permeability (2) and possibly efflux transporter expression over a large area within the affected hemisphere (15). It is still not clear how these changes might influence the delivery of drugs that could limit stroke damage and speed recovery.

## 8.2.2  Measuring Drug Permeability

Both *in vitro* and *in vivo* methods have been used to correlate the physicochemical properties of drugs with blood-brain barrier permeability. *In vitro* models measure solute transport across a confluent monolayer of cells cultured on a porous membrane. These models have included brain capillary endothelial cells in primary culture, brain capillary endothelial cell lines (rodent and human derived), and cell lines derived from dog kidney (MDCK) and human intestine (Caco-2) (16). These cells exhibit a range of paracellular permeabilities and drug efflux transporter expression levels, and none could be considered as a perfect model for the blood-brain barrier. Indeed, even for well-established epithelial cell lines and brain capillary endothelial cell cultures, measured transendothelial electrical resistances (one measure of tight junction permeability) and transporter expression levels vary from lab to lab. Certainly, for single drugs, the cell model chosen influences the permeability coefficient obtained. However, all models have been informative when comparing the permeabilities of large numbers of test compounds in attempts to understand structure-permeability relationships (see below).

*In vivo* measurements of blood-brain barrier permeability have primarily centered around rodent models, although there have been some attempts to do this in patients using pharmacological endpoints and positron emission tomography (PET) imaging. Animal studies involve injection of a drug into the vasculature and measurement of brain plasma distribution over time. Short-term measurements during which the brain vasculature is isolated and perfused with physiological saline provide kinetic measures of drug uptake and efflux. Longer term measurements of distribution in the intact animal provide both plasma and brain area under the curve (AUC) values. Corrections for drug metabolism (when using radioisotopes), drug binding to plasma and brain proteins, and drug retention within the brain vasculature are critical considerations (17). Experiments with knockout mice have been particularly revealing of the influence of drug transporters on brain/plasma distribution and have played a large part in demonstrating the importance of p-glycoprotein as major determinants of drug uptake into the brain (18). Thus, it is important to understand that differences in substrate specificity between human and rodent transporters have been documented, and these will most certainly contribute to species differences in the contributions of transporter-based efflux and uptake for specific drugs.

## 8.3   STRATEGIES TO IMPROVE CNS DRUG DELIVERY

Three general strategies have been used to increase drug delivery to the brain: transient modification of the blood-brain barrier itself, structural modification of the drug to increase permeability while retaining therapeutic potential, and packaging of drug or prodrug on or within a delivery vehicle that can pass through the barrier. Each of these approaches has achieved some measure of success, especially in animal-based experiments. However, none of these has yet provided a general solution to the problem of drug delivery to the brain, which remains a major problem in the clinic. This is especially true for antibody-based and nucleic acid-based therapies (gene therapy), which are beginning to be used to treat peripheral diseases and which hold great promise for CNS diseases once delivery is improved.

### 8.3.1   Modifying the Barrier

Targets for therapeutic modification of the barrier include both the tight junctions and the drug efflux transporters. It should be pointed out here that both tight junctions and efflux transporters serve important neuroprotective roles; this concept must be kept in mind when they are manipulated to improve drug entry. Several approaches have been used in an attempt to create a brief window in time during which tight junctions are opened to the extent that normally nonpermeable drugs can be delivered to the CNS. Initial attempts to do this employed carotid artery infusion of hyperosmotic solutions of mannitol. This treatment transiently shrinks the endothelial cells, pulls them apart at the tight junctions, and increases brain entry of chemotherapeutics and even monoclonal antibodies (19). The osmotic blood-brain barrier opening has been an element of several clinical trials in which chemotherapeutics and chemotherapeutics plus radiation were used to treat brain tumor patients; results have been generally encouraging but not spectacular. Other workers have tried to use the endothelium's own regulatory mechanisms to modulate tight junctional permeability. For example, Cereport (RMP-7), which is a bradykinin B2 receptor agonist, transiently increases the permeability of the blood-brain barrier in several animal models (20). Phase I clinical trials with Cereport combined with carboplatin showed little toxicity. Although early clinical trials with Cereport and carboplatin in brain tumor patients were promising; subsequent trials showed little or no benefit of Cereport.

p-Glycoprotein is the efflux transporter responsible for the limited blood-brain barrier penetration of a surprisingly large number of lipophilic drugs (e.g., chemotherapeutics, opioids, HIV protease inhibitors, antibiotics, immunosuppressives, and anti-epileptics). This transporter sits in the luminal membrane of brain capillary endothelial cells and uses the potential energy released from ATP splitting to pump drugs back into the blood. In animal experiments, reducing p-glycoprotein expression or inhibiting its activity can increase brain accumulation of some substrates by more than 10-fold. Several animal studies have shown dramatic reductions in volumes of transplanted tumors when specific p-glycoprotein inhibitors were administered along with chemotherapeutics that normally do not cross the blood-brain barrier (11). To date, limited clinical studies show no therapeutic advantage of inhibiting p-glycoprotein in cancer patients. Careful attention to dose levels and timing of administration for both the inhibitor and chemotherapeutic may be critical for this strategy to work.

A second general approach to the problem of transporter modulation was recently suggested by Bauer et al. (21), who targeted the intracellular signaling pathway responsible for p-glycoprotein upregulation in an animal model of epilepsy. It was shown that glutamate, which is released during seizures, signaled the upregulation of p-glycoprotein expression in brain capillaries. By blocking glutamate signaling through cyclooxygenase-2, they could abolish the increase in transporter expression caused by epileptic seizures themselves. At the time of writing, this general approach of targeting signaling to modulate transporter activity and thus increase drug access to the brain awaits confirmation in the clinic.

### 8.3.2   Modifying Drug Structure

Early attempts to understand the relationship between chemical structure and *in vivo* blood-brain barrier permeability focused on lipophilicity. Indeed, for many compounds there seems to be a simple relationship between log oil/water partition coefficient (log P) and the log brain-to-plasma concentration ratio. From these initial findings and subsequent experiments, several groups have attempted to describe the key physicochemical parameters that determine blood-brain barrier permeability. These parameters have turned out to include lipophilicity (clogP), number of hydrogen bond donors (HBD), polar surface area (PSA), and molecular size (MW) and shape; the number of hydrogen bond acceptors seems to be less critical. In drug design, each of these parameters can be modified independently, so barrier permeability actually responds to some type of aggregate score with contributions from each parameter. Nevertheless, for each practical limits exist outside of which permeability falls dramatically. In a recent review, Hitchcock and Pennington (22) suggest the following limits for drugs that must cross the blood-brain barrier: PSA < 90 Å, HBD < 3, clogP 2-5, MW < 500. Note that of the top 25 CNS drugs examined, most fell well within these limits. Note also that by these rules macromolecules (e.g., proteins, DNA and RNA), being too large and too polar, have essentially no ability to cross the blood-brain barrier.

Hitchcock and Pennington (22) also review in detail several case studies in which structurally related, small drug analogs were designed and synthesized with a view toward optimizing the physicochemical parameters for improved passive diffusion, reducing entry (to minimize CNS side-effects of peripherally acting drugs), enhancing carrier mediated uptake (through normally expressed nutrient transporters), and minimizing active efflux (modulating p-glycoprotein/drug interactions). They importantly point out that most studies of this kind only focus on one process (e.g., passive diffusion) without consideration of how structural modifications might affect the others (e.g., active efflux). Again, it must be emphasized that it is the combined effects of all these processes that determines how well or poorly a drug crosses the blood-brain barrier. Finding the right structural combination may require a sophisticated atomic balancing act.

### 8.3.3   Molecular Trojan Horses

The use of molecular "Trojan horses" to move cargos, both small molecule and macromolecule drugs, across the blood-brain barrier has been an active area of research (1, 23–25). The following three types of targeted vehicles have been used with limited success: antibodies to surface receptors, nanoparticles, and immunoliposomes. This family of technologies offers great promise in that it has the potential to deliver a systemically administered drug specifically to the CNS. Clearly, this general strategy adds substantial complexity to the drug-delivery process. For example, the release of cargo from the vehicle must be appropriate for the drug administered and its cellular target. In addition, small drugs released within brain capillary endothelial cells and the brain may be subject to efflux back into the blood. Thus, small drugs and proteins will have to be released into the CNS interstitium, and genes and RNA-based drugs will have to be delivered not just into the brain but also to the interior of their target cells.

### 8.3.4   Protein Vectors

One class of vectors used for Trojan horse technology is proteins that normally cross the brain capillary endothelium by absorptive or receptor-mediated endocytosis [e.g., cationized albumin or monoclonal antibodies to the transferin (OX26) or insulin receptors]. These proteins are conjugated to macromolecules for example by a streptavidin/biotin linkage, which is cleaved after delivery. This relatively simple approach has been used successfully to deliver many neurotrophic and growth factors to the brain with significant physiological effects. For example, intravenously injected human basic fibroblast growth factor (bFGF) conjugated to OX26 with a streptavidin/biotin linkage substantially reduced infarct volume in a rat model of stroke, whereas unconjugated bFGF was without effect (24). Other studies have shown CNS delivery of genes and antisense RNA using similar technology.

### 8.3.5 Nanoparticles

Nanoparticles are solid, colloidal matrix-like particles made of polymers: poly-butylcyanoacrylate (PBCA), polyactide homopolymers (PLA), and poly(lactide-co-glycolide) heteropolymers (PLGA). They can be made simply, are stable, and can provide sustained release of cargos to the brain and peripheral organs. Most nanoparticles used for drug delivery are 50–300 nm in diameter and cannot cross the blood-brain barrier without modification. Moreover, intravenously administered nanoparticles are rapidly cleared to the liver and reticuloendothelial system, and thus have a very short plasma half-life. Pegylation (covalent attachment of polyethylene glycol molecules) can improve the pharmacokinetic profile of the nanoparticles by reducing rates of clearance and coupling coating particles with Tween 80. Endogenous apolipoproteins improve endocytic transport of the particles with their small drug cargo (e.g., doxorubicin) across the brain capillary endothelium. Small molecule drugs can be loaded into pegylated PLA and PLGA nanoparticles, with ease of loading depending to a large extent on drug lipophlicity. Loading nanoparticles with proteins requires multiple steps that have to be fine tuned for each protein cargo (24).

### 8.3.6 Liposomes

Liposome-based Trojan horses seem to offer the greatest potential as high-capacity, multipurpose vehicles for drug delivery to the CNS. Liposomes vesicles are 50–100 nm in diameter in which an aqueous compartment is enclosed within a phospholipid membrane bilayer. The aqueous center of the liposomes can be loaded with drugs, which include small molecules, peptides, proteins, and nucleic acids [drugs, genes, and small interfering RNA (siRNA) for cell-specific gene knockdown]. Like nanoparticles, untreated liposomes are cleared rapidly from plasma into the liver and reticuloendothelial system. The pegylation of phospholipid groups improves the pharmacokinetic profile by dramatically decreasing systemic clearance by one or more orders of magnitude. Targeting liposomes to the brain involves conjugation with a protein vector that will normally cross the brain capillary endothelium by absorptive or receptor-mediated endocytosis [e.g., cationized albumin or monoclonal antibodies to the transferin (OX26) or insulin receptors]. The method of attachment of these targeting vectors to the liposomes is important, because direct bonding to the phospholipids produces interference of antibody binding by the PEG chains. A better method is to attach the proteins to a small fraction of the PEG chains themselves, which act as spacers that reduce steric interactions between PEG chains and the vector. Recent studies show substantial successes with animal models. First, a liposome-based Trojan horse was used to increase tyrosine hydroxylase gene expression in a rat model of Parkinson's disease (24). Second, a liposome-based Trojan horse that contains a specific RNA interference (RNAi) silenced epidermal growth factor receptor gene expression in a mouse model of brain cancer (intracerebrally injected human tumor cells); RNAi gene therapy doubled survival time (23). These limited successes suggest that with the right delivery system gene therapy of inherited CNS diseases may be possible in the near future.

### 8.3.7 Recent Innovations

Two examples suggest innovative, alternative strategies that may also prove to be effective for the delivery of macromolecular drugs. In one, Kumar et al. (26) used a short peptide derived from the rabies virus glycoprotein to deliver siRNA to the brain. This glycoprotein, which crosses the blood-brain barrier and binds to acetylcholine receptors on neurons, was modified by adding arginine residues to its carboxy terminus for attachment of siRNA. The intravenous injection of siRNA bound to the glycoprotein silenced genes within mouse brain. As proof of principle, an antiviral siRNA delivered via the modified rabies virus glycoprotein protected mice against fatal viral encephalitis. In the second example, Spencer and Verma (27) engineered a lentivirus vector that coded for apolipoprotein B, which is a protein secreted in the periphery (largely by the liver). The secreted protein both crossed the blood-brain barrier on the low density lipoprotein receptor and was targeted to endosomes and eventually lysosomes within neurons. This study is important for two reasons: It suggests ways to deliver protein-based drugs continuously to the CNS and to rescue certain genetic lysosomal storage disorders that target the CNS.

### ACKNOWLEDGMENTS

This research was supported in part by the Intramural Research Program of the National Institutes of Health, National Institute of Health Sciences.

### REFERENCES

1. Pardridge WM. Drug targeting to the brain. Pharm. Res. 2007;24:1733–1744.
2. Hawkins BT, Davis TP. The blood-brain barrier/neurovascular unit in health and disease. Pharmacol. Rev. 2005;57:173–185.
3. Stewart PA, Wiley MJ. Developing nervous tissue induces formation of blood-brain barrier characteristics in invading endothelial cells: a study using quail–chick transplantation chimeras. Dev. Biol. 1981;84:183–192.

4. Calabria AR, Shusta EV. A genomic comparison of in vivo and in vitro brain microvascular endothelial cells. J. Cereb. Blood Flow Metab. 2008;28:135–148.

5. Hamel E. Perivascular nerves and the regulation of cerebrovascular tone. J Appl. Physiol. 2006;100:1059–1064.

6. Iadecola C, Nedergaard M. Glial regulation of the cerebral microvasculature. Nat. Neurosci. 2007;10:1369–1376.

7. Bechmann I, Galea I, Perry VH. What is the blood-brain barrier (not)? Trends Immunol. 2007;28:5–11.

8. Hjort N, Wu O, Ashkanian M, Solling C, Mouridsen K, Christensen S, Gyldensted C, Andersen G, Ostergaard L. MRI detection of early blood-brain barrier disruption: parenchymal enhancement predicts focal hemorrhagic transformation after thrombolysis. Stroke 2008;39:1025–1028.

9. Hawkins BT, Egleton RD. Pathophysiology of the blood-brain barrier: animal models and methods. Curr. Top Dev. Biol. 2008;80:277–309.

10. Nagaraja TN, Keenan KA, Brown SL, Fenstermacher JD, Knight RA. Relative distribution of plasma flow markers and red blood cells across BBB openings in acute cerebral ischemia. Neurol. Res. 2007;29:78–80.

11. Miller DS, Bauer B, Hartz AM. Modulation of p-glycoprotein at the blood-brain barrier: opportunities to improve central nervous system pharmacotherapy. Pharmacol. Rev. 2008;60:196–209.

12. Dallas S, Miller DS, Bendayan R. Multidrug resistance-associated proteins: expression and function in the central nervous system. Pharmacol. Rev. 2006;58:140–161.

13. Gerstner ER, Fine RL. Increased permeability of the blood-brain barrier to chemotherapy in metastatic brain tumors: establishing a treatment paradigm. J. Clin. Oncol. 2007;25:2306–2312.

14. Loscher W, Potschka H. Drug resistance in brain diseases and the role of drug efflux transporters. Nat. Rev. Neurosci. 2005;6:591–602.

15. Spudich A, Kilic E, Xing H, Kilic U, Rentsch KM, Wunderli-Allenspach H, Bassetti CL, Hermann DM. Inhibition of multidrug resistance transporter-1 facilitates neuroprotective therapies after focal cerebral ischemia. Nat. Neurosci. 2006;9:487–488.

16. Liu X, Chen C, Smith BJ. Progress in brain penetration evaluation in drug discovery and development. J. Pharmacol. Exp. Ther. 2008;325:349–356.

17. Bickel U. How to measure drug transport across the blood-brain barrier. NeuroRx. 2005;2:15–26.

18. Schinkel AH. The roles of P-glycoprotein and MRP1 in the blood-brain and blood-cerebrospinal fluid barriers. Adv. Exp. Med. Biol. 2001;500:365–372.

19. Kroll RA, Neuwelt EA. Outwitting the blood-brain barrier for therapeutic purposes: osmotic opening and other means. Neurosurgery 1998;42:1083–1099; discussion 1099–1100.

20. Borlongan CV, Emerich DF. Facilitation of drug entry into the CNS via transient permeation of blood brain barrier: laboratory and preliminary clinical evidence from bradykinin receptor agonist, Cereport. Brain Res. Bull. 2003;60:297–306.

21. Bauer B, Hartz AM, Pekcec A, Toellner K, Miller DS, Potschka H. Seizure-induced up-regulation of P-glycoprotein at the blood-brain barrier through glutamate and cyclooxygenase-2 signaling. Mol. Pharmacol. 2008;73:1444–1453.

22. Hitchcock SA, Pennington LD. Structure-brain exposure relationships. J. Med. Chem. 2006;49:7559–7583.

23. Boado RJ. Blood-brain barrier transport of non-viral gene and RNAi therapeutics. Pharm. Res. 2007;24:1772–1787.

24. Jones AR, Shusta EV. Blood-brain barrier transport of therapeutics via receptor-mediation. Pharm. Res. 2007;24:1759–1771.

25. Xia CF, Zhang Y, Zhang Y, Boado RJ, Pardridge WM. Intravenous siRNA of brain cancer with receptor targeting and avidin-biotin technology. Pharm. Res. 2007;24:2309–2316.

26. Kumar P, Wu H, McBride JL, Jung KE, Kim MH, Davidson BL, Lee SK, Shankar P, Manjunath N. Transvascular delivery of small interfering RNA to the central nervous system. Nature 2007;448:39–43.

27. Spencer BJ, Verma IM. Targeted delivery of proteins across the blood-brain barrier. Proc. Natl. Acad. Sci. U.S.A. 2007;104:7594–7599.

## FURTHER READING

Abbott NJ. Dynamics of CNS barriers: evolution, differentiation, and modulation. Cell Mol. Neurobiol. 2005;25:5–23.

Abbott NJ, Ronnback L, and Hansson E. Astrocyte-endothelial interactions at the blood-brain barrier. Nat. Rev. Neurosci. 2006;7:41–53.

Begley DJ. Delivery of therapeutic agents to the central nervous system: the problems and the possibilities. Pharmacol. Ther. 2004;104:29–45.

Lok J, Gupta P, Guo S, Kim WJ, Whalen MJ, van Leyen K, Lo EH. Cell-cell signaling in the neurovascular unit. Neurochem. Res. 2007;32:2032–2045.

Reichel A. The role of blood-brain barrier studies in the pharmaceutical industry. Curr. Drug Metab. 2006;7:183–203.

Zlokovic BV. The blood-brain barrier in health and chronic neurodegenerative disorders. Neuron. 2008;57:178–201.

## SEE ALSO

Chapter 7

# 9

# PHARMACOKINETIC CONSIDERATIONS: METHODS AND SYSTEMS OF CONTROLLED DRUG DELIVERY

Zhong Zuo and Vincent H. L. Lee

*School of Pharmacy, The Chinese University of Hong Kong, Shatin, N.T., Hong Kong SAR*

Controlled drug delivery applies interdisciplinary approaches to engineer systems that improve the therapeutic value of drugs. This review addresses the biological basis of drug delivery, the pharmacokinetic/pharmacodynamic considerations important to the design of controlled release delivery systems, and the methods to fabricate them. The focus of this review will principally be on oral and transdermal applications. Systems at various stages of development for the delivery of more complex molecules, such as proteins, oligonucleotides, genes, and sRNA's will also be discussed.

The frequency of drug dosing is usually determined by the drug's duration in the body. For drugs that are inherently long lasting, once daily oral dosing is sufficient to sustain adequate drug blood levels and the desired therapeutic effect. Formulation of these drugs as conventional, immediate-release dosage forms is used for the patient. However, many drugs are not inherently long lasting and require multiple doses each day to achieve the desired therapeutic results. Multiple daily dosing often is inconvenient for the patient and can result in missed doses, made-up doses, and patient noncompliance with the therapeutic regimen. In addition, sequential therapeutic blood level peaks and valleys (troughs) are associated with each dose. The traditional approach to modifying drug release is

*Chemical Biology: Approaches to Drug Discovery and Development to Targeting Disease*, First Edition. Edited by Natanya Civjan.
© 2012 John Wiley & Sons, Inc. Published 2012 by John Wiley & Sons, Inc.

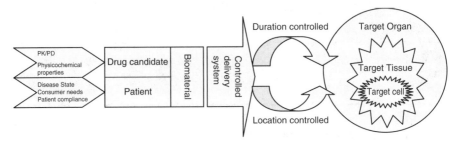

**Figure 9.1**    Scheme for the design of controlled drug delivery system.

to control the rate of drug delivery, as exemplified by extended-release tablets and capsules, which are commonly taken only once or twice daily. Modern drug delivery systems provide an additional targeted drug release dimension at a specific site in the body, spanning subcellular to organ. Thus, the term "controlled drug delivery" has a much broader meaning than just controlling the rate of drug delivery. Controlled drug delivery to the right site of action at the right time can be achieved by either 1) controlling the release rate and duration or 2) controlling the release site (i.e., localized delivery or targeted diseased organs, tissues or cells). Figure 9.1 has summarized the scheme for the design of the controlled drug delivery. The commonly used types of controlled drug delivery products in terms of its targeted diseases and specific delivery technology involved are summarized in Table 9.1.

## 9.1    GENERAL INTRODUCTION OF PHARMACOKINETICS/ PHARMACODYNAMICS

### 9.1.1    Fate of a Drug *In Vivo:* ADME Process

After administration, the fate of the drug will be determined by four key steps *in vivo*, namely, absorption (A), distribution (D), metabolism (M), and excretion (E). Effects of the four steps on the administered drug will influence the levels of drug exposure to the site of action and will eventually influence the pharmacological activities of the drug. The details of the four steps are described briefly as follows. For a more detailed discussion see Chapter 6.

*9.1.1.1    Absorption (A)*    Before a compound can exert its pharmacological effect in tissues, it has to be taken into the bloodstream usually via mucosal surfaces such as the digestive tract for intestinal absorption. In addition, uptake of drugs from the blood stream into target organs or cells needs to be ensured. However, this task is not always easy given the natural barriers that exist, like the blood-brain barrier. Extent of absorption for a drug varies with the way it has been administered. Factors such as poor compound solubility, chemical instability in the stomach, and inability to permeate the intestinal wall can all reduce the extent to which a drug is absorbed after oral administration.

**TABLE 9.1  Summary of representative products using controlled release delivery system**

| Disease state | Active ingredient | Type of controlled delivery system | Product (manufacturer) |
|---|---|---|---|
| **Endocrinology** | | | |
| Postmenstrual syndrome | Estradiol | TDD | Estraderm (Novartis), Vivelle (Novartis), Climara (Berlex) |
| Hypogonadism in males | Testosterone | TDD | Testoderm (Alza), Androderm (SmithKline Beecham) |
| Birth control | Estradiol/Norethindrone | TDD | CombiPatch (Noven/Aventis), Ortho-Evra |
| | Progesterone | IUD insert | Progestersert (Alza) |
| Hormone replacement therapy | Estrogen/Progesterone | TDD | Nuvelle TS (Ethical Holdings/Schering) |
| **Pain, inflammatory** | | | |
| Moderate/severe pain | Fentanyl | TDD | Duragesic (Janssen) |
| **Central nervous system** | | | |
| Smoking cessation | Nicotine | TDD | Habitrol (Novartis Consumer), Nicoder CQ (SmithKline Beecham Consumer) |
| Motion sickness | Scopolamine | TDD | Transderm Scop(Novartis Consumer) |
| Attention deficit hyperactivity disorder | Methylphenidate HCL | Osmotic pump | Concerta |
| **Cardiovascular disease** | | | |
| Calcium channel antagonist for hypertension, ischemic heart disease, hypertrophic cardiomyopathy, arrhythmias | Nifedipine | Diffusion, osmotic pump, erosion | Procardia, Procaria XL, Adalat CC |

*(continued)*

**TABLE 9.1** (*Continued*)

| Disease state | Active ingredient | Type of controlled delivery system | Product (manufacturer) |
|---|---|---|---|
| | Diltiazem | Diffusion, erosion | Cardizem, Cardizem CD, Cardizem SR, Discor XR, |
| | Verapamil | Diffusion, erosion | Verelan, Calan SR, Isoptin SR |
| | Felodipine | | Plendil |
| | Nicardipine | Erosion | Cardene SR |
| Nitrates for angina pectoris, ischemic heart disease, congestive heart failure. | Nitroglycerin | TDD (Monolithic and membrane controlled) | Nitrodisc (Searle), Nitro-Dur (Key Pharmaceutical), Transdermal-Nitro (Summit Medical), Deponit (Schwarz Pharma), Minitran (3 M Pharmaceuticals), etc. |
| | Isosorbide mononitrate | Matrix tablet | Imdur (Schering-Key) |
| Hypertension | Clonidine | TDD | Catapres-TTS |
| Arrhythmia | Procainamide | Matrix tablet | Procan SR (Parke Davis) |
| | Disopyramide phosphate | Multipellet | Norpace CR |
| Hypercholesterolemia | Nicotinic acid | Polygel matrix delivery system | Slo-Niacin (Upsher-Smith) |
| **Cancer** | | | |
| Recurrent ovarian cancer | Doxorubincin | Stealth liposome | Doxil, Caelyx |
| **Infectious disease** | | | |
| Fungal infection, AIDS | Amphotericin B | Liposome | AmBisome |
| **Ophthalmic disease** | | | |
| Glaucoma | Pilocarpine | Reservoir insert | Ocusert PILO (Alza) |

Absorption critically determines the compound's bioavailability. Drugs that absorb poorly when taken orally must be administered in some less desirable way, like intravenously or by inhalation.

***9.1.1.2  Distribution (D)***    Once the drug gets into the bloodstream, it will be carried to various parts in the body including different tissues and organs, as well as its site of action. Depending on the physicochemical and biological properties of the drug, its degree of distribution to various tissues and organs differs by varying extents.

***9.1.1.3  Metabolism (M)***    As soon as the drug enters the body, in addition to being distributed to other parts of the body, it begins to break down. Most small-molecule drug metabolism occurs in the liver by enzymes termed cytochrome P450 enzymes. As metabolism occurs, the parent compound is converted to metabolites, which could be pharmacologically inert or active.

***9.1.1.4  Excretion (E)***    Compounds and their metabolites need to be removed from the body via excretion, usually through the kidneys or the feces. Unless excretion is complete, accumulation of foreign substances can affect normal metabolism adversely.

### 9.1.2    Role of Transporters in ADME

The extent of absorption for a drug varies with the way it has been administered. Over the last decade, several important transporters responsible for drug uptake/disposition in various organs have been discovered. They mainly include organic anion transporter, organic cation transporter, multidrug resistant protein, and multidrug resistance-associated protein. Many transporters function in the uptake of drugs into cells, which leads to increased permeability of the drug into cells; other transporters function to export drugs out of cells, which thereby decreases the apparent permeability of the drug. Transport mechanism of drugs and substrate specificities of drug transporters in drug development have become increasingly important. Identification of drugs that are the substrates of the transporters can help to elucidate the pharmacokinetic profiles of these drugs and drug–drug interactions, as well as to improve the therapeutic safety of drugs. Therefore, research that focuses on such membrane transporters is promising, and it leads to rational design strategies for drug targeting delivery. Drug targeting to specific transporters and receptors using carrier-mediated absorption has demonstrated immense significances in the ocular drug delivery (1) and targeted drug delivery across the blood–brain barrier (2).

### 9.1.3    Concept of Pharmacodynamics, Biomarkers and Pharmacogenetics

Pharmacodynamics is the study of the mechanisms of drug action and the biological and physiological effects of these drugs. Compared with pharmacokinetics, which reveals what the body does to a drug, pharmacodynamics focuses more on what a drug does to the body.

Traditionally, the level of drug *in vivo* is assumed to be proportional to its pharmacodynamic effect. The identification of various biomarkers provides a more relevant indicator than the *in vivo* drug concentrations as far as clinical efficacy of a drug is concerned. Biomarkers enable the characterization of patient populations and the quantization of the extent to which new drugs reach intended targets, alter proposed pathophysiological mechanisms, and achieve clinical outcomes. In genomics, the biomarker challenge is to identify unique molecular signatures within complex biological mixtures that can be unambiguously correlated to biological events to validate novel drug targets and to predict drug response. Biomarkers can stratify patient populations or quantify drug benefit in primary prevention or disease-modification studies in poorly served areas such as neurodegeneration and cancer. Appropriateness of biomarkers depends on the strategy and the stage of development, as well as the nature of the medical indication. Biomarkers are perhaps most useful in the early phase of clinical development when measurement of clinical endpoints may be too time-consuming or cumbersome to provide timely proof of concept or dose-ranging information (3).

With the disclosure of the map of the human genome, the effect of human genetic variations on drug metabolism and its related clinical efficacy has become a major concern for clinical practice. Pharmacogenetics is an area of study on the genetic variation in response to drug metabolism with an emphasis on improving drug safety. The aim of drug delivery is to achieve an appropriate drug concentration at the targeted site that can elicit a desired level of response. The concentration of a drug *in vivo* is highly dependent on the ADME pharmacokinetic processes, whereas the level of response for a drug usually results from the interaction of a drug with a target receptor protein. More and more studies have revealed genetic polymorphisms in drug-metabolizing enzymes, transporters, and target receptors. The genetic basis underlying pharmacokinetic and pharmacodynamic inter-individual variability is essential in the consideration of the design of drug delivery systems (4).

## 9.2 PHARMACOKINETIC/PHARMACEUTICAL DOSAGE FORM/PHARMACODYNAMIC CONSIDERATIONS IN THE DESIGN OF A CONTROLLED DRUG DELIVERY SYSTEM

An ideal controlled-release product is expected to release the drug from the dosage form at a predetermined rate, to maintain the released drug at a sufficient concentration in the gastrointestinal fluids, and to provide sufficient gastrointestinal residence time prior to the absorption of the drug. In general, not every drug is suitable to be developed as a controlled-release product.

Traditionally, the pharmacokinetic properties of drug candidates has always played an important role in their design as a controlled release dosage form. From the pharmacokinetic point of view, the best suited drug candidates should have neither very slow nor very fast rates of absorption and excretion. Drugs with slow rates of absorption and excretion with half-lives of greater than 8 hours are usually inherently long acting, and their preparation into extended release dosage forms is not necessary. Drugs with very short half-lives (i.e., <2 hours) are poor candidates for extended-release dosage forms because of the large quantities of drug required for such a formulation. Another pharmacokinetic consideration is that the drugs prepared in extended-release forms must have good aqueous solubility, must maintain adequate residence time in the gastrointestinal tract, and must be uniformly absorbed from the gastrointestinal tract. In addition, from the classic dosage form design point of view, the drug candidate should be administered in relatively small doses because each oral dosage unit needs to maintain a sustained therapeutic blood level of the drug.

As indicated in Table 9.1, because of the nature of the duration of action for a controlled drug delivery system, another important pharmacodynamic consideration for drug candidates is that the drugs are used in the treatment of chronic rather than acute conditions. Regardless of how it is achieved, the prolonged and controlled pharmacodynamic effect is the eventual outcome expected for the controlled drug delivery system. The appropriate drug candidate should possess a reasonable therapeutic index (i.e., the median toxic dose divided by the median effective dose). It should be noted that drugs possessing a longer acting quality than indicated by their quantitative pharmacokinetic half-lives might not be good candidates for a controlled drug delivery system either. Therefore, monitoring the controlled release drug more reliably and accurately *in vivo* is becoming a challenge and the traditional pharmacokinetic/pharmacodynamic correlations as a standard component of drug development are becoming more complicated than what is usually expected (5). At the time of writing this chapter, an integrated pharmacokinetic and pharmacodynamic approach for rate-controlled drug delivery has been proposed (6). Should pharmacodynamic measurements eventually replace the traditional pharmacokinetic measurements? With the availability of various biomarkers, the measurements of related biomarkers rather than the plasma concentrations of specific drugs are believed to provide more direct and accurate monitoring of the behavior of the controlled drug delivery system *in vivo*.

## 9.3   METHODS AND SYSTEMS OF CONTROLLED DRUG DELIVERY

Natural and synthetic polymers, both erodible and nonerodible, play an important role in the fabrication of systems to control drug release. The method of combining polymers with a drug candidate, together with the manufacturing process, provides even more control on the release profile desirable of the drug candidate.

### 9.3.1   Systems to Control the Release Rate and Duration of the Drug

The rate of drug release from solid dosage forms may be modified by the technologies described below, which in general are based on 1) modifying drug dissolution by controlling access of biologic fluids to the drug through the use of barrier coatings, 2) controlling drug diffusion rates from dosage forms, and 3) chemical reactions or interactions between the drug substance or its pharmaceutical barrier and site-specific biological fluids. The commonly used systems and the related techniques are illustrated as follows.

### 9.3.1.1 Coated Beads, Granules, or Microspheres

This system aims to distribute the drug onto beads, pellets, granules, or other particulate systems. The rate at which body fluids can penetrate the coating to dissolve the drug will be adjusted by varying the thickness of the coats and the type of coating material used. The coating techniques include conventional pan-coating or air-suspension coating techniques by which a solution of the drug substance is placed onto small inert nonpareil seeds or beads made of sugar and starch or onto microcrystalline cellulose spheres. Naturally, the thicker the coat, the more resistant to penetration and the more delayed will be the drug release and dissolution. Typically, the coated beads are about 1 mm in diameter. However, the beads are combined to have three or four release groups among the more than 100 beads contained in the dosing unit. This technique provides the different desired sustained or extended release rates as well as the targeting of the coated beads to the desired segments of the gastrointestinal tract. An example of this type of dosage form is the Spansule capsule from SmithKline Beecham (King of Prussia, PA).

### 9.3.1.2 Microencapsulated Drug

Microencapsulation is a process by which solids, liquids, or even gases may be encapsulated into microscopic sized particles through the formation of thin coatings of "wall" material around the substance being encapsulated. Gelatin is a common wall-forming material, but synthetic polymers, such as polyvinyl alcohol, ethylcellulose, polyvinyl chloride, and other materials, may be used. The typical encapsulation process usually begins with the dissolving of the prospective wall material, such as gelatin, in water. The material to be encapsulated is added and the two-phase mixture is stirred thoroughly. The material to be encapsulated is then broken up into the desired particle size via addition of a solution of a second material, which is usually acacia. This additive material can concentrate the gelatin into tiny liquid droplets. These droplets (the coacervate) then form a film or coat around the particles of the substance to be encapsulated as a consequence of the extremely low interfacial tension of the residual water or solvent in the wall material so that a continuous, tight, film coating remains on the particle. The final dry microcapsules are free-flowing, discrete particles of coated material. Different rates of drug release may be obtained by changing the core to wall ratio, the polymer used for the coating, and the method of microencapsulation. One advantage of microencapsulation is that the administered dose of a drug is subdivided into small units that are spread over a large area of the gastrointestinal tract, which may enhance absorption by diminishing localized drug concentration. An example of a drug commercially available in a microencapsulated, extended-release dosage form is potassium chloride (Micro-K Extencaps; Wyeth, Madison, NJ).

***9.3.1.3   Embedding Drug in Hydrophilic or Biodegradable Matrix Systems*** By
this process, the drug substance is combined and made into granules with an
excipient material that slowly erodes in body fluids, progressively releasing the
drug for absorption. Drugs without excipients provide an immediate drug effect,
whereas drug-excipient granules provide extended drug action. The granule mix
may be tableted or placed into gelatin capsule shells for oral delivery. Hydrox-
ypropyl methylcellulose (HPMC), which is a free-flowing powder, is commonly
used to provide the hydrophilic matrix. Tablets are prepared by distributing
HPMC in the formulation thoroughly, preparing the granules by wet granulation or roller compaction, and manufacturing the tablets by compression. After
ingestion, the tablet is made wet by gastric fluid, which hydrates the polymer. A
gel layer forms around the surface of the tablet and an initial quantity of drug is
exposed and released. As water permeates further into the tablet, the thickness of
the gel layer is increased, and the soluble drug diffuses through the gel layer. As
the outer layer becomes fully hydrated, it erodes from the tablet core. If the drug
is insoluble, then it is released as such with the eroding gel layer. Thus, the rate
of drug release is controlled by the processes of diffusion and tablet erosion. An
example of a proprietary product using a hydrophilic matrix base of HPMC for
extended drug release is Oramorph SR Tablets from Roxane Laboratories, Inc.
(Columbus, OH), which contains morphine sulfate.

Scientists have been working for decades to expand the number of biodegradable polymers for clinical use, which include polyhydroxyacids, polyanhydrides,
polyorthoesters, and polyesteramides. Of these polymers, polylactide (PLA) and
poly(D,L-lactide-co-glycolide) (PLGA) are most often used because of their
relatively good biocompatibility, easily controlled biodegradability, and good
processability. In fact, they belong to one of the two classes of FDA approved
synthetic biodegradable polymers to date. These polymers can be used either
as matrix devices or as reservoirs. In matrix systems, the drug is dispersed or
dissolved in the polymer, whereas the drug is encapsulated in a biodegradable
membrane in the reservoir. The release rate of the drug is generally dependent on the degradation of the polymer. Poly(ortho)esters and polyanhydrides
erode from their surface, whereas the PLGA and PLA follow bulk erosion (i.e.,
degradation occurring throughout the whole polymer). By varying the ratios of
monomers in the copolymer mix, nearly zero-order drug release produced by
matrix degradation at steady rate can be achieved (7).

A few commercially available drug delivery systems are based on biodegradable polymers. Glidadel is a white, dime-sized wafer made up of biodegradable polyanhydride polymer [poly (1,3-bis(carb-oxyphenoxy)propane-co-sebacic
acid)] (PPCP-SA) to deliver carmustine for the treatment of brain cancer. ReGel is
an aqueous filter sterilizable ABA tri-block polymer system that consists of PLGA
and polyethylene glycol (8). Its suitability to provide sustained interleukin-2 (IL-
2) delivery for cancer immunotherapy has been reported. Atrigel is a proprietary
delivery system that contains PLA or PLGA dissolved in a biocompatible carrier
and can be used for both parenteral and site-specific drug delivery. The system
is placed in the body using conventional needles and syringes and solidifies on

contact with aqueous body fluids to form a solid implant. For a drug incorporated into such polymer solution, it becomes entrapped within the polymer matrix as it solidifies followed by being slowly released as the polymer biodegrades. Products that have already been approved by the FDA using the Atrigel technology include the Eligard (leuprolide acetate for injectable suspension) prostate cancer products that provide systemic release of leuprolide acetate for 1-, 3-, and 4-month duration and the Atridox (8.5% doxycyline) periodontal treatment product for localized subgingival delivery of doxycycline. With the clinical success of the current available products using biodegradable polymers, their additional applications in the delivery of small molecules, peptides, proteins, monoclonal antibodies, and vaccines are anticipated.

### 9.3.1.4   Embedding Drug in an Inert Plastic Matrix

By this method, the drug is granulated with an inert plastic material, such as polyethylene, polyvinyl acetate, or polymethacrylate, and the granulation is compressed into tablets. The compression of the tablet creates the matrix or plastic form that retains its shape during drug diffusion and its passage through the alimentary tract. An immediate-release portion of the drug may be compressed onto the surface of the tablet. The inert tablet matrix, which is expended of drug, is excreted with the feces. The primary example of a dosage form of this type is the Gradumet from Abbott Pharmaceutical (Abbott Park, IL).

### 9.3.1.5   Ion-exchange Resins

A solution of a cationic drug may be passed through a column that contains an ion-exchange resin, which forms a complex by the replacement of hydrogen atoms. The resin-drug complex is then washed and may be tableted, encapsulated, or suspended in an aqueous vehicle. The release of the drug is dependent on the pH and the electrolyte concentration in the gastrointestinal tract. Release is greater in the acidity of the stomach than in the less acidic environment of the small intestine. Examples of drug products of this type include hydrocodone polistirex and chlorpheniramine polistirex suspension (Tussionex Pennkinetic Extended Release Suspension; Medeva) and phentermine resin capsules (Ionamin Capsules; Pharmanex, Provo, UT).

### 9.3.1.6   Osmotic Controlled Release Systems

The pioneer oral osmotic pump drug delivery system is the Oros system, which was developed by Alza Corporation (Mountain View, CA). The system is composed of a core tablet surrounded by a semi-permeable membrane coating with a 0.4-mm diameter hole in it that was produced by a laser beam. The core tablet has two layers: one contains the drug and the other contains a polymeric osmotic agent. The system operates on the principle of osmotic pressure. When the tablet is swallowed, the semi-permeable membrane permits water to enter from the patient's stomach into the core tablet, which dissolves or suspends the drug. As pressure increases in the osmotic layer, it forces or pumps the drug solution out of the delivery orifice on the side of the tablet. Only the drug solution can pass through the hole in the tablet. The rate of inflow of water and the function of the tablet depends on the existence of an osmotic gradient between the contents of the bilayer core and the fluid in the gastrointestinal (GI) tract. Drug delivery is essentially constant as long as the osmotic gradient remains constant. The drug release rate may be altered by changing the surface area, the thickness, or the composition of the membrane, and/or by changing the diameter of the drug release orifice; the rate is not affected by gastrointestinal acidity, alkalinity, fed conditions, or GI motility. The biologically inert components of the tablet remain intact during GI transit and are eliminated in the feces as an insoluble shell. This type of osmotic system, which is termed the "gastrointestinal therapeutic system" (GITS; Pfizer, New York), is employed in the manufacture of Glucotrol XL Extended Release Tablets and Procardia XL Extended Release Tablets (Pfizer).

***9.3.1.7 Gastro-Retentive Systems***    Gastric emptying of dosage forms is an extremely variable process, and the ability to prolong and control the emptying time is a valuable asset for dosage forms, which reside in the stomach for a longer period of time than conventional dosage forms. Gastro-retentive systems can remain in the gastric region for several hours, and hence, they significantly prolong the gastric residence time of drugs. Prolonged gastric retention would therefore improve the bioavailability of drugs by reducing drug waste and by improving solubility for drugs that are less soluble in a high pH environment. It has applications also for local drug delivery to the stomach and proximal small intestines. The controlled gastric retention of solid dosage forms may be achieved by the mechanisms of mucoadhesion, flotation, sedimentation, expansion, modified shape systems, or by the simultaneous administration of pharmacological agents that delay gastric emptying. Based on these approaches, classification of floating drug delivery systems has been investigated extensively. Several examples have been reported that show the efficiency of such systems for drugs with bioavailability problems. Gastroretention helps to provide better availability of novel products with many therapeutic possibilities and substantial benefits for patients. However, because of the complexity of the pharmacokinetic and pharmacodynamic factors for certain drugs, some researchers suggest that the rationale for continuous administration obtained by controlled-release gastroretentive dosage forms should be assessed and established *in vivo* (5).

***9.3.1.8 Transdermally Controlled Drug Delivery Systems***    Because of the large surface area of the skin and its bypass of the liver as a first pass step in metabolism, many drug delivery systems have been developed that control the rate of drug delivery to the skin for subsequent absorption. Effective transdermal drug delivery systems of this type deliver uniform quantities of drug to the skin over a period of time. Technically, transdermal drug delivery systems may be classified into monolithic and membrane-controlled systems (9).

*Monolithic Systems*    The systems incorporate a drug matrix layer between backing and frontal layers. The drug-matrix layer is composed of a polymeric material in which the drug is dispersed. The polymer matrix controls the rate at which the drug is released for percutaneous absorption. In the preparation of monolithic systems, the drug and the polymer are dissolved or blended together, cast as the matrix, and dried. The gelled matrix may be produced in a sheet or cylindrical form, with individual dosage units cut and assembled between the backing and the frontal layers. Most transdermal drug delivery systems are designed to contain an excess of drug and thus drug-releasing capacity beyond the time frame recommended for replacement.

*Membrane-Controlled Transdermal Systems*   The system is designed to contain a drug reservoir, usually in liquid or gel form, a rate-controlling membrane, and backing, adhesive, and protecting layers. This system has an advantage over monolithic systems in that as long as the drug solution in the reservoir remains saturated, the release rate of drug through the controlling membrane remains constant. Membrane-controlled systems are generally prepared by preconstructing the delivery unit, filling the drug reservoir, and sealing it off, or by a process of lamination, which involves a continuous process of construction, dosing, and sealing.

The first transdermal therapeutic system designed to control the delivery of a drug to the skin for absorption was developed in 1980 by the Alza Corporation. The system, marketed as Transderm-Scop by CIBA, is a circular, flat adhesive patch designed for the continuous release of scopolamine through a rate-controlling microporous membrane. Soon after, nitroglycerin, which is a drug substance used for the treatment of angina, became another candidate for transdermal delivery. The drug has a relatively low dose, short plasma half-life, high-peak plasma levels, liver first pass metabolism, and inherent side effects when taken sublingually (a popular route for its administration). Since then, several nitroglycerin-containing systems have been developed including: Deponit (Wyeth-Ayerst), Minitran (3 M Riker, St. Paul, MN), TransdermNitro (CIBA-Geigy, Summit, NJ), Nitro-Dur (Key Pharmaceuticals, Kenilworth, NJ), Nitrocine (Schwarz Pharma, Monheim, Germany), and Nitrodisc (Searle). As indicated in Table 9.1, the application of the transdermal therapeutic system products covers various aspects with the most emphasis on its use in central nervous system disease, cardiovascular disease, and hormonal therapy.

Future developments in the field of transdermal drug delivery should address problems that relate to irritancy, sensitization, and impermeability of the stratum corneum. These hurdles currently exclude several therapeutic entities from delivery via this route. An active area of research is breaching the integrity of the stratum corneum by various means, which include small molecule penetration enhancers, microneedles, and the use of various external energy sources such as electric current and ultrasound (10, 11). The approval of lidocaine hydrochloride and epinephrine combined iontophoretic patch for localized pain treatment by FDA has invigorated the gaining interest in iontophoretic drug delivery systems for drug delivery (12). In addition, additional innovations in matrix compound and formulation will likely expand the number of candidate drugs suitable for transdermal delivery (13).

### 9.3.2 Systems to Control the Location of Drug Release: Target Drug Spatially

*9.3.2.1 Liposomes* A liposome is a spherical vesicle with a membrane composed of a phospholipids and cholesterol bilayer. Liposomes can be composed of naturally derived phospholipids with mixed lipid chains, such as egg phosphatidylethanolamine, or of pure surfactants, such as dioleolylphophtidylthanolamine. Liposomes can be made by sonicating phospholipids in water followed by extrusion. Usually, a liposome contains a core of aqueous solution, in which the hydrophilic drug substances could be dissolved. For hydrophobic substances, the drug can be dissolved into the lipid membrane.

Liposomes are used for drug delivery because of their unique properties. To deliver the molecules to sites of action, the lipid bilayer can fuse with other bilayers such as the cell membrane, which delivers the liposomal contents into the cell. For drugs that do not have diffusion capabilities across the cell membrane, formulating the drugs into a liposome would enable them to be delivered past the lipid bilayer. Liposomes can also be designed to have various ways of delivering drugs into cells. It could possess various pH's, charges, or even made to be within a particular size range such that they would be viable targets for natural macrophage phagocytosis. These liposomes may then be digested while in the macrophage's phagosome followed by releasing the drug.

Additional advances in liposome research have resulted in the generation of liposomes that can avoid detection by the body's immune system, specifically the cells of the reticuloendothelial system (RES). These liposomes are known as "stealth liposomes" and are constructed with polyethylene glycol (PEG) studding the outside of the membrane. The PEG coating, which is inert in the body, allows for a longer circulatory life for the drug. In 1995, Doxil, the liposomal doxorubicin, became the first commercially available liposomal anticancer drug. It has an enhanced circulation half-life compared with the free drug because of its surface-grafted polyethylene glycol coating. Doxil passively targets solid tumors, and once the liposomes localize to the tumor interstitial space, the cytotoxic drug is slowly released within the tumor (14). AmBisome is a sterile, nonpyrogenic lyophilized antifungal product for intravenous infusion. It consists of unilamellar bilayer liposomes that are less than 100 nm in diameter with amphotericin B intercalated within the membrane.

Injectable liposome-based drug delivery systems of anticancer drugs are the most widely used drug nanoparticle in cancer (15). In addition to a PEG coating, most stealth liposomes also have some sort of biological species attached as a ligand to the liposome to enable binding via a specific receptor for the targeted drug delivery site. These targeting ligands could be monoclonal antibodies (named as immunoliposome), vitamins, or specific antigens. Targeted liposomes can hone to nearly any cell type in the body and deliver drugs that would otherwise be systemically delivered. It is expected that the side effects of extremely toxic drugs could be drastically reduced if the drugs were only delivered to diseased tissues using such techniques (16).

***9.3.2.2 Nanoparticles***   A nanoparticle is a microscopic particle with a diameter less than 100 nm. Nanoparticles were first developed around 1970, and initially they were devised as carriers for vaccines and anticancer drugs. Nanoparticle research is currently an area of intense scientific research because of a wide variety of potential applications in biomedical, optical, and electronic fields. To enhance tumor uptake, the strategy of drug targeting was employed, and as a first important step, research focused on the development of methods to reduce the uptake of the nanoparticles by the RES cells. Simultaneously, the use of nanoparticles for ophthalmic and oral delivery was investigated (17, 18). Recent advancement of nanoparticles and nanosuspensions was caused by their application for pulmonary drug delivery (19, 20).

The size of nanoparticles generally varies from 10 to 1000 nm. The drug is usually dissolved, entrapped, encapsulated, or attached to a nanoparticle matrix, and depending on the method of preparation, nanoparticles, nanospheres, or nanocapsules can be obtained. Nanocapsules are vesicular systems in which the drug is confined to a cavity surrounded by a unique polymer membrane, whereas nanospheres are matrix systems in which the drug is physically and uniformly dispersed. In recent years, biodegradable polymeric nanoparticles using poly(D,L-lactide), PLA, poly(D,L-glycolide) (PLG), PLGA, and poly(cyanoacrylate) (PCA) have attracted considerable attention as potential drug delivery devices in view of their applications in the controlled release of drugs, their ability to target particular organs/tissues, as carriers of DNA in gene therapy, and in their ability to deliver proteins, peptides, and genes through a peroral route of administration. The PLA, PLG, and PLGA polymers have been used as controlled release formulations in parenteral and implantation drug delivery applications.

Nanomedicine is undergoing explosive development as it relates to the development of nanoparticles for enabling and improving the targeted delivery of therapeutic and diagnostic agents. The use of nanoparticles for specific delivery to malignant cancers, to the central nervous system, and across the gastrointestinal barriers has received great attention (21). The additional development of nanoparticles is highly dependent on the advancement of the material sciences. For nanotechnology rationally to generate materials useful in human therapy, it will need to progress in full recognition of all the requirements biology places on the acceptability of exogenous materials (22). In particular, nanostructured porous materials, in particular silicon-based photonic and templated materials, offer a degree of control in both the rate and the location of drug delivery that is just beginning to be recognized (23).

***9.3.2.3 Dendrimers***   Dendrimers are macromolecules characterized by their highly branched structure and globular shape. They are attractive drug carriers by virtue of the flexibility in precisely controlling their molecular size, shape, branching, length, and surface functionality. An active area of research is the development of nontoxic, biodegradable, and biocompatible dendrimers. Various kinds of dendrimer-drug conjugates have been tried using polyamidoamine, polyphenylether, and polyester dendrimers and anticancer drugs such as doxorubicin, methotrexate, and 5-fluorouracil. The hydrophobic core and hydrophilic shell of dendrimers could encapsulate hydrophobic drugs or diagnostic agents in their interior and release them slowly. With an interior of basic environment, dendrimers could also encapsulate acidic drugs, such as methotrexate.

A unique characteristic of dendrimers is that they can act as a particulate system while retaining the properties of a polymer. Dendrimers can significantly improve pharmacokinetic and pharmacodynamic properties of low molecular weight and protein-based therapeutic agents. The surface modified dendrimers with various kinds of ligands, such as sugars, folate residues, and poly (ethylene glycol), could serve as a drug carrier with targeted cell specificity. Poly (ethylene glycol) modified (PEGylation) dendrimers can generally overcome clearance by RES system. Attachments of targeting moiety on the surface of partially PEGylated dendrimer raised the possibility of a delivery system that can cross biological barriers and deliver the bioactive agent near the vicinity of its target site. Recent successes also demonstrate potential of PEGylated dendrimers in magnetic resonance imaging contrast agent and in carbonyl metallo immunoassay (24, 25).

*9.3.2.4 Erythrocytes/Cells*   Erythrocytes, which are also known as red blood cells, have been studied extensively for their potential carrier capabilities for the delivery of drugs and drug-loaded microspheres. Such drug-loaded carrier erythrocytes are prepared simply by collecting blood samples from the organism of interest, separating the erythrocytes from the plasma, entrapping the drug into the erythrocytes, and resealing the resultant cellular carriers. Hence, these carriers are called resealed erythrocytes. The overall process is based on the response of these cells to osmotic conditions. On reinjection, the drug-loaded erythrocytes serve as slowly circulating depots and target the drugs to the RES system. The advantage of erythrocytes as drug carriers is mainly caused by their biocompatibility within the host, particularly when autologous cells are used, and hence triggering an immune response is not possible.

Several methods, that include physical (e.g., electrical-pulse method), osmosis-based systems, and chemical methods (e.g., chemical perturbation of the erythrocyte membrane), can be used to load drugs or other bioactive compounds into erythrocytes.

Resealed erythrocytes have several possible applications in various fields of human and veterinary medicine. Such cells could be used as circulating carriers to disseminate a drug over a prolonged period of time in circulation or in target-specific organs, including the liver, spleen, and lymph nodes. Most drug delivery studies that use drug-loaded erythrocytes are in the preclinical phase. In a few clinical studies, successful results have been obtained.

The ability of resealed erythrocytes to accumulate selectively within RES organs makes them a useful tool for the treatment of hepatic tumors and for the treatment of parasitic diseases. Moreover, the first report of successful clinical trials of the resealed erythrocytes loaded with enzymes for replacement therapy was that of (beta)-glucocerebrosidase in the treatment of Gaucher's disease, in which the disease is characterized by an inborn deficiency of lysosomal (beta)-glucocerebrosidase in RES cells. This deficiency thereby leads to an accumulation of (beta)-glucocerebrosides in macrophages of the RES. The same concept was also extended to the delivery of biopharmaceuticals that include therapeutically significant peptides and proteins, nucleic acid-based biologicals, antigens, and vaccines (26). In addition, the possibility of using carrier erythrocytes for selective drug targeting to differentiated macrophages increases the opportunities to treat intracellular pathogens and to develop new drugs. Moreover, the availability of an application that permits the encapsulation of drugs into autologous erythrocytes has made this technology available in many clinical settings and competitive with other drug delivery systems (27).

## 9.4   SUMMARY

During the past four decades, formulations that control the rate and the period of drug delivery (i.e., time-release medications) and that target specific areas of the body for treatment have become increasingly common and complex. Because of researchers' ever-evolving understanding of the human body and because of the explosion of new and potential treatments that result from discoveries of bioactive molecules and gene therapies, pharmaceutical research hangs on the precipice of yet another great advancement. However, this next leap poses questions and challenges not only for the development of new treatments but also for the mechanisms by which to administer them.

The current methods for drug delivery exhibit specific problems that scientists are attempting to address. Therefore, the goal of all sophisticated drug delivery systems is to deploy medications intact to specifically targeted parts of the body through a medium that can control the therapy's administration by means of either a physiological or a chemical trigger. To achieve this goal, researchers are turning to advances in the worlds of microtechnology and nanotechnology. During the past decade, systems that contain polymeric materials, such as polymeric microspheres and nanoparticles, have all been shown to be effective in enhancing drug targeting specificity, lowering systemic drug toxicity, improving treatment absorption rates, and providing protection for pharmaceuticals against biochemical degradation.

Peptides, proteins, and nucleotides or DNA fragments are the new generation of drugs. They are becoming attractive because of fast development within the biotechnology field. The administration of such molecules, however, may pose problems such as sensitivity of the molecules to certain temperatures, instability of the molecules at some physiological pH values, a short plasma half-life, and a high molecular dimension, which hinders diffusive transport and makes, at the moment, the parenteral route the only possible way of administering such molecules. Controlled drug delivery that uses the development of new administration routes could be the answer to the problems in administration of these molecules. The rationale of drug delivery is to change the pharmacokinetic and pharmacodynamic properties of drugs by controlling their absorption and distribution. Rate and time of drug release at absorption site could be programmed using a so-called delivery system. Various techniques, that include chemical technologies (such as pro-drugs), biological technologies, polymers, and lipids (such as liposomes) as well as the invention of new routes of administration, have been proposed to achieve controlled drug release. In fact, it could increase drug absorption and reduce the effects of the active ingredient in those regions not requiring therapy. Drug delivery systems that allow for an effective *in vivo* release of new molecules, such as recombinant anti-idiotypic antibodies with antibiotic activity devoted to the treatment of pulmonary (tuberculosis and pneumocystosis) and mucosal (candidiasis) diseases, fall into that category (28).

## REFERENCES

1. Dey S, Anand BS, Patel J, Mitra AK. Transporters/receptors in the anterior chamber: pathways to explore ocular drug delivery strategies. Expert Opin. Biol. Ther. 2003;3:23–44.

2. Gaillard PJ, Visser CC, De Boer AG. Targeted delivery across the blood-brain barrier. Expert Opin. Drug Deliv. 2005;2:299–309.

3. Kuhlmann J. The applications of biomarkers in early clinical drug development to improve decision-making processes. Pharma Center, Bayer Health Care AG, Wuppertal, Germany. Ernst Schering Research Foundation Workshop, Volume 59, 2007. pp. 29–45.

4. Attar M, Lee VHL. Pharmacogenomic considerations in drug delivery. Pharmacogenomics 2003;4:443–461.

5. Hoffman A, Stepensky D, Lavy E, Eyal S, Klausner E, Friedman M. Pharmacokinetic and pharmacodynamic aspects of gastroretentive dosage forms. Internat. J. Pharmaceut. 2004;277:141–153.

6. Breimer DD. An integrated pharmacokinetic and pharmacodynamic approach to controlled drug delivery. J. Drug Target. 1996;3:411–415.

7. Kwon GS, Furgeson DY. Biodegradable polymers for drug delivery systems. Biomed. Polym. 2007;83–110.

8. Kissel T, Li YX, Unger F. ABA-triblock copolymers from biodegradable polyester A-blocks and hydrophilic poly (ethylene oxide) B-blocks as a candidate for in situ forming hydrogel delivery systems for proteins. Adv. Drug Deliv. Rev. 2002;54:99–134.

9. Moore L, Chien WY. Transdermal drug delivery: a review of pharmaceutics, pharmacokinetics, and pharmacodynamics. Crit. Rev. Ther. Drug Carrier Sys. 1988;4:285–349.

10. Gill HS, Prausnitz MR. Coated microneedles for transdermal delivery. J. Control. Rel. 2007;117:227–237.

11. McKinnie J. Transdermal market: the future of transdermal drug delivery relies on active patch technology. Drug Deliv. Tech. 2006;6:54–58.

12. Dixit N, Bali V, Baboota S, Ahuja A, Ali J. Iontophoresis - an approach for controlled drug delivery: a review. Curr. Drug Deliv. 2007;4:1–10.

13. Hadgraft J, Lane ME. Passive transdermal drug delivery systems: recent considerations and advances. Am. J. Drug Deliv. 2006;4:153–160.

14. Allen TM, Cheng WWK, Hare JI, Laginha KM. Pharmacokinetics and pharmacodynamics of lipidic nano-particles in cancer. Anti-Cancer Agents Med. Chem. 2006;6:513–523.

15. Gabizon AA, Shmeeda H, Zalipsky S. Pros and cons of the liposome platform in cancer drug targeting. J. Liposome Res. 2006;16:175–183.

16. Gabizon AA. Applications of liposomal drug delivery systems to cancer therapy. Nanotech. Cancer Ther. 2007;595–611.

17. Zimmer A, Kreuter J. Microspheres and nanoparticles used in ocular delivery systems. Adv. Drug Deliv. Rev. 1995;16:61–73.

18. Vandervoort J, Ludwig A. Ocular drug delivery: nanomedicine applications. Nanomedicine 2007:(1):11–21.

19. Ozeki T, Okada H. Nanoparticle-containing microspheres for pulmonary drug delivery. Yakuzaigaku 2006;66:249–253.

20. Chow AHL, Tong HHY, Chattopadhyay P, Shekunov BY. Particle Engineering for Pulmonary Drug Delivery. Pharm. Res. 2007;24:411–437.

21. Emerich DF, Thanos CG. Targeted nanoparticle-based drug delivery and diagnosis. J. Drug Target. 2007;15:163–183.

22. Petrak K. Nanotechnology and site-targeted drug delivery. J. Biomat. Sci. 2006;17:1209–1219.

23. Cunin F, Li Y, Sailor MJ. Nanodesigned pore-containing systems for biosensing and controlled drug release. Bio. Biomed. Nanotech. 2006;3:213–222.

24. Bai SH, Thomas C, Rawat A, Ahsan F. Recent progress in dendrimer-based nanocarriers. Crit. Rev. Ther. Drug Carrier Sys. 2006;23:437–495.

25. Patri AK, Kukowska-Latallo JF, Baker Jr. Targeted drug delivery with dendrimers: comparison of the release kinetics of covalently conjugated drug and non-covalent drug inclusion complex. Adv. Drug Delivery Rev. 2005;57:2203–2214.

26. Hamidi M, Zarrin A, Foroozesh M, Mohammadi-Samani S. Applications of carrier erythrocytes in delivery of biopharmaceuticals. J. Control. Rel. 2007;118:145–160.

27. Rossi L, Serafini S, Pierige F, Antonelli A, Cerasi A, Fraternale A, Chiarantini L, Magnani M. Erythrocyte -based drug delivery. Exp. Opin. Drug Deliv. 2005;2:311–322.

28. Conti S, Polonelli L, Frazzi R, Artusi M, Bettini R, Cocconi D, Colombo P. Controlled delivery of biotechnological products. Curr. Pharm. Biotech. 2000;1:313–323.

## FURTHER READING

Katz B, Rosenberg A, Frishman WH. Controlled-release drug delivery systems in cardiovascular medicine. Am. Heart J. 1995;129:359–368.

Mayer PR. Pharmacodynamic and pharmacokinetic considerations in controlled drug delivery. Control. Drug Deliv. 1997; 589–595.

Suzuki H, Nakai D, Seita T, Sugiyama Y. Design of a drug delivery system for targeting based on pharmacokinetic consideration. Adv. Drug Deliv. Rev. 1996;19:335–357.

# 10

# PHARMACEUTICALS: NATURAL PRODUCTS AND NATURAL PRODUCT MODELS

SHEO B. SINGH

*Merck Research Laboratories, Rahway, New Jersey*

Natural products have played a vital role in the treatment of human ailments for thousands of years and continue to play a big role in the modern discovery of new agents for the treatment of diseases today. In certain therapeutic areas, natural products account for almost all key modern medicine used today. Many drugs are formulated and used directly as they are found in nature, some are derived directly from natural products by semisynthesis, and others are modeled after natural products. In this overview, examples of the pharmaceutically important natural products have been summarized.

## 10.1 INTRODUCTION

Natural product preparations have played a vital role in empiric treatment of ailments for thousands of years in many advanced civilizations and continue to play a significant role even today in various parts of the world. The use of plants and preparations derived from plants has been the basis for the sophisticated medical treatments in Chinese, Indian, and Egyptian civilizations for many thousands of years. These medical applications have been documented in the Chinese *Materia Medica* (1100 BC), in the Indian *Ayurveda* (1000 BC), and in

*Chemical Biology: Approaches to Drug Discovery and Development to Targeting Disease*, First Edition.
Edited by Natanya Civjan.
© 2012 John Wiley & Sons, Inc. Published 2012 by John Wiley & Sons, Inc.

Egyptian medicine as early as 2900 BC. Plant preparations continued to be the basis of medical treatments in the ancient Western world as well. This knowledge migrated through Greece to Western Europe, including England, during the ancient period and led to its formal codification in the United Kingdom and to the publication of the *London Pharmacopoeia* in 1618.

The isolation of strychnine (**1**), morphine (**2**), atropine (**3**), colchicine (**4**), and quinine (**5**) in the early 1800s from the commonly used plants and their use for the treatment of certain ailments might constitute the early idea of "pure" compounds as drugs. E. Merck isolated and commercialized morphine (**2**) as the first pure natural product for the treatment of pain (1–3). Preparations of the Willow tree have been used as a painkiller for a long period in traditional medicine. Isolation of salicylic acid (**6**) as the active component followed by acetylation produced the semisynthetic product called "Aspirin" (**7**) that was commercialized by Bayer in 1899 for the treatment of arthritis and pain (4).

Strychnine (**1**)     Morphine (**2**)     Atropine (**3**)

Colchicine (**4**)     Quinine (**5**)     Salicylic acid (**6**)     Aspirin (**7**)

The World Health Organization estimates that herbal and traditional medicines, derived mostly from plants, constitute primary health care for ~80% of the world population even today. The compounds produced by plants play significant roles in the treatment of diseases for the rest of the 20% of populations that are fortunate to use modern medicine. About 50% of the most prescribed drugs in the United States consist of natural products or their semisynthetic derivatives, or they were modeled after natural products. "Curare," the crude extract from the South American plant, *Chondodendron tomentosum*, and the derived purified compound tubocurarine, has been used as anesthetic in surgery until recently. Purified digitoxin, as well as the crude extracts that contain digitalis glycosides from foxglove plant, *Digitalis lanata*, is used as cardiotonic even today.

Drugs derived from microbial fermentations have played perhaps a bigger role in the modern drug discovery and have revolutionized the practice of medicine, which leads to saving human lives. Although the contribution of purified natural products as single agent drugs is significant in almost all therapies, their contribution in the treatment of bacterial infection is perhaps most critical (5). Natural products constitute drugs or leads to all but three classes of antibiotics. The discovery of microbial natural products-based antibiotics began with the serendipitous observation by Fleming in 1929 that bacterial growth was prevented by the growth of *Penicillium notatum*. Although this discovery was highly publicized and very important, it took over 10 years before the active material, penicillin, was purified and structurally elucidated by Florey and Chain in early 1940s. Subsequent commercialization was very quick, driven largely by the medical needs of World War II. Penicillin was one of the first broad-spectrum antibiotics that treated bacterial infection and saved millions of lives. Fleming, Florey, and Chain were awarded the Nobel Prize in 1945 for their efforts on penicillin. The success of penicillin led to unparalleled efforts by government, academia, and the pharmaceutical industry to focus drug-discovery efforts based on the newfound "microbial" sources for the discovery of natural products beyond plants. However, initial efforts were mostly focused on the discovery of antibiotic compounds from fermentations of a variety of microorganisms of not only fungal origin by also soil-dwelling prokaryotes (e.g., *Streptomyces* spp.), which led to the discovery by 1962 of almost all novel classes of antibiotic scaffolds that are being used today. The antibiotic discovery effort was performed largely by Fleming's method of detection of antibacterial activity on petri plates. Zones of inhibition of bacterial strains on agar plates were measured after applying whole broth or extracts obtained from microbial ferments (5). As newer biological assays and screening techniques became available in the 1960s, microbial sources, along with plant and marine sources, started to be used for screening against other therapeutic targets, which led to the discovery of leads and drugs in those areas. Examples of these discoveries will be discussed with the target areas. As time progressed, improved technologies in biology and chemistry helped with the popularization of natural products; natural product extracts became part of the screening resource in most large pharmaceutical houses from 1960 through the 1980s until their de-emphasis in the early 1990s. Therefore, natural product extracts became popular sources for the screening against purified enzymes and receptors, an occurrence that led to the identification of many nonantibiotic natural products that have revolutionized the practice of medicine, saved countless human lives, improved quality of life, and perhaps helped increase life expectancy for humans.

## 10.2   ANTIBACTERIAL AGENTS

Natural products contribute to over 80% of all antibiotics that are in clinical practice today. Natural products contribute to all but three classes of antibiotics (3).

   Penicillin was the first β-lactam and the first broad-spectrum antibiotic discovered that started the "Golden age" (1940–1962) of antibiotics. The structure of penicillin contains a thiazolidine ring that is fused to a β-lactam ring. The existence and stability of the β-lactam ring was highly controversial at the time despite the availability of a single crystal X-ray structure of one of the penicillins. Penicillin G (**8**) was the first penicillin that was clinically used. Penicillin

G was converted easily by either chemical or biochemical means to 6-amino-penicillanic acid (**9**), which became the lead for the semisynthetic modifications that led to the synthesis of various penicillin derivatives. Some early derivatives (e.g., amoxicillin **10**) are still in clinical use. Penicillins (general structure **11**) bind to penicillin-binding proteins and inhibit the bacterial cell wall. Penicillins became targets of β-lactamases that opened the β-lactam ring and abolished the antibacterial effectiveness of these compounds (5, 6).

Cephalosporin C (**12**), a second class of β-lactam antibiotics, was first discovered from *Cephalosporium acremonium*, isolated from a sewer outfall of Sardinia, Italy in 1948. Although Cephalosporin C was less active than penicillin G, it was less prone to β-lactamase action and therefore attracted a lot of attention that led to the development of five generations of orally active clinical agents (e.g., cephalexin, **13** and general structure, **14**) (5, 6).

| Penicillin G (**8**) | 6-Amino-penicillin acid (**9**) | Amoxicillin (**10**) | Penicillins (**11**) |

| Cephalosporin C (**12**) | Cephalexin (**13**) | Cephalosporins (**14**) |

Continued search for even better antibiotics led to the discovery of the highly potent and broadest-spectrum antibiotic thienamycin (**15**), the third class of the β-lactams, called carbapenems, in which the sulfur atom of the thiazolidine ring was replaced by a methylene group. Thienamycin was produced by *Streptomyces cattleya* (7). The primary amine group of thienamycin self-catalyzes the opening of the β-lactam ring, which leads to the concentration-dependent instability that poses a serious challenge for the fermentation-based production of the compound. The Merck group stabilized the compound by replacing the primary amine with an aminomethylidineamino group and synthesized imipenem (**16**) (8). They developed a highly efficient total synthesis that remains in commercial use today. Imipenem was approved for clinical use 23 years ago in 1985, but it remains one of the most important broad-spectrum hospital antibiotics in the market today. Like other β-lactams, several generations of carbapenems (general structure **17**) have been approved for clinical use in recent years (9).

As resistance to β-lactam antibiotics increased because of the expression of a variety of β-lactamases, many groups focused their efforts on discovering compounds that could be more reactive to β-lactamases without having significant intrinsic antibiotic activities of their own and pharmacokinetic properties that would be similar to β-lactam antibiotics. This focus led to the discoveries of clavulanic acid (**18**) and monobactam sulfazecins (**19**). Nature effectively stabilized the latter monobactam structure by the addition of a *N*-sulfamic acid. The β-lactamase inhibitor clavulanic acid was combined with amoxicillin, which led to the development of a potent and successful antibacterial agent, Augmentin® (GSK, Surrey, UK) (10). Chemical modifications of the monobactam produced aztreonam (**20**), a clinical agent with a narrow spectrum but significantly improved activity against Gram-negative pathogens, particularly *Pseudomonas aeruginosa* (11).

Thienamycin (**15**)          Imipenem (**16**)          Carbapenems (**17**)

Clavulanic acid (**18**)          sulfazecin (**19**)          Aztreonam (**20**)

Immediately after the discovery of penicillin, Waksman started efforts on soil-dwelling bacteria and discovered the first of the aminoglycosides, streptomycin (**21**) from *Streptomyces griseus*, in 1943 (6). Subsequently, a series of aminoglycosides was isolated. These aminoglycosides are potent broad-spectrum antibiotics and are potent inhibitors of protein synthesis. Unfortunately, nephrotoxicity limited their wider use, and they are used mainly for treatment of infections caused by Gram-negative bacteria. Continued efforts to screen prokaryotic organisms led to the discovery of the phenyl propanoids (chloramphenicol, **22**) and tetracyclines. The latter is a major class of tetracyclic polyketides that were discovered from various species of *Streptomyces* spp. Although the parent tetracycline (**23**) was not used as an antibiotic to a great extent, the chloro derivative (Chlortetracycline **24**), oxytetracycline (**25**), and minocycline (**26**) are clinical agents. This class of compound suffered from the selection for rapid resistance via efflux mechanism that limited their use (6). Recently, however, chemical modifications of the A-ring yielded compounds that overcame the efflux pump and lead to the development of tigecycline (**27**) as an effective broad spectrum antibiotic (12).

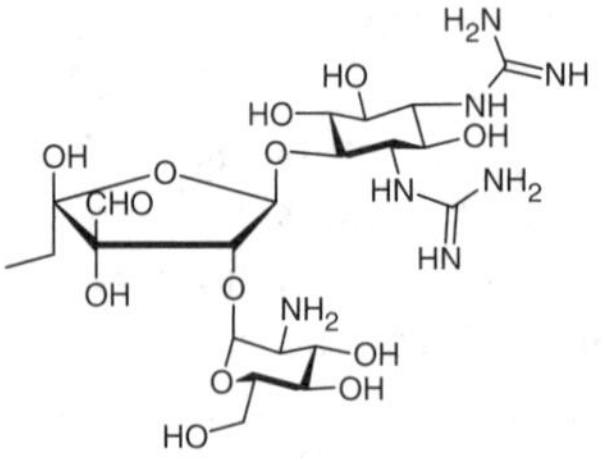

Streptomycin (**21**)

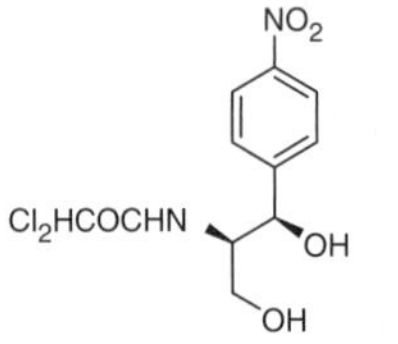

Chloramphenicol (**22**)

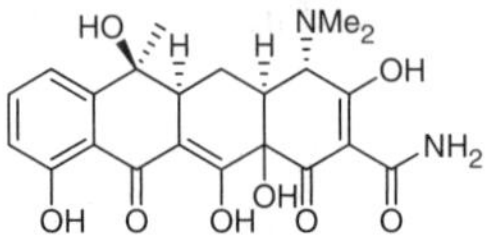

Tetracycline (**23**)

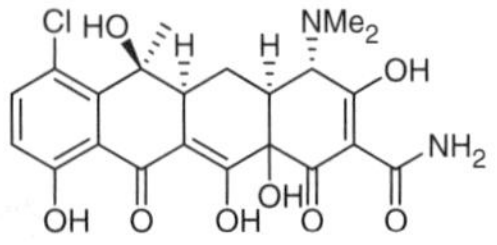

Chlortetracycline (**24**)

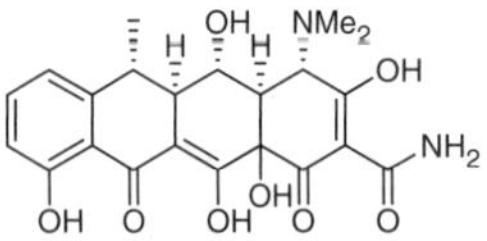

Doxycycline (**25**)

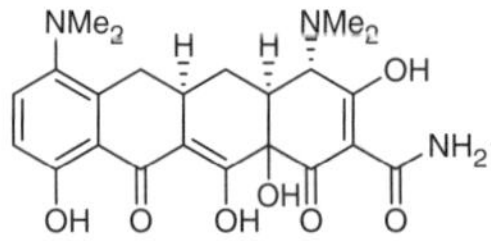

Minocycline (**26**)

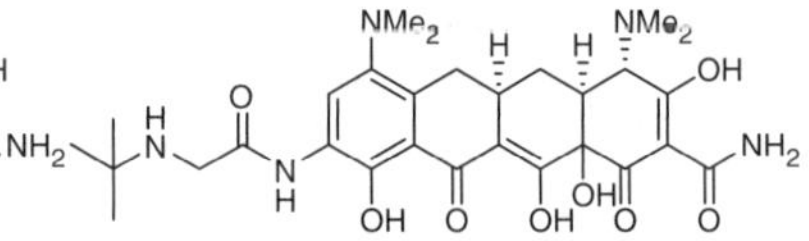

Tigecycline (**27**)

Another large class of orally active protein syntheis inhibitor antibiotics that were produced by *Streptomyces* spp. is represented by 14-membered lactones generically called macrolides, exemplified by the first member, erythromycin (**28**) (6). Chemical modifications of this class of compounds led to many clinical agents such as the aza derivatives, azithromycin (**29**) and ketolide (telithromycin, **30**) (13, 14). Mupirocin (pseudomonic acid, **31**) is another protein synthesis inhibitor that was isolated from *Pseudomonas fluorescens* and is used only as a topical agent (6).

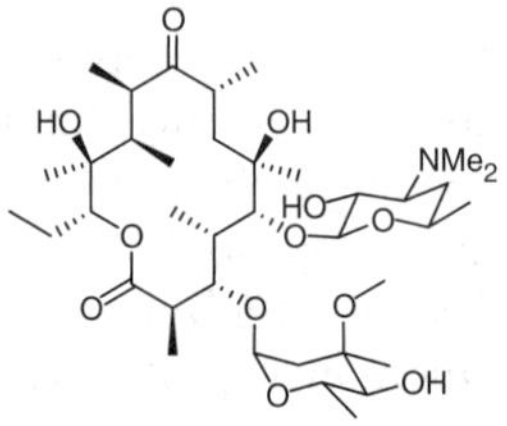

Erythromycin (**28**)

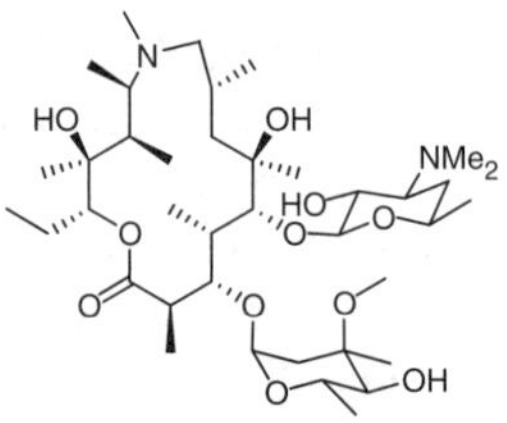

Azithromycin (**29**)

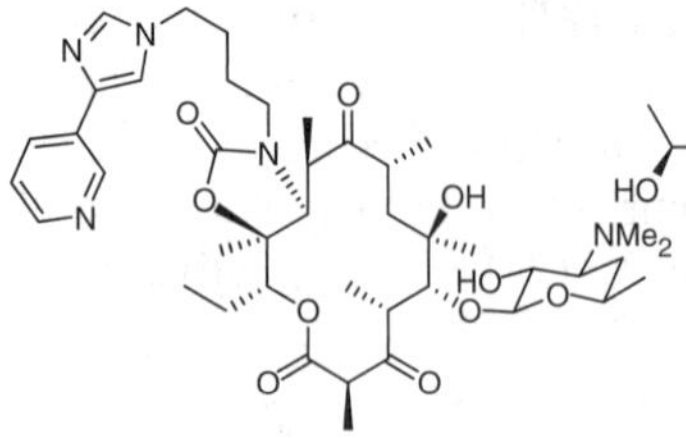

Telithromycin (**30**)

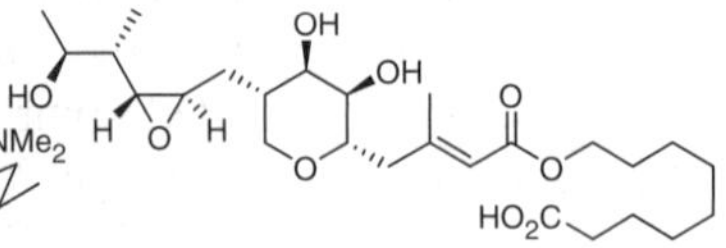

Mupirocin (**31**)

Vancomycin (**32**), a glycopeptide produced by *Streptomyces orientalis*, is a key Gram-positive antibiotic, originally discovered in 1954, and remains a critical antibiotic in clinical practice even today for the treatment of Gram-positive bacterial infections (6). Teicoplanin (**33**), a related glycopeptide produced by *Streptomyces teicomyceticus*, is a newer antibiotic that complements vancomycin in the clinic but is not effective against vancomycin-resistant bacteria. Ramoplanin (**34**) represents another glycopeptide that is larger in molecular size and structurally different from vancomycin and teicoplanin; it is in the late stages of clinical development for treatment of Gram-positive bacterial infections. Glycopeptides inhibit the bacterial cell wall. Daptomycin (**35**), a cyclic lipopeptide produced by *Streptomyces roseosporus*, is one of the newest members of antibiotics approved for the clinical practice as a broad-spectrum Gram-positive agent. It works by depolariztion of the bacterial cell membrane (14). Streptogramins were discovered in the early 1960s but were used for humans only recently when a 70/30 mixture of dalfopristin (**36**) and quinupristin (**37**) with the trade name Synercid® King Pharmaceuticals, Bristol, NJ; was developed for the treatment of drug-resistant Gram-positive bacterial infections (15).

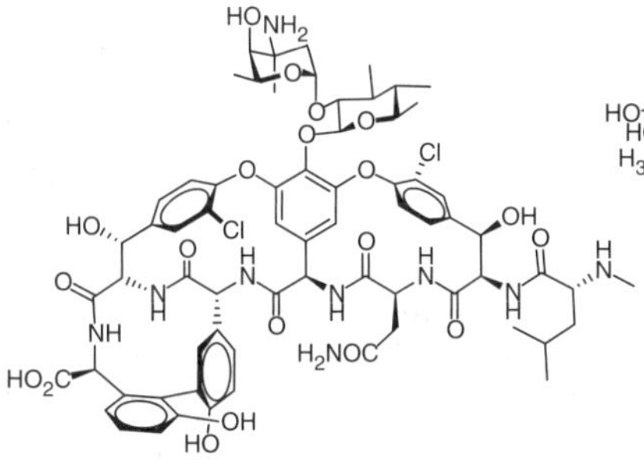

Vancomycin (**32**)

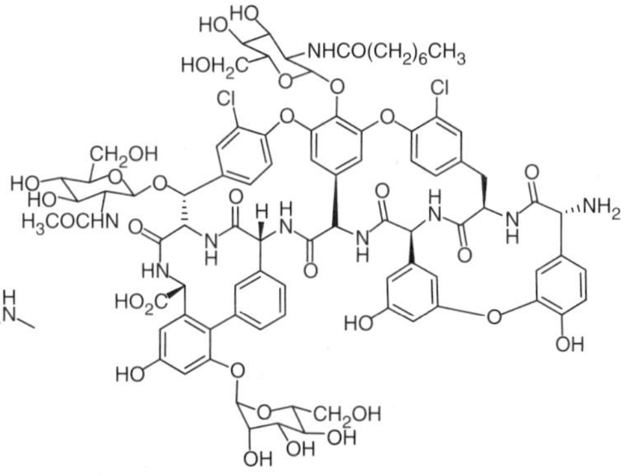

Teicoplaninn (**33**)

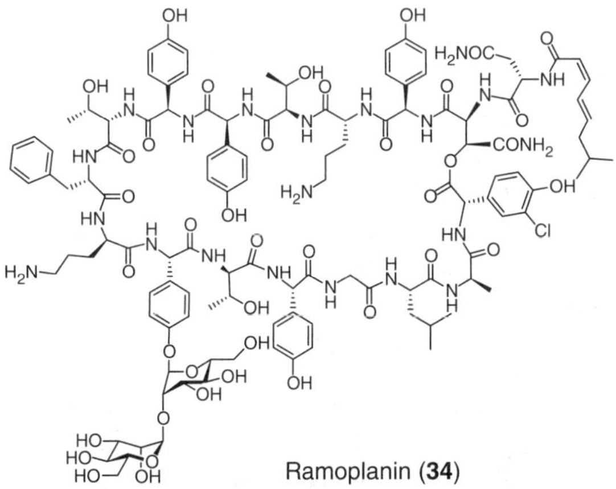

Ramoplanin (**34**)

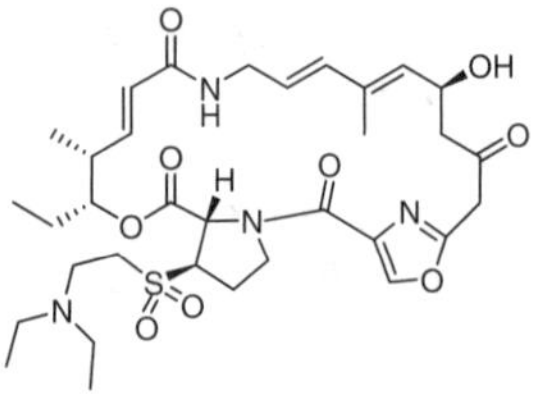

Dalfopristin (36)

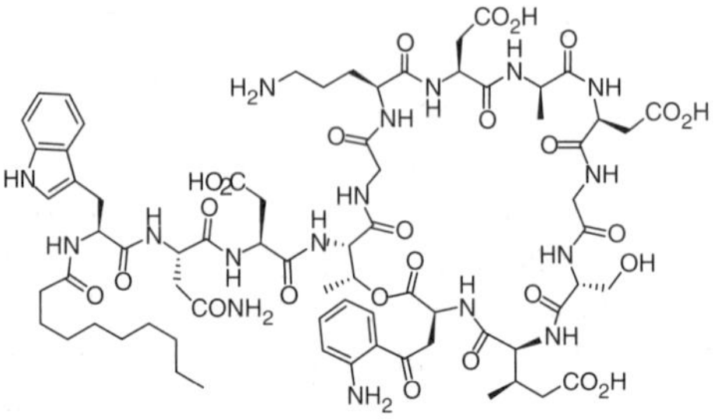

Daptomycin (35)

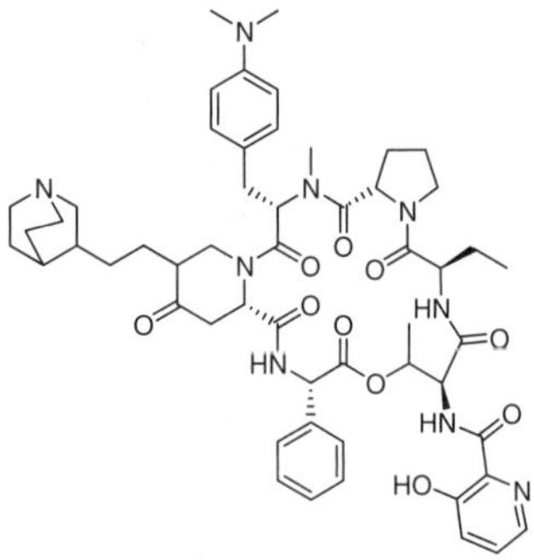

Quinupristin (37)

Antibiotics from natural sources range from compounds with small molecular size (e.g., thienamycin) to large peptides (e.g., ramoplanin). They generally possess complex architectural scaffolds and densely deployed functional groups, which affords the maximal number of interactions with molecular targets and often leads to exquisite selectivity for killing pathogens versus the host. This function is nicely illustrated by vancomycin binding to its target. Vancomycin has five hydrogen bond contacts with the D-Ala-D-Ala terminal end of peptidoglycan. Resistant organisms modify the terminal D-Ala with D-lactate, which leads to loss of one hydrogen bond and a 1000-fold drop in binding affinity and loss of antibiotic activity (see Fig. 10.1) (6).

## 10.3 ANTIFUNGAL AGENTS

Significant similarities in the fungal and mammalian cellular processes result in very few fungal-specific drug targets that lead to the development of only a few quality and safe antifungal agents. Amphotericin B (38), a natural product that consists of a polyene lactone, is a highly effective broad-spectrum antifungal agent unfortunately with a very limited safety margin. Glucan synthesis was identified as a fungal-specific target that could be inhibited by a series of cyclic peptides called echinocandins, which were identified in the 1970s as having potent antifungal activities (16, 17). This identification provided impetus for the discovery and development of new related lipopeptides that led to the identification of pneumocandins from *Glarea lozoyensis* (e.g., pneumocandin Bo, 39). Chemical

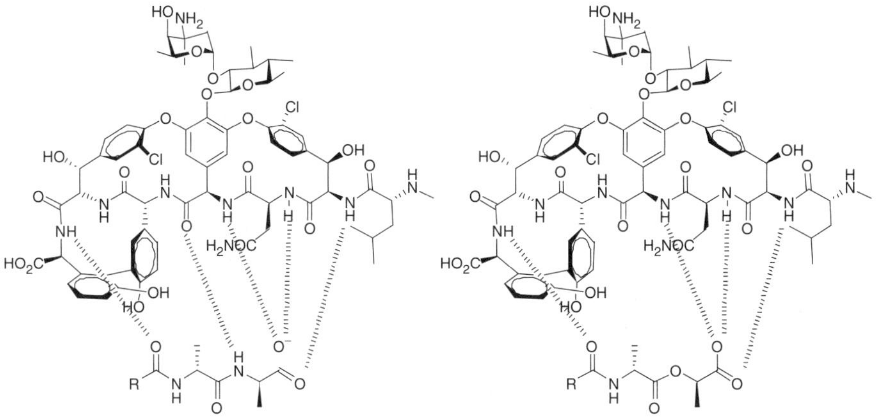

N-Acyl-D-Ala-D-Ala N-Acyl-D-Ala-D-Lactate

**Figure 10.1** Vancomycin binding to the active site D-Ala-D-Ala of peptidoglycan (left panel) and D-Ala-D-Lactate (right panel).

modifications at two sites of pneumocandin Bo led to the synthesis of caspofungin (**40**), which was the first in the class of glucan synthesis inhibitors; it is a "potent," highly effective, and safe antifungal agent approved for serious fungal infections in hospitals. Side chain replacements of the related cyclic peptide

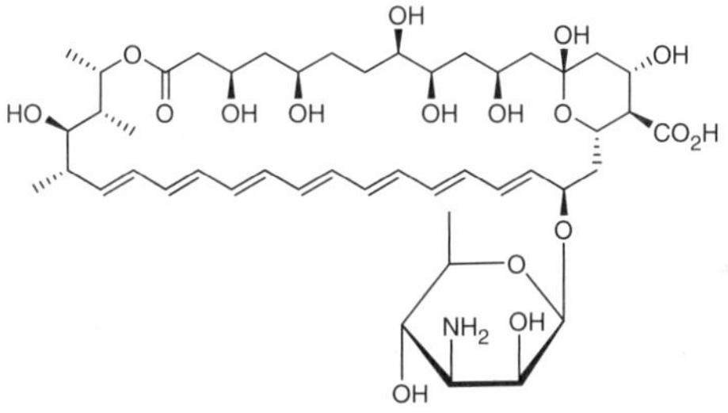

Amphotericin B (**38**)

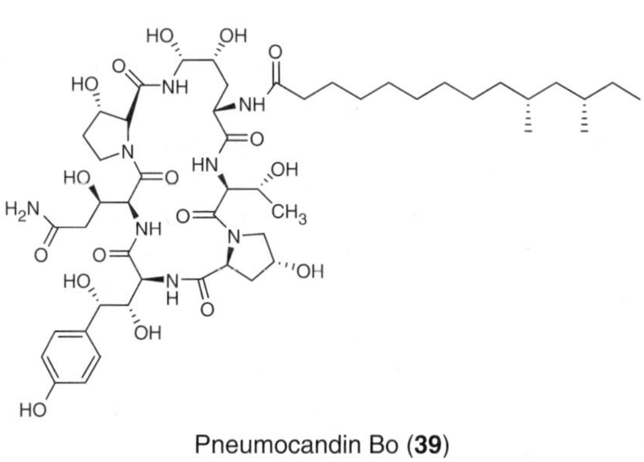

Pneumocandin Bo (**39**)

Caspofungin (**40**)

FR901379 (isolated from the fungus *Coleophoma empetri*) and echinocandin B (isolated from *Aspergillus nidulans*) led to two additional clinical agents of this class, micafungin (**41**) and anidulafungin (**42**), respectively. (16, 17)

Micafungin (**41**)                                             Anidulafungin (**42**)

## 10.4  ANTIMALARIALS

Quinine (**5**) isolated from Cinchona bark was one of the first antimalarials discovered, and it became a model for the discovery and development of some of the most successful antimalarial agents, chloroquine and its successors. However, the development of resistance by the malarial parasite *Plasmodium falciparum* for these drugs has rendered them ineffective. Artemisinin (**43**), a sesquiterpene peroxide originally isolated from a Chinese herb *Artemisia annua* in 1972 as an antimalarial agent, was chemically modified to a derivative, artemether (**44**), which is a very effective and widely used antimalarial agent (18). Unfortunately, limited supply of this plant-derived compound rendered it inaccessible for wider use. Biosynthetic genes of artimisinin have been identified and successfully transfected to an heterologous host, *Escherichia coli*. This method has allowed the production of an intermediate, amorphadiene (**45**) and artemisinic acid (**46**), which could be transformed chemically to artmether and potentially could relieve the strain of supply and could provide wider availability (19–21).

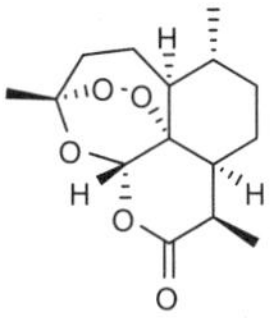

Artemisinin (**43**) 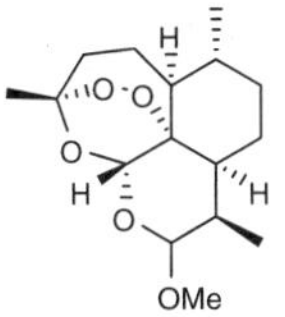　　Artemether (**44**) 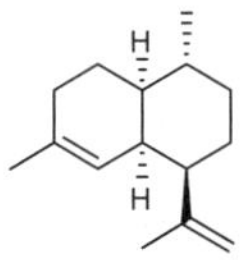　　Amorphadiene (**45**) 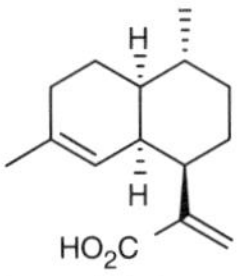　　Artemisinic acid (**46**)

## 10.5　ANTIVIRALS

Most antiviral agents are based on nucleoside structures and have their origin from spongouridine (**47**) and spongothymidine (**48**) that were isolated from marine sponges in the 1950s by Bergmann and his coworkers (22–24). These natural nucleosides possessed sugars other than ribose and deoxyribose and provided rationale for the substitution of the sugars in the antiviral nucleosides with various sugar mimics, including linear polar groups that led to the synthesis of Ara-A (**49**) and acyclovir (**50**). HIV protease inhibitors were developed from pepstatin (**51**), a pepsin inhibitor produced by various fungal species. Pepstatin possesses as the structural component statine a β-hydroxy-γ-amino acid that mimics the transition-state intermediate of the hydrolytic reaction catalyzed by the proteases (25). This structure became the foundation for the rational peptidomimetics and design of all HIV protease inhibitors, for example, indinavir (Crixivan®, Merck & Co., Inc., Whitehouse Station, NJ; **52**) and others (26).

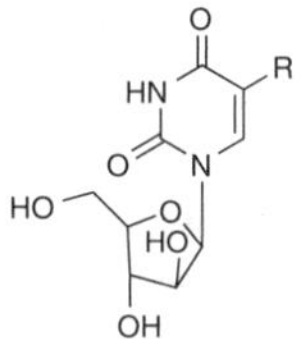

Spongouridine, R = H (**47**)　　Ara-A (**49**) 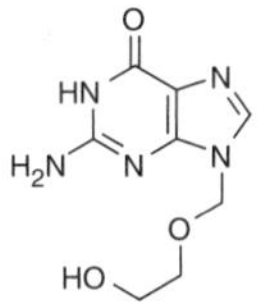　　Acyclovir (**50**)
Spongothymidine, R = Me (**48**)

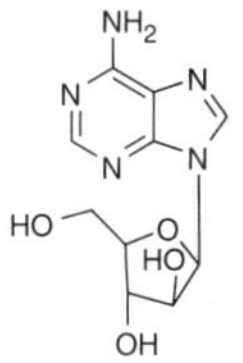

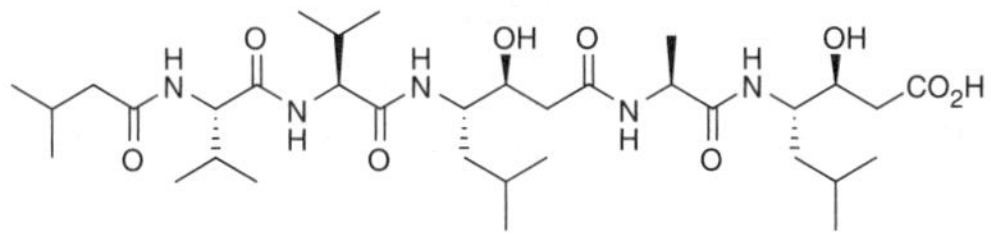

Pepstatin (**51**)

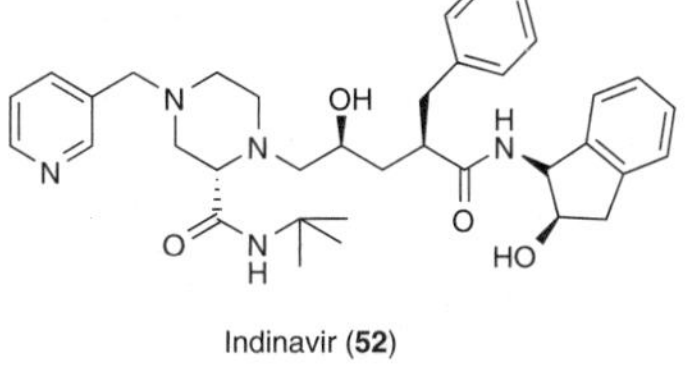

Indinavir (**52**)

## 10.6  ANTIPAIN AGENTS

Use of the opium poppy (*Papaver somniferum*) to ameliorate pain dates back thousands of years, and the active metabolite morphine (**2**) was isolated first from its extracts in 1806 followed by codeine (**53**) in 1832 (27, 28). Morphine and its derivatives are agonists of opiate receptors in the central nervous system and are some of the most effective pain relievers known and prescribed for postoperative pain. Morphine and codeine differ by substitution by methyl ether. Unfortunately, addictive properties of these compounds limit their use. Efforts have been made to reduce the addictive properties of morphine, which resulted in a semisynthetic derivative buprenorphine (**54**) (29). This compound is 25 to 50 times more potent than morphine with lower addictive potential and has been indicated for use by morphine addicts.

Codeine (53)                        Buprenorphine (54)

Conotoxins, a class of 10 to 35 amino acid-containing peptides produced by cone snails to intoxicate their prey, were isolated and characterized by Olivera and coworkers (30). They are a novel class of analgesics that helped identify the target and blocking of $N$-type $Ca^{+2}$ channels. One compound, Ziconotide (**55**), was synthesized and developed as a treatment for severe chronic pain (31, 32).

H2N-CKGKGAKCSRLMYDCCTGSCRSGKC-CONH2

Ziconotide (55)

The alkaloid epibatidine (**56**) was discovered from the skin of an Ecuadorian poison frog (*Epipedobates tricolor*), and its potent analgesic activity was demonstrated as early as in 1974 by Daly and coworkers (33–35). The paucity of the material delayed the structure elucidation and was only accomplished after the invention of newer and more sensitive NMR techniques in 1980s. Once the structure was elucidated as chloronicotine derivative, it was synthesized and was shown to antagonize nicotinic receptors in neurons. It did not show any specificity with similar receptors in other tissues and lacked a therapeutic index of any clinical value. However it served as a model for designing compounds with desired specificity and resulted in the synthesis of ABT594 (**57**), which is an agonist of the nicotinic acetylcholine receptor, is about 50-fold more potent than morphine without addictive properties, and was under advanced clinical development until its discontinuance in 2003 (36–39).

Epibatidine (**56**)     ABT594 (**57**)

## 10.7 ANTIOBESITY

Lipstatin (**58**), comprised of a 3,5-dihydroxy-2-hexyl hexadeca-7,9-dienoic acid with cyclization of the hydroxy group at C-3 with the carboxylic acid to form a β-lactone and esterification of the hydroxy group at C-5 with a $N$-formyl leucine, was isolated from *Streptomyces toxytricini*. It inhibits gastric and pancreatic lipase and blocks intestinal absorption of lipids (40, 41). Reduction of the olefins led to the synthesis of tetrahydrolipstatin, which was approved in 1999 as Xenical (**59**) for the treatment of obesity (42–44).

Lipstatin (**58**)     Orlistat (**59**)

## 10.8 ALZHEIMER'S AGENTS

Galantamine (**60**) is a tetracyclic alkaloid that was isolated originally from *Galanthus nivalis* and subsequently from *Narcissus* spp. (45). It was approved by trade name Reminyl® Johnson & Johnson, New Brunswick, NJ for the treatment of Alzheimer's disease. That galantamine, a selective inhibitor of acetylcholinesterase, was confirmed by X-ray structural characterization of galantamine bound to a plant acetylcholinesterase (46, 47).

Galantamine (**60**)

## 10.9 ANTINEOPLASTIC AGENTS

Natural products have played a much bigger role, perhaps second only to antibacterial agents, in the discovery and development of anticancer agents either directly as drugs or leads to drugs (48). In fact, they contribute 64% of all approved cancer drugs (3). Taxol® BMS, NY (**61**, paclitaxel) is unarguably the most successful anticancer drug in clinical use. It is a taxane diterpenoid that was isolated from the Pacific yew *Taxus brevifolia* in 1967 as a cytotoxic agent but not pursued as a development candidate until its novel mode of action was determined in 1979 as a stabilizer of microtubule assembly (49–52). The discovery of the mechanism of action led to the United States National Cancer Institute (NCI) committing significant resources to the large-scale production, eventual clinical development, and approval of Taxol® (paclitaxel) by U.S. Food and Drug Administration in 1992 for treatment of breast, lung, and ovarian cancer. Another important plant product, camptothecin (**62**), an alkaloid from *Camptotheca acuminata*, was discovered in 1966, also by the Wall and Wani group (53). The development of this compound was also hampered by the lack of knowledge of the mechanism of action and most importantly by its extremely poor water solubility. Determination of the mechanism of action as an inhibitor of topoisomerases-I led to significant efforts both by NCI and the pharmaceutical industry, which resulted in chemical modification of the structure, introduction of water-solubilizing groups (e.g., amino), and development of several derivatives as anticancer agents exemplified by topotecan (**63**) and irinotecan (**64**) (54, 55).

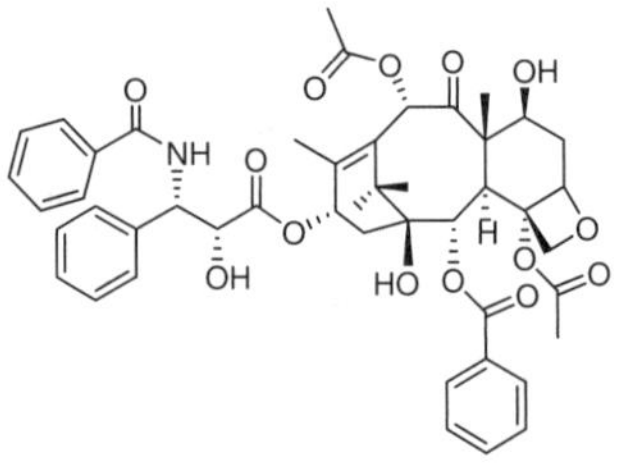

Taxol (**60**)

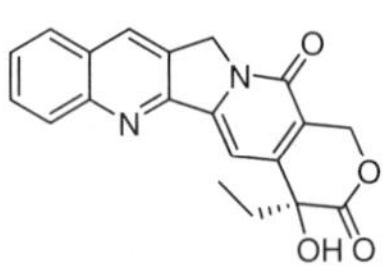

Camptothecin (**61**)

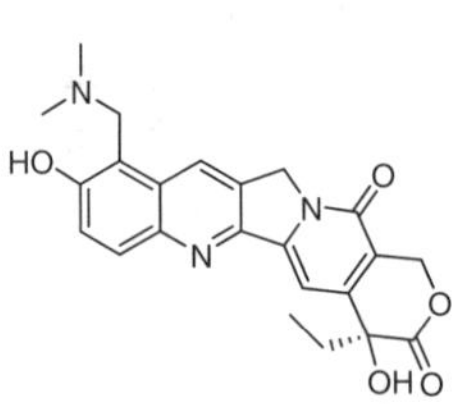

Topotecan (**62**)

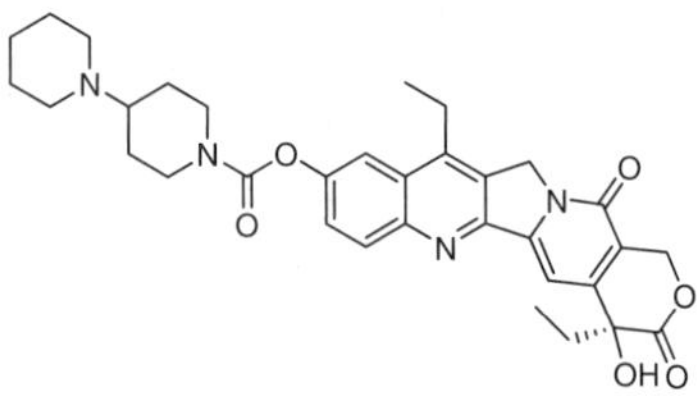

Irinotecan (**63**)

The Vinca alkaloids, vinblastine (**64**) and vincristine (**65**), isolated from *Catharanthus roseus*, have contributed significantly to the understanding and treatment of cancer (56, 57). These compounds bind to tubulin and inhibit cell division by inhibiting mitosis; they were perhaps the best-known anticancer agents before Taxol®. Chemical modifications of vinblastine led to the clinical agents vinorelbine (**66**) (58) and vindesine (**67**) (48 one, 59–61). Podophyllotoxin (**68**) was isolated from various species of the genus *Podophyllum* spp. as an anticancer agent. Chemical modifications of the naturally occurring epimer, epipodophyllo-toxin (**69**), led to the synthesis and development of etoposide (**70**) and teniposide (**71**) as clinical agents (48, 54, 62–65).

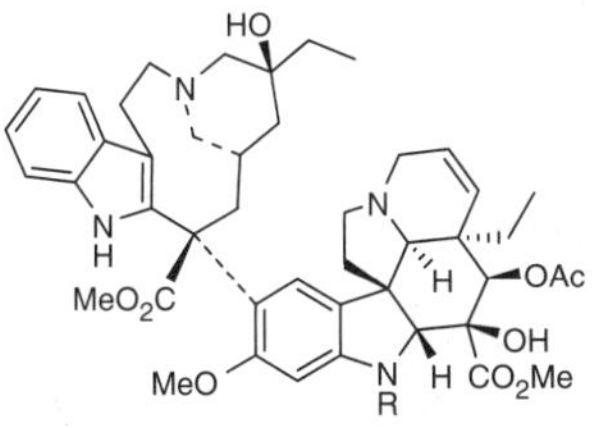

Vinblastine, R = Me(**64**)
Vincristine, R = CHO(**65**)

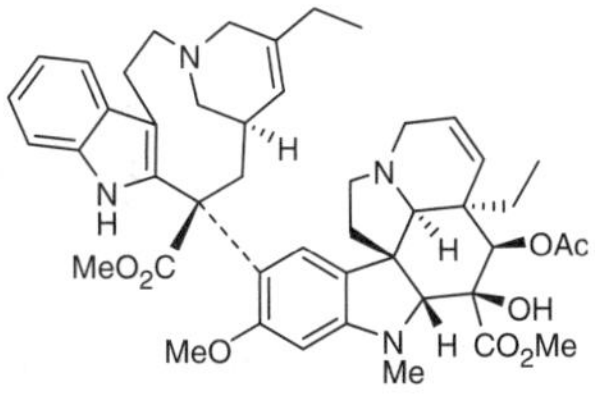

Vinorelbine (**66**)

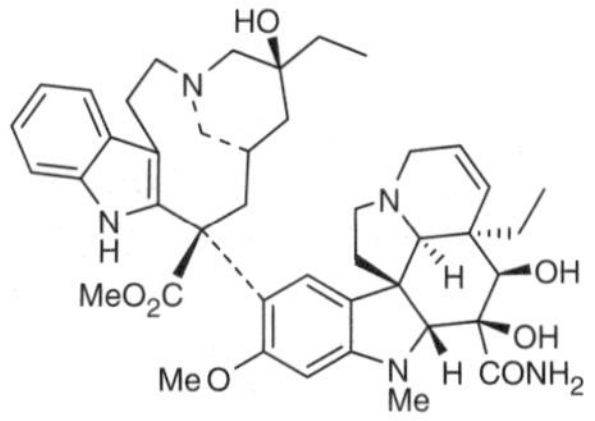

Vindesine (**67**)

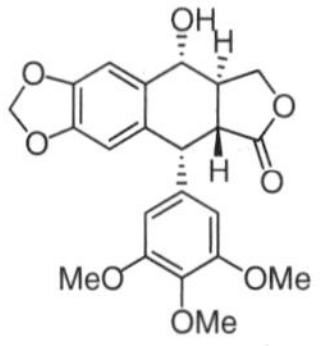

Podophyllotoxin (**68**) 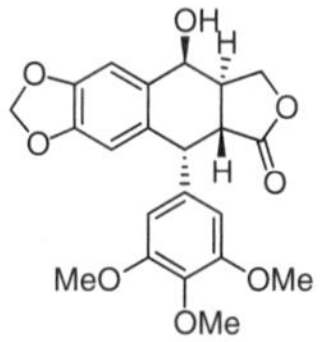     Epipodophyllotoxin (**69**)

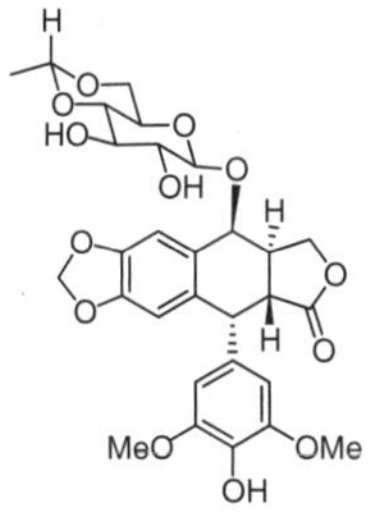

Etoposide(**70**) 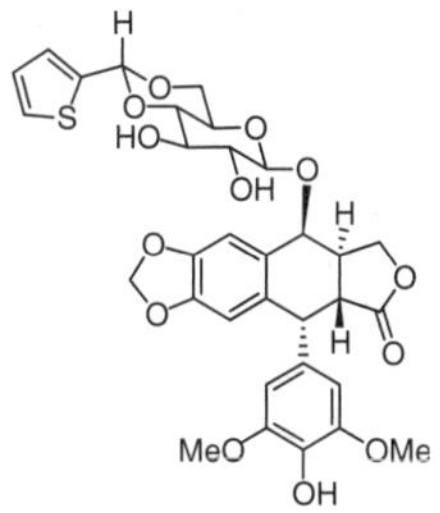     Teniposide (**71**)

Combretastatin A4 phosphate (**72**) is a phosphate prodrug of combretastatin A4, a *cis*-stilbene, isolated from *Combretum caffrum* (66). Combretastatin A4 is one of the many combretastatins that inhibits tubulin polymerization (67), shows efficacy against solid tumor, is a vascular targeting agent that blocks the blood supply to solid tumors, and is in Phase II/III clinical development for the treatment of various types of tumors as a vascular targeting agent (68–71).

Combretastatin A4
phosphate (**72**)

Microbial sources have been a very rich source for cancer chemotherapeutic agents. Of particular note is the *Streptomyces* spp., which has been responsible for the production of many approved anticancer agents that are in clinical practice. These agents are represented by highly diverse structural classes exemplified by the anthracycline family (e.g., doxorubicin, **73**) (72–74), actinomycin family (e.g., dactinomycin, **74**), glycopeptides family (e.g., bleomycins A2 and B2, **75** and **76**) (75), and mitomycin family (e.g., mitomycin C, **77**) (72, 76). All these compounds specifically interact with DNA for their mode of action.

Doxorubicin (**73**)                    Dactinomycin (**74**)

Mitomycin C (**77**)

Bleomycin A₂, R=CH₂CH₂CH₂S+(CH₃)₂ (**75**)
Bleomycin B₂, R=CH₂CH₂CH₂CH₂NHC(NH)NH₂, (**76**)

Staurosporine (**78**) produced by *Streptomyces* spp. is a potent inhibitor of protein kinase C (77–79). This compound inhibits many other kinases with almost equal potency and has become a great tool for the study of kinases. Lack of selectivity for protein kinase C has significantly hampered the development of this compound. Recently, however, several compounds derived from this lead have entered in the clinic for potential treatment of cancer. These include 7-deoxystaurosporine (**79**) and CGP41251 (**80**) (80, 81). CGP41251 shows multiple modes of action including inhibition of angiogenesis *in vivo*.

Staurosporine,R = H (**78**)              CGP41251 (**80**)
7-Hydrxystaurosporine, R =OH (**79**)

Microbial sources other than *Streptomyces* spp. have also provided highly interesting and structurally diverse compounds. Discovery of epothilones from myxobacterial strains by a German group (82) and the Merck group (83, 84) constitute a breakthrough discovery. The Merck group used an assay that mimicked Taxol® at the active site for the screening of natural products that led to the isolation of epothilones A (**81**) and B (**82**). The discovery of a unique structural class, interesting biological activity, and clinically proven mode of action drew significant attention from the scientific community and led to a variety of approaches, including combinatorial biosynthesis, chemical modifications, and total synthesis, that permitted preparation of many derivatives with improved potency and drug-like properties. A series of these compounds have entered human clinical trials, and many are in the late stages of development. Epothilone discovery and development has been reviewed (85).

Recent pursuit of marine microbial sources led to the isolation of salinosporamide A (**83**). It is a β-lactone produced by the marine bacteria *Salinispora tropica* and is a proteasome inhibitor (86). Mechanistically, it works by specific covalent modification of the target. This compound has entered human clinical development for treatment of multiple myeloma (87–89).

Epothilone A, R = H (**81**)              Salinosporamide A (**83**)
Epothilone B, R = Me (**82**)

Although use of marine microbial sources for the discovery of natural products is a somewhat recent phenomenon, marine natural products from higher species have contributed tremendously to the discovery of novel architecturally complex compounds as anticancer agent leads with one, Ecteinascidin-743, now approved

in the European Union for treatment of sarcoma. The discovery of natural products derived from marine sources exploded in the 1970s not only because of
increased level of NCI funding but also because of technological advancements
in the techniques for collection of specimens, chemical isolation, and structural
elucidation of low amounts of compounds initially isolated. Because of the fear
of a limited supply of marine sources for large-scale production, marine natural
products have remained the exclusive purview of academia except for a small
Spanish pharmaceutical company, PharmaMar, which collaborates closely with
academia and governments. Bryostatins are among the most interesting marine
natural products known. They were isolated from the bryozoan *Bugula neritina*.
They are a series of polyketide macro lactones represented here by the major
congener bryostatin I (**84**), which is a modulator of protein kinase C and has
been subjected to several human clinical trials (90, 91). At the time of writing,
combination studies have been recommended for Phase I and Phase II trials,
mainly under the auspices of the NCI. Modeling studies along with a diligent
chemical design approach has led to the synthesis of a simplified analog **85** that
has been shown to be equally active as bryostatin I in most *in vitro* studies (92,
93). This compound stands a better chance of being produced at larger scale by
total synthesis.

Bryostatin 1(**84**)                                    (**85**)

Dolastatins are a class of peptides comprised of mostly nonribosomal amino
acids. They were isolated from a sea hare *Dolabella auricularia* (94). Dolastatin-
10 (**86**) is one of the most potent and the best-studied members (95, 96). It exerts
it antitumor effect by inhibiting tubulin polymerization and binds at the vinca
alkaloid binding site (97, 98). Dolastatin-10 has been studied in Phase II human
clinical trials but was discontinued because of lack of efficacy (99). Auristatin PE
(**87**), a synthetic analog, seems to be more promising and, at the time of writing,
is being studied in Phase II human trials (100).

Dolastatin-10 (**86**)                                    Auristatin PE (**87**)

Discodermolide (**88**) was isolated from *Discodermia dissoluta* by using a P388 cell line toxicity bioassay; later it was determined that it stabilized microtubule assembly better than Taxol®, and it drew a lot of attention as an anticancer agent (101, 102). Its development, like that of many other complex marine natural products, was hampered because of the lack of ample supply of the material required for the clinical studies. In this case, the supply problem was overcome by the synthetic efforts of the Novartis process group. They synthesized it on a large enough scale to allow clinical studies. Unfortunately, its development seems to have been halted because of toxicity at Phase I (103, 104).

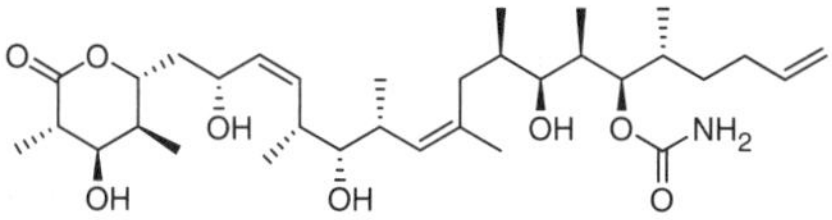

Discodermolide (**88**)

Several other novel, structurally and mechanistically diverse marine natural products have entered various preclinical and clinical studies. One of these products is ecteinascidin-743 (**89** ET-743), isolated from the tunicate *Ecteinscidea turbinata* (105, 106) and recently approved for the treatment of sarcoma in the European Union as the first "direct-from-the-sea" drug. Total synthesis and methods developed during the total synthesis allowed the preparation of a simpler analog phthalascidin (**90**) with comparable activities (107–109).

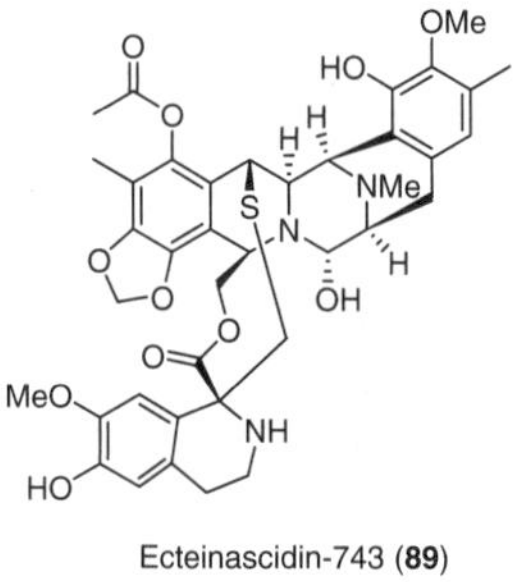

Ecteinascidin-743 (**89**)

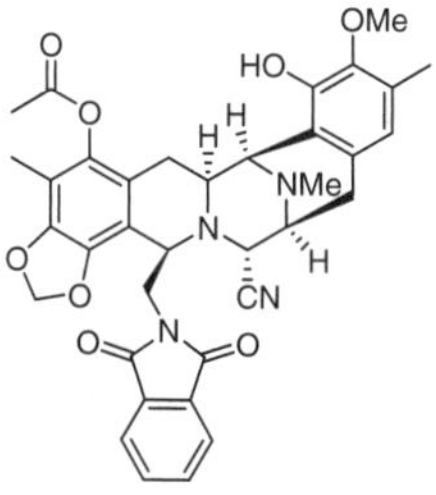

Phthalascidin (**90**)

Hemiasterlin (**91**), a tripeptide isolated from a sponge that was chemically modified to HTI-286 (**92**), which binds to the vinca binding site of tubulin, depolarizes microtubules; it entered into clinical development but was apparently dropped (110, 111).

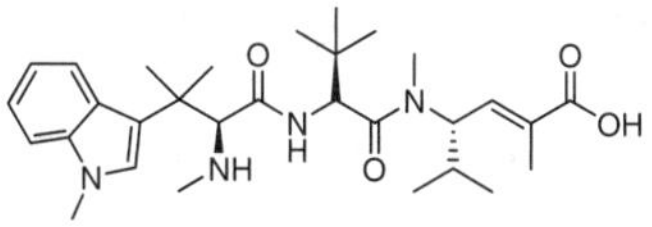

Hemiasterlin (**91**)

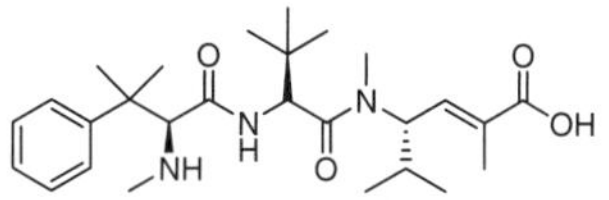

HTI-286 (**92**)

One difficulty with the cytotoxic agents that are used for the treatment of cancer is the differentiation of cytotoxicity between target tumor cells and normal cells. In an innovative approach, the Wyeth group took advantage of a tumor cell-specific drug delivery mechanism of antibodies. They conjugated calicheamicin (**93**), perhaps the best described member of the ene-diyne class of highly cytotoxic antitumor antibiotics produced by Actinomycetes, with recombinant humanized IgG$_4$ kappa antibody and developed Mylotarg$^®$ Wyeth, Madison, NJ (**94**), which binds to CD33 antigens expressed on the surface of leukemia blasts. Mylotarg$^®$ is an effective and less toxic treatment of myeloid leukemia (112–115).

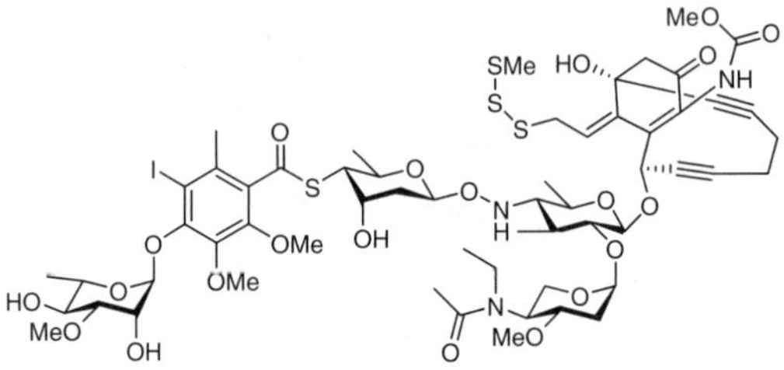

Calicheamicin (**93**)

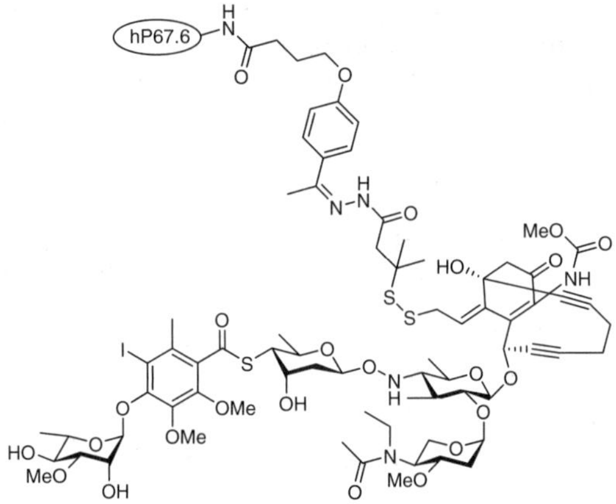

Mylotarg$^R$ (**94**)

## 10.10   IMMUNOSUPPRESSANT AGENTS

Natural products represent essentially all clinically used immunosuppressant agents. These agents collectively have made organ transplant possible. Cyclosporin (**95**) is an N-methyl cyclic peptide and originally was isolated from the fungus *Trichoderma polysporum* as an antifungal agent; almost immediately, the inhibition of T-cell proliferation and *in vivo* immunosuppressive properties were discovered and led to the development and approval of this molecule as a highly effective immunosuppressive agent (116–118). Natural products isolation of related compounds allowed the discovery of new congeners with reduced or no immunosuppressive activity in favor of antifungal and various other biological activities (e.g., antiparasitic activity) and asthma (119, 120). FK506 (**96**), a macrocyclic lactone, discovered from *Streptomyces tsukubaensis* as an immunosuppressive agent (121–125), was approved for clinical use for organ transplant as Tacrolimus® Astellas, Tokyo, Japan (126). Rapamycin (**97**), another macrocyclic lactone that is a very potent immunosuppressive agent, was approved as Sirolimus® Wyeth, Madison, NJ for clinical use for transplant rejection (127, 128). Rapamycin was isolated from *Streptomyces hygroscopicus* (129, 130). Mechanistically, all three of these compounds bind to their specific intracellular receptors, immunophilins, and the resulting complexes target the protein phosphatase, calcineurin (cyclosprin and FK506), and mammalian target of rapamycin (mTOR) to exert their immunosuppressive effects (131–133). Rapamycin and FK506 have played significant roles in studies of signal transduction and identification of various targets for other therapeutic applications, such as mTOR. Mycophenolic acid (**98**) originally was isolated from various species of *Penicillium*, and its antifungal activity has been known since 1932 (134). Mycophenolate was approved for acute rejection of kidney transplant (135, 136).

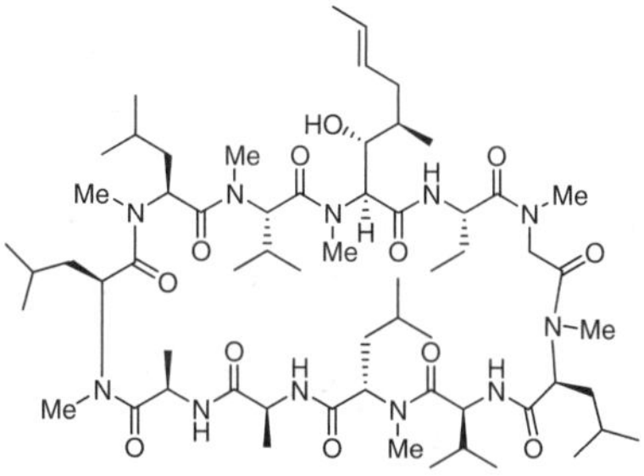

Cyclosporin A (**95**)

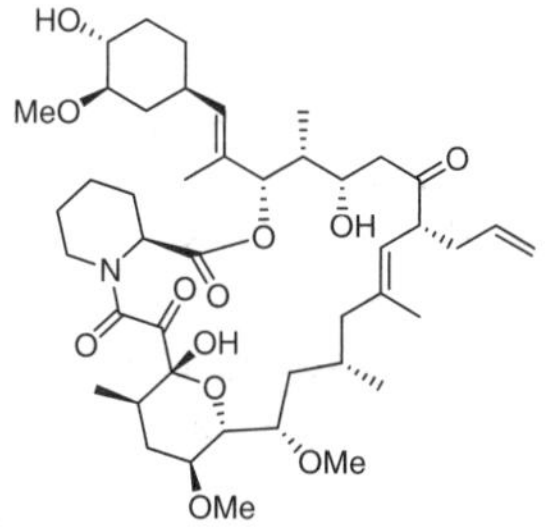

Tacrolimus^R (FK506) (**96**)

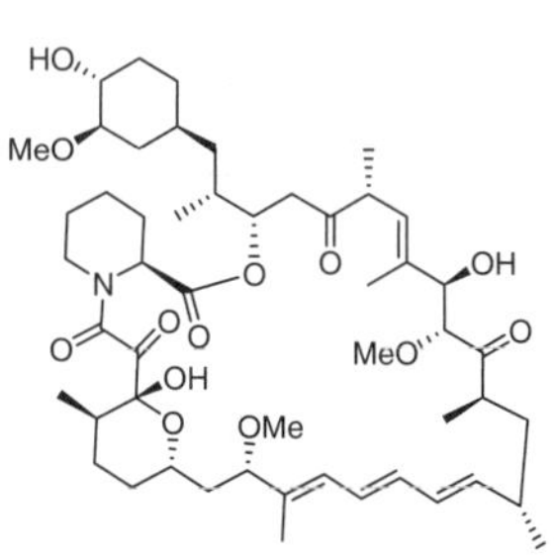

Rapamycin (**97**)

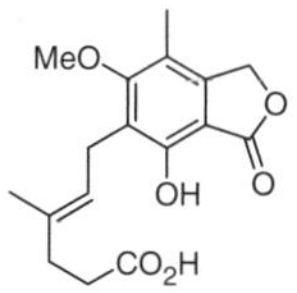

Mycophenolic acid (**98**)

## 10.11 CARDIOVASCULAR AGENTS

The biggest impact made by natural products in the treatment of cardiovascular diseases is undoubtedly associated with the discovery of the first of the HMG CoA reductase inhibitors by enzyme-based screening of microbial extracts that led to the isolation (from *Aspergillus terreus*) and characterization of mevinolin (lovastatin **99**), which is a homolog of compactin (**100**) that was discovered earlier (137, 138). These compounds possess a lipophilic hexahydrodecalin, a 2-methylbutanoate side chain, and a β-hydroxy-δ-lactone connected to the decalin unit with a two-carbon linker. These compounds are potent inhibitors of HMG CoA reductase, the rate-limiting enzyme of cholesterol biosynthesis, and inhibit the synthesis of cholesterol in the liver. Lovastatin (Mevacor®) Merck & Co., Inc., Whitehouse Station, NJ; was the first compound approved for lowering cholesterol in humans and became the cornerstone of all cholesterol-lowering agents generically called "statins." The modification of 2-methylbutanoate to 2,2-dimethylbutanoate led to the semisynthetic derivative, simvastatin (Zocor®, Merck & Co., Inc., Whitehouse Station, NJ; **101**), the second and more effective agent approved for human use (139). Hydroxylation of compactin by biotransformation led to pravastatin (Pravachol®, BMS, NY **102**) (139). The key pharmacophore of the statins is the β-hydroxy-δ-lactone or open acid. As the importance and value of cholesterol lowering to human pathophysiology became clearer, the search for additional cholesterol-lowering agents became more prominent and led to the discovery and development of several other clinical agents. All these compounds retained nature's gift of the pharmacophore, β-hydroxy-δ-lactone (or open acid), with replacement of the decalin unit of the natural products with a variety of aromatic lipophilic groups that resulted in fluvastin (**103**) (140), Atorvastatin (**104**) (141), Cerivastatin (**105**, withdrawn from the clinic) (142), Rosuvastatin (**106**) (143), and Pitavastatin (**107**). The statins have had tremendous impact in improvement of overall human health and quality of life because of the lowering of low-density lipoprotein (LDL) particles, which leads to a reduction in the incidence of coronary heart disease; arguably, they are the most successful class of medicines.

Lovastatin (**99**)    Mevastatin (**100**)    Simvastatin (**101**)    Pravastatin (**102**)

Fluvastin (**103**)    Atorvastatin (**104**)    Cerivastatin (**105**)    Rosuvastatin (**106**)

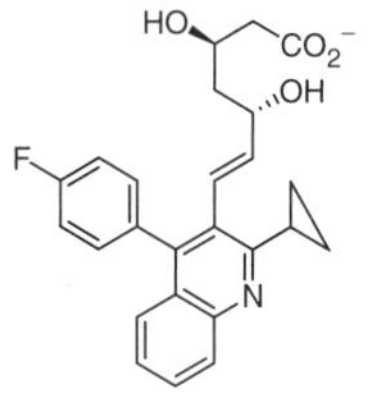

Pitavasatin (**107**)

Ephedrine (**108**), isolated from the Chinese plant *Ephedra sinaica*, was approved as one of the first bronchodilators and cardiovascular agents. This discovery led to a variety of such antihypertensive agents including β-blockers (4).

Angiotensin-converting enzyme (ACE) converts angiotensin I to angiotensin II, and its inhibition has led to several very successful, clinically useful antihypertensive agents. Although these inhibitors are of synthetic origin, the original lead was modeled after a nonapeptide, teprotide (**109**). This peptide was isolated from snake (viper, *Bothrops jararaca*) venom by Ondetti et al. It had antihypertensive activity in the clinic by parenteral administration (138, 144, 145) but was devoid of oral activity. Ondetti and coworkers worked diligently, and, recognizing that ACE was a metallo-enzyme, they visualized the binding of a smaller snake-venom peptide SQ20475 (**110**) with ACE; they modeled an acyl-proline with a sulfhydryl substitution at the zinc binding site, which led to the design and synthesis of captopril (**111**) as an orally active highly effective antihypertensive clinical agent. Additional application of the rational design by Patchett and coworkers led to the synthesis of enalapril (**112**) and other clinically relevant oral ACE inhibitors (138).

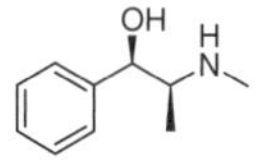

Ephedrine (**108**)

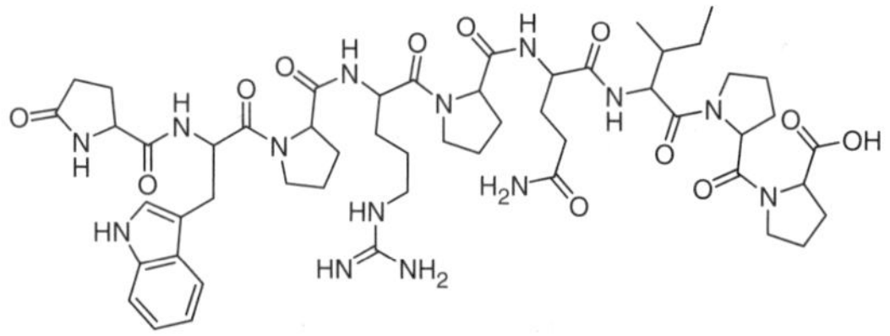

Teprotide (**109**)

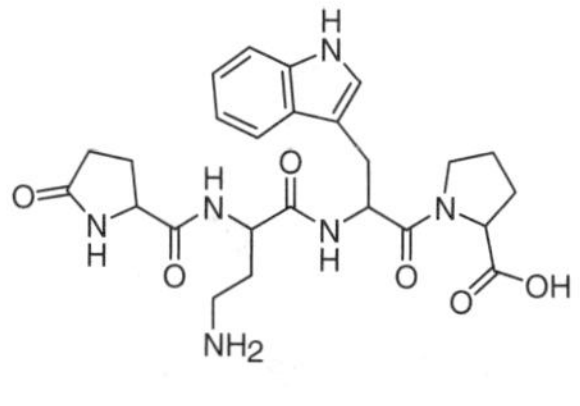

SQ 20575 (**110**)

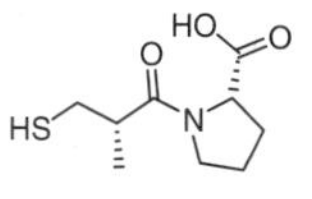

Captropril (**111**)

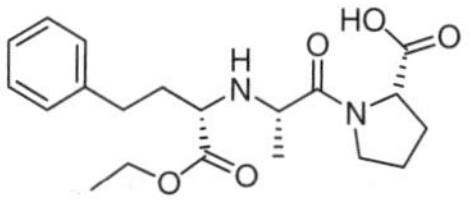

Enalapril (**112**)

## 10.12  ANTIPARASITIC AGENTS

Avermectins (**113**, **114**) are a series of macrocyclic lactones that are broad-spectrum, highly potent, glutamate-gated, chloride channel-modulator antiparasitic agents produced by *Streptomyces avermitilis* (146, 147). Ivermectin, 23,24-dihydroavermectin $B_{1a}/B_{1b}$ (**115**), was the first product approved in the mid-1980s for treatment of intestinal parasites in domesticated and farm animals, and it remains the standard of care (148, 149). The remarkable activity of ivermectin against *Onchocerca volvulus*, the causative parasitic agent of onchocerciasis (river blindness), led to clinical development and the approval of Mectizan® Merck & Co., Inc., Whitehouse Station, NJ; for the treatment of such diseases. These parasitic diseases have debilitated millions of people in many countries in Africa and South America. Because Mectizan® is a very effective treatment, Merck is providing this drug free of cost to all people in need as a part of the "Mectizan Donation Program," which has had tremendous impact on the health and quality of life of people affected by these diseases (150).

Spinosyns were discovered from the fermentation broth of *Saccharopolyspora spinosa* by screening for mortality of blowfly larvae, and a mixture of spinosyns A (**116**) and D (**117**) was approved and used successfully as a crop protection and an antiparasitic animal health agent. (151) Nodulisporic acids are an indole diterpenoid class discovered from various species of *Nodulisporium* as orally active antiflea and antitick agents for dogs and cats (152, 153). The most active of the series is nodulisporic acid A (**118**), which selectively modulates the activity of insect-specific glutamate-gated chloride channels (153).

Avemectin $B_{1a}$, R = $CH(CH_3)CH_2CH_3$, $\Delta^{23,24}$ (**113**)
Avermecin $B_{1b}$, R = $CH(CH_3)_2$, $\Delta^{23,24}$ (**114**)
Ivermectin R = $CH(CH_3)CH_2CH_3$ + $CH(CH_3)_2$, (**115**)

Spinosyn A, R = H (**116**)
Spinosyn D, R = Me (**117**)

Nodulisporic Acid A (**118**)

## 10.13   PHARMACEUTICAL MODELS

The roles played by natural products as models for design and development of pharmaceutical agents are too many to cover in this overview. A few examples are illustrated during the discussions of specific disease areas above. For example, a marine sponge-derived nucleoside was the precursor for various nucleoside-based antiviral agents, pepstatin for renin and HIV protease inhibitors, snake venom peptide for ACE inhibitors, lovastatin and compactin for all statins, and ephedrine for many painkillers and β-blockers. Below are a few critical examples that have played a big role in defining leads for some therapeutic areas but have not resulted in a drug yet.

Asperlicin (**119**) was isolated from *Aspergillus alliaceus* as a weak cholecystokinin A receptor (CCK-A) antagonist by using CCK receptor binding screening assays (154). It is a competitive antagonist of CCK-A (but not CCK-B) but did not have sufficient potency or oral activity to qualify as a drug candidate. In a remarkable strategy, medicinal chemists simplified the molecule to a benzodiazepine core of asperlicin, which led to the synthesis of potent, safe, and orally active analogs (**120** and **121**) with selectivity for either CCK-A (**120**) or CCK-B (**121**) receptors. They entered human clinical trials but were abandoned because of lack of efficacy (155, 156). The benzodiazepine scaffold was coined as "privileged structures" by Evans et al (155). This discovery is a beautiful demonstration of how a natural product became a model for the CCK program and played a pivotal role in defining the entire field (138).

Asperlicin (**119**)            MK-329(**120**)            L-365,260 (**121**)

The second example is apicidin (**122**), which is a cyclic tetrapeptide isolated from a fungus *Fusarium pallidoroseum* by using an empiric antiprotozoal screen (157, 158). It showed potent inhibition of apicomplexan protozoa including the malarial parasite *Plasmodium falciparum* and coccidiosis parasite *Eimeria* spp. It was effective *in vivo* against reducing malaria parasite infection in a mouse model (157) and exhibited strong activity against tumor cell lines (159, 160). Cyclic tetrapeptides with a terminal epoxy-ketone were known to be effective cytotoxic agents before the discovery of apicidin, but the pharmacophore was associated with the epoxy-ketone group (e.g., Trapoxin B, **123**) with covalent modification as a mode of action. Apicidin does not contain the epoxy-ketone but showed potent antitumor activity (161). The mode of action of apicidin was shown to be the inhibition of histone deacetylase (HDAC) (157). The amino-*oxo*-decanoic acid (L-Aoda) mimics the acetylated lysine residue and positions itself at the zinc-binding site of HDAC (162). Chemical modification of apicidin with retention of the ethyl or methyl ketone led to the synthesis of small dipeptides (e.g., **124**) that retained the HDAC and tumor cell line inhibitory activities with significant reduction of inhibition of normal cells (161).

Apicidin (**122**)          Trapoxin B (**123**)          **124**

In summary, nature has provided a great set of molecules with enormous chemical diversity that has contributed to the treatment of many human diseases. Nature continues to amaze us with novel chemical diversity with unimaginable biological activity and target specificity, as illustrated by the recent discovery of the fatty acid synthesis inhibitor antibiotic platensimycin (**125**) (163, 164) and platencin (**126**) (165, 166). The former shows exquisite selectivity for FabF, whereas the latter compound is a balanced inhibitor of both condensing enzymes, FabF and FabH.

Platensimycin (**125**)          Platencin (**126**)

# REFERENCES

1. Butler M. The role of natural product chemistry in drug discovery. J. Nat. Prod. 2004;67:2141–2153.

2. Clardy J, Walsh CT. Lessons from natural molecules. Nature 2004;432:829–836.

3. Newman DJ, Cragg GM, Snader KM. Natural products as sources of new drugs over the period 1981–2002. J. Nat. Prod. 2003;66:1022–1037.

4. Newman DJ, Cragg GM, Snader KM. The influence of natural products upon drug discovery. Nat. Prod. Rep. 2000;17:215–234.

5. Singh SB, Barrett JF. Empirical antibacterial drug discovery–foundation in natural products. Biochem. Pharmacol. 2006;71:1006–1015.

6. Walsh CT. Antibiotics: Actions, Origin, Resistance. 2003. ASM Press, Washington, DC.

7. Albers-Schoenberg G, Arison BH, Hensens OD, Hirshfield J, Hoogsteen K, Kaczka EA, Rhodes RE, Kahan JS, Kahan FM, et al. Structure and absolute configuration of thienamycin. J. Am. Chem. Soc. 1978;100:6491–6499.

8. Salzmann TN, Ratcliffe RW, Christensen BG, Bouffard FA. A stereocontrolled synthesis of+-thienamycin. J. Am. Chem. Soc. 1980;102:6161–6163.

9. Zhanel GG, Wiebe R, Dilay L, Thomson K, Rubinstein E, Hoban DJ, Noreddin AM, Karlowsky JA. Comparative review of the carbapenems. Drugs 2007;67:1027–1052.

10. White AR, Kaye C, Poupard J, Pypstra R, Woodnutt G, Wynne B. Augmentin amoxicillin/clavulanate in the treatment of community-acquired respiratory tract infection: a review of the continuing development of an innovative antimicrobial agent. J. Antimicrob. Chemother. 2004;53:3–20.

11. Guay DR, Koskoletos C. Aztreonam, a new monobactam antimicrobial. Clin. Pharm. 1985;4:516–526.

12. Rubinstein E, Vaughan D. Tigecycline: a novel glycylcycline. Drugs 2005;65: 1317–1336.

13. Yassin HM, Dever LL. Telithromycin: a new ketolide antimicrobial for treatment of respiratory tract infections. Expert. Opin. Investig. Drugs 2001;10:353–367.

14. Woodford N. Novel agents for the treatment of resistant Gram-positive infections. Expert. Opin. Investig. Drugs 2003;12:117–137.

15. Bonfiglio G, Furneri PM. Novel streptogramin antibiotics. Expert. Opin. Investig. Drugs 2001;10:185–198.

16. Georgopapadakou NH. Update on antifungals targeted to the cell wall: focus on beta-1,3-glucan synthase inhibitors. Expert. Opin. Investig. Drugs 2001;10:269–280.

17. Denning DW. Echinocandin antifungal drugs. Lancet 2003;362:1142–1151.

18. Vroman JA, Alvim-Gaston M, Avery MA. Current progress in the chemistry, medicinal chemistry and drug design of artemisinin based antimalarials. Curr. Pharm. Des. 1999;5:101–138.

19. Ro DK, Paradise EM, Ouellet M, Fisher KJ, Newman KL, Ndungu JM, Ho KA, Eachus RA, Ham TS, Kirby J, Chang MC, Withers ST, Shiba Y, Sarpong R, Keasling JD. Production of the antimalarial drug precursor artemisinic acid in engineered yeast. Nature 2006;440:940–943.

20. Withers ST, Keasling JD. Biosynthesis and engineering of isoprenoid small molecules. Appl. Microbiol. Biotechnol. 2007;73:980–990.

21. Chang MC, Eachus RA, Trieu W, Ro DK, Keasling JD. Engineering Escherichia coli for production of functionalized terpenoids using plant P450s. Nat. Chem. Biol. 2007;3:274–277.

22. Bergmann W, Feeney RJ. Contributions to the study of marine products. XXXII. The nucleosides of sponges I. J. Org. Chem. 1951;16:981–987.

23. Bergmann W, Feeney RJ. The isolation of a new thymine pentoside from sponges. J. Am. Chem. Soc. 1950;72:2809–2810.

24. Bergmann W, Burke DC. Contributions to the Study of Marine Products. XL. The nucleosides of sponges. IV. Spongosine. J. Org. Chem. 1956;21:226–228.

25. Wiley RA, Rich DH. Peptidomimetics derived from natural products. Med. Res. Rev. 1993;13:327–384.

26. Dorsey BD, McDonough C, McDaniel SL, Levin RB, Newton CL, Hoffman JM, Darke PL, Zugay-Murphy JA, Emini EA, Schleif WA, Olsen DB, Stahlhut MW, Rutkowski CA, Kuo LC, Lin JH, Chen IW, Michelson SR, Holloway MK, Huff JR, Vacca JP. Identification of MK-944a: a second clinical candidate from the hydroxylaminepentanamide isostere series of HIV protease inhibitors. J. Med. Chem. 2000;43:3386–3399.

27. Terr CE, Pellens M. The Opium Problem. 1928. Bureau of Social Hygeine, New York:

28. Eddy NB, Friebel H, Hahn KJ, Halbach H. Codeine and its alternates for pain and cough relief. 2. Alternates for pain relief. Bull. World Health Organ. 1969;40:1–53.

29. Heel RC, Brogden RN, Speight TM, Avery GS. Buprenorphine: a review of its pharmacological properties and therapeutic efficacy. Drugs 1979;17:81–110.

30. Walker CS, Steel D, Jacobsen RB, Lirazan MB, Cruz LJ, Hooper D, Shetty R, DelaCruz RC, Nielsen JS, Zhou LM, Bandyopadhyay P, Craig AG, Olivera BM. The T-superfamily of conotoxins. J. Biol. Chem. 1999;274:30664–30671.

31. Heading CE. Ziconotide Elan Pharmaceuticals. IDrugs 2001;4:339–350.

32. Miljanich GP. Ziconotide: neuronal calcium channel blocker for treating severe chronic pain. Curr. Med. Chem. 2004;11:3029–3040.

33. Spande TF, Garraffo HM, Edwards MW, Yeh HJC, Pannell L, Daly JW. Epibatidine: a novel chloropyridylazabicycloheptane with potent analgesic activity from an Ecuadoran poison frog. J. Am. Chem. Soc. 1992;114:3475–3478.

34. Daly JW. The chemistry of poisons in amphibian skin. Proc. Natl. Acad. Sci. U.S.A. 1995;92:9–13.

35. Badio B, Daly JW. Epibatidine, a potent analgetic and nicotinic agonist. Mol. Pharmacol. 1994;45:563–9.

36. Holladay MW, Bai H, Li Y, Lin NH, Daanen JF, Ryther KB, Wasicak JT, Kincaid JF, He Y, Hettinger AM, Huang P, Anderson DJ, Bannon AW, Buckley MJ, Campbell JE, Donnelly-Roberts DL, Gunther KL, Kim DJ, Kuntzweiler TA, Sullivan JP, Decker MW, Arneric SP. Structure-activity studies related to ABT-594, a potent nonopioid analgesic agent: effect of pyridine and azetidine ring substitutions on nicotinic acetylcholine receptor binding affinity and analgesic activity in mice. Bioorg. Med. Chem. Lett. 1998;8:2797–2802.

37. Decker MW, Rueter LE, Bitner RS. Nicotinic acetylcholine receptor agonists: a potential new class of analgesics. Curr. Top. Med. Chem. 2004;4:369–384.

38. Decker MW, Meyer MD, Sullivan JP. The therapeutic potential of nicotinic acetylcholine receptor agonists for pain control. Expert. Opin. Investig. Drugs 2001;10:1819–1830.

39. Decker MW, Meyer MD. Therapeutic potential of neuronal nicotinic acetylcholine receptor agonists as novel analgesics. Biochem. Pharmacol. 1999;58:917–923.

40. Hochuli E, Kupfer E, Maurer R, Meister W, Mercadal Y, Schmidt K. Lipstatin, an inhibitor of pancreatic lipase, produced by Streptomyces toxytricini. II. Chemistry and structure elucidation. J Antibiot. (Tokyo) 1987;40:1086–1091.

41. Weibel EK, Hadvary P, Hochuli E, Kupfer E, Lengsfeld H. Lipstatin an inhibitor of pancreatic lipase, produced by Streptomyces toxytricini. I. Producing organism, fermentation, isolation and biological activity. J. Antibiot. (Tokyo) 1987;40:1081–1085.

42. McNeely W, Benfield P. Orlistat. Drugs 1998;56:241–9;250.

43. Mancino JM. Orlistat: Current issues for patients with type 2 diabetes. Curr. Diab. Rep. 2006;6:389–394.

44. Henness S, Perry CM. Orlistat: a review of its use in the management of obesity. Drugs 2006;66:1625–1656.

45. Miyazaki Y, Godaishi K. Experimental Cultivation of the Plants Containing Galanthamine at Izu. 1 General Growth of Shokiran Lycoris Aurea Herb., Natsuzuisen L. Squamigera Maxim., Snowflake Leucojum Aestivum L., and Snowdrop Galanthus Nivalis L., 1961 to 1962. Eisei. Shikenjo. Hokoku. 1963;81:172–176.

46. Thomsen T, Kewitz H. Selective inhibition of human acetylcholinesterase by galanthamine in vitro and in vivo. Life Sci. 1990;46:1553–1558.

47. Greenblatt HM, Kryger G, Lewis T, Silman I, Sussman JL. Structure of acetylcholinesterase complexed with –galanthamine at 2.3 A resolution. FEBS Lett. 1999;463:321–326.

48. Lee KH. Novel antitumor agents from higher plants. Med. Res. Rev. 1999;19:569–596.

49. Wani MC, Taylor HL, Wall ME, Coggon P, McPhail AT. Plant antitumor agents. VI. The isolation and structure of taxol, a novel antileukemic and antitumor agent from Taxus brevifolia. J. Am. Chem. Soc. 1971;93:2325–2327.

50. Schiff PB, Horwitz SB. Taxol assembles tubulin in the absence of exogenous guanosine 5'-triphosphate or microtubule-associated proteins. Biochemistry 1981;20:3247–3252.

51. Schiff PB, Horwitz SB. Taxol stabilizes microtubules in mouse fibroblast cells. Proc. Natl. Acad. Sci. U.S.A. 1980;77:1561–1565.

52. Schiff PB, Fant J, Horwitz SB. Promotion of microtubule assembly in vitro by taxol. Nature 1979;277:665–667.

53. Wall ME, Wani MC, Cook CE, Palmer KH, McPhail AT, Sim GA. Plant Antitumor Agents. I. The Isolation and Structure of Camptothecin, a Novel Alkaloidal Leukemia and Tumor Inhibitor from Camptotheca acuminata. J. Am. Chem. Soc. 1966;88:3888–3890.

54. Srivastava V, Negi AS, Kumar JK, Gupta MM, Khanuja SP. Plant-based anticancer molecules: a chemical and biological profile of some important leads. Bioorg. Med. Chem. 2005;13:5892–5908.

55. Sandler A. Irinotecan therapy for small-cell lung cancer. Oncology 2002;16:419–438.

56. Neuss N, Gorman M, Svoboda GH, Maciak G, Beer CT. Vinca alkaloids. III. Characterization of leurosine and vindaleukoblastine, new alkaloids from Vinca rosea Linn. J. Am. Chem. Soc. 1959;81:4754–4755.

57. Moncrief JW, Lipscomb WN. Structures of leurocristine vincristine and vincaleukoblastine.1 X-Ray analysis of leurocristine methiodide. J. Am. Chem. Soc. 1965;87:4963–4964.

58. Hochster HS, Vogel CL, Burman SL, White R. Activity and safety of vinorelbine combined with doxorubicin or fluorouracil as first-line therapy in advanced breast cancer: a stratified phase II study. Oncologist 2001 6:269–277.

59. Barnett CJ, Cullinan GJ, Gerzon K, Hoying RC, Jones WE, Newlon WM, Poore GA, Robison RL, Sweeney MJ, et al. Structure-activity relationships of dimeric Catharanthus alkaloids. 1. Deacetyl vinblastine amide vindesine sulfate. J. Med. Chem. 1978;21:88–96.

60. Sorensen JB, Hansen HH. Is there a role for vindesine in the treatment of non-small cell lung cancer? Invest. New Drugs 1993;11:103–133.

61. Sorensen JB, Osterlind K, Hansen HH. Vinca alkaloids in the treatment of non-small cell lung cancer. Cancer Treat. Rev. 1987;14:29–51.

62. Ruckdeschel JC. Etoposide in the management of non-small cell lung cancer. Cancer 1991;67:2533.

63. Gordaliza M, Castro MA, del Corral JM, Feliciano AS. Antitumor properties of podophyllotoxin and related compounds. Curr. Pharm. Des. 2000;6:1811–1839.

64. Baldwin EL, Osheroff N. Etoposide, topoisomerase II and cancer. Curr. Med. Chem. Anticancer Agents 2005;5:363–372.

65. You Y. Podophyllotoxin derivatives: current synthetic approaches for new anticancer agents. Curr. Pharm. Des. 2005;11:1695–1717.

66. Pettit GR, Singh SB, Hamel E, Lin CM, Alberts DS, Garcia-Kendall D. Isolation and structure of the strong cell growth and tubulin inhibitor combretastatin A-4. Experientia 1989;45:209–211.

67. Lin CM, Singh SB, Chu PS, Dempcy RO, Schmidt JM, Pettit GR, Hamel E. Interactions of tubulin with potent natural and synthetic analogs of the antimitotic agent combretastatin: a structure-activity study. Mol. Pharmacol. 1988;34:200–208.

68. Kirwan IG, Loadman PM, Swaine DJ, Anthoney DA, Pettit GR, Lippert JW, 3rd, Shnyder SD, Cooper PA, Bibby MC. Comparative preclinical pharmacokinetic and metabolic studies of the combretastatin prodrugs combretastatin A4 phosphate and A1 phosphate. Clin. Cancer Res. 2004;10:1446–1453.

69. Nabha SM, Mohammad RM, Dandashi MH, Coupaye-Gerard B, Aboukameel A, Pettit GR, Al-Katib AM. Combretastatin-A4 prodrug induces mitotic catastrophe in chronic lymphocytic leukemia cell line independent of caspase activation and polyADP-ribose polymerase cleavage. Clin. Cancer Res. 2002;8:2735–2741.

70. Chaplin DJ, Pettit GR, Hill SA. Anti-vascular approaches to solid tumour therapy: evaluation of combretastatin A4 phosphate. Anticancer Res. 1999;19:189–195.

71. Dorr RT, Dvorakova K, Snead K, Alberts DS, Salmon SE, Pettit GR. Antitumor activity of combretastatin-A4 phosphate, a natural product tubulin inhibitor. Invest. New Drugs 1996;14:131–137.

72. Phillips DR, White RJ, Cullinane C. DNA sequence-specific adducts of adriamycin and mitomycin C. FEBS Lett. 1989;246:233–240.

73. Skorobogaty A, White RJ, Phillips DR, Reiss JA. Elucidation of the DNA sequence preferences of daunomycin. Drug. Des. Deliv. 1988;3:125–151.

74. Skorobogaty A, White RJ, Phillips DR, Reiss JA. The 5'-CA DNA-sequence preference of daunomycin. FEBS Lett. 1988;227:103–106.

75. Hecht SM. Bleomycin: new perspectives on the mechanism of action. J. Nat. Prod. 2000;63:158–168.

76. White RJ, Durr FE. Development of mitoxantrone. Invest. New Drugs 1985;3: 85–93.

77. Sasaki Y, Seto M, Komatsu K, Omura S. Staurosporine, a protein kinase inhibitor, attenuates intracellular Ca2 + -dependent contractions of strips of rabbit aorta. Eur. J. Pharmacol. 1991;202:367–372.

78. Omura S. The expanded horizon for microbial metabolites–a review. Gene 1992;115:141–149.

79. Omura S, Sasaki Y, Iwai Y, Takeshima H. Staurosporine, a potentially important gift from a microorganism. J. Antibiot. (Tokyo) 1995;48:535–548.

80. Thavasu P, Propper D, McDonald A, Dobbs N, Ganesan T, Talbot D, Braybrook J, Caponigro F, Hutchison C, Twelves C, Man A, Fabbro D, Harris A, Balkwill F. The protein kinase C inhibitor CGP41251 suppresses cytokine release and extracellular signal-regulated kinase 2 expression in cancer patients. Cancer Res. 1999;59:3980–3984.

81. Propper DJ, McDonald AC, Man A, Thavasu P, Balkwill F, Braybrooke JP, Caponigro F, Graf P, Dutreix C, Blackie R, Kaye SB, Ganesan TS, Talbot DC, Harris AL, Twelves C. Phase I and pharmacokinetic study of PKC412, an inhibitor of protein kinase C. J. Clin. Oncol. 2001;19:1485–1492.

82. Gerth K, Bedorf N, Hofle G, Irschik H, Reichenbach H. Epothilons A and B: antifungal and cytotoxic compounds from Sorangium cellulosum Myxobacteria. Production, physico-chemical and biological properties. J. Antibiot. (Tokyo) 1996;49:560–563.

83. Bollag DM. Epothilones: novel microtubule-stabilising agents. Expert. Opin. Investig. Drugs 1997;6:867–873.

84. Bollag DM, McQueney PA, Zhu J, Hensens O, Koupal L, Liesch J, Goetz M, Lazarides E, Woods CM. Epothilones, a new class of microtubule-stabilizing agents with a taxol-like mechanism of action. Cancer Res. 1995;55:2325–2333.

85. Altmann KH, Pfeiffer B, Arseniyadis S, Pratt BA, Nicolaou KC. The Chemistry and Biology of Epothilones-The Wheel Keeps Turning. ChemMedChem 2007;2:396–423.

86. Feling RH, Buchanan GO, Mincer TJ, Kauffman CA, Jensen PR, Fenical W. Salinosporamide A: a highly cytotoxic proteasome inhibitor from a novel microbial source, a marine bacterium of the new genus salinispora. Angew. Chem. Int. Ed. Engl. 2003;42:355–357.

87. Groll M, Huber R, Potts BC. Crystal structures of Salinosporamide A NPI-0052 and B NPI-0047 in complex with the 20S proteasome reveal important consequences of beta-lactone ring opening and a mechanism for irreversible binding. J. Am. Chem. Soc. 2006;128:5136–5141.

88. Macherla VR, Mitchell SS, Manam RR, Reed KA, Chao TH, Nicholson B, Deyanat-Yazdi G, Mai B, Jensen PR, Fenical WF, Neuteboom ST, Lam KS, Palladino MA, Potts BC. Structure-activity relationship studies of salinosporamide A NPI-0052, a novel marine derived proteasome inhibitor. J. Med. Chem. 2005;48:3684–3687.

89. Chauhan D, Catley L, Li G, Podar K, Hideshima T, Velankar M, Mitsiades C, Mitsiades N, Yasui H, Letai A, Ovaa H, Berkers C, Nicholson B, Chao TH, Neuteboom ST, Richardson P, Palladino MA, Anderson KC. A novel orally active proteasome inhibitor induces apoptosis in multiple myeloma cells with mechanisms distinct from Bortezomib. Cancer Cell 2005;8:407–419.

90. Pettit GR, Herald CL, Doubek DL, Herald DL, Arnold E, Clardy J. Isolation and structure of bryostatin 1. J. Am. Chem. Soc. 1982;104:6846–6848.

91. Hennings H, Blumberg PM, Pettit GR, Herald CL, Shores R, Yuspa SH. Bryostatin 1, an activator of protein kinase C, inhibits tumor promotion by phorbol esters in SENCAR mouse skin. Carcinogenesis 1987;8:1343–1346.

92. Baryza JL, Brenner SE, Craske ML, Meyer T, Wender PA. Simplified analogs of bryostatin with anticancer activity display greater potency for translocation of PKCdelta-GFP. Chem. Biol. 2004;11:1261–1267.

93. Wender PA, Hinkle KW, Koehler MF, Lippa B. The rational design of potential chemotherapeutic agents: synthesis of bryostatin analogues. Med. Res. Rev. 1999;19:388–407.

94. Pettit GR. Th dolastatins. Fortschr Chem. Org. Naturst. 1997;70:1–79.

95. Pettit GR, Kamano Y, Herald CL, Tuinman AA, Boettner FE, Kizu H, Schmidt JM, Baczynskyj L, Tomer KB, Bontems RJ. The isolation and structure of a remarkable marine animal antineoplastic constituent: dolastatin 10. J. Am. Chem. Soc. 1987;109:6883–6885.

96. Pettit GR, Singh SB, Hogan F, Lloyd-Williams P, Herald DL, Burkett DD, Clewlow PJ. Antineoplastic agents. Part 189. The absolute configuration and synthesis of natural –dolastatin 10. J. Am. Chem. Soc. 1989;111:5463–5465.

97. Bai R, Pettit GR, Hamel E. Dolastatin 10, a powerful cytostatic peptide derived from a marine animal. Inhibition of tubulin polymerization mediated through the vinca alkaloid binding domain. Biochem. Pharmacol. 1990;39:1941–1949.

98. Bai RL, Pettit GR, Hamel E. Binding of dolastatin 10 to tubulin at a distinct site for peptide antimitotic agents near the exchangeable nucleotide and vinca alkaloid sites. J. Biol. Chem. 1990;265:17141–17149.

99. Saad ED, Kraut EH, Hoff PM, Moore DF, Jr., Jones D, Pazdur R, Abbruzzese JL. Phase II study of dolastatin-10 as first-line treatment for advanced colorectal cancer. Am. J. Clin. Oncol. 2002;25:451–453.

100. Patel S, Keohan ML, Saif MW, Rushing D, Baez L, Feit K, DeJager R, Anderson S. Phase II study of intravenous TZT-1027 in patients with advanced or metastatic soft-tissue sarcomas with prior exposure to anthracycline-based chemotherapy. Cancer 2006;107:2881–2887.

101. Gunasekera SP, Gunasekera M, Longley RE, Schulte GK. Discodermolide: a new bioactive polyhydroxylated lactone from the marine sponge Discodermia dissoluta. J. Org. Chem. 1990;55:4912–4915.

102. ter Haar E, Kowalski RJ, Hamel E, Lin CM, Longley RE, Gunasekera SP, Rosenkranz HS, Day BW. Discodermolide, a cytotoxic marine agent that stabilizes microtubules more potently than taxol. Biochemistry 1996;35:243–250.

103. Mickel SJ. Toward a commercial synthesis of+-discodermolide. Curr. Opin. Drug Discov. Devel. 2004;7:869–881.

104. Mickel SJ, Niederer D, Daeffler R, Osmani A, Kuesters E, Schmid E, Schaer K, Gamboni R, Chen W, Loeser E, Kinder FR, Konigsberger K, Prasad K, Ramsey TM, Repic O, Wang R-M, Florence G, Lyothier I, Paterson I. Large-scale synthesis of the anti-cancer marine natural product + -discodermolide. Part 5: linkage of fragments C and Finale. Org. Process Res. Dev. 2004;8:122–130.

105. Wright AE, Forleo DA, Gunawardana GP, Gunasekera SP, Koehn FE, McConnell OJ. Antitumor tetrahydroisoquinoline alkaloids from the colonial ascidian Ecteinascidia turbinata. J. Org. Chem. 1990;55:4508–4512.

106. Rinehart KL, Holt TG, Fregeau NL, Stroh JG, Keifer PA, Sun F, Li LH, Martin DG. Ecteinascidins 729, 743, 745, 759A, 759B, and 770: potent antitumor agents from the Caribbean tunicate Ecteinascidia turbinata. J. Org. Chem. 1990;55:4512–4515.

107. Corey EJ, Gin DY, Kania RS. Enantioselective total synthesis of ecteinascidin 743. J. Am. Chem. Soc. 1996;118:9202–9203.

108. Martinez EJ, Corey EJ. A new, more efficient, and effective process for the synthesis of a key pentacyclic intermediate for production of ecteinascidin and phthalascidin antitumor agents. Org. Lett. 2000;2:993–996.

109. Martinez EJ, Owa T, Schreiber SL, Corey EJ. Phthalascidin, a synthetic antitumor agent with potency and mode of action comparable to ecteinascidin 743. Proc. Natl. Acad. Sci. U.S.A. 1999;96:3496–3501.

110. Anderson HJ, Coleman JE, Andersen RJ, Roberge M. Cytotoxic peptides hemiasterlin, hemiasterlin A and hemiasterlin B induce mitotic arrest and abnormal spindle formation. Cancer Chemother. Pharmacol. 1997;39:223–226.

111. Loganzo F, Discafani CM, Annable T, Beyer C, Musto S, Hari M, Tan X, Hardy C, Hernandez R, Baxter M, Singanallore T, Khafizova G, Poruchynsky MS, Fojo T, Nieman JA, Ayral-Kaloustian S, Zask A, Andersen RJ, Greenberger LM. HTI-286, a synthetic analogue of the tripeptide hemiasterlin, Is a potent antimicrotubule agent that circumvents P-Glycoprotein-mediated resistance in vitro and in Vivo. Cancer Res. 2003;63:1838–1845.

112. Maiese WM, Lechevalier MP, Lechevalier HA, Korshalla J, Kuck N, Fantini A, Wildey MJ, Thomas J, Greenstein M. Calicheamicins, a novel family of antitumor antibiotics: taxonomy, fermentation and biological properties. J. Antibiot. (Tokyo) 1989;42:558–563.

113. Giles F, Estey E, O'Brien S. Gemtuzumab ozogamicin in the treatment of acute myeloid leukemia. Cancer 2003;98:2095–2104.

114. Lee MD, Dunne TS, Siegel MM, Chang CC, Morton GO, Borders DB. Calichemicins, a novel family of antitumor antibiotics. 1. Chemistry and partial structure of calichemicin.gamma.1I. J. Am. Chem. Soc. 1987;109:3464–3466.

115. Lee MD, Dunne TS, Chang CC, Ellestad GA, Siegel MM, Morton GO, McGahren WJ, Borders DB. Calichemicins, a novel family of antitumor antibiotics. 2. Chemistry and structure of calichemicin.gamma.1I. J. Am. Chem. Soc. 1987;109:3466–3468.

116. Ruegger A, Kuhn M, Lichti H, Loosli HR, Huguenin R, Quiquerez C, von Wartburg A. Cyclosporin A, a peptide metabolite from trichoderma polysporum link ex pers. Rifai, with a remarkable immunosuppressive activity. Helv. Chim. Acta 1976;59:1075–1092.

117. Wenger RM. Pharmacology of cyclosporin sandimmune. II. Chemistry. Pharmacol. Rev. 1990;41:243–247.

118. Zenke G, Baumann G, Wenger R, Hiestand P, Quesniaux V, Andersen E, Schreier MH. Molecular mechanisms of immunosuppression by cyclosporins. Ann. N. Y. Acad. Sci. 1993;685:330–335.

119. Bua J, Ruiz AM, Potenza M, Fichera LE. In vitro anti-parasitic activity of Cyclosporin A analogs on Trypanosoma cruzi. Bioorg. Med. Chem. Lett. 2004;14:4633–4637.

120. Eckstein JW, Fung J. A new class of cyclosporin analogues for the treatment of asthma. Expert. Opin. Investig. Drugs 2003;12:647–653.

121. Kino T, Hatanaka H, Miyata S, Inamura N, Nishiyama M, Yajima T, Goto T, Okuhara M, Kohsaka M, Aoki H, et al. FK-506, a novel immunosuppressant isolated from a Streptomyces. II. Immunosuppressive effect of FK-506 in vitro. J. Antibiot. (Tokyo) 1987;40:1256–1265.

122. Kino T, Inamura N, Sakai F, Nakahara K, Goto T, Okuhara M, Kohsaka M, Aoki H, Ochiai T. Effect of FK-506 on human mixed lymphocyte reaction in vitro. Transplant. Proc. 1987;19:36–39.

123. Ochiai T, Nagata M, Nakajima K, Suzuki T, Sakamoto K, Enomoto K, Gunji Y, Uematsu T, Goto T, Hori S, et al. Studies of the effects of FK506 on renal allografting in the beagle dog. Transplantation 1987;44:729–733.

124. Kino T, Goto T. Discovery of FK-506 and update. Ann. N. Y. Acad. Sci. 1993;685:13–21.

125. Tanaka H, Kuroda A, Marusawa H, Hatanaka H, Kino T, Goto T, Hashimoto M, Taga T. Structure of FK506, a novel immunosuppressant isolated from Streptomyces. J. Am. Chem. Soc. 1987;109:5031–5033.

126. Spencer CM, Goa KL, Gillis JC. Tacrolimus. An update of its pharmacology and clinical efficacy in the management of organ transplantation. Drugs 1997;54:925–975.

127. Kahan BD, Camardo JS. Rapamycin: clinical results and future opportunities. Transplantation 2001;72:1181–1193.

128. Camardo J. The Rapamune era of immunosuppression 2003: the journey from the laboratory to clinical transplantation. Transplant. Proc. 2003;35:18–24.

129. Baker H, Sidorowicz A, Sehgal SN, Vezina C. Rapamycin AY-22,989, a new antifungal antibiotic. III. In vitro and in vivo evaluation. J. Antibiot. (Tokyo) 1978;31:539–545.

130. Sehgal SN, Baker H, Vezina C. Rapamycin AY-22,989, a new antifungal antibiotic. II. Fermentation, isolation and characterization. J. Antibiot. (Tokyo) 1975;28:727–732.

131. Schreiber SL. Chemistry and biology of the immunophilins and their immunosuppressive ligands. Science 1991;251:283–287.

132. Ho S, Clipstone N, Timmermann L, Northrop J, Graef I, Fiorentino D, Nourse J, Crabtree GR. The mechanism of action of cyclosporin A and FK506. Clin. Immunol. Immunopathol. 1996;80:40–45.

133. Schreiber SL, Crabtree GR. The mechanism of action of cyclosporin A and FK506. Immunol. Today 1992;13:136–142.

134. Campbell IM, Calzadilla CH, McCorkindale NJ. Some new metabolites related to mycophenolic acid. Tetrahedron Lett. 1966:5107–5111.

135. Behrend M. Mycophenolate mofetil Cellcept. Expert Opin. Investig. Drugs 1998;7:1509–1519.

136. Sollinger HW. Mycophenolate mofetil for the prevention of acute rejection in primary cadaveric renal allograft recipients. U.S. Renal Transplant Mycophenolate Mofetil Study Group. Transplantation 1995;60:225–232.

137. Alberts AW, Chen J, Kuron G, Hunt V, Huff J, Hoffman C, Rothrock J, Lopez M, Joshua H, Harris E, Patchett A, Monaghan R, Currie S, Stapley E, Albers-Schonberg G, Hensens O, Hirshfield J, Hoogsteen K, Liesch J, Springer J. Mevinolin: a highly potent competitive inhibitor of hydroxymethylglutaryl-coenzyme A reductase and a cholesterol-lowering agent. Proc. Natl. Acad. Sci. U.S.A. 1980;77:3957–3961.

138. Patchett AA. 2002 Alfred Burger Award Address in Medicinal Chemistry. Natural products and design: interrelated approaches in drug discovery. J. Med. Chem. 2002;45:5609–5616.

139. Coukell AJ, Wilde MI. Pravastatin. A pharmacoeconomic review of its use in primary and secondary prevention of coronary heart disease. Pharmacoeconomics 1998;14:217–236.

140. Plosker GL, Wagstaff AJ. Fluvastatin: a review of its pharmacology and use in the management of hypercholesterolaemia. Drugs 1996;51:433–459.

141. Roth BD. The discovery and development of atorvastatin, a potent novel hypolipidemic agent. Prog. Med. Chem. 2002;40:1–22.

142. Plosker GL, Dunn CI, Figgitt DP. Cerivastatin: a review of its pharmacological properties and therapeutic efficacy in the management of hypercholesterolaemia. Drugs 2000;60:1179–1206.

143. Chong PH, Yim BT. Rosuvastatin for the treatment of patients with hypercholesterolemia. Ann. Pharmacother. 2002;36:93–101.

144. Ondetti MA, Williams NJ, Sabo EF, Pluscec J, Weaver ER, Kocy O. Angiotensin-converting enzyme inhibitors from the venom of Bothrops jararaca. Isolation, elucidation of structure, and synthesis. Biochemistry 1971;10:4033–4039.

145. Ondetti MA, Rubin B, Cushman DW. Design of specific inhibitors of angiotensin-converting enzyme: new class of orally active antihypertensive agents. Science 1977;196:441–444.

146. Burg RW, Miller BM, Baker EE, Birnbaum J, Currie SA, Hartman R, Kong YL, Monaghan RL, Olson G, Putter I, Tunac JB, Wallick H, Stapley EO, Oiwa R, Omura S. Avermectins, new family of potent anthelmintic agents: producing organism and fermentation. Antimicrob, Agents Chemother. 1979;15:361–367.

147. Cully DF, Vassilatis DK, Liu KK, Paress PS, Van der Ploeg LH, Schaeffer JM, Arena JP. Cloning of an avermectin-sensitive glutamate-gated chloride channel from Caenorhabditis elegans. Nature 1994;371:707–711.

148. Davies HG, Green RH. Avermectins and milbemycins. Nat. Prod. Rep. 1986;3:87–121.

149. Chabala JC, Mrozik H, Tolman RL, Eskola P, Lusi A, Peterson LH, Woods MF, Fisher MH, Campbell WC, Egerton JR, Ostlind DA. Ivermectin, a new broad-spectrum antiparasitic agent. J. Med. Chem. 1980;23:1134–1136.

150. Goa KL, McTavish D, Clissold SP. Ivermectin. A review of its antifilarial activity, pharmacokinetic properties and clinical efficacy in onchocerciasis. Drugs 1991;42:640–658.

151. Kirst HA, Creemer LC, Naylor SA, Pugh PT, Snyder DE, Winkle JR, Lowe LB, Rothwell JT, Sparks TC, Worden TV. Evaluation and development of spinosyns to control ectoparasites on cattle and sheep. Curr. Top. Med. Chem. 2002;2:675–699.

152. Ondeyka JG, Helms GL, Hensens OD, Goetz MA, Zink DL, Tsipouras A, Shoop WL, Slayton L, Dombrowski AW, Polishook JD, Ostlind DA, Tsou NN, Ball RG, Singh SB. Nodulisporic Acid A, a novel and potent insecticide from a nodulisporium sp. Isolation, structure determination, and chemical transformations. J. Am. Chem. Soc. 1997;119:8809–8816.

153. Meinke PT, Smith MM, Shoop WL. Nodulisporic acid: its chemistry and biology. Curr. Top. Med. Chem. 2002;2:655–674.

154. Chang RS, Lotti VJ, Monaghan RL, Birnbaum J, Stapley EO, Goetz MA, Albers-Schonberg G, Patchett AA, Liesch JM, Hensens OD, et al. A potent nonpeptide cholecystokinin antagonist selective for peripheral tissues isolated from Aspergillus alliaceus. Science 1985;230:177–179.

155. Evans BE, Bock MG, Rittle KE, DiPardo RM, Whitter WL, Veber DF, Anderson PS, Freidinger RM. Design of potent, orally effective, nonpeptidal antagonists of the peptide hormone cholecystokinin. Proc. Natl. Acad. Sci. U.S.A. 1986;83:4918–4922.

156. Bock MG, DiPardo RM, Evans BE, Rittle KE, Whitter WL, Veber DE, Anderson PS, Freidinger RM. Benzodiazepine gastrin and brain cholecystokinin receptor ligands: L-365,260. J. Med. Chem. 1989;32:13–16.

157. Darkin-Rattray SJ, Gurnett AM, Myers RW, Dulski PM, Crumley TM, Allocco JJ, Cannova C, Meinke PT, Colletti SL, Bednarek MA, Singh SB, Goetz MA, Dombrowski AW, Polishook JD, Schmatz DM. Apicidin: a novel antiprotozoal agent that inhibits parasite histone deacetylase. Proc. Natl. Acad. Sci. U.S.A. 1996;93:13143–13147.

158. Singh SB, Zink DL, Liesch JM, Mosley RT, Dombrowski AW, Bills GF, Darkin-Rattray SJ, Schmatz DM, Goetz MA. Structure and chemistry of apicidins, a class of novel cyclic tetrapeptides without a terminal alpha-keto epoxide as inhibitors of histone deacetylase with potent antiprotozoal activities. J. Org. Chem. 2002;67:815–825.

159. Han JW, Ahn SH, Park SH, Wang SY, Bae GU, Seo DW, Kwon HK, Hong S, Lee HY, Lee YW, Lee HW. Apicidin, a histone deacetylase inhibitor, inhibits proliferation of tumor cells via induction of p21WAF1/Cip1 and gelsolin. Cancer Res. 2000;60:6068–6074.

160. Kim MS, Son MW, Kim WB, In Park Y, Moon A. Apicidin, an inhibitor of histone deacetylase, prevents H-ras-induced invasive phenotype. Cancer Lett. 2000;157:23–30.

161. Jones P, Altamura S, Chakravarty PK, Cecchetti O, De Francesco R, Gallinari P, Ingenito R, Meinke PT, Petrocchi A, Rowley M, Scarpelli R, Serafini S, Steinkuhler C. A series of novel, potent, and selective histone deacetylase inhibitors. Bioorg. Med. Chem. Lett. 2006;16:5948–5952.

162. Remiszewski SW. Recent advances in the discovery of small molecule histone deacetylase inhibitors. Curr. Opin. Drug Discov. Devel. 2002;5:487–499.

163. Singh SB, Jayasuriya H, Ondeyka JG, Herath KB, Zhang C, Zink DL, Tsou NN, Ball RG, Basilio A, Genilloud O, Diez MT, Vicente F, Pelaez F, Young K, Wang J. Isolation, structure, and absolute stereochemistry of platensimycin, a broad spectrum antibiotic discovered using an antisense differential sensitivity strategy. J. Am. Chem. Soc. 2006;128:11916–11920.

164. Wang J, Soisson SM, Young K, Shoop W, Kodali S, Galgoci A, Painter R, Parthasarathy G, Tang YS, Cummings R, Ha S, Dorso K, Motyl M, Jayasuriya H, Ondeyka J, Herath K, Zhang C, Hernandez L, Allocco J, Basilio A, Tormo JR, Genilloud O, Vicente F, Pelaez F, Colwell L, Lee SH, Michael B, Felcetto T, Gill C, Silver LL, Hermes JD, Bartizal K, Barrett J, Schmatz D, Becker JW, Cully D, Singh SB. Platensimycin is a selective FabF inhibitor with potent antibiotic properties. Nature 2006;441:358–361.

165. Wang J, Kodali S, Lee SH, Galgoci A, Painter R, Dorso K, Racine F, Motyl M, Hernandez L, Tinney E, Colletti SL, Herath K, Cummings R, Salazar O, Gonzalez

I, Basilio A, Vicente F, Genilloud O, Pelaez F, Jayasuriya H, Young K, Cully DF, Singh SB. Discovery of platencin, a dual FabF and FabH inhibitor with in vivo antibiotic properties. Proc. Natl. Acad. Sci. U.S.A 2007;104:7612–7616.

166. Jayasuriya H, Herath KB, Zhang C, Zink DL, Basilio A, Genilloud O, Diez MT, Vicente F, Gonzalez I, Salazar O, Pelaez F, Cummings R, Ha S, Wang J, Singh SB. Isolation and structure of platencin: a FabH and FabF dual inhibitor with potent broad-spectrum antibiotic activity. Angew. Chem. Int. Ed. Engl. 2007;46:4684–4688.

# PART II

## CHEMICAL BIOLOGY TO UNDERSTAND AND TARGET DISEASE

# 11

# THE ROLE OF CHEMICAL BIOLOGY IN UNDERSTANDING AND TREATING DISEASE—ARE SMALL MOLECULE "CORRECTORS" THE WAY OF THE FUTURE?

DAVID SELWOOD

*Wolfson Institute for Biomedical Research, University College London, London*

In this section, we take on a diverse group of diseases, including those related to mental health, metabolism, and protein trafficking and misfolding, and show how chemical biology has impacted the study of these diseases. An intriguing theme emerges from these reviews: the influence of genetics on the understanding of disease. It has been 12 years since the publication of the human genome, and the early hype surrounding its publication was not translated into instant gains in the understanding of diseases and the search for new molecularly targeted medicines. Now we have genomewide association studies (GWAS), pointing to either outright genetic causes for disease or associations that, when combined with environmental factors, can lead to disease. Perhaps it is the combination of this knowledge with the advent of a new type of therapeutic agent, small-molecule "correctors" of protein function, which will yield the biggest gains. Small molecules such as the Vertex cystic fibrosis drug offer a precisely tailored genetic medicine for the treatment of cystic fibrosis, correcting aberrant protein function at the level of the gene product.

In simpler organisms, the effect of genetic changes is already evident and plays a major role in the design and development of new HIV therapies

*Chemical Biology: Approaches to Drug Discovery and Development to Targeting Disease*, First Edition.
Edited by Natanya Civjan.
© 2012 John Wiley & Sons, Inc. Published 2012 by John Wiley & Sons, Inc.

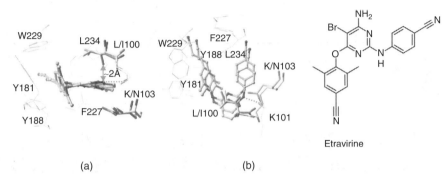

(a)                                    (b)

Etravirine

**Figure 11.1**    The extraordinary ability of etravirine to adjust to changes in protein conformation caused by resistant HIV mutations is shown. The rotatable bonds linked through the linking oxygen and nitrogen atoms from the central phenyl core allow wiggling (a) and jiggling (b). Figure adapted from K. Das, J.D. Bauman, A.D. Clark, Y.V. Frenkel, P.J. Lewi, A.J. Shatkin, et al. Proc. Natl. Acad. Sci. 2008;105:1466–1471 Copyright (2008) National Academy of Sciences, U.S.A.

(see Chapter 16). HIV is uniquely variable as a result of its inaccurate reverse transcriptase enzyme, which leads to a number of "mistakes" in the HIV genome, with a consequent variability and an advantage for the virus in producing resistant forms. In this case, new drugs have to be designed to be able to adjust to these new mutants, and this creates unique challenges for the medicinal chemist. Perhaps these are best illustrated by the new non-nucleoside reverse transcriptase inhibitors such as etravirine (1). In this case, the design of the molecule allows a limited and precisely defined flexibility to accommodate the required adjustments (Fig. 11.1).

Such second- or third-generation non-nucleoside reverse transcriptase inhibitors are a step change in terms of their clinical usefulness from the early inhibitors such as nevarapine where the conformationally inflexible molecule was unable to adjust to mutations in the protein, and resistance developed rapidly.

Many of the lessons learnt from the successful battle to develop new HIV medicines are being applied in cancer therapy. The newly discovered Braf kinase inhibitor Zelboraf (vermurafenib) is another example of cutting-edge medicinal chemistry. This kinase is strongly linked to the progression of malignant melanoma (Fig. 11.2). In this case, the mutation in B-Raf of V600E is the key oncogenic event. A system of fragment-based chemical design was instituted at Plexicon, Inc. to identify chemical scaffolds likely to have optimal activity against the B-Raf kinase. Although the V600E mutation is located away from the active site of B-Raf, vermurafenib is 10 times more potent at inhibiting the V600E form. This molecule has spectacular Absorption, distribution, metabolism and excretion (ADME) characteristics reaching plasma levels of 86 μM in humans (2). Such molecules should represent the future of healthcare, being highly selective yet simple to manufacture and to produce.

These advances in the fields of HIV and cancer provide the direction for other major human diseases. First, the genetic understanding is overlaid onto the

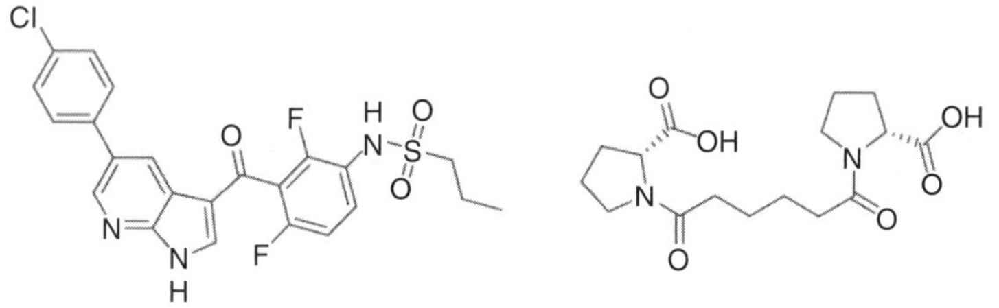

**Vermurafenib (PLX4032)**                    **CPHPC (Ro 63-8695)**

**Figure 11.2**   The B-raf kinase inhibitor vermurafenib and the SAP protein dimerizer CPHPC.

biochemical and chemical knowledge of the pathology and signaling pathways. Specific molecular changes at the protein level can then lead to the design of new molecules.

The design of small molecules to fit enzyme active sites or receptors is well accepted, but the ability of small molecules to alter protein function or "correct" the function of a misfolded protein is less well appreciated. Amyloidosis is a set of conditions in which amyloid protein aggregates are deposited in organs or tissues. The best known of these diseases is Alzheimer's disease where amyloid deposits occur in the brain. These deposits have a general β-sheet structure. Defects in normal proteins can be caused by (inherited) mutations and result in misfolding (see Chapter 19). Alzheimer's is characterized by the deposits of aggregates of a relatively small peptide of only 42 amino acids (amyloid beta, Aβ). Intriguingly, a link between diabetes and Alzheimer's disease has now been shown, which has led to the term *type 3 diabetes* where it was shown that soluble amyloid oligomers can bind to the insulin receptor. Soluble amyloid oligomers are known to bind to several proteins including the cyclophilin D (PPIF) of mitochondria (3). Only further studies will show if these interactions are intergral to in the disease process. Many small molecule inhibitors of Aβ aggregation have been investigated, usually targeted toward a specific subregion of the Aβ peptide (4). To date none of these have been successful in clinical trials. A number of biological therapies also target Aβ and can reduce the formation of amyloid deposits. As an example, serum amyloid protein (SAP) binds to amyloid deposits, and antibodies to SAP can clear amyloids from the body if the circulating SAP is removed first by the small-molecule dimerizer of SAP, CPHPC (Fig. 11.3). The action of CPHPC on SAP may be a special case, but such novel ideas exemplify the range of possible solutions to a given therapeutic demand. For Alzheimer's disease, the range of therapies currently under trial would imply that the amyloid hypothesis of Alzheimer's will be confirmed or rejected in the next few years. A small-molecule drug is already available for one of the other amyloidoses. Transthyretin is involved with the transport of retinol in association with retinol-binding protein and thyroxine in both blood and the central nervous system (CNS). Specific mutations tend to destabilize

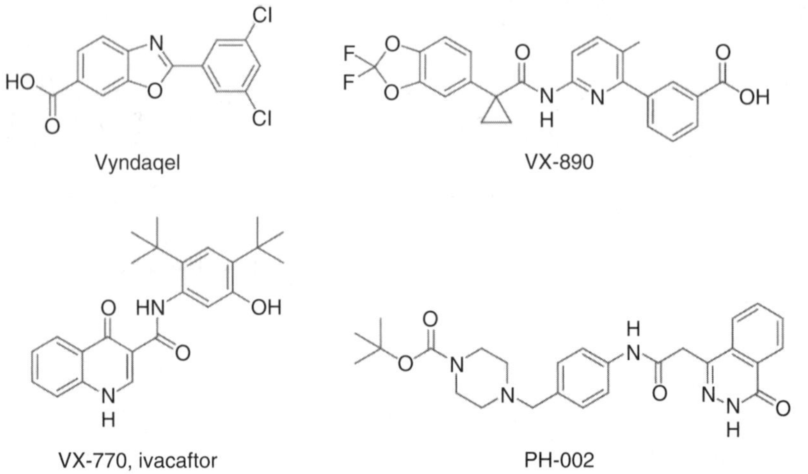

**Figure 11.3** Chemical structure of correctors of protein function. The amyloidosis treatment vyndagel, and the cystic fibrosis molecules VX-890 and VX 770. PH-002 prevents intraprotein domain interactions in apoE4, a target for Alzheimer's disease.

the protein and also favor monomer formation over the normal tetramer. This monomer is thought to be the amyloid/fibril-forming species. Thus, the V30M mutation in transthyretin causes transthyretin familial amyloid polyneuropathy, a rare progressive disease leading to loss of motor control and sensory loss. In a remarkable piece of chemical biology design, Kelly and coworkers identified small molecules that can bind to the thyroxine sites on TTR and stabilize the tetramer (5). This work led to the discovery of tafamidis (Vyndaqel), recently approved by the European Commission in November 2011 for the treatment of this disease. Currently, the only alternative treatment is a liver transplant, a vastly expensive procedure.

An overview of protein-trafficking diseases is provided in Chapter 20. This field overlaps with that of protein folding because defects in protein trafficking can be a result of a misfolding event. The best known example is that of the cystic fibrosis transmembrane conductance regulator (CFTR) protein, an ABC transporter of chloride and thiocyanate anions. Many mutations in the CFTR gene have been linked to cystic fibrosis. Here, the ΔF508 CFTR mutant (phenylalanine deleted at position 508) prevents the protein from being trafficked to the cell membrane. As a result, a thick mucus builds up in several organs, notably the lungs. ΔF508 is present in most cystic fibrosis cases worldwide (6). The mucus interferes with normal function and encourages bacterial infections. Research on small-molecule chaperones or correctors has been going on for several years. The most advanced of the ΔF508 CFTR correctors is the VX-809 compound. VX-809 improves folding and processing and enhances chloride transport (7). It is now being combined with another small-molecule corrector targeting a different cystic fibrosis mutation G551D, a mutation found in only 4% of sufferers. This molecule Kalydeco (ivacaftor, VX-770) has been approved by the U.S. Food and

Drug Administration (FDA) (Jan 2012). The discovery of these molecules is all the more remarkable in that it was done without any structural information on the target—just expert medicinal chemistry brilliantly applied using the latest understanding of favorable drug characteristics (8).

The fact that small molecules can target the single mutated proteins resulting from genetic defects could now be applied to other areas, such as diseases of mitochondria (see Chapter 22); several mitochondrial diseases are caused by mutations to mitochondrial proteins but do not appear as yet to be the targets of small molecule correctors or rescue agents. Some proteins that affect mitochondrial function have been studied, however. Apolipoprotein E4 (apoE4), the major genetic risk factor for late onset Alzheimer disease, is a case in point. In this protein, the E4 allele has amino acid substitutions that promote intraprotein domain interactions of the C and N termini of apoE4. This domain interaction is thought to be key in the disease process. In contrast, the E2 and E3 alleles with slightly different sequences do not promote this domain interaction. By utilizing a FRET assay, Mahley and coworkers were able to identify molecules that inhibited this domain interaction in apoE4 (9). This resulted in the identification of PH-002, a molecule that could reverse the decrease in mtCOX1 and improve mitochondrial motility. Although simply a lead, probably requiring much further optimization, PH-002 provides another potential example of a genetically tailored future medicine for Alzheimer's.

Lysosomal disorders (see Chapter 23) are often caused by a specific genetic mutation. Fabry disease, an X-linked storage disorder of lysosomes, is caused by genetic defects in the gene for α-galactosidase A (10). This results in a deficiency of this enzyme and accumulation of neutral glycosphingolipids such as globotri-azoylceramide in the lysosomes. The only treatment available currently for the disease is enzyme replacement therapy. A small molecule chaperone can have two beneficial effects here: it can stabilize the mutated protein, increasing the activity and stability to allow it to carry out its normal function, and it can stabilize the exogenously delivered recombinant therapy to improve its half-life and delivery to the tissues (11). The small-molecule migalastat (1-deoxygalactonojirimycin, Amigal, AT1001) achieves both these goals. Migalastat is a potent inhibitor of α-galactosidase A, but at subinhibitory concentrations, it acts as a stabilizer or chaperone and can enhance the activity (12). Amicus Therapeutics, the company that produces migalastat, is now developing two related compounds, duvoglustat and afegostat, for other genetic lysosomal disorders, Pompe's disease and Gaucher disease (Fig. 11.4). In Gaucher disease, innovative efforts are ongoing to produce new agents such as dideoxyiminoxylitols through thiol-ene chemistry (13).

Sometimes, the genetic link with disease has been more difficult to establish. In Chapter 13 on disease mechanisms in chronic obstructive pulmonary disease (COPD), it is clear that COPD is a complex inflammatory disease overwhelmingly occurring among smokers. COPD occurs in 25% of smokers. Here, the only firmly established genetic link is with the *SERPINA1* gene encoding the α-1 antitrypsin protein, a major plasma protease inhibitor with its main action on

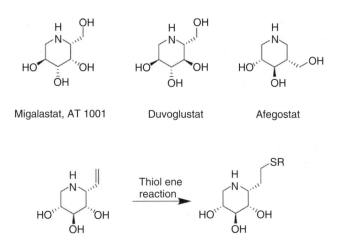

**Figure 11.4**   Small-molecule correctors of lysosomal diseases.

neutrophil elastase, which degrades the elastin of alveolar walls, but this link is only present in 2% of COPD (14). The most common (Z) mutation K342E causes the loss of a normal salt bridge, resulting in the intracellular aggregation of the protein and lower protein levels (15). Although there are no reports of small-molecule correctors or stabilizers for the α-1 antitrypsin protein, this would seem to be a logical target for this approach.

In Chapter 17 on allergy and asthma, multiple inflammatory molecules are described with multiple small-molecule approaches to therapy. This highlights the traditional approach to therapies but genetic studies have also shown significant causal relationships with disease (16). GWAS studies highlight multiple genes that may be involved with asthma; however, asthma is a heterogeneous disease and far away from the diseases with a defect in a single gene. A gene consistently linked with asthma is *IL13* where polymorphisms have been proposed to affect the gene function (17). Of course, where polymorphisms are in the promoter region of genes, a small molecule corrector approach will not be possible.

In many ways, osteoarthritis (see Chapter 15) presents a similar complicated picture, but some accepted gene associations have already been implicated in the disease. For example, the gene *GDF5* codes for a member of the TGF β-signaling pathway (18). Again, polymorphisms may be in the untranslated region (19) or in a region where a functional effect on the protein is realized (20).

The biological mechanisms of metabolic diseases, a grouping of diabetes, lipid disorders, and hypertension, are discussed in Chapter 21. These diseases collectively amount to a global epidemic fueled by a modern Western lifestyle consisting of inactivity, poor diet, and smoking. Any detailed description is beyond the scope of this commentary; however, it is possible to focus on type 2 diabetes (21) where some 12 genes have been identified where susceptibility loci exist. For one of these, the ZnT8/*SLC30A8* gene, encoding a zinc transporter linked to insulin polymorphisms can affect the $Zn^{2+}$ content of cells and hence the functioning of insulin (22). In this case, a specific R325W mutation thought to be located at the interface between protein subunits causes a defect in zinc transport.

Three reviews bring us the state of the art in the understanding and potential treatments of depression (Chapter 14), anxiety (Chapter 12), and schizophrenia (Chapter 18, see also Reference 23). The contrast between the fields is also stark with the understanding of schizophrenia in terms of genetics being arguably well ahead. Surely, if there is a field where chemical biology is to make an impact, it is in mental illness where small chemicals—neurotransmitters—govern response between neurones on the millisecond timescale. In schizophrenia, we have reached a point where the etiology of the disease is largely understood. Now proposed to be a neurodevelopmental disease, schizophrenia shows abnormalities in defined brain regions and an altered brain chemistry. Most recently, a number of genetic causes or risk factors have been highlighted, and one of protein target catechol O-methyltransferase (COMT), is has been studied in detail, with a specific V108M polymorphism implicated. COMT is responsible for the metabolism of dopamine, and Valine substitution increases the stability of the protein and hence increases the degradation of dopamine. If we examine the crystal structure of human COMT (pdb: 3BWM), we can see that the Valine residue occurs at the base of an $\alpha$ helix (Fig. 11.5). Current therapies such as entacapone (COMTan) are inhibitors and are designed to interact at the substrate site. Entacapone is currently used to treat Parkinson's disease while a related molecule, tolcapone, is under trial for schizophrenia (24). Could a more specific therapy be designed that would target the Valine 108 form directly? Such investigations require new types of assays of protein function (i.e., destabilizers rather than inhibitors or activators).

Perhaps one of the most intriguing associations with schizophrenia is that of the *DISC1* gene (Disrupted in Schizophrenia 1). Originally, *DISC1* mutations were identified in a Scottish family where mental illness was prevalent (25). *DISC1* is associated with other mental disorders such as bipolar disorder, in addition to schizophrenia. The translated DISC1 protein expressed in multiple splice variant forms (26), has multiple binding interactions with other proteins (27) and is involved with neuronal development (28). The known DISC1 protein variations have been collated with their function and reveal a wide array of changes to protein–protein interaction sites, phosphorylation sites, or sites likely to alter the structure of the protein (27). A traditional approach to finding new medicines would identify which mediators are upregulated by the genetic changes and seek to develop inhibitors for these mediators. The interplay between DISC1 protein and GSK3β is a case in point where lentivirus knockdown of DISC 1 protein expression in the dentate gyrus gave rise to schizophrenia-like behaviors. These behavioral phenotypes could be rescued by a GSK3β inhibitor (27). Such inhibitors might make useful drugs, but if the GSK3β binding site on DISC1 protein were amenable to a "corrector" molecule, then it might be possible to restore normal function rather than to modulate a separate signaling system. In order to do this, sophisticated phenotypic assays would be required. DISC1

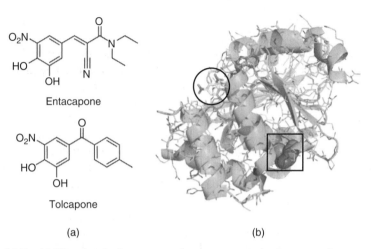

Entacapone

Tolcapone

(a)                                                          (b)

**Figure 11.5** (a) The chemical structure of entacapone and tolcapone substrate analogs of dopamine that inhibit COMT and (b) a ribbon representation of the crystal structure of COMT, showing the position of the V108M mutation (spheres). A substrate analog is shown as a light grey stick model (shown with circle) and $S$-adenosylmethionine as a dark grey stick model (shown with box).

protein is not amenable to simple cloning and expression, and there are no crystal structures available.

## 11.1   THE POTENTIAL FOR SMALL-MOLECULE CORRECTORS

We cannot yet know how widespread the use of small-molecule "correctors" will become. Clearly, not every mutational defect in a protein will be amenable to such a strategy. New assays will need to be established to allow identification of molecules that can precisely modulate a protein's activity or interaction with a separate protein. Conventional enzyme-based or receptor-binding assays may prove to be of limited use. It is likely that most of these assays will need to be phenotypic and backed up by sophisticated genetically based animal models. The potential of such medicines is enormous, however. Not only will they be targeted at the root cause of a disease rather than the symptoms, but they will (once the patent life is completed) also be cheap and affordable medicines promising sustainable healthcare for the future.

## REFERENCES

1. Das K, Bauman JD, Clark AD, Frenkel YV, Lewi PJ, Shatkin AJ, et al. High-resolution structures of HIV-1 reverse transcriptase/TMC278 complexes: strategic flexibility explains potency against resistance mutations. Proc. Natl. Acad. Sci. 2008;105:1466–1471.

2. Bollag G, Hirth P, Tsai J, Zhang J, Ibrahim PN, Cho H, et al. Clinical efficacy of a RAF inhibitor needs broad target blockade in BRAF-mutant melanoma. Nature 2010;467:596–599.

3. Du H, Guo L, Fang F, Chen D, Sosunov AA, McKhann GM, et al. Cyclophilin D deficiency attenuates mitochondrial and neuronal perturbation and ameliorates learning and memory in Alzheimer's disease. Nat. Med. 2008;14:1097–1105.

4. Nie Q, Du XG, Geng MY. Small molecule inhibitors of amyloid [beta] peptide aggregation as a potential therapeutic strategy for Alzheimer's disease. Acta. Pharmacol. Sin. 2011;32:545–551.

5. Razavi H, Palaninathan SK, Powers ET, Wiseman RL, Purkey HE, Mohamedmohaideen NN, et al. Benzoxazoles as transthyretin amyloid fibril inhibitors: synthesis, evaluation, and mechanism of action. Angew. Chem. 2003;115:2864–2867.

6. Bobadilla JL, Macek M, Fine JP, Farrell PM. Cystic fibrosis: a worldwide analysis of CFTR mutations—correlation with incidence data and application to screening. Hum. Mutat. 2002;19:575–606.

7. Van Goor F, Hadida S, Grootenhuis PDJ, Burton B, Stack JH, Straley KS, et al. Correction of the F508del-CFTR protein processing defect in vitro by the investigational drug VX-809. Proc. Natl. Acad. Sci. 2011;108:18843–18848.

8. Van Goor F, Hadida S, Grootenhuis PDJ, Burton B, Cao D, Neuberger T, et al. Rescue of CF airway epithelial cell function in vitro by a CFTR potentiator, VX-770. Proc. Natl. Acad. Sci. 2009;106:18825–18830.

9. Chen H-K, Liu Z, Meyer-Franke A, Brodbeck J, Miranda RD, McGuire JG, et al. Small molecule structure correctors abolish detrimental effects of apolipoprotein E4 in cultured neurons. J. Biol. Chem. 2012;287:5253–5266.

10. Wu X, Katz E, Valle MCD, Mascioli K, Flanagan JJ, Castelli JP, et al. A pharmacogenetic approach to identify mutant forms of α-galactosidase a that respond to a pharmacological chaperone for Fabry disease. Hum. Mutat. 2011;32:965–977.

11. Asano N, Ishii S, Kizu H, Ikeda K, Yasuda K, Kato A, et al. In vitro inhibition and intracellular enhancement of lysosomal α-galactosidase~A activity in Fabry lymphoblasts by 1-deoxygalactonojirimycin and its derivatives. Eur. J. Biochem. 2000;267:4179–4186.

12. Fan J-Q, Ishii S, Asano N, Suzuki Y. Accelerated transport and maturation of lysosomal [alpha]-galactosidase A in Fabry lymphoblasts by an enzyme inhibitor. Nat. Med. 1999;5:112–115.

13. Goddard-Borger ED, Tropak MB, Yonekawa S, Tysoe C, Mahuran DJ, Withers SG. rapid assembly of a library of lipophilic iminosugars via the thiol–ene reaction yields promising pharmacological chaperones for the treatment of Gaucher disease. J. Med. Chem. 2012;55:2737–2745.

14. Castaldi PJ, Cho MH, Cohn M, Langerman F, Moran S, Tarragona N, et al. The COPD genetic association compendium: a comprehensive online database of COPD genetic associations. Hum. Mol. Genet. 2010;19:526–534.

15. Brantly M, Courtney M, Crystal R. Repair of the secretion defect in the Z form of alpha 1-antitrypsin by addition of a second mutation. Science 1988;242:1700–1702.

16. Meyers DA. Genetics of asthma and allergy: What have we learned? J. Allergy Clin. Immunol. 2010;126:439–446.

17. Beghé B, Hall IP, Parker SG, Moffatt MF, Wardlaw A, Connolly MJ, et al. Polymorphisms in IL13 pathway genes in asthma and chronic obstructive pulmonary disease. Allergy 2010;65:474–481.

18. Sandell LJ. Etiology of osteoarthritis: genetics and synovial joint development. Nat. Rev. Rheumatol. 2012;8:77–89.

19. Egli RJ, Southam L, Wilkins JM, Lorenzen I, Pombo-Suarez M, Gonzalez A, et al. Functional analysis of the osteoarthritis susceptibility-associated GDF5 regulatory polymorphism. Arthritis Rheum. 2009;60:2055–2064.

20. Seemann P, Brehm A, König J, Reissner C, Stricker S, Kuss P, et al. Mutations in GDF5 reveal a key residue mediating BMP inhibition by NOGGIN. PLoS Genet. 2009;5:e1000747.

21. Voight BF, Scott LJ, Steinthorsdottir V, Morris AP, Dina C, Welch RP, et al. Twelve type 2 diabetes susceptibility loci identified through large-scale association analysis. Nat. Genet. 2010;42:579–589.

22. Nicolson TJ, Bellomo EA, Wijesekara N, Loder MK, Baldwin JM, Gyulkhandanyan AV, et al. Insulin storage and glucose homeostasis in mice null for the granule zinc transporter ZnT8 and studies of the type 2 diabetes-associated variants. Diabetes 2009;58:2070–2083.

23. Perrone-Bizzozero, NI and Bullock, W.M. Schizophrenia: Biological Mechanisms. Wiley Encyclopedia of Chemical Biology. 2008.

24. Bitsios P, Roussos P. Tolcapone, COMT polymorphisms and pharmacogenomic treatment of schizophrenia. Pharmacogenomics 2011;12:559–566.

25. Chubb JE, Bradshaw NJ, Soares DC, Porteous DJ, Millar JK. The DISC locus in psychiatric illness. Mol. Psychiatry 2008;13:36–64.

26. Nakata K, Lipska BK, Hyde TM, Ye T, Newburn EN, Morita Y, et al. DISC1 splice variants are upregulated in schizophrenia and associated with risk polymorphisms. Proc. Natl. Acad. Sci. 2009;106:15873–15878.

27. Soares DC, Carlyle BC, Bradshaw NJ, Porteous DJ. DISC1: structure, function, and therapeutic potential for major mental illness. ACS Chem. Neurosci. 2011;2:609–632.

28. Mao Y, Ge X, Frank CL, Madison JM, Koehler AN, Doud MK, et al. Disrupted in schizophrenia 1 regulates neuronal progenitor proliferation via modulation of GSK3 $\beta/\beta$-catenin signaling. Cell 2009;136:1017–1031.

# 12

# ANXIETY DISORDERS

MIKLOS TOTH

*Department of Pharmacology, Weill Medical College of Cornell University, New York, New York*

Fear and anxiety can be a normal adaptive reaction to help cope with stress in the short term, but when the emotional, cognitive, and physical manifestations are long lasting, extreme, and disproportionate to threat, whether real or perceived, anxiety is maladaptive and has become a disabling disorder. Anxiety disorders may be deconstructed to elementary behaviors/symptoms that can be conceptualized as quantitative characters determined by the combined effects of several risk genes and nongenetic factors (e.g., early-life adversity). Progress in neurogenetics, molecular and cellular neuroscience, and neuroimaging is beginning to yield significant insights of how genetic and non-genetic factors contribute to specific manifestations of anxiety disorders. The aim of this overview is to summarize and integrate the current knowledge on anxiety-related macromolecular pathways and mechanisms initiated by genetic risk and environmental factors. These pathways interact with each other, often during specific periods of development, and could lead to alterations in the formation and function of neuronal circuits that encode emotional behavior.

Anxiety is a state characterized by feelings of fear, apprehension, and worry. Emotionally, anxiety causes a sense of dread or panic, and behaviorally, it can be associated with both voluntary and involuntary behaviors such as escaping or avoiding the source of anxiety. Also, anxiety is associated with specific physical manifestations including increased heart rate and blood pressure. Although anxiety is clinically different from depression, the two disorders are often comorbid. Indeed, it is generally difficult to find individuals with pure anxiety and pure depression. Here we focus on anxiety exclusively but will mention and discuss

*Chemical Biology: Approaches to Drug Discovery and Development to Targeting Disease*, First Edition.
Edited by Natanya Civjan.

depression when the separation is not clear. Anxiety, similarly to other psychiatric disorders, has a significant genetic basis, and it is relatively stable during lifetime (trait anxiety). Recent studies implicated candidate genes, each with a small contribution to the risk of anxiety disorders (1). Several of these candidate genes encode proteins that regulate neurotransmitter synthesis and metabolism or correspond to neurotransmitter receptors. Studies implicate an equally important role for environmental factors, including early-life adversity and maternal care, in the development of anxiety and anxiety-like behavior in animals (2–4). Some of these environmental effects also converge on neurotransmitter systems. Because the risk genes or environmental factors individually represent only a small contribution to anxiety, the current view is that interactions between several risk genes and environmental effects are necessary to lead to the symptoms of anxiety. This overview focuses on molecular systems related to anxiety such as neurotransmitters, their receptor, and, when known, downstream intracellular signaling pathways that ultimately regulate the development and/or function of neuronal networks that underlie the behavioral anxiety response.

## 12.1  NEUROTICISM AND ANXIETY

Neuroticism is a personality trait characterized by an enduring tendency to experience negative emotional states. Personality traits are underlying characteristics of an individual that can explain the major dimensions of human behavior. Neuroticism represents a continuum between emotional stability and instability, and most people fall in between the extremes (5–9). Neuroticism has a wide individual variability, but it is relatively stable in individuals over time (10). People with a high level of neuroticism respond more poorly to stress and are more likely to interpret ordinary situations as threatening and frustrating. In addition, autonomic arousal is an integral part of neuroticism characterized by increased heart rate and blood pressure, cold hands, sweating, and muscular tension. Extroversion and openness, two other personality traits, are also part of the NEO Personality Inventory (NEO-PI) and the revised (R) NEO-PI (11), and Gray and McNaughton (12) argued that anxiety proneness is primarily captured by measures of neuroticism, together with a smaller contribution from the dimension of extroversion. Specifically, rotating the dimensions of neuroticism and extroversion by 45°, two new dimensions, anxiety (N+ , E−) and impulsivity (N+ , E+ ), were proposed (12).

Although neuroticism is not a disease *per se*, it predisposes individuals to anxiety disorders (12, 13). Neuroticism is a vulnerability factor for all forms of anxiety (14–16). A system established by the *Diagnostic and Statistical Manual for Psychiatric Disorders in the United States*, currently in its 4th edition (DSM-IV) text revision (TR) (American Psychiatric Association, 2000), sets the boundary at which a particular level of behavior becomes an anxiety disorder—a level often based on the number and the duration of symptoms. DSM is a categorical system based on the qualitative separation of disease states from the state of well-being. The DSM-IVTR category of anxiety disorders currently includes

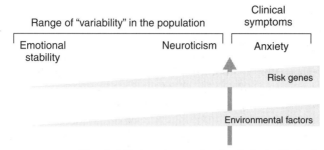

**Figure 12.1**   The quantitative trait model of neuroticism and anxiety.

generalized anxiety disorder (GAD), simple phobia, posttraumatic stress disorder (PTSD), panic disorder, social phobia, and obsessive compulsive disorder (OCD) as discrete anxiety disorders. The International Classification of Diseases-10 (IC-10) is a similar system, but it is less frequently used in research (17).

Although DSM-IV and IC-10 are widely used for the clinical diagnosis of anxiety disorders, these categorical models are not based on the underlying biological pathophysiology and may not be optimal for the identification of genetic and environmental factors involved in anxiety disorders. The main reason is the complexity and heterogeneity of anxiety disorders that makes the association of genetic and/or environmental effects with overt anxiety phenotypes difficult. An alternative approach is to deconstruct anxiety disorders to elementary behavioral manifestations or specific symptoms that may represent less complex biological traits and/or environmental influences. The rationale is that a specific phenotype that consists of few elementary behaviors is likely linked to a more limited number of genes and environmental effects than complex disease phenotypes. These elementary behaviors are primarily state-independent (manifest in an individual regardless of whether illness is active). Similar to this principle is the term "endophenotype," although it usually refers to an internal phenotype along the pathway between the genotype and disease. Evidence suggests that elementary behaviors/endophenotypes of psychiatric disease may be understood as quantitative traits. This model is based on the notion that risk gene variants carried by a given individual in combination with various nongenetic factors such as early-life adversity and nutrition produce anxiety symptoms with variable onset and severity depending on the strengths of the genetic and environmental factors (Fig. 12.1).

## 12.2   GENES, THEIR PROTEIN PRODUCTS, AND ASSOCIATED BIOLOGICAL PATHWAYS IMPLICATED IN ANXIETY AND ANXIETY-LIKE BEHAVIOR

Using the techniques of quantitative behavior genetics, it became clear that roughly 20–60% of the variation in most personality traits has a genetic base, and broad personality traits are under polygenic influence (18, 19). Similarly, genetic epidemiological studies estimate that heritability in anxiety varies between 23 and 50% (1, 20, 21). Genome-wide linkage studies have been performed by using Eysenck personality questionnaire (EPQ) (22, 23) to identify chromosomal regions associated with neuroticism. A two-point and multipoint nonparametric regression identified 1q, 4q, 7p, 8p, 11q, 12q, and 13q (24), whereas another similar study using multipoint, nonparametric allele sharing and regression identified 1q, 3centr, 6q, 11q, and 12p (25), which confirms some links in the previous study. Still another study using a broad anxiety definition instead of the DSM-IV classification identified a linkage at chromosome 14 between 99 and 115 cM (26), and this finding replicated a linkage for a broad anxiety phenotype in a clinically based study (27). However, these studies have not yet identified specific genes.

So far, the candidate gene approach has been more productive than linkage and association studies in implicating gene polymorphisms related to neuroticism and anxiety disorders. The candidate gene approach relies on prior biochemical studies that implicate various molecules, primarily neurotransmitters, their receptors, and signaling in anxiety. Anxiety disorders traditionally have been viewed as disturbances in neurotransmitter systems including the serotonin (5-HT), gamma-aminobutyric acid (GABA), and corticotropin releasing hormone (CRH) systems, among others. Many of these neurotransmitters and their receptors have also been identified as sites of action for anxiolytic drugs. Neurotransmitter systems and corresponding neurobiological pathways that are well established in anxiety both in human studies and animal experiments are discussed below.

Animal studies, especially rodent studies, significantly contributed to the current knowledge on how genes, the environment, and their interaction may produce anxiety. As people with a high level of neuroticism respond more poorly to stress, certain inbred, selectively bred (28, 29), and knockout rodent strains (see below) have increased emotional reactions to stress (emotionality). These behavioral responses include avoidance, hypoactivity/freezing, and autonomic arousal among others. Although substantial similarities are found between human and murine stress responses, complex anxiety phenotypes cannot be reproduced in animals. Nevertheless, emotional stress reactions in animals represent relatively simple behaviors that are evolutionarily conserved and quantifiable. A multitude of tests can measure emotionality in rodents (30–34). Although all behavioral tests use a novel environment and/or fearful situation to produce avoidance, hypoactivity/freezing, and/or autonomic arousal, it seems that the underlying pathophysiology is test specific. Indeed, data with recombinant inbred mouse strains indicate that open field and elevated plus maze behaviors are linked to specific but partly overlapping sets of quantitative trait loci (35, 36). Furthermore, even these relatively simple behaviors can be dissected to more elementary behaviors with a smaller assembly of quantitative trait loci (36). However, it is not clear currently if mouse QTL data can be directly extrapolated to humans.

Many recently developed genetically manipulated mouse strains exhibit increased emotionality, often referred to as anxiety-like behavior. Most strains have constitutive genetic inactivation, but some strains are also available with a conditional allele. Although constitutive gene inactivation may elicit compensatory processes that can complicate the phenotype, this occurrence is not necessarily a disadvantage. Indeed, genetic risk in humans is also constitutive and present from early life. Probably more important is that polymorphisms in risk genes do not cause a complete functional loss; therefore, it may be more appropriate to analyze the behavioral phenotype of mice with a heterozygous inactivation of the risk gene.

### 12.2.1 The 5-HT Pathway and Associated Genes Involved in Anxiety

5-HT has long been associated with emotion and anxiety (37). 5-HT is synthesized from tryptophan by the rate-limiting enzyme tryptophan hydroxylase (TPH) in serotonergic neurons in the raphe. Release of 5-HT is controlled by the $5\text{-HT}_{1A}$ and $5\text{-HT}_{1B}$ autoreceptors located at the somatodendritic compartment and axon terminals, respectively. In addition, the synaptic and extracellular levels of 5-HT are regulated by the 5-HT transporter (5-HTT). Genetic risk for anxiety has been associated with all of these macromolecules.

*12.2.1.1 Tryptophan Hydroxylase* The two isoforms of TPH are as follows: 1) the classical TPH isoform, now termed TPH1, that is detected in the periphery, especially in the duodenum and in blood but not in the brain, and 2) the relatively recently identified TPH2 expressed exclusively in the brain (38). Several common single nucleotide polymorphisms are found around the transcriptional control region of TPH2. T allele carriers of a functional polymorphism in the upstream regulatory region of TPH2 (G-703 T) were found to be overrepresented in individuals with anxiety-related personality traits (39, 40). In agreement with these data, T carriers have been shown to exhibit relatively greater activity in the amygdala than do G-allele homozygotes to affective facial expressions (41). Activation of the amygdala, indicated by increased blood flow, is a typical reaction to stress and to unpleasant and potentially harmful stimuli.

*12.2.1.2 The Serotonin Transporter* The transporter removes 5-HT after its release into the synaptic cleft and returns it to the presynaptic terminal, where it is metabolized by monoaminoxidases or stored in secretory vesicles and results in the termination of postsynaptic serotonergic effects (42). 5-HTT is the target of the selective serotonin reuptake inhibitors (SSRIs) that have been shown to be effective in certain anxiety disorders and depression (43).

The 5-HTT transporter belongs to the family of $Na^+/Cl^-$-dependent transporters, which shows a certain degree of similarity with the GABA and the dopamine transporters (42, 44). The 5-HTT has 12 transmembrane domains (TM) with a large extracellular loop between TMs 3 and 4. Both the N- and C-termini are located within the cytoplasm. Growing evidence suggests that 5-HTT forms a homomultimer in the plasma membrane, although most studies suggest an autonomous function for each monomer (45). Amino acid substitution experiments indicate the importance of TM1 as an important contributor to substrate, ion, and inhibitor interactions (46). In addition to TM1, TM3 has also been shown to play a role in substrate and inhibitor binding (47). It has been demonstrated that Tyr-95 and Ile-172 in transmembrane segments 1 and 3 of human 5-HTT interact to establish the high-affinity site for antidepressants (48). Although SSRIs have a rapid action on the transporter, their anxiolytic and antidepressant actions are delayed. It is believed that one factor involved in this delay is the activation of 5-HT$_{1A}$ autoreceptors on serotonergic neurons that effectively reduces neuronal firing and 5-HT release rebalancing extracellular 5-HT levels (43). As the SSRI treatment is prolonged, the 5-HT$_{1A}$ autoreceptor desensitizes, and firing activity is restored. The notion that an adaptive change underlies, at least partly, the delayed therapeutic effect of SSRIs is supported by the acceleration of the anxiolytic and antidepressant response by the concomitant administration of the 5-HT$_{1A}$ autoreceptor antagonist pindolol (49). Once 5-HT levels are significantly increased during chronic SSRI administration, it is assumed that the elevated 5-HT levels act on specific 5-HT receptors to elicit anxiolytic and antidepressant effects. Although at least 13 different types of 5-HT receptors exist in mammals, a report using a specific behavioral test (novelty induced suppression of feeding) showed that the presence of postsynaptic 5-HT$_{1A}$ receptors is necessary for the anxiolytic effect of the SSRI fluoxetine (50).

Lesch et al. demonstrated that a functional 5-HTT promoter polymorphism is associated with the factor "neuroticism" in the revised NEO personality inventory (51). Specifically, individuals who carry the *s/s* (short promoter repeat) alleles of the 5-HTT, as compared with individuals with *s/l* (long) or *l/l* alleles, have increased neuroticism. Extension of these genetic studies to anxiety disorders by the same authors showed no differences in 5-HTT genotype-distribution between anxiety patients and comparison subjects, but among anxiety patients, carriers of the *s/s* alleles exhibited higher neuroticism scores (52). Several later reports found association between the 5-HTT promoter polymorphism and anxiety-related traits (53–55). Also, meta-analyses of several prior studies found a small but significant association between 5-HTT polymorphism and some but not all measures of neuroticism/anxiety (56–58). Interestingly, imaging data found a significant increase in amygdala activity in subjects who carry the *s* allele, which indicates a functional link between 5-HTT polymorphism and anxiety (59). Consistent with these data, genetic inactivation of the 5-HTT gene in mice results in an increased anxiety-like phenotype (60).

The *s* allele is associated with reduced 5-HTT mRNA expression *in vitro* and in lymphoblasts (51), but functional imaging studies and postmortem samples generally show no effect of the 5-HTT genotype on 5-HTT expression in adult subjects (61, 62). To resolve this discrepancy, it has been hypothesized that the *s* allele acts primarily during development and less so in adult brain and would primarily affect brain development leading to anxiety in later life (Fig. 12.2). Consistent with this idea, pharmacological blockade of the 5-HTT in rats and mice during early postnatal life (that mostly corresponds to the third trimester in human) results in changes in emotional behavior (63, 64). However, it is not

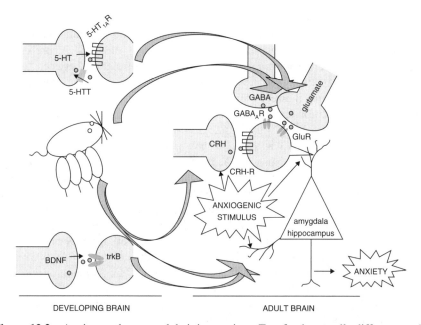

DEVELOPING BRAIN          ADULT BRAIN

**Figure 12.2** Anxiety pathways and their interactions. Two fundamentally different mechanisms involved in anxiety and anxiety-like behavior are proposed. The first mechanism is developmental, and its consequences are manifested later in life. Both genetic and environmental factors linked to anxiety can have a developmental origin, as illustrated by the example of the *s* allele of the 5-HTT, the deficiency in the $5HT_{1A}$ receptor, and variability in maternal care. The second possible mechanism is not developmental but based on "acute" or current molecular abnormalities that result in anxiety or anxiety-like behavior. Examples include deficiencies in GABAergic neurotransmission and abnormalities in the central CRH system, which have the direct behavioral output of anxiety. It is also proposed that the developmental mechanisms lead to anxiety by converging on the "acute" mechanisms.

known how increased 5-HT levels during development lead to lifelong anxiety nor what 5-HT receptors are involved.

*12.2.1.3  The 5-HT$_{1A}$ Receptor*   The G-protein-coupled 5-HT$_{1A}$ receptor emerged as another candidate gene in anxiety as early studies indicated a deficit in the receptor in panic disorder patients (65, 66). These data seem to be consistent with the anxiolytic effect of partial 5-HT$_{1A}$ receptor agonists in the treatment of generalized anxiety disorder, but the pharmacological inhibition of the receptor in animals does not elicit anxiety (67). Then in 1998, it was shown that the genetic inactivation of the 5-HT$_{1A}$ receptor in mice results in enhanced anxiety-like behaviors (68–70) alongside reduced immobility in the forced swim test (70) and tail suspension test (68, 69). Anxiety-related tests in these and follow-up studies included open field, elevated plus maze, zero maze, novelty-induced suppression of feeding, and some fear conditioning paradigms (68–71). More recently, human studies strengthen the association between a 5-HT$_{1A}$ receptor deficiency and certain forms of anxiety. A preliminary neuroimaging study reported a significant negative correlation between 5-HT$_{1A}$ receptor binding in the dorsolateral prefrontal, anterior cingulate, parietal, and occipital cortices and indirect measures of anxiety in healthy volunteers (72). Then, reduced receptor levels were reported in the anterior cingulate, posterior cingulate, and raphe by positron tomography in patients with panic disorder (73). No such association was found in posttraumatic stress disorder (74).

5-HT$_{1A}$ receptors are expressed at both postsynaptic locations in 5-HT target areas (including the amygdala, hippocampus, and cortex) and presynaptic sites on 5-HT neurons in the raphe nuclei as somatodendritic autoreceptors. Because autoreceptors control neuronal firing and consequently 5-HT release, and because an increase in extracellular 5-HT levels during development (see previous section) has been implicated in anxiety, it initially was believed that the anxiety-like phenotype of 5-HT$_{1A}$ receptor-deficient mice was because of increased 5-HT release. However, basal 5-HT levels are not altered, as measured by *in vivo* microdialysis, in 5-HT$_{1A}$ receptor-deficient mice, presumably because of the compensatory action of presynaptic 5-HT$_{1B}$ receptors (75–77), and expression of 5-HT$_{1A}$ receptors in forebrain regions rescued the anxiety phenotype of 5-HT$_{1A}$ receptor knockout mice (78), which suggests that the behavioral phenotype results from the absence of postsynaptic rather than the presynaptic 5-HT$_{1A}$ receptors. The anxiety-like phenotype of 5-HT$_{1A}$ receptor-deficient mice has been shown to be associated with genetic background-specific perturbations in the prefrontal cortex GABA-glutamate system and in GABA$_A$ receptor subunit expression in this region as well as in the amygdala (79, 80) (Fig. 12.2). Our recent studies implicate hippocampal abnormalities in some manifestations of the anxiety-like phenotype in 5-HT$_{1A}$ receptor-deficient mice (unpublished data). Presumably the lack of 5-HT$_{1A}$ receptors during development at these postsynaptic areas leads to neuronal network abnormalities that underlie the exaggerated behavioral responses in anxiogenic environments.

Considering the many studies with 5-HTT polymorphism (see above), it is surprising that relatively little is known about 5-HT$_{1A}$ receptor alleles in the context of anxiety. Although the common genetic polymorphism ($-1019$ C/G) in the promoter region of the 5-HT$_{1A}$ receptor gene has been associated with panic disorder (81), more is known about the link of this polymorphism to depression/suicide (82) (but see Reference 83) and to antidepressant treatment response (84, 85). Nevertheless, the $-1019$ C/G polymorphism has been found to have functional effects on gene expression. Specifically, the 5-HT$_{1A}$ G(-1019) allele fails to bind Deaf-1 and Hes5, two transcriptional repressors, in raphe-derived cells leading to upregulation of autoreceptor expression. Consistent with these data, an increase in 5-HT$_{1A}$ autoreceptor expression in individuals with the G/G genotype has been observed (86, 87). However, as mentioned earlier, autoreceptors may not be implicated in anxiety-like phenotype, at least not in the mouse model. Deaf-1 has an opposite effect in neurons that model the postsynaptic site as it enhances 5-HT$_{1A}$ receptor promoter activity, and this enhancement is attenuated in the presence of the G allele (88). This cell-specific activity of Deaf-1 in 5-HT$_{1A}$ gene regulation could account for the region-specific decrease of receptor level in postsynaptic 5-HT$_{1A}$ receptor described in anxiety and mood disorders (see above).

### 12.2.2   The GABA Pathway

Abnormalities in GABA levels have been noted in subjects with anxiety disorders. Magnetic resonance spectroscopy studies have shown lower occipital cortex GABA concentrations in subjects with panic disorder (89) and with major depression (90), compared with healthy controls. GABA is synthesized from glutamate by glutamic acid decarboxylase (GAD). Its two isoforms, GAD65 and GAD67, are products of two independently regulated genes, *GAD2* and *GAD1*, respectively. GAD65 is responsible for the synthesis of a small pool of GABA associated with nerve terminals and synaptic vesicles and can be rapidly activated in times of high GABA demand (91, 92). In contrast, GAD67 is responsible for the majority of GABA synthesis, and it is important in developmental processes (92). GABA interacts with the ionotropic GABA$_A$ and metabotropic GABA$_B$ receptors. GABA$_A$ receptors in particular have been associated with anxiety as they control neuronal excitability. The GABA$_A$ receptor is a pentameric ion channel composed of subunits from seven families ($\alpha$1-6, $\beta$1-3, $\gamma$1-3, $\delta$, $\epsilon$, $\theta$, and $\rho$1-3)(93). Receptors that express $\alpha$1 and $\alpha$2, together with a $\beta$ subunit and the $\gamma$2 subunit, are the predominant forms in the central nervous system.

*12.2.2.1   GAD and Anxiety*   GABA neurotransmission has been linked to anxiety disorders, and therefore genes that encode GAD are reasonable candidate susceptibility genes for these conditions. In a multivariate structural equation model, several of the six single-nucleotide polymorphisms tested in the *GAD1* region (encoding GAD67) formed a common high-risk haplotype that contributed to individual differences in neuroticism and impacted susceptibility across a range of anxiety disorders, including generalized anxiety disorder, panic disorder, agoraphobia, and social phobia, and also major depression (94). No such association was found with single-nucleotide polymorphisms in *GAD2* (encoding GAD65) in this study. Nevertheless, genetic inactivation of GAD65 results in anxiety-like behavior in mice (95). These mice also show increased seizure sensitivity, but otherwise they show no overt developmental phenotype (91). It is not possible to study anxiety in GAD67$^{-/-}$ mice because they die of severe cleft palate during the first morning after birth (92). The less severe and more specific phenotype of GAD65$^{-/-}$ mice is probably because the overall GABA content is normal in GAD65$^{-/-}$ tissues and only the K$^+$ stimulated GABA release is reduced, whereas in GAD67$^{-/-}$ mice a severe depletion of GABA occurs.

*12.2.2.2   GABA$_A$ Receptors and Anxiety*   A deficit in GABA$_A$ receptors has been identified in the hippocampus and parahippocampus of patients with panic disorder and generalized anxiety disorders (96–98). Also, $GABA_A$ receptor antagonists elicit anxiety in patients with panic disorder, which suggests an underlying deficit in receptor function in these individuals (99).

Animal studies suggest that a reduction in $GABA_A$ receptors that contain $\alpha_4$ subunits is associated with withdrawal-induced anxiety (100, 101). Genetic inactivation of some other subunit genes has a similar effect. For example, heterozygote $\gamma_2^{+/-}$ mice have reduced numbers of $GABA_A$ receptors and display anxiety phenotype in the elevated plus maze and the dark-light box tests (100, 102). In addition, $\gamma_2^{+/-}$ mice show increased responses in the passive avoidance paradigm. These behavioral alterations are associated with a lower single channel conductance, a pronounced deficit of functional receptors, and a reduction in $\alpha_2$/gephyrin containing postsynaptic $GABA_A$ receptor clusters in cortex, hippocampus, and thalamus. Transgenic mice overexpressing either the mouse $\gamma_{2L}$ or $\gamma_2$ s subunit of the $GABA_A$ receptor showed no difference in anxiety-related behavior as compared with wild-type littermates (103). Compensation at the level of $GABA_A$ receptor subunit expression and assembly often occurs when subunit expression is disturbed, which may explain the lack of phenotype in these mice. In contrast to the $\gamma_2$ subunit, the deletion of the also abundant $\alpha_1$ subunit (104) does not increase anxiety or elicit other behavioral abnormalities (105–107), probably because lack of the $\alpha_1$ subunit is compensated and substituted by other $\alpha$ subunits (105). Although $\alpha_2$ subunit-deficient mice have been generated and a point mutation in this subunit (H101 R) abolishes the anxiolytic effect of diazepam (108, 109), it is not entirely clear whether these mice have a change in anxiety-related behavior. Mice deficient for the $\alpha_2$ subunit show a faster habituation to a novel environment, which is not a typical measure of anxiety (110). As mentioned earlier, a strain of $5\text{-}HT_{1A}$ receptor$^{-/-}$ mice have reduced expression of both the $\alpha_1$ and $\alpha_2$ subunits that could explain, at least partly, their increased anxiety phenotype (79) (Fig. 12.2). "Knock-in" mice in the $\alpha_5$ subunit (H105 R) display enhanced trace fear conditioning to threat cues (102). More analysis showed that these knockin mice exhibit a 33% reduction in hippocampal (CA1 and CA3) $\alpha_5$ receptor subunits (102); thus, the phenotype may be simply because of a partial knockout. Finally, the genetic inactivation of $\beta_2$, a predominant $\beta$ subunit, resulted in a more than 50% reduction in the total number of $GABA_A$ receptors and increased locomotor activity in open field, which suggests that these receptors may control motor activity (106). Taken together, these studies in general support the notion that a deficiency in $GABA_A$ receptors results in anxiety and anxiety-like behavior, but it is difficult to assign specific subunit-containing receptors to anxiety.

## 12.2.3   The BDNF Pathway in Anxiety

BDNF is a secretory protein that belongs to the neurotrophin family. A large proportion of neuronal BDNF is secreted in the proform (proBDNF), which is subsequently converted to mature (m) BDNF by extracellular proteases such as plasmin or matrix metalloproteinases (111). A functional single nucleotide polymorphism that produces a valine-(Val)-to-methionine (Met) substitution at aa. 66 in pro-BDNF, first described as altering the intracellular trafficking and activity-dependent secretion of BDNF (112), has been studied extensively in association studies. The outcome of these studies has been variable. One study found that the Met 66 may be a risk allele for anxious temperament as measured by the Tridimensional Personality Questionnaire (113). Another study however concluded that the Val 66 allele is associated with greater neuroticism (114), whereas still another has shown no difference in neuroticism between the Val and Met genotypes (115). A more definitive outcome of the Val–Met polymorphism was reported with mice. Because the mouse does not have the Met allele, it was generated by knockin (116). BDNF$^{Met/Met}$ mice exhibited increased anxiety-related behaviors in a variety of behavioral paradigms including the elevated plus maze, open field, and novelty-induced hypophagia tests (116). Although BDNF expression in BDNF$^{Met/Met}$ mice is equivalent to that in BDNF$^{Val/Val}$ mice, a ~30% deficit in activity-dependent release of BDNF from neurons occurs. This deficit suggests that the anxiety phenotype of BDNF$^{Met/Met}$ mice can be linked to the activity-dependent release of BDNF. Similarly to the BDNF$^{Met/Met}$ mice, BDNF$^{+/-}$ mice have an anxiety-like phenotype and have BDNF levels lower than normal (by 50%); thus, it seems that a partial deficit in BDNF is sufficient to elicit anxiety. The receptor for m-BDNF is trkB, a receptor tyrosine kinase (117). Consistent with the increased anxiety-like phenotype of BDNF$^{+/-}$, conditional BDNF$^{-/-}$, and BDNF$^{Met/Met}$ mice (116, 118), transgenic mice overexpressing trkB in postmitotic neurons in a pattern similar to that of the endogenous receptor display less anxiety in the elevated plus maze test (119).

Similarly to the serotonin-related genes and their polymorphisms, the Met allele of BDNF causes developmental brain abnormalities. Humans heterozygous for the Met allele have smaller hippocampal volumes (120) and perform poorly on hippocampal-dependent memory tasks (112, 121). Consistent with these data, BDNF$^{+/Met}$ or BDNF$^{Met/Met}$ mice have a significant decrease in hippocampal volume as compared with WT mice (116). Also, a significant decrease in dendritic complexity in dentate gyrus neurons occurs in BDNF$^{+/Met}$ and BDNF$^{Met/Met}$ mice. These data raise the possibility that the Met allele contributes to anxiety by modulating brain development (Fig. 12.2).

### 12.2.4   The Central CRH–CRH-R Pathway and Anxiety

CRH is a 41-amino acid neuropeptide in mammals (122), and it is an important mediator of the central stress response (122, 123). The biological function of CRH is determined by the aminated end of the peptide's C-terminus that binds to the extracellular binding pocket of the receptors CRH-R1 and -R2, whereas its N-terminus contacts other sites on the receptor to initiate signaling (124, 125). CRH-R1 in the anterior pituitary is thought to be the subtype through which hypothalamic CRH primarily initiates its effect on pituitary ACTH release (123, 126). ACTH then induces the secretion of corticosteroids in the adrenal. CRH via CRH-Rs, however, has functions outside the hypothalamic–pituitary–adrenal

(HPA) axis, in particular in the amygdala. Indeed, data support the notion that extrahypothalamic CRH, presumably via the central noradrenergic systems, is significantly involved in anxiety (127), whereas dysregulation of the HPA axis via hypothalamic CRH seems to be more characteristic for depression (128).

*12.2.4.1 CRH* An association between behavioral inhibition and three single nucleotide polymorphisms in the CRH gene, including one in the coding sequence of the gene, was found in children at risk for panic disorder (129). Behavioral inhibition is a trait that involves the tendency to display fearful, avoidant, or shy behavior in novel situations. In animal experiments, the central administration of CRH produces behavioral effects that correlate with anxiety, such as reduced exploration in a novel environment or enhanced fear response (130). Also, an anxiety-like phenotype has been described in transgenic mice overexpressing CRH (131, 132). However, mice with a deleted CRH gene, although they had significantly decreased basal corticosterone levels, showed no anxiety (133, 134). One possible explanation is that the central CRH system is redundant. Indeed, CRH-like peptides, urocortin 1–3 (that also bind CRH-Rs), are present in the CNS (135–138). Two groups have generated mice with a deletion of the urocortin 1 gene (139, 140), but only one of these studies found an increased anxiety-like phenotype (139). At the time of writing, no obvious explanation is found for the significant difference observed between the two urocortin 1-deficient mice. Finally, deletion of the CRH binding protein, that normally binds and inactivates CRH, resulted in increased anxiety (141). The authors hypothesized that the inactivation of CRH–BP may increase the "free" or unbound levels of CRH or urocortin that lead to anxiety.

As described earlier, hypothalamic CRH regulates the HPA axis. Some reports have indicated the dysregulation of the HPA axis in PTSD and panic disorders. Small but significant decreases in plasma cortisol levels and increased HPA axis sensitivity to low glucocorticoid negative feedback signals have been reported in PTSD (142). In contrast, other studies showed persistent increases in salivary cortisol levels in pediatric PTSD patients (143) and significantly greater CRH-induced ACTH and cortisol responses in women with chronic PTSD (144). In panic patients, abnormal HPA axis regulation, including increased basal cortisol secretion and overnight hypercortisolemia, have also been documented (129). These HPA axis changes, however, may not be specific for anxiety disorders but rather reflect the presence of comorbid depression (145).

*12.2.4.2 CRH-Rs* No significant association was found between an intron 2 polymorphism in CRH-R1 and the "neuroticism" dimension of personality as assessed by the Revised NEO Personality Inventory in healthy Japanese subjects (146). Similarly, three CRH-R2 gene polymorphisms had no association with panic disorder in another study (147).

The role of CRH-Rs in anxiety-like behavior has been studied extensively by using knockout mice. Whereas mice lacking CRH-R1 display decreased anxiety in the light–dark box and the elevated plus maze (148–150), CRH-R2-deficient mice, generated independently by three groups, exhibit varying degrees of anxiety-related behavior. In one study, increased anxiety was reported in the elevated plus maze and open field but not in the light–dark box test (151); another study found anxious behavior in both the elevated plus maze and light–dark box, but only in males (152). However, this latter study showed an increased time spent by the knockout mice in the center of an open field, which is more consistent with reduced anxiety. Still another report found no significant change in anxiety behavior in the elevated plus maze or open field (153). More recently, a mouse with a conditional deletion of the CRH-R1 in forebrain, hippocampus, and the amygdala, but with normal expression in the anterior pituitary, showed markedly reduced anxiety-like behavioral responses in two avoidance tests, and it exhibited normal ACTH and corticosterone secretion to stress (154).

Taken together, some of behavioral data obtained with various CRH and CRH-R knockout mice suggest that extrahypothalamic CRH and/or urocortin mediate a dual modulation of anxiety behavior. Activation of CRH-R1 seems to be anxiogenic, whereas activation of CRH-R2 is anxiolytic. Therefore, it may not be surprising that dual CRH-R1/2 knockout mice have only a subtle behavioral phenotype (155).

*12.2.4.3   CRH Signaling and Anxiety*   CRH-R1/2 are coupled to several signaling pathways, including the adenyl cyclase-protein kinase A (PKA)-CREB and the ERK-mitogen-activated protein kinase (MAPK) pathways (156). These pathways can also be linked to anxiety. For example, genetic inactivation of adenylyl cyclase type 8 results in reduced anxiety-like behavior (157). Mice with mutations of the PKA RIIβ subunit and CREB also exhibit abnormal anxiety responses (158–160), and activation of CREB in amygdala produces anxiety-like behavior (160). Thus, anxiety-like responses may be initiated and regulated by Gs-coupled CRH receptor signaling, at least partly, via the cyclic AMP–PKA–CREB pathway. The involvement of the ERK–MAPK pathway can also be implicated in CRH-related anxiety-like behavior. ERK1/2 is strongly activated in hippocampal CA1 and CA3 pyramidal cells and basolateral amygdala by the intracerebroventricular administration of CRH (161), and CRH-induced phosphorylation of ERK1/2 was absent in mice with a conditional knockout of forebrain and limbic CRH-R1 exhibiting a low level of anxiety (162). One may hypothesize that processing of anxiogenic stimuli is altered in the amygdala, hippocampus, or other relevant brain region as a result of abnormal CRH signaling that consecutively leads to the behavioral manifestations of anxiety (Fig. 12.2).

## 12.3   ENVIRONMENTAL FACTORS AND RELATED BIOLOGICAL PATHWAYS INVOLVED IN ANXIETY

### 12.3.1   Stressful Life Events and Anxiety

Although most frequently associated with depression (163), stressful life events have also been linked to anxiety disorders (164–166). Kendler et al. attempted to determine if anxiety, specifically GAD symptoms and depression, are associated with different dimensions (humiliation, entrapment, loss, and danger) of stressful life events associated with high contextual threat. Onset of GAD symptoms was predicted by higher ratings of loss and danger, whereas depression was associated with the combination of humiliation and loss, which indicates that event dimensions that predispose to pure GAD episodes versus pure depression can be distinguished with moderate specificity (164). Unfortunately, little is known about the molecular and cellular mechanisms elicited by stressful life events and how these events predispose an individual to anxiety or depression. However, animal studies provided mechanistic insights of how early-life events can contribute to the development of adult anxiety (discussed below).

### 12.3.2   Maternal Care and Adult Life Anxiety in Rodents

Brief "handling" of rat pups results in a lifelong decrease to behavioral and endocrine effects of stress, whereas animals separated from their mothers/litters for longer periods of time, for example, for several hours, exhibit increased anxiety (167). Later studies determined that the critical effect of short-term handling is the increase in maternal care (licking and grooming) after the return of the pups to the nest (3).

More studies showed that rat pups nursed by mothers selected for either a high or a low level of licking and grooming (LG) and arched-back nursing (ABN) exhibit a decreased and increased level of anxiety-like behavior in adult life (open field, novelty-suppressed feeding, and shock-probe burying assays), respectively (3, 168–170). Offspring of high LG–ABN mothers (as well as briefly handled pups) show increased glucocorticoid feedback sensitivity, increased hippocampal GR mRNA expression, and decreased hypothalamic CRH mRNA levels (171). Because the activity of the hypothalamic CRH and the function of the HPA axis may not be directly related to anxiety-like behavior (128, 154), it is not clear whether these mechanisms can explain the anxiety-related behavioral consequences of maternal care. Perhaps more plausible mechanisms are the numerous neurochemical changes related to differences in maternal care. For example, rat pups of high LG–ABN dams show altered $GABA_A$ receptor subunit expression in the amygdala, locus coeruleus, medial prefrontal cortex, and hippocampus that could contribute to their reduced anxiety-like behavior as compared with pups from low LG–ABN dams (172, 173) (Fig. 12.2). In addition to the GABAergic system, other potential factors mediating the environmental effects include the glutamatergic system and neurotrophins such as BDNF. Liu et al. found that increased LG–ABN of offspring resulted in increased hippocampal mRNA expression of NR2A and NR2B NMDA receptor subunits at postnatal day 8, a change that was sustained into adulthood (174). Also, increased levels of BDNF, but not NGF or NT-3, mRNA were observed in the dorsal hippocampus of 8-day old high LG–ABN pups (174). Neuronal network changes that could directly explain the behavioral consequences of maternal care have also been found as adult offspring of high LG–ABN dams show increased hippocampal synaptogenesis as compared with low LG–ABN offspring, and it has been found that this change could be normalized when low LG–ABN pups are cross-fostered to high LG–ABN dams (174) (Fig. 12.2).

## 12.4   INTERACTION BETWEEN GENETIC AND ENVIRONMENTAL FACTORS

Although the interaction of genes and environment in shaping behavior is well accepted, direct experimental evidence to support their role in the pathogenesis of psychiatric diseases has been difficult to obtain. However, association studies with the 5-HTT polymorphism have indicated, at least in depressive disorders, that genetic and environmental factors act together, enhancing the phenotype beyond the level established by either factor alone. In 2003, Caspi et al. reported that carriers of the $s$, transcriptionally less active, allele of 5-HTT are more likely to develop depression after stressful life events or childhood abuse than individuals homozygous for the $l$ allele (175). The actual interpretation of these data was that the $l$ allele moderates the environmental effect. The majority of replication studies have provided results consistent with such effect (176). It seems, however, that this environment × gene interaction is specific for depressive symptomatology as it was not identified for generalized anxiety disorder or anxiety symptoms (177, 178). However, the combined effects of early-life adversity (nursery rearing as opposed to maternal rearing) and 5-HTT polymorphism in anxiety behavior have been found in primates. Rhesus monkeys have a 5-HTT polymorphism similar to that of the human, and it was shown that although both mother- and nursery-reared heterozygote ($l/s$) animals demonstrate increased affective responding (a measure of temperament) relative to $l/l$ homozygotes (a genotype effect), nursery-reared, but not mother-reared, $l/s$ infants exhibited lower orientation scores (a measure of visual orientation, visual tracking, and attentional capabilities) than their $l/l$ counterparts (an environment × genotype interaction) (179). Also, monkeys with deleterious early-rearing experiences could be differentiated by genotype in cerebrospinal fluid concentrations of the 5-HT metabolite, 5-hydroxyindoleacetic acid, whereas monkeys reared normally were not (180).

In rodents, several environment × gene interaction studies related to anxiety behavior have been conducted. For example, early-life handling or cross-fostering of genetically highly neophobic BALB/c mice to less neophobic C57BL/6 mice equalizes both the behavioral and the benzodiazepine receptor expression differences between these two strains (170, 181–184) and, thus, indicates the moderating effect of the environment on genetic disposition to anxiety-like behavior. Also, $5HT_{1A}$ receptor-deficient mice have increased ultrasonic vocalization (USV) when reared by $5HT_{1A}$ receptor$^{-/-}$ instead of $5HT_{1A}$ receptor$^{+/+}$ dams (185), which indicates that the maternal environment, presumably in interaction with the offspring genotype, alters offspring behavior. However, contrary to the expectation, $5HT_{1A}R^{+/-}$ offspring reared by $5HT_{1A}R^{-/-}$ mothers have decreased measures of anxiety in the elevated plus maze as adults when compared with $5HT_{1A}R^{+/-}$ offspring raised by $5HT_{1A}R^{+/+}$ dams. Although it is difficult to consolidate these conflicting results, it seems that the level of anxiety in $5HT_{1A}$ receptor-deficient mice can be altered by the maternal environment.

In addition to early environmental influences, later-life or adult environment can also influence the expression of a genetic effect on emotionality, as demonstrated in mouse models. Lack of the nociceptin/orphanin FQ gene leads to an enhanced anxiety phenotype in mice (186, 187), but the degree of this phenotype is dependent on environmental influences such as social interactions. Ouagazzal et al. found that homozygous mutant animals, when housed alone, performed similarly to their wild-type controls on tests of emotional reactivity. Enhanced emotionality became apparent only when the singly-housed animals were introduced to group housing (five animals/cage) that induced greater levels of aggression and increased anxiety responses (188).

## 12.5 CONVERGENCE OF ANXIETY-RELATED PATHWAYS AND MECHANISMS

Two fundamentally different mechanisms associated with anxiety and anxiety-like behavior seem to exist: one that has a developmental origin with the cause and the adult manifestations of the phenotype separated in time and another mechanism that presents itself in "acute" settings (Fig. 12.2). Typical examples for developmental anxiety are those caused by the $s$ allele of the 5-HTT, the deficiency in the $5HT_{1A}$ receptor and BDNF, and the variability in maternal care. All of these produce their effect in adults only when present during prenatal and/or early postnatal life. Indeed, genetic inactivation of the $5HT_{1A}$ receptor during adult life is not accompanied by anxiety. Because the individual genetic risk factors mentioned above have subtle effects, a combination of them may be necessary to lead to a significant level of anxiety. For example, $BDNF^{+/-}$ and $5\text{-}HTT^{+/-}$ double mutants show a more pronounced anxiety phenotype as compared with singly heterozygous mice (189).

In other anxiety forms and models, developmental mechanisms do not seem to play a role. A typical example is represented by a deficient GABAergic neurotransmission (as a result of less GABA availability or altered composition of the $GABA_A$ receptor) that results in reduced neuronal inhibition and increased excitation. The direct behavioral output of these neuronal and neuronal network changes is anxiety (Fig. 12.2). Another example is the central CRH system as its pharmacological manipulation acutely alters anxiety levels. Because many developmentally relevant anxiety genes and environmental effects and their corresponding mechanisms (see above) modulate GABAergic and glutamatergic transmission as well as central CRH signaling, it is possible that the developmental mechanisms converge on and are interconnected sequentially to the "acute" mechanisms that ultimately lead to the anxiety behavior (Fig. 12.2).

In summary, it seems that anxiety in many cases has a developmental origin whether it is elicited by genetic polymorphisms or environmental effects. Because individual genetic influences have small effects, it is believed that multiple genetic risk genes, together with adverse environment, are required to have a large enough impact to cause anxiety. Although some early signs of anxiety may be manifested during childhood, anxiety becomes more apparent in predisposed individuals as the brain and behavior mature during adolescence. Although relatively little is known about how early developmental mechanisms lead to anxiety in later life, data suggest that long-lasting alterations in neurotransmitter systems (GABA, glutamate, and CRH) and/or the morphology and function of neuronal networks (amygdala, hippocampus, etc.) are involved. Additional environmental influences during adolescence and adulthood can increase the incidence and severity of anxiety.

## REFERENCES

1. Hettema JM, Neale MC, Kendler KS. A review and meta-analysis of the genetic epidemiology of anxiety disorders. Am. J. Psych. 2001;158:1568–1578.
2. Caspi A, Moffitt TE. Gene-environment interactions in psychiatry: joining forces with neuroscience. Nat. Rev. Neurosci. 2006;7:583–590.
3. Francis DD, Meaney MJ. Maternal care and the development of stress responses. Curr. Opin. Neurobiol. 1999;9:128–134.
4. Nemeroff CB. Early-Life Adversity, CRF dysregulation, and vulnerability to mood and anxiety disorders. Psychopharmacol. Bull. 2004;38(suppl 1): 14–20.
5. Costa PTJ, McCrae RR. Normal personality assessment in clinical practice: The NEO Personality Inventory. Psych. Assess. 1992;4:5–13.
6. McAdams DP. The five-factor model in personality: a critical appraisal. J. Personal. 1992;60:329–361.
7. McCrae RR, Costa PT Jr. Personality trait structure as a human universal. Am. Psychol. 1997;52:509–516.
8. John OP. The "Big Five" factor taxonomy: dimensions of personality in the natural language and in questionnaires. In: Handbook of Personality: Theory and Research. Pervin LA, ed. 1990. Guilford, New York. pp. 66–100.
9. Eysenck HJ. Biological basis of personality. Nature 1963;199:1031–1034.
10. Soldz S, Vaillant GE. The Big Five personality traits and the life course: A 45-year longitudinal study. J. Res. Personal. 1999;33:208–232.
11. Costa PTJ, McCrae RR. Normal personality assessment in clinical practice: the NEO Personality Inventory. Psych. Assess. 1992;4:5–13.
12. Gray JA, McNaughton N. The Neuropsychology of Anxiety. 2000. Oxford University Press, Oxford, UK.
13. Cloninger CR. A systematic method for clinical description and classification of personality variants. A proposal. Arch. Gen. Psychiatry 1987;44:573–588.
14. Clark LA, Watson D, Mineka S. Temperament, personality, and the mood and anxiety disorders. J. Abnorm. Psychol. 1994;103:103–116.
15. Hettema JM, Prescott CA, Kendler KS. Genetic and environmental sources of covariation between generalized anxiety disorder and neuroticism. Am. J. Psychiatry 2004;161:1581–1587.

16. Hayward C, Killen JD, Kraemer HC, Taylor CB. Predictors of panic attacks in adolescents. J. Am. Acad. Child Adolesc. Psychiatry 2000;39:207–214.

17. World Health Organization. The ICD-10 Classification of Mental and Behavioral Disorders - Clinical Description and Diagnostic Guidelines. 1992. World Health Organization, Geneva, Switzerland.

18. Jang KL, McCrae RR, Angleitner A, Riemann R, Livesley WJ. Heritability of facet-level traits in a cross-cultural twin sample: support for a hierarchical model of personality. J. Pers. Soc. Psychol. 1998;74:1556–1565.

19. Loehlin JC, Horn JM, Willerman L. Heredity, environment, and personality change: evidence from the Texas Adoption Project. J. Pers. 1990;58:221–243.

20. Middeldorp CM, Birley AJ, Cath DC, Gillespie NA, Willemsen G, Statham DJ, de Geus EJ, Andrews JG, van Dyck R, Beem AL, et al. Familial clustering of major depression and anxiety disorders in Australian and Dutch twins and siblings. Twin Res. Hum. Genet. 2005;8:609–615.

21. Hettema JM, Prescott CA, Myers JM, Neale MC, Kendler KS. The structure of genetic and environmental risk factors for anxiety disorders in men and women. Arch. Gen. Psychiatry 2005;62:182–189.

22. Eysenck HJIE. Biological dimensions of personality. In: Handbook of Personality: Theory and Research. Pervin LA, ed. 1990. Guilford, New York. pp. 244–276.

23. Eysenck HJ. The Biological Basis of Personality. 1967. Charles C. Thomas, Springfield, IL.

24. Fullerton J, Cubin M, Tiwari H, Wang C, Bomhra A, Davidson S, Miller S, Fairburn C, Goodwin G, Neale MC, et al. Linkage analysis of extremely discordant and concordant sibling pairs identifies quantitative-trait loci that influence variation in the human personality trait neuroticism. Am. J. Hum. Genet. 2003;72:879–890.

25. Neale BM, Sullivan PF, Kendler KS. A genome scan of neuroticism in nicotine dependent smokers. Am. J. Med. Genet. B Neuropsychiatr. Genet. 2005;132:65–69.

26. Middeldorp CM, Hottenga JJ, Slagboom PE, Sullivan PF, de Geus EJ, Posthuma D, Willemsen G, Boomsma DI. Linkage on chromosome 14 in a genome-wide linkage study of a broad anxiety phenotype. Mol. Psychiatry 2008;13:84–89.

27. Kaabi B, Gelernter J, Woods SW, Goddard A, Page GP, Elston RC. Genome scan for loci predisposing to anxiety disorders using a novel multivariate approach: strong evidence for a chromosome 4 risk locus. Am. J. Hum. Genet. 2006;78:543–553.

28. Berrettini WH, Harris N, Ferraro TN, Vogel WH. Maudsley reactive and non-reactive rats differ in exploratory behavior but not in learning. Psychiatr. Genet. 1994;4:91–94.

29. Liebsch G, Linthorst AC, Neumann ID, Reul JM, Holsboer F, Landgraf R. Behavioral, physiological, and neuroendocrine stress responses and differential sensitivity to diazepam in two Wistar rat lines selectively bred for high- and low-anxiety-related behavior. Neuropsychopharmacology 1998;19:381–396.

30. Hogg S. A review of the validity and variability of the elevated plus-maze as an animal model of anxiety. Pharmacol. Biochem. Behav. 1996;54:21–30.

31. Bourin M, Hascoet M. The mouse light/dark box test. Eur. J. Pharmacol. 2003;463:55–65.

32. Blanchard DC, Griebel G, Blanchard RJ. The Mouse Defense Test Battery: pharmacological and behavioral assays for anxiety and panic. Eur. J. Pharmacol. 2003;463:97–116.

33. Belzung C. Rodent models of anxiety-like behaviors: are they predictive for compounds acting via non-benzodiazepine mechanisms? Curr. Opin. Invest. Drugs 2001;2:1108–1111.

34. Uys JD, Stein DJ, Daniels WM, Harvey BH. Animal models of anxiety disorders. Curr. Psychiatry Rep. 2003;5:274–281.

35. Henderson ND, Turri MG, DeFries JC, Flint J. QTL analysis of multiple behavioral measures of anxiety in mice. Behav. Genet. 2004;34:267–293.

36. Flint J, Corley R, DeFries JC, Fulker DW, Gray JA, Miller S, Collins AC. A simple genetic basis for a complex psychological trait in laboratory mice. Science 1995;269:1432–1435.

37. Lucki I. The spectrum of behaviors influenced by serotonin. Biol. Psychiatry 1998;44:151–162.

38. Walther DJ, Peter JU, Bashammakh S, Hortnagl H, Voits M, Fink H, Bader M. Synthesis of serotonin by a second tryptophan hydroxylase isoform. Science 2003;299:76.

39. Mossner R, Freitag CM, Gutknecht L, Reif A, Tauber R, Franke P, Fritze J, Wagner G, Peikert G, Wenda B, et al. The novel brain-specific tryptophan hydroxylase-2 gene in panic disorder. J. Psychopharmacol. 2006;20:547–552.

40. Reuter M, Kuepper Y, Hennig J. Association between a polymorphism in the promoter region of the TPH2 gene and the personality trait of harm avoidance. Int. J. Neuropsychopharmacol. 2007;10:401–404.

41. Brown SM, Peet E, Manuck SB, Williamson DE, Dahl RE, Ferrell RE, Hariri AR. A regulatory variant of the human tryptophan hydroxylase-2 gene biases amygdala reactivity. Mol. Psychiatry 2005;10:884–888, 805.

42. Amara SG, Kuhar MJ. Neurotransmitter transporters: recent progress. Annu. Rev. Neurosci. 1993;16:73–93.

43. Blier P, Pineyro G, el Mansari M, Bergeron R, de Montigny C. Role of somatodendritic 5-HT autoreceptors in modulating 5-HT neurotransmission. Ann. N. Y. Acad. Sci. 1998;861:204–216.

44. Blakely RD, Berson HE, Fremeau RT Jr, Caron MG, Peek MM, Prince HK, Bradley CC. Cloning and expression of a functional serotonin transporter from rat brain. Nature 1991;354:66–70.

45. Kilic F, Rudnick G. Oligomerization of serotonin transporter and its functional consequences. Proc. Natl. Acad. Sci. U. S. A. 2000;97:3106–3111.

46. Henry LK, Adkins EM, Han Q, Blakely RD. Serotonin and cocaine-sensitive inactivation of human serotonin transporters by methanethiosulfonates targeted to transmembrane domain I. J. Biol. Chem. 2003;278:37052–37063.

47. Chen JG, Sachpatzidis A, Rudnick G. The third transmembrane domain of the serotonin transporter contains residues associated with substrate and cocaine binding. J. Biol. Chem. 1997;272:28321–28327.

48. Henry LK, Field JR, Adkins EM, Parnas ML, Vaughan RA, Zou MF, Newman AH, Blakely RD. Tyr-95 and Ile-172 in transmembrane segments 1 and 3 of human serotonin transporters interact to establish high affinity recognition of antidepressants. J. Biol. Chem. 2006;281:2012–2023.

49. Blier P, Ward NM. Is there a role for 5-HT1A agonists in the treatment of depression? Biol. Psychiatry 2003;53:193–203.

50. Santarelli L, Saxe M, Gross C, Surget A, Battaglia F, Dulawa S, Weisstaub N, Lee J, Duman R, Arancio O, et al. Requirement of hippocampal neurogenesis for the behavioral effects of antidepressants. Science 2003;301:805–809.

51. Lesch KP, Bengel D, Heils A, Sabol SZ, Greenberg BD, Petri S, Benjamin J, Muller CR, Hamer DH, Murphy DL. Association of anxiety-related traits with a polymorphism in the serotonin transporter gene regulatory region. Science 1996;274:1527–1531.

52. Jacob CP, Strobel A, Hohenberger K, Ringel T, Gutknecht L, Reif A, Brocke B, Lesch KP. Association between allelic variation of serotonin transporter function and neuroticism in anxious cluster C personality disorders. Am. J. Psychiatry 2004;161:569–572.

53. Katsuragi S, Kunugi H, Sano A, Tsutsumi T, Isogawa K, Nanko S, Akiyoshi J. Association between serotonin transporter gene polymorphism and anxiety-related traits. Biol. Psychiatry 1999;45:368–370.

54. Mazzanti CM, Lappalainen J, Long JC, Bengel D, Naukkarinen H, Eggert M, Virkkunen M, Linnoila M, Goldman D. Role of the serotonin transporter promoter polymorphism in anxiety-related traits. Arch. Gen. Psychiatry 1998;55:936–940.

55. Melke J, Landen M, Baghei F, Rosmond R, Holm G, Bjorntorp P, Westberg L, Hellstrand M, Eriksson E. Serotonin transporter gene polymorphisms are associated with anxiety-related personality traits in women. Am. J. Med. Genet. 2001;105:458–463.

56. Sen S, Burmeister M, Ghosh D. Meta-analysis of the association between a serotonin transporter promoter polymorphism (5-HTTLPR) and anxiety-related personality traits. Am. J. Med. Genet. B Neuropsychiatr. Genet. 2004;127:85–89.

57. Schinka JA, Busch RM, Robichaux-Keene N. A meta-analysis of the association between the serotonin transporter gene polymorphism (5-HTTLPR) and trait anxiety. Mol. Psychiatry 2004;9:197–202.

58. Munafo MR, Clark T, Flint J. Does measurement instrument moderate the association between the serotonin transporter gene and anxiety-related personality traits? A meta-analysis. Mol. Psychiatry 2005;10:415–419.

59. Hariri AR, Mattay VS, Tessitore A, Kolachana B, Fera F, Goldman D, Egan MF, Weinberger DR. Serotonin transporter genetic variation and the response of the human amygdala. Science 2002;297:400–403.

60. Murphy DL, Li Q, Engel S, Wichems C, Andrews A, Lesch KP, Uhl G. Genetic perspectives on the serotonin transporter. Brain Res. Bull. 2001;56:487–494.

61. Mann JJ, Huang YY, Underwood MD, Kassir SA, Oppenheim S, Kelly TM, Dwork AJ, Arango V. A serotonin transporter gene promoter polymorphism (5-HTTLPR) and prefrontal cortical binding in major depression and suicide. Arch. Gen. Psychiatry 2000;57:729–738.

62. Parsey RV, Hastings RS, Oquendo MA, Hu X, Goldman D, Huang YY, Simpson N, Arcement J, Huang Y, Ogden RT, et al. Effect of a triallelic functional polymorphism of the serotonin-transporter-linked promoter region on expression of serotonin transporter in the human brain. Am. J. Psychiatry 2006;163:48–51.

63. Ansorge MS, Zhou M, Lira A, Hen R, Gingrich JA. Early-life blockade of the 5-HT transporter alters emotional behavior in adult mice. Science 2004;306:879–881.

64. Maciag D, Simpson KL, Coppinger D, Lu Y, Wang Y, Lin RC, Paul IA. Neonatal antidepressant exposure has lasting effects on behavior and serotonin circuitry. Neuropsychopharmacology 2006;31:47–57.

65. Lesch KP, Wiesmann M, Hoh A, Muller T, Disselkamp-Tietze J, Osterheider M, Schulte HM. 5-HT1A receptor-effector system responsivity in panic disorder. Psychopharmacology (Berl) 1992;106:111–117.

66. Lopez JF, Chalmers DT, Little KY, Watson SJ. A.E. Bennett Research Award. Regulation of serotonin1A, glucocorticoid, and mineralocorticoid receptor in rat and human hippocampus: implications for the neurobiology of depression. Biol. Psychiatry 1998;43:547–573.

67. Millan MJ. The neurobiology and control of anxious states. Prog. Neurobiol. 2003;70:83–244.

68. Ramboz S, Oosting R, Amara DA, Kung HF, Blier P, Mendelsohn M, Mann JJ, Brunner D, Hen R. Serotonin receptor 1A knockout: an animal model of anxiety-related disorder. Proc. Natl. Acad. Sci. U. S. A. 1998;95:14476–14481.

69. Heisler LK, Chu HM, Brennan TJ, Danao JA, Bajwa P, Parsons LH, Tecott LH. Elevated anxiety and antidepressant-like responses in serotonin 5-HT1A receptor mutant mice. Proc. Natl. Acad. Sci. U. S. A. 1998;95:15049–15054.

70. Parks CL, Robinson PS, Sibille E, Shenk T, Toth M. Increased anxiety of mice lacking the serotonin1A receptor. Proc. Natl. Acad. Sci. U. S. A. 1998;95:10734–10739.

71. Sibille E, Pavlides C, Benke D, Toth M. Genetic inactivation of the Serotonin(1A) receptor in mice results in downregulation of major GABA(A) receptor alpha subunits, reduction of GABA(A) receptor binding, and benzodiazepine-resistant anxiety. J. Neurosci. 2000;20:2758–2765.

72. Tauscher J, Bagby RM, Javanmard M, Christensen BK, Kasper S, Kapur S. Inverse relationship between serotonin 5-HT(1A) receptor binding and anxiety: a ((11)C)WAY-100635 PET investigation in healthy volunteers. Am. J. Psychiatry 2001;158:1326–1328.

73. Neumeister A, Bain E, Nugent AC, Carson RE, Bonne O, Luckenbaugh DA, Eckelman W, Herscovitch P, Charney DS, Drevets WC. Reduced serotonin type 1A receptor binding in panic disorder. J. Neurosci. 2004;24:589–591.

74. Bonne O, Bain E, Neumeister A, Nugent AC, Vythilingam M, Carson RE, Luckenbaugh DA, Eckelman W, Herscovitch P, Drevets WC, et al. No change in serotonin type 1A receptor binding in patients with posttraumatic stress disorder. Am. J. Psychiatry 2005;162:383–385.

75. He M, Sibille E, Benjamin D, Toth M, Shippenberg T. Differential effects of 5-HT1A receptor deletion upon basal and fluoxetine-evoked 5-HT concentrations as revealed by in vivo microdialysis. Brain Res. 2001;902:11–17.

76. Parsons LH, Kerr TM, Tecott LH. 5-HT(1A) receptor mutant mice exhibit enhanced tonic, stress-induced and fluoxetine-induced serotonergic neurotransmission. J. Neurochem. 2001;77:607–617.

77. Bortolozzi A, Amargos-Bosch M, Toth M, Artigas F, Adell A. In vivo efflux of serotonin in the dorsal raphe nucleus of 5-HT1A receptor knockout mice. J. Neurochem. 2004;88:1373–1379.

78. Gross C, Zhuang X, Stark K, Ramboz S, Oosting R, Kirby L, Santarelli L, Beck S, Hen R. Serotonin1A receptor acts during development to establish normal anxiety-like behaviour in the adult. Nature 2002;416:396–400.

79. Bailey SJ, Toth M. Variability in the benzodiazepine response of serotonin 5-HT1A receptor null mice displaying anxiety-like phenotype: evidence for genetic modifiers in the 5-HT-mediated regulation of GABA(A) receptors. J. Neurosci. 2004;24:6343–6351.

80. Bruening S, Oh E, Hetzenauer A, Escobar-Alvarez S, Westphalen RI, Hemmings HC Jr, Singewald N, Shippenberg T, Toth M. The anxiety-like phenotype of 5-HT receptor null mice is associated with genetic background-specific perturbations in the prefrontal cortex GABA-glutamate system. J. Neurochem. 2006;99:892–899.

81. Strobel A, Gutknecht L, Rothe C, Reif A, Mossner R, Zeng Y, Brocke B, Lesch KP. Allelic variation in 5-HT1A receptor expression is associated with anxiety- and depression-related personality traits. J. Neural Transm. 2003;110:1445–1453.

82. Lemonde S, Turecki G, Bakish D, Du L, Hrdina PD, Bown CD, Sequeira A, Kushwaha N, Morris SJ, Basak A, et al. Impaired repression at a 5-hydroxytryptamine 1A receptor gene polymorphism associated with major depression and suicide. J. Neurosci. 2003;23:8788–8799.

83. Hettema JM, An SS, van den Oord EJ, Neale MC, Kendler KS, Chen X. Association study between the serotonin 1A receptor (HTR1A) gene and neuroticism, major depression, and anxiety disorders. Am. J. Med. Genet. B Neuropsychiatr. Genet. 2007.

84. Lemonde S, Du L, Bakish D, Hrdina P, Albert PR. Association of the C(-1019)G 5-HT1A functional promoter polymorphism with antidepressant response. Int. J. Neuropsychopharmacol. 2004;7:501–506.

85. Serretti A, Artioli P, Lorenzi C, Pirovano A, Tubazio V, Zanardi R. The C(-1019)G polymorphism of the 5-HT1A gene promoter and antidepressant response in mood disorders: preliminary findings. Int J Neuropsychopharmacol 2004;7:453–60.

86. David SP, Murthy NV, Rabiner EA, Munafo MR, Johnstone EC, Jacob R, Walton RT, Grasby PM. A functional genetic variation of the serotonin (5-HT) transporter affects 5-HT1A receptor binding in humans. J. Neurosci. 2005;25:2586–2590.

87. Parsey RV, Oquendo MA, Ogden RT, Olvet DM, Simpson N, Huang YY, Van Heertum RL, Arango V, Mann JJ. Altered serotonin 1A binding in major depression: a (carbonyl-C-11)WAY100635 positron emission tomography study. Biol. Psychiatry 2006;59:106–113.

88. Czesak M, Lemonde S, Peterson EA, Rogaeva A, Albert PR. Cell-specific repressor or enhancer activities of Deaf-1 at a serotonin 1A receptor gene polymorphism. J. Neurosci. 2006;26:1864–1871.

89. Goddard AW, Mason GF, Almai A, Rothman DL, Behar KL, Petroff OA, Charney DS, Krystal JH. Reductions in occipital cortex GABA levels in panic disorder detected with 1h-magnetic resonance spectroscopy. Arch. Gen. Psychiatry 2001;58:556–561.

90. Sanacora G, Gueorguieva R, Epperson CN, Wu YT, Appel M, Rothman DL, Krystal JH, Mason GF. Subtype-specific alterations of gamma-aminobutyric acid and glutamate in patients with major depression. Arch. Gen. Psychiatry 2004;61:705–713.

91. Asada H, Kawamura Y, Maruyama K, Kume H, Ding R, Ji FY, Kanbara N, Kuzume H, Sanbo M, Yagi T, et al. Mice lacking the 65kDa isoform of glutamic acid decarboxylase (GAD65) maintain normal levels of GAD67 and GABA in their brains but are susceptible to seizures. Biochem. Biophys. Res. Commun. 1996;229:891–895.

92. Asada H, Kawamura Y, Maruyama K, Kume H, Ding RG, Kanbara N, Kuzume H, Sanbo M, Yagi T, Obata K. Cleft palate and decreased brain gamma-aminobutyric acid in mice lacking the 67-kDa isoform of glutamic acid decarboxylase. Proc. Natl. Acad. Sci. U. S. A. 1997;94:6496–6499.

93. Mohler H, Fritschy JM, Rudolph U. A new benzodiazepine pharmacology. J. Pharmacol. Exp. Ther. 2002;300:2–8.

94. Hettema JM, An SS, Neale MC, Bukszar J, van den Oord EJ, Kendler KS, Chen X. Association between glutamic acid decarboxylase genes and anxiety disorders, major depression, and neuroticism. Mol. Psychiatry 2006;11:752–762.

95. Kash SF, Tecott LH, Hodge C, Baekkeskov S. Increased anxiety and altered responses to anxiolytics in mice deficient in the 65-kDa isoform of glutamic acid decarboxylase. Proc. Natl. Acad. Sci. U. S. A. 1999;96:1698–1703.

96. Schlegel S, Steinert H, Bockisch A, Hahn K, Schloesser R, Benkert O. Decreased benzodiazepine receptor binding in panic disorder measured by IOMAZENIL-SPECT. A preliminary report. Eur. Arch. Psychiatry Clin. Neurosci. 1994;244:49–51.

97. Kaschka W, Feistel H, Ebert D. Reduced benzodiazepine receptor binding in panic disorders measured by iomazenil SPECT. J. Psychiatr. Res. 1995;29:427–434.

98. Tiihonen J, Kuikka J, Rasanen P, Lepola U, Koponen H, Liuska A, Lehmusvaara A, Vainio P, Kononen M, Bergstrom K, et al. Cerebral benzodiazepine receptor binding and distribution in generalized anxiety disorder: a fractal analysis. Mol. Psychiatry 1997;2:463–471.

99. Nutt DJ, Glue P, Lawson C, Wilson S. Flumazenil provocation of panic attacks. Evidence for altered benzodiazepine receptor sensitivity in panic disorder. Arch. Gen. Psychiatry 1990;47:917–925.

100. Essrich C, Lorez M, Benson JA, Fritschy JM, Luscher B. Postsynaptic clustering of major GABAA receptor subtypes requires the gamma 2 subunit and gephyrin. Nat. Neurosci. 1998;1:563–571.

101. Smith SS, Gong QH, Hsu FC, Markowitz RS, ffrench-Mullen JM, Li X. GABA(A) receptor alpha4 subunit suppression prevents withdrawal properties of an endogenous steroid. Nature 1998;392:926–930.

102. Crestani F, Lorez M, Baer K, Essrich C, Benke D, Laurent JP, Belzung C, Fritschy JM, Luscher B, Mohler H. Decreased GABAA-receptor clustering results in enhanced anxiety and a bias for threat cues. Nat. Neurosci. 1999;2:833–839.

103. Wick MJ, Radcliffe RA, Bowers BJ, Mascia MP, Luscher B, Harris RA, Wehner JM. Behavioural changes produced by transgenic overexpression of gamma2L and gamma2S subunits of the GABAA receptor. Eur. J. Neurosci. 2000;12:2634–2638.

104. McKernan RM, Whiting PJ. Which GABAA-receptor subtypes really occur in the brain? Trends Neurosci. 1996;19:139–143.

105. Kralic JE, O'Buckley TK, Khisti RT, Hodge CW, Homanics GE, Morrow AL. GABA(A) receptor alpha-1 subunit deletion alters receptor subtype assembly, pharmacological and behavioral responses to benzodiazepines and zolpidem. Neuropharmacology 2002;43:685–694.

106. Sur C, Wafford KA, Reynolds DS, Hadingham KL, Bromidge F, Macaulay A, Collinson N, O'Meara G, O Howell, Newman R, et al. Loss of the major GABA(A) receptor subtype in the brain is not lethal in mice. J. Neurosci. 2001;21:3409–3418.

107. Vicini S, Ferguson C, Prybylowski K, Kralic J, Morrow AL, Homanics GE. GABA(A) receptor alpha1 subunit deletion prevents developmental changes of inhibitory synaptic currents in cerebellar neurons. J. Neurosci. 2001;21:3009–3016.

108. Low K, Crestani F, Keist R, Benke D, Brunig I, Benson JA, Fritschy JM, Rulicke T, Bluethmann H, Mohler H, et al. Molecular and neuronal substrate for the selective attenuation of anxiety. Science 2000;290:131–134.

109. Reynolds DS, McKernan RM, Dawson GR. Anxiolytic-like action of diazepam: which GABA(A) receptor subtype is involved? Trends Pharmacol. Sci. 2001;22: 402–403.

110. Boehm SL 2nd, Ponomarev I, Jennings AW, Whiting PJ, Rosahl TW, Garrett EM, Blednov YA, Harris RA. gamma-Aminobutyric acid A receptor subunit mutant mice: new perspectives on alcohol actions. Biochem. Pharmacol. 2004;68:1581–1602.

111. Teng HK, Teng KK, Lee R, Wright S, Tevar S, Almeida RD, Kermani P, Torkin R, Chen ZY, Lee FS, et al. ProBDNF induces neuronal apoptosis via activation of a receptor complex of p75NTR and sortilin. J. Neurosci. 2005;25:5455–5463.

112. Egan MF, Kojima M, Callicott JH, Goldberg TE, Kolachana BS, Bertolino A, Zaitsev E, Gold B, Goldman D, Dean M, et al. The BDNF val66met polymorphism affects activity-dependent secretion of BDNF and human memory and hippocampal function. Cell 2003;112:257–269.

113. Jiang X, Xu K, Hoberman J, Tian F, Marko AJ, Waheed JF, Harris CR, Marini AM, Enoch MA, Lipsky RH. BDNF variation and mood disorders: a novel functional promoter polymorphism and Val66Met are associated with anxiety but have opposing effects. Neuropsychopharmacology 2005;30:1353–1361.

114. Sen S, Nesse RM, Stoltenberg SF, Li S, Gleiberman L, Chakravarti A, Weder AB, Burmeister M. A BDNF coding variant is associated with the NEO personality inventory domain neuroticism, a risk factor for depression. Neuropsychopharmacology 2003;28:397–401.

115. Tsai SJ, Hong CJ, Yu YW, Chen TJ. Association study of a brain-derived neurotrophic factor (BDNF) Val66Met polymorphism and personality trait and intelligence in healthy young females. Neuropsychobiology 2004;49:13–16.

116. Chen ZY, Jing D, Bath KG, Ieraci A, Khan T, Siao CJ, Herrera DG, Toth M, Yang C, McEwen BS, et al. Genetic variant BDNF (Val66Met) polymorphism alters anxiety-related behavior. Science 2006;314:140–143.

117. Huang EJ, Reichardt LF. Trk receptors: roles in neuronal signal transduction. Annu. Rev. Biochem. 2003;72:609–642.

118. Rios M, Fan G, Fekete C, Kelly J, Bates B, Kuehn R, Lechan RM, Jaenisch R. Conditional deletion of brain-derived neurotrophic factor in the postnatal brain leads to obesity and hyperactivity. Mol. Endocrinol. 2001;15:1748–1757.

119. Koponen E, Voikar V, Riekki R, Saarelainen T, Rauramaa T, Rauvala H, Taira T, Castren E. Transgenic mice overexpressing the full-length neurotrophin receptor trkB exhibit increased activation of the trkB-PLCgamma pathway, reduced anxiety, and facilitated learning. Mol. Cell Neurosci. 2004;26:166–181.

120. Pezawas L, Verchinski BA, Mattay VS, Callicott JH, Kolachana BS, Straub RE, Egan MF, Meyer-Lindenberg A, Weinberger DR. The brain-derived neurotrophic factor val66met polymorphism and variation in human cortical morphology. J. Neurosci. 2004;24:10099–10102.

121. Hariri AR, Goldberg TE, Mattay VS, Kolachana BS, Callicott JH, Egan MF, Weinberger DR. Brain-derived neurotrophic factor val66met polymorphism affects human memory-related hippocampal activity and predicts memory performance. J. Neurosci. 2003;23:6690–6694.

122. Vale W, Spiess J, Rivier C, Rivier J. Characterization of a 41-residue ovine hypothalamic peptide that stimulates secretion of corticotropin and beta-endorphin. Science 1981;213:1394–1397.

123. Bale TL, Vale WW. CRF and CRF receptors: role in stress responsivity and other behaviors. Annu. Rev. Pharmacol. Toxicol. 2004;44:525–557.

124. Grace CR, Perrin MH, DiGruccio MR, Miller CL, Rivier JE, Vale WW, Riek R. NMR structure and peptide hormone binding site of the first extracellular domain of a type B1 G protein-coupled receptor. Proc. Natl. Acad. Sci. U. S. A. 2004;101:12836–12841.

125. Hoare SR, Sullivan SK, Schwarz DA, Ling N, Vale WW, Crowe PD, Grigoriadis DE. Ligand affinity for amino-terminal and juxtamembrane domains of the corticotropin releasing factor type I receptor: regulation by G-protein and nonpeptide antagonists. Biochemistry 2004;43:3996–4011.

126. Dautzenberg FM, Hauger RL. The CRF peptide family and their receptors: yet more partners discovered. Trends Pharmacol. Sci. 2002;23:71–77.

127. Steckler T, Holsboer F. Corticotropin-releasing hormone receptor subtypes and emotion. Biol. Psychiatry 1999;46:1480–1508.

128. Arborelius L, Owens MJ, Plotsky PM, Nemeroff CB. The role of corticotropin-releasing factor in depression and anxiety disorders. J. Endocrinol. 1999;160:1–12.

129. Smoller JW, Yamaki LH, Fagerness JA, Biederman J, Racette S, Laird NM, Kagan J, Snidman N, Faraone SV, Hirshfeld-Becker D, et al. The corticotropin-releasing hormone gene and behavioral inhibition in children at risk for panic disorder. Biol. Psychiatry 2005;57:1485–1492.

130. Sutton RE, Koob GF, Le Moal M, Rivier J, Vale W. Corticotropin releasing factor produces behavioural activation in rats. Nature 1982;297:331–333.

131. Stenzel-Poore MP, Heinrichs SC, Rivest S, Koob GF, Vale WW. Overproduction of corticotropin-releasing factor in transgenic mice: a genetic model of anxiogenic behavior. J. Neurosci. 1994;14:2579–2584.

132. Stenzel-Poore MP, Duncan JE, Rittenberg MB, Bakke AC, Heinrichs SC. CRH overproduction in transgenic mice: behavioral and immune system modulation. Ann. N. Y. Acad. Sci. 1996;780:36–48.

133. Dunn AJ, Swiergiel AH. Behavioral responses to stress are intact in CRF-deficient mice. Brain Res. 1999;845:14–20.

134. Muglia L, Jacobson L, Majzoub JA. Production of corticotropin-releasing hormone-deficient mice by targeted mutation in embryonic stem cells. Ann. N. Y. Acad. Sci. 1996;780:49–59.

135. Vaughan J, Donaldson C, Bittencourt J, Perrin MH, Lewis K, Sutton S, Chan R, Turnbull AV, Lovejoy D, Rivier C, et al. Urocortin, a mammalian neuropeptide related to fish urotensin I and to corticotropin-releasing factor. Nature 1995;378:287–292.

136. Lewis K, Li C, Perrin MH, Blount A, Kunitake K, Donaldson C, Vaughan J, Reyes TM, Gulyas J, Fischer W, et al. Identification of urocortin III, an additional member of the corticotropin-releasing factor (CRF) family with high affinity for the CRF2 receptor. Proc. Natl. Acad. Sci. U. S. A. 2001;98:7570–7575.

137. Reyes TM, Lewis K, Perrin MH, Kunitake KS, Vaughan J, Arias CA, Hogenesch JB, Gulyas J, Rivier J, Vale WW, et al. Urocortin II: a member of the corticotropin-releasing factor (CRF) neuropeptide family that is selectively bound by type 2 CRF receptors. Proc. Natl. Acad. Sci. U. S. A. 2001;98:2843–2848.

138. Hsu SY, Hsueh AJ. Human stresscopin and stresscopin-related peptide are selective ligands for the type 2 corticotropin-releasing hormone receptor. Nat. Med. 2001;7:605–611.

139. Vetter DE, Li C, Zhao L, Contarino A, Liberman MC, Smith GW, Marchuk Y, Koob GF, Heinemann SF, Vale W, et al. Urocortin-deficient mice show hearing impairment and increased anxiety-like behavior. Nat. Genet. 2002;31:363–369.

140. Wang X, Su H, Copenhagen LD, Vaishnav S, Pieri F, Shope CD, Brownell WE, De Biasi M, Paylor R, Bradley A. Urocortin-deficient mice display normal stress-induced anxiety behavior and autonomic control but an impaired acoustic startle response. Mol. Cell Biol. 2002;22:6605–6610.

141. Karolyi IJ, Burrows HL, Ramesh TM, Nakajima M, Lesh JS, Seong E, Camper SA, Seasholtz AF. Altered anxiety and weight gain in corticotropin-releasing hormone-binding protein-deficient mice. Proc. Natl. Acad. Sci. U. S. A. 1999;96:11595–11600.

142. Yehuda R. Current status of cortisol findings in post-traumatic stress disorder. Psychiatr. Clin. North Am. 2002;25:341–368, vii.

143. Carrion VG, Weems CF, Ray RD, Glaser B, Hessl D, Reiss AL. Diurnal salivary cortisol in pediatric posttraumatic stress disorder. Biol. Psychiatry 2002;51:575–582.

144. Rasmusson AM, Lipschitz DS, Wang S, Hu S, Vojvoda D, Bremner JD, Southwick SM, Charney DS. Increased pituitary and adrenal reactivity in premenopausal women with posttraumatic stress disorder. Biol. Psychiatry 2001;50:965–977.

145. Young EA, Breslau N. Saliva cortisol in posttraumatic stress disorder: a community epidemiologic study. Biol. Psychiatry 2004;56:205–209.

146. Tochigi M, Kato C, Otowa T, Hibino H, Marui T, Ohtani T, Umekage T, Kato N, Sasaki T. Association between corticotropin-releasing hormone receptor 2 (CRHR2) gene polymorphism and personality traits. Psychiatry Clin. Neurosci. 2006;60:524–526.

147. Tharmalingam S, King N, De Luca V, Rothe C, Koszycki D, Bradwejn J, Macciardi F, Kennedy JL. Lack of association between the corticotrophin-releasing hormone receptor 2 gene and panic disorder. Psychiatr. Genet. 2006;16:93–97.

148. Moreau JL, Kilpatrick G, Jenck F. Urocortin, a novel neuropeptide with anxiogenic-like properties. Neuroreport 1997;8:1697–1701.

149. Contarino A, Dellu F, Koob GF, Smith GW, Lee KF, Vale W, Gold LH. Reduced anxiety-like and cognitive performance in mice lacking the corticotropin-releasing factor receptor 1. Brain Res. 1999;835:1–9.

150. Smith GW, Aubry JM, Dellu F, Contarino A, Bilezikjian LM, Gold LH, Chen R, Marchuk Y, Hauser C, Bentley CA, et al. Corticotropin releasing factor receptor 1-deficient mice display decreased anxiety, impaired stress response, and aberrant neuroendocrine development. Neuron 1998;20:1093–1102.

151. Bale TL, Contarino A, Smith GW, Chan R, Gold LH, Sawchenko PE, Koob GF, Vale WW, Lee KF. Mice deficient for corticotropin-releasing hormone receptor-2 display anxiety-like behaviour and are hypersensitive to stress. Nat. Genet. 2000;24:410–414.

152. Kishimoto T, Radulovic J, Radulovic M, Lin CR, Schrick C, Hooshmand F, Hermanson O, Rosenfeld MG, Spiess J. Deletion of crhr2 reveals an anxiolytic role for corticotropin-releasing hormone receptor-2. Nat. Genet. 2000;24:415–419.

153. Coste SC, Kesterson RA, Heldwein KA, Stevens SL, Heard AD, Hollis JH, Murray SE, Hill JK, Pantely GA, Hohimer AR, et al. Abnormal adaptations to stress and impaired cardiovascular function in mice lacking corticotropin-releasing hormone receptor-2. Nat. Genet. 2000;24:403–409.

154. Muller MB, Zimmermann S, Sillaber I, Hagemeyer TP, Deussing JM, Timpl P, Kormann MS, Droste SK, Kuhn R, Reul JM, et al. Limbic corticotropin-releasing hormone receptor 1 mediates anxiety-related behavior and hormonal adaptation to stress. Nat. Neurosci. 2003;6:1100–1107.

155. Bale TL, Picetti R, Contarino A, Koob GF, Vale WW, Lee KF. Mice deficient for both corticotropin-releasing factor receptor 1 (CRFR1) and CRFR2 have an impaired stress response and display sexually dichotomous anxiety-like behavior. J. Neurosci. 2002;22:193–199.

156. Hauger RL, Risbrough V, Brauns O, Dautzenberg FM. Corticotropin releasing factor (CRF) receptor signaling in the central nervous system: new molecular targets. CNS Neurol Disord. Drug Targets 2006;5:453–479.

157. Schaefer ML, Wong ST, Wozniak DF, Muglia LM, Liauw JA, Zhuo M, Nardi A, Hartman RE, Vogt SK, Luedke CE, et al. Altered stress-induced anxiety in adenylyl cyclase type VIII-deficient mice. J. Neurosci. 2000;20:4809–4820.

158. Fee JR, Sparta DR, Knapp DJ, Breese GR, Picker MJ, Thiele TE. Predictors of high ethanol consumption in RIIbeta knock-out mice: assessment of anxiety and ethanol-induced sedation. Alcohol Clin. Exp. Res. 2004;28:1459–1468.

159. Barrot M, Olivier JD, Perrotti LI, DiLeone RJ, Berton O, Eisch AJ, Impey S, Storm DR, Neve RL, Yin JC, et al. CREB activity in the nucleus accumbens shell controls gating of behavioral responses to emotional stimuli. Proc. Natl. Acad. Sci. U. S. A. 2002;99:11435–11440.

160. Carlezon WA Jr, Duman RS, Nestler EJ. The many faces of CREB. Trends Neurosci. 2005;28:436–445.

161. Sananbenesi F, Fischer A, Schrick C, Spiess J, Radulovic J. Mitogen-activated protein kinase signaling in the hippocampus and its modulation by corticotropin-releasing factor receptor 2: a possible link between stress and fear memory. J. Neurosci. 2003;23:11436–11443.

162. Refojo D, Echenique C, Muller MB, Reul JM, Deussing JM, Wurst W, Sillaber I, Paez-Pereda M, Holsboer F, Arzt E. Corticotropin-releasing hormone activates ERK1/2 MAPK in specific brain areas. Proc. Natl. Acad. Sci. U. S. A. 2005; 102:6183–6188.

163. Kendler KS, Karkowski LM, Prescott CA. Causal relationship between stressful life events and the onset of major depression. Am. J. Psychiatry 1999;156:837–841.

164. Kendler KS, Hettema JM, Butera F, Gardner CO, Prescott CA. Life event dimensions of loss, humiliation, entrapment, and danger in the prediction of onsets of major depression and generalized anxiety. Arch. Gen. Psychiatry 2003;60:789–796.

165. Faravelli C, Pallanti S. Recent life events and panic disorder. Am. J. Psychiatry 1989;146:622–626.

166. Finlay-Jones R, Brown GW. Types of stressful life event and the onset of anxiety and depressive disorders. Psychol. Med. 1981;11:803–815.

167. Meaney MJ, Diorio J, Francis D, Widdowson J, LaPlante P, Caldji C, Sharma S, Seckl JR, Plotsky PM. Early environmental regulation of forebrain glucocorticoid receptor gene expression: implications for adrenocortical responses to stress. Dev. Neurosci. 1996;18:49–72.

168. Meaney MJ. Maternal care, gene expression, and the transmission of individual differences in stress reactivity across generations. Ann. Rev. Neurosci. 2001; 24:1161–1192.

169. Menard JL, Champagne DL, Meaney MJ. Variations of maternal care differentially influence 'fear' reactivity and regional patterns of cFos immunoreactivity in response to the shock-probe burying test. Neuroscience 2004;129:297–308.

170. Caldji C, Tannenbaum B, Sharma S, Francis D, Plotsky PM, Meaney MJ. Maternal care during infancy regulates the development of neural systems mediating the expression of fearfulness in the rat. Proc. Natl. Acad. Sci. U. S. A. 1998;95:5335–5340.

171. Liu D, Diorio J, Tannenbaum B, Caldji C, Francis D, Freedman A, Sharma S, Pearson D, Plotsky PM, Meaney MJ. Maternal care, hippocampal glucocorticoid receptors, and hypothalamic-pituitary-adrenal responses to stress. Science 1997;277:1659–1662.

172. Caldji C, Diorio J, Meaney MJ. Variations in maternal care alter GABA(A) receptor subunit expression in brain regions associated with fear. Neuropsychopharmacology 2003;28:1950–1959.

173. Caldji C, Diorio J, Anisman H, Meaney MJ. Maternal behavior regulates benzodiazepine/GABAA receptor subunit expression in brain regions associated with fear in BALB/c and C57BL/6 mice. Neuropsychopharmacology 2004;29:1344–1352.

174. Liu D, Diorio J, Day JC, Francis DD, Meaney MJ. Maternal care, hippocampal synaptogenesis and cognitive development in rats. Nat. Neurosci. 2000;3:799–806.

175. Caspi A, Sugden K, Moffitt TE, Taylor A, Craig IW, Harrington H, McClay J, Mill J, Martin J, Braithwaite A, et al. Influence of life stress on depression: moderation by a polymorphism in the 5-HTT gene. Science 2003;301:386–389.

176. Uher R, McGuffin P. The moderation by the serotonin transporter gene of environmental adversity in the aetiology of mental illness: review and methodological analysis. Mol. Psychiatry. 2007.

177. Kendler KS, Kuhn JW, Vittum J, Prescott CA, Riley B. The interaction of stressful life events and a serotonin transporter polymorphism in the prediction of episodes of major depression: a replication. Arch. Gen. Psychiatry 2005;62:529–535.

178. Taylor SE, Way BM, Welch WT, Hilmert CJ, Lehman BJ, Eisenberger NI. Early family environment, current adversity, the serotonin transporter promoter polymorphism, and depressive symptomatology. Biol. Psychiatry 2006;60:671–676.

179. Champoux M, Bennett A, Shannon C, Higley JD, Lesch KP, Suomi SJ. Serotonin transporter gene polymorphism, differential early rearing, and behavior in rhesus monkey neonates. Mol. Psychiatry 2002;7:1058–1063.

180. Bennett AJ, Lesch KP, Heils A, Long JC, Lorenz JG, Shoaf SE, Champoux M, Suomi SJ, Linnoila MV, Higley JD. Early experience and serotonin transporter gene variation interact to influence primate CNS function. Mol. Psychiatry 2002;7:118–122.

181. Crawley JN, Belknap JK, Collins A, Crabbe JC, Frankel W, Henderson N, Hitzemann RJ, Maxson SC, Miner LL, Silva AJ, et al. Behavioral phenotypes of inbred mouse strains: implications and recommendations for molecular studies. Psychopharmacology (Berl) 1997;132:107–124.

182. Anisman H, Zaharia MD, Meaney MJ, Merali Z. Do early-life events permanently alter behavioral and hormonal responses to stressors? Int. J. Dev. Neurosci. 1998;16:149–164.

183. Zaharia MD, Kulczycki J, Shanks N, Meaney MJ, Anisman H. The effects of early postnatal stimulation on Morris water-maze acquisition in adult mice: genetic and maternal factors. Psychopharmacology (Berl) 1996;128:227–239.

184. Francis DD, Szegda K, Campbell G, Martin WD, Insel TR. Epigenetic sources of behavioral differences in mice. Nat. Neurosci. 2003;6:445–446.

185. Weller A, Leguisamo AC, Towns L, Ramboz S, Bagiella E, Hofer M, Hen R, Brunner D. Maternal effects in infant and adult phenotypes of 5HT1A and 5HT1B receptor knockout mice. Dev. Psychobiol. 2003;42:194–205.

186. Koster A, Montkowski A, Schulz S, Stube EM, Knaudt K, Jenck F, Moreau JL, Nothacker HP, Civelli O, Reinscheid RK. Targeted disruption of the orphanin FQ/nociceptin gene increases stress susceptibility and impairs stress adaptation in mice. Proc. Natl. Acad. Sci. U. S. A. 1999;96:10444–10449.

187. Reinscheid RK, Civelli O. The orphanin FQ/nociceptin knockout mouse: a behavioral model for stress responses. Neuropeptides 2002;36:72–76.

188. Ouagazzal AM, Moreau JL, Pauly-Evers M, Jenck F. Impact of environmental housing conditions on the emotional responses of mice deficient for nociceptin/orphanin FQ peptide precursor gene. Behav. Brain Res. 2003;144:111–117.

189. Ren-Patterson RF, Cochran LW, Holmes A, Sherrill S, Huang SJ, Tolliver T, Lesch KP, Lu B, Murphy DL. Loss of brain-derived neurotrophic factor gene allele exacerbates brain monoamine deficiencies and increases stress abnormalities of serotonin transporter knockout mice. J. Neurosci. Res. 2005;79:756–771.

## FURTHER READING

Caspi A, Moffitt TE. Gene-environment interactions in psychiatry: joining forces with neuroscience. Nat. Rev. Neurosci. 2006;7:583–590.

Leonardo ED, Hen R. Anxiety as a developmental disorder. Neuropsychopharmacology 2008;33:134–140.

Hauger RL, Risbrough V, Brauns O, Dautzenberg FM. Corticotropin releasing factor (CRF) receptor signaling in the central nervous system: new molecular targets. CNS Neurol. Disord. Drug Targets 2006;5:453–479.

Martinowich K, Manji H, Lu B. New insights into BDNF function in depression and anxiety. Nat. Neurosci. 2007;10:1089–1093.

# 13

# CHRONIC OBSTRUCTIVE PULMONARY DISEASE (COPD)

PETER J. BARNES

*National Heart and Lung Institute, Imperial College, London, United Kingdom*

Airway obstruction in chronic obstructive pulmonary disease (COPD) is caused by narrowing of small airways as a result of inflammation and fibrosis and the disruption of their alveolar attachments as a result of emphysema. COPD is characterized by a complex inflammatory disease process that increases as the disease progresses, which leads to increasing airflow limitation. Many inflammatory cells and mediators have now been implicated in the pathogenesis of COPD. Increased numbers of macrophages, neutrophils, T-lymphocytes (particularly CD8[+] cells) and B-lymphocytes have been observed, as well as the release of multiple inflammatory mediators (lipids, chemokines, cytokines, growth factors). Macrophages seem to play an important role in orchestrating the inflammatory process, which includes the recruitment of neutrophils and T-cells into small airways and lung parenchyma and the secretion of proteinases that lead to emphysema. A high level of oxidative and nitrative stress may amplify this inflammation through the reduction in histone deacetylase-2, which also results in corticosteroid resistance.

Chronic obstructive pulmonary disease (COPD) has now become a major global epidemic, and it is predicted to become the third leading cause of death and fifth leading cause of disability over the next decade (1). The increase in COPD is a particular problem in developing nations; in developed countries, is the only common cause of death that is increasing. COPD now has a world-wide prevalence of over 10% in men and is increasing toward this figure in women (2). Because of the enormous burden of disease and escalating health-care costs, there is now

*Chemical Biology: Approaches to Drug Discovery and Development to Targeting Disease*, First Edition.
Edited by Natanya Civjan.
© 2012 John Wiley & Sons, Inc. Published 2012 by John Wiley & Sons, Inc.

renewed interest in the underlying cellular and molecular mechanisms of COPD (3) and a search for new therapies (4). The definition of COPD adopted by the Global initiative on Obstructive Lung Disease encompasses the idea that COPD is a chronic inflammatory disease, and much recent research has focused on the nature of this inflammatory response (5). COPD is an obstructive disease of the lungs that slowly progresses over many decades leading to death from respiratory failure unless patients die of comorbidities such as heart disease and lung cancer before this stage. Although the most common cause of COPD is chronic cigarette smoking, some patients, particularly in developing countries, develop the disease from inhalation of wood smoke from biomass fuels or other inhaled irritants (2). However, only about 25% of smokers develop COPD, which suggests that genetic or host factors may predispose patients to its development, although these factors have not yet been identified. The disease is relentlessly progressive, and only smoking cessation reduces the rate of decline in lung function; as the disease becomes more severe, there is less effect of smoking cessation, and lung inflammation persists.

## 13.1  COPD AS AN INFLAMMATORY DISEASE

The progressive airflow limitation in COPD is caused by two major pathological processes: remodeling and narrowing of small airways and destruction of the lung parenchyma with consequent destruction of the alveolar attachments of these airways as a result of emphysema (Fig. 13.1). This disease results in diminished lung recoil, higher resistance to flow and closure of small airways at higher lung volumes during expiration, which traps air in the lung. This trapped air leads to the characteristic hyperinflation of the lungs, which causes the sensation of dyspnea and limits exercise capacity. The major symptom of COPD is shortness of breath on exertion. Both the small airway remodeling and narrowing and the emphysema are caused by chronic inflammation in the lung periphery. Quantitative studies have shown that the inflammatory response in small airways and lung parenchyma increases as the disease progresses (6). A specific pattern of inflammation in COPD airways and lung parenchyma is observed with increased numbers of macrophages, T-lymphocytes, with predominance of CD8$^+$ (cytotoxic) T-cells, and in more severe disease B-lymphocytes with increased numbers of neutrophils in the lumen (3). The inflammatory response in COPD involves both innate and adaptive immune responses. Multiple inflammatory mediators are increased in COPD and are derived from inflammatory cells and structural cells of the airways and lungs (7). A similar pattern of inflammation is observed in smokers without airflow limitation, but in COPD, this inflammation is amplified even more during acute exacerbations of the disease, which are usually precipitated by bacterial and viral infections (Fig. 13.2). The molecular basis of this amplification of inflammation is not yet understood but may be, at least in part, determined by genetic factors. Cigarette smoke and other irritants in the respiratory tract may activate surface macrophages and airway epithelial cells to release

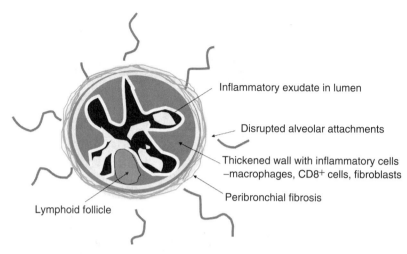

Inflammatory exudate in lumen

Disrupted alveolar attachments

Thickened wall with inflammatory cells
−macrophages, CD8$^+$ cells, fibroblasts

Peribronchial fibrosis

Lymphoid follicle

**Figure 13.1** Small airways in COPD patients. The airway wall is thickened and infiltrated with inflammatory cells, predominately macrophages and CD8$^+$ lymphocytes, with increased numbers of fibroblasts. In severe COPD, lymphoid follicles are observed, which consist of a central core of B-lymphocytes, surrounded by T-lymphocytes and are thought to indicate chronic exposure to antigens (bacterial, viral, or autoantigens). Similar changes are also reported in larger airways. The lumen is often filled with an inflammatory exudate and mucus. Peribronchial fibrosis occurs, and it results in progressive and irreversible narrowing of the airway. Airway smooth muscle may be increased slightly.

chemotactic factors that then attract circulating leukocytes into the lungs. Among chemotactic factors, chemokines predominate and therefore play a key role in orchestrating the chronic inflammation in COPD lungs and its amplification during acute exacerbations (8). These events might be the initial inflammatory events that occur in all smokers. However in smokers who develop COPD, this inflammation progresses into a more complicated inflammatory pattern of adaptive immunity and involves T- and B-lymphocytes and possibly dendritic cells along with a complicated interacting array of cytokines and other mediators (9).

### 13.1.1 Differences from Asthma

Histopathological studies of COPD show a predominant involvement of peripheral airways (bronchioles) and lung parenchyma, whereas asthma involves inflammation in all airways (particularly proximal airways) but usually without involvement of the lung parenchyma (10). In COPD, the bronchioles become narrow, with fibrosis and infiltration with macrophages and T-lymphocytes, along with destruction of lung parenchyma and an increased number of macrophages and T-lymphocytes, with a greater increase in CD8$^+$ (cytotoxic) than CD4$^+$ (helper) cells (6) (Fig. 13.3). Bronchial biopsies show similar changes with an infiltration of macrophages and CD8$^+$ cells and an increased

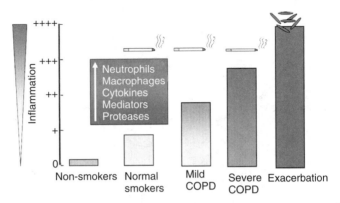

**Figure 13.2** Amplification of lung inflammation in COPD. Normal smokers have a mild inflammatory response, which represents the normal (probably protective) reaction of the respiratory mucosa to chronic inhaled irritants. In COPD, this same inflammatory response is markedly amplified, and this amplification increases as the disease progresses. It is increased even more during exacerbations triggered by infective organisms. The molecular mechanisms of this amplification are currently unknown, but they may be determined by genetic factors or possibly latent viral infection. Oxidative stress is an important amplifying mechanism and may increase the expression of inflammatory genes through impairing the activity of histone deacetylase 2 (HDAC2), which is needed to switch off inflammatory genes.

number of neutrophils in patients with severe COPD. Bronchoalveolar lavage (BAL) fluid and induced sputum demonstrate a marked increase in macrophages and neutrophils. In contrast to asthma, eosinophils are not prominent except during exacerbations or when patients have concomitant asthma (10).

## 13.2  INFLAMMATORY CELLS

For many years it was believed that the inflammatory reaction in the lungs of smokers consisted of neutrophils and macrophages and that proteinases from these cells were responsible for the lung destruction in COPD. More recently it has been recognized that there is a prominent T-cell infiltration in the lungs of patients with COPD, with a predominance of CD8+ (cytotoxic) T-cells, although CD4+ (helper) T-cells are also numerous. Although abnormal numbers of inflammatory cells have been documented in COPD, the relationship between these cell types and the sequence of their appearance and their persistence are not yet understood in detail (3). Most studies have been cross-sectional based on selection of patients with different stages of the disease, and comparisons have been made between smokers without airflow limitation (normal smokers) and those with COPD who have smoked a similar amount. No serial studies have been conducted, and selection biases (such as selecting tissue from patients suitable for lung volume reduction surgery) may give misleading results. Nonetheless,

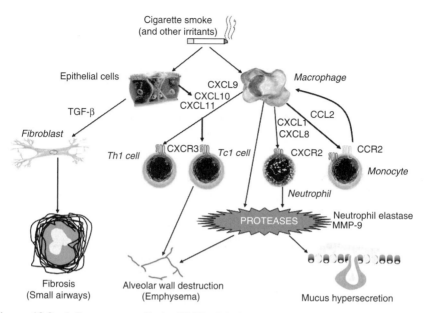

**Figure 13.3** Inflammatory cells in COPD. Inhaled cigarette smoke and other irritants activate epithelial cells and macrophages to release several chemotactic factors that attract inflammatory cells to the lungs, including CC-chemokine ligand 2 (CCL2), which acts on CC-chemokine receptor 2 (CCR2) to attract monocytes, CXC-chemokine ligand 1 (CXCL1) and CXCL8, which act on CCR2 to attract neutrophils and monocytes (which differentiate into macrophages in the lungs) and CXCL9, CXCL10, and CXCL11, which act on CXCR3 to attract T-helper 1 (Th1) cells and type 1 cytotoxic T-cells (Tc1 cells). These inflammatory cells together with macrophages and epithelial cells release proteases, such as MMP-9, which cause elastin degradation and emphysema. Neutrophil elastase also causes mucus hypersecretion. Epithelial cells and macrophages also release TGF-β, which stimulates fibroblast proliferation and results in fibrosis in the small airways.

a progressive increase in the numbers of inflammatory cells in small airways and lung parenchyma are observed as COPD becomes more severe, even though the patients with most severe obstruction have stopped smoking for many years (6). This finding indicates the existence of some mechanisms that perpetuate the inflammatory reaction in COPD. This characteristic is in contrast to many other chronic inflammatory diseases, such as rheumatoid arthritis and interstitial lung diseases, in which the inflammation tends to diminish in severe disease. The inflammation of COPD lungs involves both innate immunity (neutrophils, macrophages, eosinophils, mast cells, NK cells, γδ-T-cells, and dendritic cells) and adaptive immunity (T and B cells).

### 13.2.1 Epithelial Cells

Epithelial cells are activated by cigarette smoke to produce inflammatory mediators, which include tumor necrosis factor (TNF)-α, interleukin (IL)-1β,

IL-6, granulocyte-macrophage colony-stimulating factor (GM-CSF), and CXCL8 (IL-8). Epithelial cells in small airways may be an important source of transforming growth factor (TGF)-$\beta$, which then induces local fibrosis. Vascular endothelial growth factor (VEGF) seems to be necessary to maintain alveolar cell survival and blockade of VEGF receptors in rats induces apoptosis of alveolar cells and an emphysema-like pathology, which may be mediated via the sphingolipid ceramide (11). Airway epithelial cells are also important in defense of the airways, with mucus production from goblet cells, and secretion of antioxidants, antiproteases and defensins. It is possible that cigarette smoke and other noxious agents impair these innate and adaptive immune responses of the airway epithelium, which increases susceptibility to infection. The airway epithelium in chronic bronchitis and COPD often shows squamous metaplasia, which may result from increased proliferation of basal airway epithelial cells, but the nature of the growth factors involved in epithelial cell proliferation, cell cycle, and differentiation in COPD are not yet known. Epithelial growth factor receptors (EGFR) show increased expression in airway epithelial cells of smokers and may contribute to basal cell proliferation, which results in squamous metaplasia and an increased risk of bronchial carcinoma (12).

### 13.2.2 Neutrophils

Increased numbers of activated neutrophils are found in sputum and BAL fluid of patients with COPD (13), yet neutrophil levels are increased relatively little in the airways or lung parenchyma. This finding may reflect their rapid transit through the airways and parenchyma. The role of neutrophils in COPD is not yet clear; however, neutrophil numbers in induced sputum are correlated with COPD disease severity (13) and with the rate of decline in lung function. Smoking has a direct stimulatory effect on granulocyte production and release from the bone marrow and survival in the respiratory tract, which is possibly mediated by GM-CSF and G-CSF released from lung macrophages. Smoking may also increase neutrophil retention in the lung. Neutrophil recruitment to the airways and parenchyma involves adhesion to endothelial cells and E-selectin, which is upregulated on endothelial cells in the airways of COPD patients. Adherent neutrophils then migrate into the respiratory tract under the direction of neutrophil chemotactic factors. Several chemotactic signals have the potential for neutrophil recruitment in COPD, which include leukotriene (LT)B$_4$, CXCL8, and related CXC chemokines, including CXCL1 (GRO-$\alpha$) and CXCL5 (ENA-78), which are increased in COPD airways (14). These mediators may be derived form alveolar macrophages T-cells and epithelial cells, but the neutrophil itself may be a major source of CXCL8. Neutrophils from the circulation marginate in the pulmonary circulation and adhere to endothelial cells in the alveolar wall before passing into the alveolar space. The neutrophils recruited to the airways of COPD patients are activated because increased concentrations of granule proteins, such as myeloperoxidase and human neutrophil lipocalin, are found in the sputum supernatant (15). Neutrophils secrete serine proteases, which include neutrophil elastase, cathepsin

G, and proteinase-3, as well as matrix metalloproteinase (MMP)-8 and MMP-9, which may contribute to alveolar destruction. Neutrophils have the capacity to induce tissue damage through the release of serine proteases and oxidants. However, whereas neutrophils have the capacity to cause elastolysis, this ability is not a prominent feature of other pulmonary diseases in which chronic airway neutrophilia is even more prominent, including cystic fibrosis and bronchiectasis. This comparison suggests that other factors are involved in the generation of emphysema. Indeed, neutrophils are not a prominent feature of parenchymal inflammation in COPD. It is likely that airway neutrophilia is more linked to mucus hypersecretion in chronic bronchitis. Serine proteases from neutrophils, which include neutrophil elastase, cathepsin G, and proteinase-3, are all potent stimulants of mucus secretion from submucosal glands and goblet cells in the epithelium. A marked increase in neutrophil numbers is observed in the airways in acute exacerbations of COPD, which accounts for the increased purulence of sputum. This finding may reflect increased production of neutrophil chemotactic factors, which include $LTB_4$ and CXCL8 (16, 17)

### 13.2.3  Macrophages

Macrophages seem to play a pivotal role in the pathophysiology of COPD and can account for most of the known features of the disease (18) (Fig. 13.4).

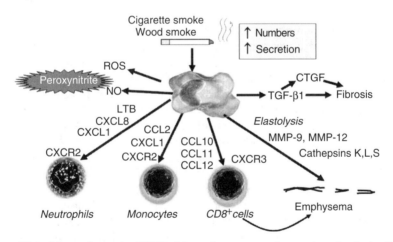

**Figure 13.4**  Macrophages in COPD. Macrophages may play a pivotal role in COPD as they are activated by cigarette-smoke extract and secrete many inflammatory proteins that may orchestrate the inflammatory process in COPD. Neutrophils may be attracted by CXCL8, CXCL1, and $LTB_4$, monocytes by CCL2, and $CD8^+$ lymphocytes by CXCL10 and CXCL11. Release of elastolytic enzymes, such as MMP and cathepsins, cause elastolysis, and release of TGF-β1 and connective tissue growth factor (CTGF). Macrophages also generate ROS and NO, which together form peroxynitrite and may contribute to steroid resistance.

A marked increase (5–10-fold) in the numbers of macrophages in airways, lung parenchyma, BAL fluid, and sputum in patients with COPD. A careful morphometric analysis of macrophage numbers in the parenchyma of patients with emphysema showed a 25-fold increase in the numbers of macrophages in the tissue and alveolar space compared with normal smokers (19). Furthermore, macrophages are localized to sites of alveolar wall destruction in patients with emphysema, and a correlation is observed between macrophage numbers in the parenchyma and severity of emphysema (20). Macrophages may be activated by cigarette smoke extract to release inflammatory mediators, which includes TNF-α, CXCL8, and other CXC chemokines; CCL2 (MCP-1); $LTB_4$; and reactive oxygen species. This release is a cellular mechanism that links smoking with inflammation in COPD. Alveolar macrophages also secrete elastolytic enzymes, which include MMP-2; MMP-9; MMP-12; cathepsins K, L, and S; and neutrophil elastase taken up from neutrophils (21). Alveolar macrophages from patients with COPD and with exposure to cigarette smoke secrete more inflammatory proteins and have a greater elastolytic activity at baseline than those from normal smokers (21). Macrophages demonstrate this difference even when maintained in culture for 3 days, and therefore they seem to be intrinsically different from the macrophages of normal smokers and nonsmoking normal control subjects (21). The predominant elastolytic enzyme secreted by alveolar macrophages in COPD patients is MMP-9. Most inflammatory proteins that are upregulated in COPD macrophages are regulated by the transcription factor nuclear factor-κB (NF-κB), which is activated in alveolar macrophages of COPD patients, particularly during exacerbations (22).

The increased numbers of macrophages in smokers and COPD patients may be caused by increased recruitment of monocytes from the circulation in response to the monocyte-selective chemokines CCL2 and CXCL1, which are increased in sputum and BAL of patients with COPD (14). Monocytes from patients with COPD show a greater chemotactic response to GRO-α than cells from normal smokers and nonsmokers, but this finding is not explained by an increase in CXCR2 (23). Interestingly, whereas all monocytes express CCR2, which is the receptor for CCL2, only ∼30% of monocytes express CXCR2. It is possible that these CXCR2-expressing monocytes transform into macrophages that are more inflammatory. Macrophages also release the chemokines CXCL9, CXCL10, and CXCL11, which are chemotactic for CD8+ Tc1 and CD4+ Th1 cells, via interaction with the chemokine receptor CXCR3 expressed on these cells (24).

The increased numbers of macrophages in COPD are mainly caused by increased recruitment of monocytes, as macrophages have a very low proliferation rate in the lungs. Macrophages have a long survival time so the macrophage level is difficult to measure directly. However, in macrophages from smokers, a markedly increased expression of the antiapoptotic protein Bcl-$X_L$ and increased expression of p21[CIP/WAF1] is observed in the cytoplasm (25). This finding suggests that macrophages may have a prolonged survival in smokers and patients with COPD. Once activated, macrophages will increase production of reactive oxygen species, nitric oxide, and lysosomal enzymes and will increase secretion of many cytokines, which include TNFα, IL-1β, IL-6, CXCL8, and IL-18, among others. Activated macrophages are aimed at the more efficient killing of organisms and promote inflammation mainly by TNFα, IL-1β, and short-lived lipid mediators.

Corticosteroids are ineffective in suppressing inflammation, which include cytokines, chemokines, and proteases, in patients with COPD (26). *In vitro*, the release of CXCL8, TNF-α and MMP-9 macrophages from normal subjects and normal smokers are inhibited by corticosteroids, whereas corticosteroids are ineffective in macrophages from patients with COPD (27). The reasons for resistance to corticosteroids in COPD and to a lesser extent macrophages from smokers may be the marked reduction in activity of histone deacetylase-2 (HDAC2) (28), which is recruited to activated inflammatory genes by glucocorticoid receptors to switch off inflammatory genes. The reduction in HDAC activity in macrophages is correlated with increased secretion of cytokines like TNF-α and CXCL8 and reduced response to corticosteroids. The reduction of HDAC activity on COPD patients may be mediated through oxidative stress and peroxynitrite formation (29).

### 13.2.4 Eosinophils

Although eosinophils are the predominant leukocyte in asthma, their role in COPD is much less certain. Increased numbers of eosinophils have been described in the airways and BAL of patients with stable COPD, whereas others have not found increased numbers in airway biopsies, BAL, or induced sputum. The presence of eosinophils in patients with COPD predicts a response to corticosteroids and may indicate coexisting asthma (30). Increased numbers of eosinophils have been reported in bronchial biopsies and BAL fluid during acute exacerbations of chronic bronchitis (31). Surprisingly, the levels of eosinophil basic proteins in induced sputum are as elevated in COPD, as in asthma, despite the absence of eosinophils, which suggests that they may have degranulated and are no longer recognizable by microscopy (15). Perhaps this finding is caused by the high levels of neutrophil elastase that have been shown to cause degranulation of eosinophils.

### 13.2.5 Dendritic Cells

Dendritic cells play a central role in the initiation of the innate and adaptive immune response, and it is believed that they provide a link between them (32). The airways and lungs contain a rich network of dendritic cells that are localized near the surface, so that they are located ideally to signal the entry of foreign substances that are inhaled. Dendritic cells can activate a variety of other inflammatory and immune cells, which include macrophages and neutrophils, as well as T- and B-lymphocytes, so dendritic cells may play an important role in the pulmonary response to cigarette smoke and other inhaled noxious agents. However, an increase in dendritic cells is not observed in the airways of COPD patients in contrast to asthma patients (33)

## 13.3   T-LYMPHOCYTES

An increase in the total numbers of T-lymphocytes is observed in lung parenchyma as well as in peripheral and central airways of patients with COPD; a greater increase is observed in CD8$^+$ than CD4$^+$ cells (6, 24). A correlation is observed between the numbers of T-cells and the amount of alveolar destruction and the severity of airflow obstruction. Furthermore, the only significant difference in the inflammatory cell infiltrate in asymptomatic smokers and smokers with COPD is an increase in T-cells, mainly CD8$^+$, in patients with COPD. An increase in the absolute number of CD4$^+$ T-cells, albeit in smaller numbers, is evidenced in the airways of smokers with COPD, and these cells express activated STAT-4, which is a transcription factor that is essential for activation and commitment of the Th1 lineage, and IFN-$\gamma$.

The ratio of CD4$^+$:CD8$^+$ cells are reversed in COPD. Most T-cells in the lung in COPD are of the Tc1 and Th1 subtypes (24). A marked increase is observed in T-cells in the walls of small airways in patients with severe COPD, and the T-cells are formed into lymphoid follicles, which surround B-lymphocytes (6).

The mechanisms by which CD8$^+$, and to a lesser extent CD4$^+$ cells, accumulate in the airways and parenchyma of patients with COPD is not yet understood (34). However, homing of T-cells to the lung must depend on some initial activation (only activated T-cells can home to the organ source of antigenic products), then adhesion and selective chemotaxis. CD4$^+$ and CD8$^+$ T-cells in the lung of COPD patients show increased expression of CXCR3, which is a receptor activated by the chemokines CXCL9, CXCL10, and CXCL11, all of which are increased in COPD (35). Increased expression of CXCL10 by bronchiolar epithelial cells is observed, and it could contribute to the accumulation of CD4$^+$ and CD8$^+$ T-cells, which preferentially express CXCR3 (36) (Fig. 13.5). CD8$^+$ cells are typically increased in airway infections, and it is possible that the chronic colonization of the lower respiratory tract of COPD patients by bacterial and viral pathogens is responsible for this inflammatory response. It is possible that cigarette-induced lung injury may uncover previously sequestered autoantigens, or cigarette smoke itself may damage lung interstitial and structural cells and make them antigenic (37). The role of increased numbers of CD4$^+$ cells in COPD, particularly in severe disease, is also unknown (19); however, it is now clear that T-cell help is required for the priming of cytotoxic T-cell responses, for maintaining CD8$^+$ T-cell memory, and for ensuring CD8$^+$ T-cell survival. It is also possible that CD4$^+$ T-cells have immunological memory and play a role in perpetuating the inflammatory process in the absence of cigarette smoking. In a mouse model of cigarette-induced emphysema, there is a predominance of T-cells that are directly related to the severity of emphysema (38).

The role of T-cells in the pathophysiology of COPD is not yet certain, although they have the potential to produce extensive damage in the lung. CD8$^+$ cells have the capacity to cause cytolysis and apoptosis of alveolar epithelial cells through release of perforins, granzyme-B, and TNF-$\alpha$ (39, 40). An association between CD8$^+$ cells and apoptosis of alveolar cells is observed in emphysema

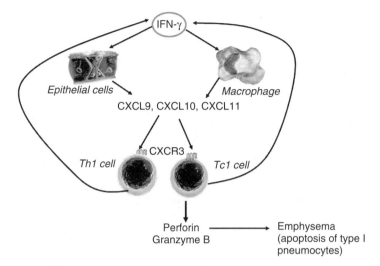

**Figure 13.5** T-lymphocytes in COPD. Epithelial cells and macrophages are stimulated by interferon-γ (IFNγ) to release the chemokines CXCL9, CXCL10, and CXCL11, which together act on CXC-chemokine receptor 3 (CXCR3) expressed on Th1 cells and Tc1 cells to attract them into the lungs. Tc1 cells, through the release of perforin and granzyme B, induce apoptosis of type 1 pneumocytes, which thereby contributes to emphysema. IFNγ released by Th1 and Tc1 cells then stimulates release of CXCR3 ligands, which results in a persistent inflammatory activation.

(41). Apoptotic cells are powerful sources of antigenic material that could reach the DC and perpetrate the T-cell response. In addition, CD8+ T-cells also produce several cytokines of the Tc1 phenotype, which include TNF-α, lymphotoxin, and IFN-γ, and evidence suggests that CD8+ in the lungs of COPD patients expresses IFN-γ (42). All these cytokines would enhance the inflammatory reaction in the lung besides the direct killing by CD8+ cells. COPD has been considered an autoimmune disease triggered by smoking, as previously suggested (37), and the presence of highly activated oligoclonal T-cells in emphysema patients supports this conclusion (43). Evidence suggests that anti-elastin antibodies exist in experimental models of COPD and in COPD patients (44). In addition to activated Th1 cells, some evidence indicates an increase in Th2 cells that express IL-4 in BAL fluid of COPD patients in COPD patients (45).

## 13.4   MEDIATORS OF INFLAMMATION

Many inflammatory mediators have now been implicated in COPD, which include lipids, free radicals, cytokines, chemokines, and growth factors (7). These mediators are derived from inflammatory and structural cells in the lung and interact with each other in a complex manner.

### 13.4.1   Lipid Mediators

The profile of lipid mediators in exhaled breath condensates of patients with COPD shows an increase in prostaglandins and leukotrienes (46). A significant increase in $PGE_2$ and $F_{2\alpha}$ and an increase in $LTB_4$ but not cysteinyl-leukotrienes is observed. This increase is a different pattern to that observed in asthma, in which increases in thromboxane and cysteinyl-leukotrienes have been shown. The increased production of prostanoids in COPD is likely to be secondary to the induction of cyclo-oxygenase-2 (COX2) by inflammatory cytokines, and increased expression of COX2 is found in alveolar macrophages of COPD patients. $LTB_4$ concentrations are also increased in induced sputum, and concentrations of $LTB_4$ are increased even more in sputum and exhaled breath condensate during acute exacerbations (16). $LTB_4$ is a potent chemoattractant of neutrophils, which acts through high-affinity $BLT_1$-receptors. A $BLT_1$-receptor antagonist reduces the neutrophil chemotactic activity of sputum by approximately 25% (47). $BLT_1$-receptors have been identified on T-lymphocytes, and evidence indicates that $LTB_4$ is involved in recruitment of T-cells.

### 13.4.2   Oxidative Stress

Oxidative stress occurs when reactive oxygen species (ROS) are produced in excess of the antioxidant defense mechanisms and result in harmful effects, which include damage to lipids, proteins, and DNA. Increasing evidence suggests that oxidative stress is an important feature in COPD (48). Inflammatory and structural cells that are activated in the airways of patients with COPD produce ROS, such as neutrophils, eosinophils, macrophages, and epithelial cells. Superoxide anions ($\cdot O_2^-$) are generated by NADPH oxidase and converted to hydrogen peroxide ($H_2O_2$) by superoxide dismutases. $H_2O_2$ is then dismuted to water by catalase. $\cdot O_2^-$, and $H_2O_2$ may interact in the presence of free iron to form the highly reactive hydroxyl radical ($\cdot OH$). $\cdot O_2^-$ may also combine with NO to form peroxynitrite, which also generates $\cdot OH$. Oxidative stress leads to the oxidation of arachidonic acid and the formation of a new series of prostanoid mediators called isoprostanes, which may exert significant functional effects, such as bronchoconstriction and plasma exudation (49) (Fig. 13.6).

The normal production of oxidants is counteracted by several antioxidant mechanisms in the human respiratory tract (48). The major intracellular antioxidants in the airways are catalase, SOD, and glutathione, which is formed by the enzyme γ-glutamyl cysteine synthetase, and glutathione synthetase. In the lung, intracellular antioxidants are expressed at relatively low levels and are not induced by oxidative stress, whereas the major antioxidants are extracellular. Extracellular antioxidants, particularly glutathione peroxidase, are markedly upregulated in response to cigarette smoke and oxidative stress. Extracellular antioxidants also include the dietary antioxidants vitamin C (ascorbic acid) and vitamin E (α-tocopherol), uric acid, lactoferrin, and extracellular superoxide dismutase, which is highly expressed in human lung, but its role in COPD is not yet clear.

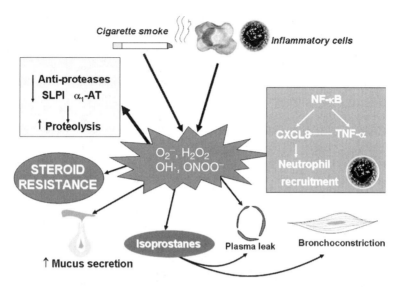

**Figure 13.6**   Oxidative stress in COPD. Oxidative stress plays a key role in the patho-physiology of COPD and amplifies the inflammatory and destructive process. ROS from cigarette smoke or from inflammatory cells (particularly macrophages and neutrophils) result in several damaging effects in COPD, which include decreased antiprotease defenses, such as $\alpha1$-antitrypsin (AT) and secretory leukoprotease inhibitor (SLPI), activation of NF-$\kappa$B resulting in increased secretion of the cytokines CXCL8 and TNF-$\alpha$s, increased production of isoprostanes, and direct effects on airway function. In addition, recent evidence suggests that oxidative stress induces steroid resistance.

ROS have several effects on the airways and parenchyma and increase the inflammatory response. ROS activate NF-$\kappa$B, which switches on multiple inflammatory genes resulting in amplification of the inflammatory response. The molecular pathways by which oxidative stress activates NF-$\kappa$B have not been fully elucidated, but several redox-sensitive steps must be followed in the activation pathway. Oxidative stress results in activation of histone acetyltransferase activity, which opens up the chromatin structure and is associated with increased transcription of multiple inflammatory genes (50). Exogenous oxidants may also be important in worsening airway disease. Considerable evidence suggests increased oxidative stress in COPD (48). Cigarette smoke itself contains a high concentration of ROS. Inflammatory cells, such as activated macrophages and neutrophils, also generate ROS, as discussed above. Several markers of oxidative stress may be detected in the breath, and several studies have demonstrated increased production of oxidants, such as $H_2O_2$, 8-isoprostane, and ethane, in exhaled air or breath condensates, particularly during exacerbations (16).

The increased oxidative stress in the lung epithelium of a COPD patient may play an important pathophysiological role in the disease by amplifying the inflammatory response in COPD. This increase may reflect the activation of NF-$\kappa$B and AP-1, which then induce a neutrophilic inflammation via increased expression of CXC chemokines, TNF-$\alpha$, and MMP-9. Oxidative stress may also impair the function of antiproteases such as $\alpha_1$-antitrypsin and SLPI, and thereby accelerates the breakdown of elastin in lung parenchyma. Corticosteroids are much less effective in COPD than in asthma and do not reduce the progression or mortality of the disease. Alveolar macrophages from patients with COPD show a marked reduction in responsiveness to the anti-inflammatory effects of corticosteroids, compared with cells from normal smokers and nonsmokers (27). In patients with COPD, a marked reduction in activity of HDAC and reduced expression of HDAC2 is observed in alveolar macrophages and peripheral lung tissue (28), which is correlated with increased expression of inflammatory cytokines and a reduced response to corticosteroids. This finding may result directly or indirectly from oxidative stress and is mimicked by the effects of $H_2O_2$ in cell lines (51).

### 13.4.3   Nitrative Stress

The increase in exhaled NO is less marked in COPD than in asthma, partly because cigarette smoking reduces exhaled NO. Recently exhaled NO has been partitioned into central and peripheral portions and this shows reduced NO in the bronchial fraction but increased NO in the peripheral fraction, which includes lung parenchyma and small airways (52). The increased peripheral NO in COPD patients may reflect increased expression of inducible NO synthase in epithelial cells and macrophages of patients with COPD (53). NO and superoxide anions combine to from peroxynitrite, which nitrates certain tyrosine residues in protein, and increased expression of 3-nitrotyrosine is observed in peripheral lung and macrophages of COPD patients (53). Tyrosine nitration of HDAC2 may lead it impaired activity and degradation of this enzyme, which results in steroid resistance (51).

### 13.4.4   Inflammatory Cytokines

Cytokines are the mediators of chronic inflammation, and several have been implicated in COPD (7). An increase in concentration of TNF-$\alpha$ is observed in induced sputum in stable COPD with an additional increase during exacerbations (13, 17). TNF-$\alpha$ production from peripheral blood monocytes is also increased in COPD patients and has been implicated in the cachexia and skeletal muscle apoptosis found in some patients with severe disease. TNF-$\alpha$ is a potent activator of NF-$\kappa$B, which may amplify the inflammatory response. Unfortunately anti-TNF therapies have not proved to be effective in COPD patients. IL-1$\beta$ and IL-6 are other proinflammatory cytokines that may amplify the inflammation in COPD and may be important for systemic circulation.

### 13.4.5   Chemokines

Chemokines are small chemotactic cytokines that play a key role in the recruitment and activation if inflammatory cells through specific chemokine receptors.

Several chemokines have now been implicated in COPD and have been of particular interest since chemokine receptors are G-protein coupled receptors, for which small molecule antagonists have now been developed (8). CXCL8 concentrations are increased in induced sputum of COPD patients and increase even more during exacerbations (13, 17). CXCL8 is secreted from macrophages, T-cells, epithelial cells, and neutrophils. CXCL8 activates neutrophils via low-affinity specific receptors CXCR1, and is chemotactic for neutrophils via high-affinity receptors CXCR2, which are also activated by related CXC chemokines, such as CXCL1. CXCL1 concentrations are markedly elevated in sputum and in BAL fluid of COPD patients, and this chemokine may be more important as a chemoattractant than CXCL8, acting via CXCR2 that are expressed on neutrophils and monocytes (14). CXCL1 induces significantly more chemotaxis of monocytes of COPD patient compared with those of normal smokers, and it may reflect increased turnover and recovery of CXCR2 in monocytes of COPD patients (23). CXCL5 shows a marked increase in expression in airway epithelial cells during exacerbations of COPD; this increase is accompanied by a marked up regulation of epithelial CXCR2.

CCL2 is increased in concentration in COPD sputum and BAL fluid (14) and plays a role in monocyte chemotaxis via activation of CCR2. CCL2 seems to cooperate with CXCL1 in recruiting monocytes into the lungs. The chemokine CCL5 (RANTES) is also expressed in airways of COPD patients during exacerbations and activates CCR5 on T cells and CCR3 on eosinophils, which may account for the increased eosinophils and T-cells in the wall of large airways that have been reported during exacerbations of chronic bronchitis. As discussed above, CXCR3 are upregulated on Tc1 and Th1 cells of COPD patients with increased expression of their ligands CXCL9, CXCL10, and CXCL11.

### 13.4.6 Growth Factors

Several growth factors have been implicated in COPD and mediate the structural changes that are found in the airways. TGF-$\beta$1 is expressed in alveolar macrophages and airway epithelial cells of COPD patients, and it is released from epithelial cells of small airways. TGF-$\beta$ is released in a latent from and activated by various factors, which include MMP-9. TGF-$\beta$ may play an important role in the characteristic peribronchiolar fibrosis of small airways, either directly or through the release of connective tissue growth factor (Fig. 13.7). TGF-$\beta$ downregulates $\beta_2$-adrenergic receptors by inhibiting gene transcription in human cell lines, and it may reduce the bronchodilator response to $\beta$-agonists in airway smooth muscle. Alveolar macrophages produce TGF-$\alpha$ in greater amounts than TGF-$\beta$, which may be a major endogenous activator of EGFR that plays a key role in regulating mucus secretion in response to many stimuli, which include cigarette smoke. Cigarette smoke activates TNF-$\alpha$-converting enzyme on airway epithelial cells, which results in the shedding of TGF-$\alpha$ and the activation of EGFR, resulting in increased mucus secretion (54) (Fig. 13.8).

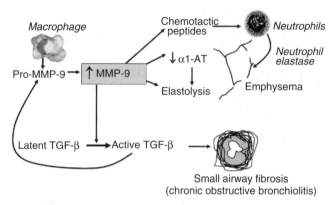

**Figure 13.7** TGF-β in COPD. TGF-β is released in a latent form that may be activated by MMP-9. It may then cause fibrosis directly through effects on fibroblasts or indirectly via the release of CTGF. TGF-β may also downregulate β$_2$-adrenoceptors on cells such as airway smooth muscle to diminish the bronchodilator response to β-agonists.

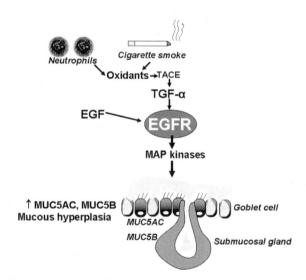

**Figure 13.8** Epidermal growth factor receptors (EGFR) in COPD. EGFR play a key role in the regulation of mucus hypersecretion, with increased expression of mucin genes (MUC5AC, MUCB) and differentiation of goblet cells and hyperplasia of mucus-secreting cells. These effects are mediated via the activation of mitogen-activated protein (MAP) kinases. EGFR are activated by TGF-α, which is in turn activated by tumor necrosis factor-α converting enzyme (TACE), activated via release of oxidants from cigarette smoke and neutrophils. EGFR may also be activated by EGF.

VEGF is a major regulator of vascular growth and is likely to be involved in the pulmonary vascular remodeling that occurs as a result of hypoxic pulmonary vasoconstriction in severe COPD. Increased expression of VEGF is observed in pulmonary vascular smooth muscle of patients with mild and moderate COPD, but paradoxically a reduction in is indicated expression in severe COPD with emphysema. Inhibition of VEGF receptors using a selective inhibitor induces apoptosis of alveolar endothelial cells in rats, which results in emphysema; this finding seems to be driven by oxidative stress. In addition, VEGF is also an important proinflammatory cytokine produced by epithelial and endothelial cells, macrophages, and activated T-cells, which acts by increasing endothelial cell permeability, by inducing expression of endothelial adhesion molecules and via its ability to act as a monocyte chemoattractant. VEGF also stimulates the expression of CXCL10 and its receptor CXCR3. Thus VEGF is likely an intermediary between cell-mediated immune inflammation and the associated angiogenesis reaction.

## 13.5 CONCLUSIONS

In summary, cigarette smoke exposure induces a florid inflammatory response in the lung that involves structural and inflammatory cells and a large array of inflammatory mediators. The interaction of these complex steps eventually leads to airway remodeling and obstruction and emphysema, albeit in only about 25% of chronic smokers. Of interest, the main difference between smokers who develop COPD and the ones who do not seems to be the presence of an adaptive immune response with $CD8^+$, $CD4^+$, and B-cells, which express obvious signs of being activated effector cells. It is likely that genetic and epigenetic factors (such as histone acetylation) are involved in determining the progression of the inflammatory cascade, as it is supported by animal models, where different strains seem to have different sensitivities to cigarette smoke. COPD is a complex inflammatory disease, and the interactions between different inflammatory cells and mediators are still uncertain. More research into these mechanisms is needed to identify novel targets that may lead to the discovery of more effective therapies that can prevent disease progression and reduce the high mortality of this common disease.

## REFERENCES

1. Barnes PJ. Chronic obstructive pulmonary disease: A growing but neglected epidemic. PLoS Med. 2007;4:e112.
2. Mannino DM, Buist AS. Global burden of COPD: risk factors, prevalence, and future trends. Lancet 2007;370:765–773.
3. Barnes PJ, Shapiro SD, Pauwels RA. Chronic obstructive pulmonary disease: molecular and cellular mechanisms. Eur. Respir. J. 2003;22:672–688.

4. Barnes PJ, Hansel TT. Prospects for new drugs for chronic obstructive pulmonary disease. Lancet 2004;364:985–996.

5. Rabe KF, Hurd S, Anzueto A, Barnes PJ, Buist SA, Calverley P, Fukuchi Y, Jenkins C, Rodriguez-Roisin R, van Weel C, Zielinski J. Global strategy for the diagnosis, management, and prevention of COPD - 2006 Update. Am. J. Respir. Crit. Care. Med. 2007;176:532–555.

6. Hogg JC, Chu F, Utokaparch S, Woods R, Elliott WM, Buzatu L, Cherniack RM, Rogers RM, Sciurba FC, Coxson HO, Pare PD. The nature of small-airway obstruction in chronic obstructive pulmonary disease. New Engl. J. Med. 2004;350:2645–2653.

7. Barnes PJ. Mediators of chronic obstructive pulmonary disease. Pharm. Rev. 2004;56:515–548.

8. Donnelly LE, Barnes PJ. Chemokine receptors as therapeutic targets in chronic obstructive pulmonary disease. Trends Pharmacol. Sci. 2006;27:546–553.

9. Barnes PJ. Immunology of asthma and chronic obstructive pulmonary disease. Nat. Immunol. Rev. 2008;8:183–192.

10. Fabbri LM, Romagnoli M, Corbetta L, Casoni G, Busljetic K, Turato G, Ligabue G, Ciaccia A, Saetta M, Papi A. Differences in airway inflammation in patients with fixed airflow obstruction due to asthma or chronic obstructive pulmonary disease. Am. J. Respir. Crit. Care Med. 2003;167:418–424.

11. Petrache I, Natarajan V, Zhen L, Medler TR, Richter AT, Cho C, Hubbard WC, Berdyshev EV, Tuder RM. Ceramide upregulation causes pulmonary cell apoptosis and emphysema-like disease in mice. Nat. Med. 2005;11:491–498.

12. Franklin WA, Veve R, Hirsch FR, Helfrich BA, Bunn PA, Jr. Epidermal growth factor receptor family in lung cancer and premalignancy. Semin.Oncol. 2002;29:3–14.

13. Keatings VM, Collins PD, Scott DM, Barnes PJ. Differences in interleukin-8 and tumor necrosis factor-a in induced sputum from patients with chronic obstructive pulmonary disease or asthma. Am. J. Respir. Crit. Care Med. 1996;153:530–534.

14. Traves SL, Culpitt S, Russell REK, Barnes PJ, Donnelly LE. Elevated levels of the chemokines GRO-a and MCP-1 in sputum samples from COPD patients. Thorax 2002;57:590–595.

15. Keatings VM, Barnes PJ. Granulocyte activation markers in induced sputum: Comparison between chronic obstructive pulmonary disease, asthma and normal subjects. Am. J. Respir. Crit. Care. Med. 1997;155:449–453.

16. Biernacki WA, Kharitonov SA, Barnes PJ. Increased leukotriene B4 and 8-isoprostane in exhaled breath condensate of patients with exacerbations of COPD. Thorax 2003;58:294–298.

17. Aaron SD, Angel JB, Lunau M, Wright K, Fex C, Le Saux N, Dales RE. Granulocyte inflammatory markers and airway infection during acute exacerbation of chronic obstructive pulmonary disease. Am. J. Respir. Crit. Care Med. 2001;163:349–355.

18. Barnes PJ. Macrophages as orchestrators of COPD. J. COPD 2004;1:59–70.

19. Retamales I, Elliott WM, Meshi B, Coxson HO, Pare PD, Sciurba FC, Rogers RM, Hayashi S, Hogg JC. Amplification of inflammation in emphysema and its association with latent adenoviral infection. Am. J. Respir. Crit. Care Med. 2001;164:469–473.

20. Meshi B, Vitalis TZ, Ionescu D, Elliott WM, Liu C, Wang XD, Hayashi S, Hogg JC. Emphysematous lung destruction by cigarette smoke. The effects of latent adenoviral infection on the lung inflammatory response. Am. J. Respir. Cell. Mol. Biol. 2002;26:52–57.

21. Russell RE, Thorley A, Culpitt SV, Dodd S, Donnelly LE, Demattos C, Fitzgerald M, Barnes PJ. Alveolar macrophage-mediated elastolysis: Roles of matrix metalloproteinases, cysteine, and serine proteases. Am. J. Physiol. Lung Cell Mol. Physiol 2002;283:L867–L873.

22. Caramori G, Romagnoli M, Casolari P, Bellettato C, Casoni G, Boschetto P, Fan CK, Barnes PJ, Adcock IM, Ciaccia A, Fabbri LM, Papi A. Nuclear localisation of p65 in sputum macrophages but not in sputum neutrophils during COPD exacerbations. Thorax 2003;58:348–351.

23. Traves SL, Smith SJ, Barnes PJ, Donnelly LE. Specific CXC but not CC chemokines cause elevated monocyte migration in COPD: a role for CXCR2. J. Leukoc.Biol. 2004;76:441–450.

24. Grumelli S, Corry DB, Song L-X, Song L, Green L, Huh J, Haccken J, Espada R, Bag R, Lewis DE, Kheradmand F. An immune basis for lung parenchymal destruction in chronic obstructive pulmonary disease and emphysema. PLoS Med. 2004;1:75–83.

25. Tomita K, Caramori G, Lim S, Ito K, Hanazawa T, Oates T, Chiselita I, Jazrawi E, Chung KF, Barnes PJ, Adcock IM. Increased p21CIP1/WAF1 and B cell lymphoma leukemia-xL expression and reduced apoptosis in alveolar macrophages from smokers. Am. J. Respir. Crit. Care. Med. 2002;166:724–731.

26. Keatings VM, Jatakanon A, Worsdell YM, Barnes PJ. Effects of inhaled and oral glucocorticoids on inflammatory indices in asthma and COPD. Am. J. Respir. Crit. Care. Med. 1997;155:542–548.

27. Culpitt SV, Rogers DF, Shah P, de Matos C, Russell RE, Donnelly LE, Barnes PJ. Impaired inhibition by dexamethasone of cytokine release by alveolar macrophages from patients with chronic obstructive pulmonary disease. Am. J. Respir. Crit. Care Med. 2003;167:24–31.

28. Ito K, Ito M, Elliott WM, Cosio B, Caramori G, Kon OM, Barczyk A, Hayashi M, Adcock IM, Hogg JC, Barnes PJ. Decreased histone deacetylase activity in chronic obstructive pulmonary disease. N. Engl. J. Med. 2005;352:1967–1976.

29. Barnes PJ. Reduced histone deacetylase in COPD: clinical implications. Chest 2006;129:151–155.

30. Papi A, Romagnoli M, Baraldo S, Braccioni F, Guzzinati I, Saetta M, Ciaccia A, Fabbri LM. Partial reversibility of airflow limitation and increased exhaled NO and sputum eosinophilia in chronic obstructive pulmonary disease. Am. J. Respir. Crit. Care Med. 2000;162:1773–1777.

31. Zhu J, Qiu YS, Majumdar S, Gamble E, Matin D, Turato G, Fabbri LM, Barnes N, Saetta M, Jeffery PK. Exacerbations of Bronchitis: Bronchial eosinophilia and gene expression for interleukin-4, interleukin-5, and eosinophil chemoattractants. Am. J. Respir. Crit. Care Med. 2001;164:109–116.

32. Hammad H, Lambrecht BN. Recent progress in the biology of airway dendritic cells and implications for understanding the regulation of asthmatic inflammation. J.Allergy Clin. Immunol. 2006;118:331–336.

33. Rogers AV, Adelroth E, Hattotuwa K, Dewar A, Jeffery PK. Bronchial mucosal dendritic cells in smokers and ex-smokers with COPD: An electron microscopic study. Thorax 2008;63:108–114.

34. Barnes PJ, Cosio MG. Characterization of T lymphocytes in chronic obstructive pulmonary disease. PLoS Med. 2004;1:25–27.

35. Costa C, Rufino R, Traves SL, Lapa E Silva JR, Barnes PJ, Donnelly LE. CXCR3 and CCR5 chemokines in the induced sputum from patients with COPD. Chest 2008;133:26–33.

36. Saetta M, Mariani M, Panina-Bordignon P, Turato G, Buonsanti C, Baraldo S, Bellettato CM, Papi A, Corbetta L, Zuin R, Sinigaglia F, Fabbri LM. Increased expression of the chemokine receptor CXCR3 and its ligand CXCL10 in peripheral airways of smokers with chronic obstructive pulmonary disease. Am. J. Respir. Crit. Care Med. 2002;165:1404–1409.

37. Cosio MG, Majo J, Cosio MG. Inflammation of the airways and lung parenchyma in COPD: role of T cells. Chest 2002;121:160S–165S.

38. Takubo Y, Guerassimov A, Ghezzo H, Triantafillopoulos A, Bates JH, Hoidal JR, Cosio MG. Alpha1-antitrypsin determines the pattern of emphysema and function in tobacco smoke-exposed mice: parallels with human disease. Am. J. Respir. Crit. Care Med. 2002;166:1596–1603.

39. Hashimoto S, Kobayashi A, Kooguchi K, Kitamura Y, Onodera H, Nakajima H. Upregulation of two death pathways of perforin/granzyme and FasL/Fas in septic acute respiratory distress syndrome. Am. J. Respir. Crit Care Med. 2000;161: 237–243.

40. Chrysofakis G, Tzanakis N, Kyriakoy D, Tsoumakidou M, Tsiligianni I, Klimathianaki M, Siafakas NM. Perforin expression and cytotoxic activity of sputum CD8⁺ lymphocytes in patients with COPD. Chest 2004;125:71–76.

41. Majo J, Ghezzo H, Cosio MG. Lymphocyte population and apoptosis in the lungs of smokers and their relation to emphysema. Eur. Respir. J. 2001;17:946–953.

42. Cosio MG T-lymphocytes. In: Chronic Obstructive Pulmonary Disease: Cellular and Molecular Mechanisms. Barnes PJ, ed. 2005. Taylor & Francis Group, New York.

43. Sullivan AK, Simonian PL, Falta MT, Mitchell JD, Cosgrove GP, Brown KK, Kotzin BL, Voelkel NF, Fontenot AP. Oligoclonal CD4⁺ T cells in the lungs of patients with severe emphysema. Am. J. Respir. Crit. Care Med. 2005;172:590–596.

44. Lee SH, Goswami S, Grudo A, Song LZ, Bandi V, Goodnight-White S, Green L, Hacken-Bitar J, Huh J, Bakaeen F, Coxson HO, Cogswell S, Storness-Bliss C, Corry DB, Kheradmand F. Antielastin autoimmunity in tobacco smoking-induced emphysema. Nat. Med. 2007;13:567–569.

45. Barczyk A, Pierzchala W, Kon OM, Cosio B, Adcock IM, Barnes PJ. Cytokine production by bronchoalveolar lavage T lymphocytes in chronic obstructive pulmonary disease. J. Allergy. Clin. Immunol. 2006;117:1484–1492.

46. Montuschi P, Kharitonov SA, Ciabattoni G, Barnes PJ. Exhaled leukotrienes and prostaglandins in COPD. Thorax 2003;58:585–588.

47. Beeh KM, Kornmann O, Buhl R, Culpitt SV, Giembycz MA, Barnes PJ. Neutrophil chemotactic activity of sputum from patients with COPD: role of interleukin 8 and leukotriene B4. Chest 2003;123:1240–1247.

48. Rahman I. Oxidative stress in pathogenesis of chronic obstructive pulmonary disease: Cellular and molecular mechanisms. Cell Biochem. Biophys. 2005;43:167–188.

49. Montuschi P, Barnes PJ, Roberts LJ. Isoprostanes: Markers and mediators of oxidative stress. FASEB J. 2004;18:1791–1800.

50. Tomita K, Barnes PJ, Adcock IM. The effect of oxidative stress on histone acetylation and IL-8 release. Biochem. Biophys. Res. Comm. 2003;301:572–577.

51. Ito K, Tomita T, Barnes PJ, Adcock IM. Oxidative stress reduces histone deacetylase (HDAC)2 activity and enhances IL-8 gene expression: Role of tyrosine nitration. Biochem. Biophys. Res. Commun. 2004;315:240–245.

52. Brindicci C, Ito K, Resta O, Pride NB, Barnes PJ, Kharitonov SA. Exhaled nitric oxide from lung periphery is increased in COPD. Eur. Respir. J. 2005;26:52–59.

53. Ricciardolo FL, Caramori G, Ito K, Capelli A, Brun P, Abatangelo G, Papi A, Chung KF, Adcock I, Barnes PJ, Donner CF, Rossi A, Di Stefano A. Nitrosative stress in the bronchial mucosa of severe chronic obstructive pulmonary disease. J. Allergy Clin. Immunol. 2005;116:1028–1035.

54. Shao MX, Nakanaga T, Nadel JA. Cigarette smoke induces MUC5AC mucin over-production via tumor Necrosis factor-α converting enzyme in human airway epithelial (NCI-H292) cells. Am. J. Physiol. Lung Cell Mol. Physiol. 2004;287:L420–427.

# 14

# DEPRESSION

GLEN B. BAKER AND NICHOLAS D. MITCHELL

*Neurochemical Research Unit, Department of Psychiatry, University of Alberta, Edmonton, Alberta, Canada*

Although the etiology of depression remains to be elucidated, our knowledge of the neurobiology and biochemistry of this mental disorder has been updated increasingly. Early research focused on the biogenic amines norepinephrine (NE; noradrenaline) and 5-hydroxytryptamine (5-HT; serotonin). The biogenic amine (monoamine) hypothesis of depression states that depression is the result of a functional deficiency of NE and/or 5-HT at central synapses. However, it soon became obvious that other factors were also important in the neurobiology of depression. Dopamine is thought to be involved in various aspects, which include reward and locomotion. The amino acids gamma-aminobutyric acid (GABA) and glutamate, their receptors, and various neuroactive steroids that act as allosteric modulators at GABA-A and/or NMDA glutamate receptors have been proposed to have an important role. Stress and the hypothalamic-pituitary-adrenal (HPA) axis play a central role, and it has been proposed that increased secretion of corticotropin releasing factor (CRF) may be critical in producing the symptoms of depression. Various other neurochemicals have also been implicated in depression, and in several cases they have links to the HPA axis. Abnormal levels of Substance P have been reported in depression, which led to development of potential antidepressants that interact with neurokinin receptors. Abnormalities in the immune system, which presumably involve cytokines, have been reported in depression. Cell loss seems to occur in some brain areas (e.g., hippocampus) in depression, and such loss can be produced by stress and prevented by antidepressants (which also increase expression of various neurotrophic and transcription factors). Agonists at melatonin receptors have been proposed in recent years as effective antidepressants.

*Chemical Biology: Approaches to Drug Discovery and Development to Targeting Disease*, First Edition. Edited by Natanya Civjan.
© 2012 John Wiley & Sons, Inc. Published 2012 by John Wiley & Sons, Inc.

The possible involvement of the various neurochemicals mentioned above in depression will be reviewed, and some other neurochemicals that are being examined will also be mentioned.

## 14.1  INTRODUCTION

Depression (referring here to major depressive disorder in the Diagnostic and Statistical Manual of the American Psychiatric Association, volume IV; DSM-IV) is a common, chronic, functionally limiting clinical syndrome that likely reflects a heterogenous group of underlying disorders (1, 2). In the DSM-IV (DSM-IV-TR is the most recent version), characteristic symptoms of a major depressive episode listed include five or more of the following symptoms, which have been present nearly every day in most cases during the same 2-week period and represent a change from previous functioning: 1) depressed mood; 2) markedly diminished interest or pleasure in all, or almost all, activities; 3) significant weight loss or gain; 4) insomnia or hypersomnia; 6) fatigue or loss of energy; 7) feelings of worthlessness or excessive or inappropriate guilt; 8) diminished ability to think or concentrate, or indecisiveness; and 9) recurrent thoughts of death or suicidal ideation without a specific plan, or a suicide attempt or a specific plan for committing suicide. Note: of the five symptoms, at least one is either 1) or 2) above. The lifetime prevalence of depression in the United States has been estimated at 5–12% in men and 10–25% in women (2). Depression has been shown to impact individual quality of life significantly (3), carry a heavy economic burden (4), and affect brain structure and function (5). Despite research into many potential causative factors, a putative etiology for depression remains elusive (6, 7). Research in the twentieth century initially focused on a decreased functional availability of the monoamine neurotransmitters norepinephrine (NE) and 5-hydroxytryptamine (5-HT) as the possible agents of depressive illness, and it was supported in part by the pharmacological properties of available antidepressants. However, a growing body of literature suggests that multiple contributory chemical systems, many of which are intertwined, are involved in producing the depressive phenotype (8–10).

## 14.2  BIOGENIC AMINES

Endogenous biogenic amines in the brain include catecholamines [NE (noradrenaline, NA), dopamine (DA), epinephrine (adrenaline)] 5-HT, histamine, and the so-called trace amines ($\beta$-phenylethylamine, tyramine, tryptamine, and octopamine). These amines have in common a arylalkylamine structure, and all have been implicated in the etiology of one or more psychiatric disorders and/or in therapeutic and/or adverse effects of drugs used to treat such disorders. In this review on depression, the focus in the case of biogenic amines will be on

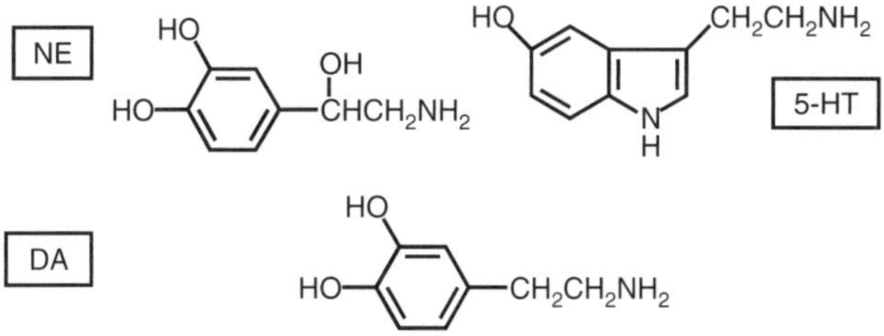

**Figure 14.1**  Structures of NE, 5-HT, and DA.

5-HT, NE, and DA, although epinephrine and histamine and trace amines have also been implicated (see Section 14.5).

Early studies on the origins of depression focused on the monoamine neurotransmitters NE (a catecholamine) and 5-HT (an indoleamine) (for reviews, see References 7, 11, and 12) (Fig. 14.1). The antihypertensive and antipsychotic medication reserpine depletes central stores of NE, 5-HT, and DA and is known to produce symptoms of depression in some patients (13). Iproniazid is an antitubercular drug that was noted to produce mood elevation in tuberculosis patients and was shown to be an inhibitor of monoamine oxidase (MAO), which is a major catabolic enzyme for monoamines, including NE, 5-HT, and DA (another catecholamine neurotransmitter). In addition, studies in animals with drugs that caused depletions of NE and/or 5-HT demonstrated changes in locomotor activity, sleep, and sexual activity; these symptoms are similar to physiological symptoms observed in depressed patients. These observations led to catecholamine (14) and indoleamine (15, 16) hypotheses, which are now generally combined into the biogenic amine (monoamine) hypothesis that states that depression is the result of a functional deficiency of NE and/or 5-HT at specific synapses in the central nervous system (12). In addition to the effects of the MAO inhibitors (several of which became approved antidepressants), the hypothesis was supported by the actions of the tricyclic antidepressants (TCAs), which inhibit the reuptake of NE and 5-HT back into nerve terminals. Such reuptake is a major inactivation mechanism for the catecholamines and 5-HT, and inhibition of this process leaves more NE and 5-HT available in the synaptic cleft between neurons to interact with postsynaptic receptors (Fig. 14.2). Indeed, most antidepressants developed subsequently inhibit NE and/or 5-HT reuptake and/or act on presynaptic receptors that affect synthesis and release of biogenic amine neurotransmitters (see Fig. 14.2).

### 14.2.1  Serotonin (5-Hydroxytryptamine, 5-HT)

Regulation of mood, sleep, and aggression have all been shown to involve the serotoninergic system (17–19), and most antidepressant drugs currently being used inhibit 5-HT reuptake and/or act on 5-HT receptors (of which there are several subtypes). 5-HT is produced centrally from the amino acid tryptophan,

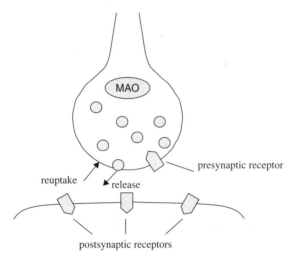

**Figure 14.2** Sites of actions of antidepressants at the synapse (presynaptic nerve terminal, postsynaptic nerve cell, and synaptic cleft between the two are shown). MAO is present in mitochondria, and the shaded circles represent synaptic vesicles that contain neurotransmitter amines.

and depressed mood can be induced experimentally by acute tryptophan depletion in healthy individuals. This effect is accentuated in those with a family history of depression (18–21). Similarly, depressive relapse can be initiated in individuals treated with MAO inhibitors or selective serotonin reuptake (SSRI) inhibitors by depleting tryptophan (22, 23).

A major problem with the biogenic amine hypothesis of depression is the discrepancy in the time course between relatively rapid biochemical and pharmacologic effects (e.g., inhibition of reuptake, inhibition of MAO) of the antidepressants and their clinical beneficial effects, which often require administration for 2–3 weeks or longer. This discrepancy led to the studies in both laboratory animals (receptor-binding studies and electrophysiological investigations) and humans (postmortem studies and *in vivo* neuroimaging studies) on possible dysregulation of receptors for the biogenic amines in depression and effects of antidepressants on that dysregulation (24–28). It has been suggested that antidepressants normalize the density and/or function of 5-HT1 and 5-HT2 receptors, but there are inconsistencies in the literature in this regard. Electroconvulsive shock in rats (an approximate animal model of electroconvulsive therapy in humans) seems to cause an opposite effect on 5-HT2 receptor regulation to that of several antidepressants (24, 25, for a review see Reference 26).

The 5-HT transporter protein (5-HTT) is critical for the reuptake of 5-HT into the presynaptic neuron. Decreased 5-HTT binding has been reported in depressed patients both in postmortem samples and in functional imaging studies (28, 29). Two common alleles of the gene that encode this protein have been identified, with the short (*s*) form of the allele being less active, which results in decreased transcription and reduced expression of 5-HTT. Studies have suggested that the *s* allele occurs more frequently in depressed patients as well as in suicide victims, and homozygous individuals are more likely to have family histories of depression. This finding seems paradoxical, as the *s* allele of the gene results in decreased serotonin reuptake, which thereby increases the duration of 5-HT in the synapse. It has been suggested that this observation may be explained by the lifelong duration of the genetic polymorphism compared with the acute effect of medication administration (28, 29).

### 14.2.2   Norepinephrine (Noradrenaline)

Investigations have also been conducted on the regulation of NE receptors in depression and after chronic administration of antidepressants (11, 30–32). In animal models, induced chronic stress has been used as a model for depression. Chronic stress has been reported to result in an increase in presynaptic inhibitory $\alpha_{2A}$ adrenergic autoreceptors. Long-term exposure to drugs that block the $\alpha_2$ receptor result in an increased levels of synaptic NE and has been shown to counter the effects of chronic stress in rats (11, 31). In humans and laboratory animals, contradictory evidence exists as to the role of noradrenergic receptors in depression (31, 32). Down regulation of $\beta1$-adrenoreceptor density occurs after chronic administration of some types of antidepressants, which coincides with clinical effect, but it does not occur with all antidepressants (25, 32). It is not clear whether $\beta$-adrenoreceptor density is increased in victims of suicide; presynaptic $\alpha_{2A}$-adrenoreceptors have been reported to be increased in postmortem brains of depressed individuals who died by suicide compared with controls (30, 31).

NE is synthesized from the amino acid tyrosine, which can be depleted in the brain by administering a competitive inhibitor of its production, $\alpha$-methyl-*para*-tyrosine. Tyrosine depletion has been shown to reverse antidepressant responses in individuals treated with antidepressants that inhibit NE reuptake, which suggests a direct role of this neurotransmitter in the course of depression (31).

### 14.2.3   Dopamine

Although the focus on monoamine neurotransmitters in depression has been on NE and 5-HT, it is generally considered that DA (Fig. 14.1) also plays a role, and readers are referred to several comprehensive papers that review basic science and clinical evidence for the involvement of DA (11, 33–35). Motivation, psychomotor speed, concentration, and anhedonia have all been linked to dopaminergic circuits. Most DA-producing neurons have nuclei located in the brain stem. Projections from dopaminergic neurons form three primary paths to the cortical and subcortical structures: The nigrostriatal pathway is involved in motor planning and execution; the mesocortical pathway is concerned with concentration and executive functions; and the mesolimbic pathway is important in motivation, pleasure, and reward. Mesolimbic DA dysfunction is observed in animal models of depression, with subsequent antidepressant use causing enhanced dopamine signaling (11, 33–35). Decreased levels of DA in the nucleus accumbens of rats are associated with a decreased response to rewards. Animals that experience learned helplessness have decreased levels of DA in the caudate nucleus and nucleus accumbens; exposure to antidepressant medications has been reported to increase the level of DA in the same regions (11, 33, 34).

In humans, genetic studies have shown that particular polymorphisms of the $D_3$ and $D_4$ DA receptors are associated with depressed phenotypes. In studies of CSF concentrations of homovanillic acid (HVA), which is a metabolite of DA, lower levels than controls have been reported in some depressed patients. Conversely, depressed individuals with psychotic symptoms have been reported to have increased CSF levels of HVA and DA compared with controls (33).

## 14.3  CLASSES OF ANTIDEPRESSANTS

The three neurotransmitter monoamines mentioned above do not operate independently, but rather they interact with one another and are influenced by various central and peripheral biological processes. The actions of most commercially available antidepressant medications continue to target the monoamine neurotransmitter systems. The structures of these antidepressants are shown in Fig. 14.3. The MAO inhibitors (tranylcypromine, phenelzine, and moclobemide) prevent metabolic breakdown of the amines; the TCAs (represented by imipramine and desipramine in Fig. 14.3, although several other structurally related TCAs are available) are inhibitors of reuptake of NE and 5-HT back into the presynaptic neuron, which results in increased levels of these amines in the synaptic cleft; the tetracyclic maprotiline is a NE reuptake inhibitor; bupropion and its metabolites inhibit reuptake of NE and DA; trazodone inhibits reuptake of 5-HT and acts on α adrenergic receptors; the selective serotonin reuptake inhibitors (SSRIs) are, as their name suggests, very selective for inhibiting serotonin (5-HT) reuptake into nerve terminals; venlafaxine, duloxetine, and milnacipran are inhibitors of NE and 5-HT reuptake; mirtazapine inhibits 5-HT reuptake and blocks presynaptic α2 adrenergic receptors; and reboxetine is a selective NE reuptake inhibitor. These medications have been shown to produce their pharmacological effects on NE and/or 5-HT very quickly (within hours or minutes), but clinical effects lag, which becomes apparent only after days to weeks. Up to 80% of patients respond to currently available therapies; however, only 50% of patients achieve complete remission (7). These phenomena suggest that biogenic amines alone are not sufficient targets for research into the etiology and treatment of depression. In recent years, extensive research has focused on other possible causative factors, and the literature in this area will now be reviewed.

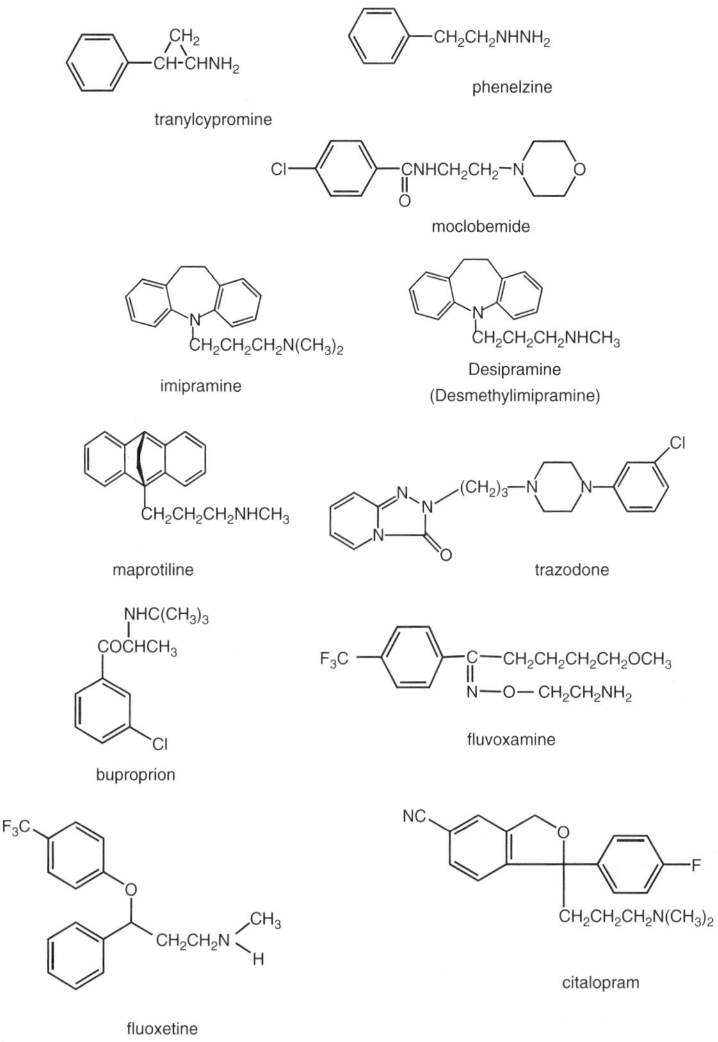

**Figure 14.3** Structures of antidepressants.

## 14.4 BEYOND BIOGENIC AMINES

### 14.4.1 γ-Aminobutyric Acid (GABA) and Glutamate

The amino acids GABA and glutamic acid (glutamate) (Figs. 14.4 and 14.5, respectively) are major inhibitory and excitatory neurotransmitters, respectively, in the central nervous system, and a requisite balance between the two operates

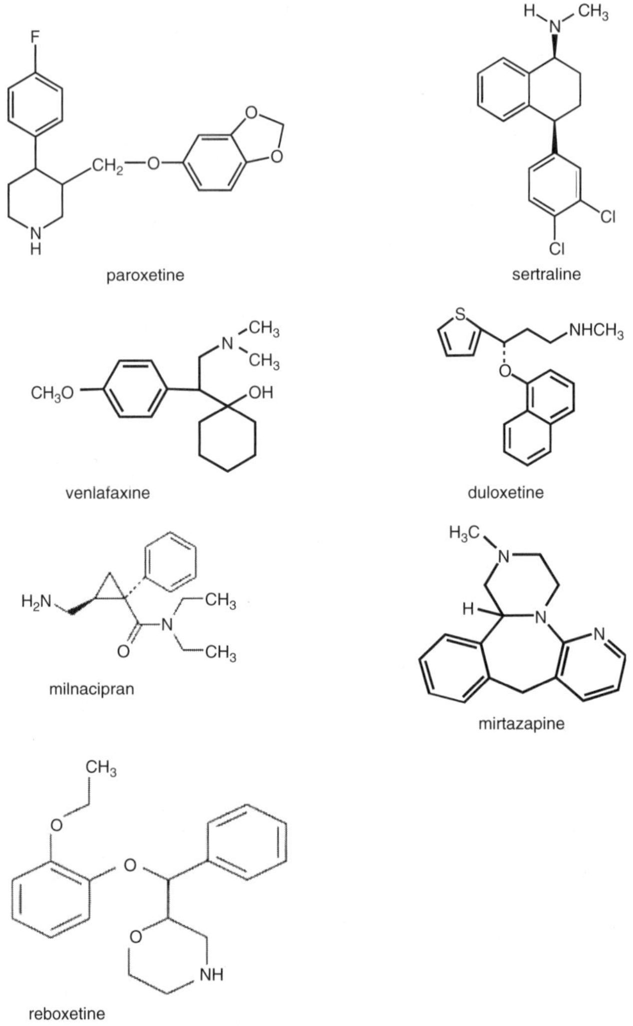

paroxetine

sertraline

venlafaxine

duloxetine

milnacipran

mirtazapine

reboxetine

**Figure 14.3**   (*Continued*)

in normal brain. Aberrations in the functions of one or both of these neuro-transmitters have been implicated in the pathogenesis of several neurological and psychiatric disorders, which include depression (36–38).

Modulators of *alpha*-amino-3-hydroxy-5-methyl-4-isoxazolepropionic acid (AMPA) and N-methyl-D-aspartate (NMDA) ionotropic (associated directly with ion channels) glutamate receptors as well as metabotropic (linked to G-proteins and second messenger systems) glutamate receptors have been investigated for antidepressant properties (39–42). NMDA receptor antagonists have shown effects similar to other antidepressants in animal models of depression. In

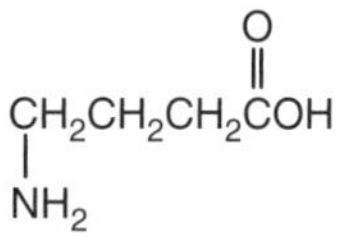

$$\begin{array}{c} O \\ \parallel \\ CH_2CH_2CH_2COH \\ | \\ NH_2 \end{array}$$

**Figure 14.4**   Structure of GABA.

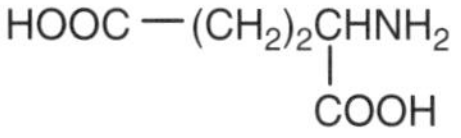

$$HOOC - (CH_2)_2\underset{\underset{COOH}{|}}{C}HNH_2$$

**Figure 14.5**   Structure of glutamic acid.

humans, plasma levels of glutamate have been reported to be correlated with the severity of depression (43), and NMDA receptor abnormalities have been observed in postmortem brain tissue of suicide victims and individuals with depression. Clinical trials of intravenously administered ketamine, which is an NMDA receptor antagonist, have produced rapid diminishment of depressive symptoms after a single administration (41). Concern has been raised over the psychotogenic side effects of NMDA receptor antagonists; however, memantine, which is a lower-affinity NMDA receptor antagonist without these side effects, has been investigated and did not show clinical improvement (42). Less evidence exists for the role of the AMPA receptor in depression. Tricyclic antidepressants have affinity for this receptor, and in a rat model of depression, AMPA receptor density increases with chronic antidepressant treatment (44).

Decreased GABAergic effects have also been associated with depression. *In vivo* evidence of GABAergic dysfunction in patients with depression includes decreased levels of GABA in the CSF, plasma, and occipital cortex (45, 46). Premenstrual dysphoric disorder (PMDD), which is a condition of depressive symptoms prior to menstruation, has been associated with a reduced variability in cortical GABA levels across the menstrual cycle (47). The MAO inhibitor antidepressant phenelzine has been shown to also cause marked increases in brain levels of GABA when administered to rats (48, 49). Using magnetic resonance spectroscopy, SSRI antidepressants have been demonstrated to increase brain levels of GABA in humans (50).

Both GABA and glutamate are important beyond their direct effects as neurotransmitters. They interact with several other neurotransmitter and neuromodulatory systems and have effects on the regulation of the hypothalamic-pituitary-adrenal (HPA) axis, which may contribute to hyperactivity in this circuit in depression. The HPA axis will be discussed later in this chapter.

### 14.4.2   Neuroactive Steroids

In recent years, a great deal of interest has focused on the possible involvement of several so-called neuroactive steroids (Fig. 14.6) in the etiology and pharmacotherapy of a variety of neurologic and psychiatric disorders, which

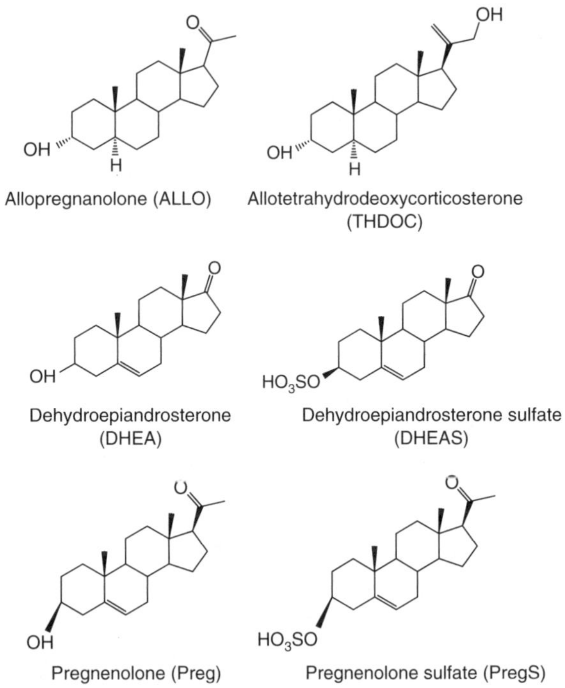

**Figure 14.6**  Structure of some neuroactive steroids.

include depression (51–53). These compounds are rapid-acting steroids that act as positive or negative allosteric modulators at many receptors, which include GABA-A and NMDA receptors. These steroids include pregananolone, pregnenolone, allopregnanolone, isoallopregnanolone, 3α,5β-tetrahydroprogesterone (THP), 3β,5α-THP, 3α,5α-tetrahydrodeoxycorticosterone (THDOC), dehydroepiandrosterone (DHEA), DHEA sulphate, and pregnenolone sulfate. Allopregnanolone and THDOC are strong positive modulators at the GABA receptor and are potent anxiolytics (Note: Most antidepressants drugs also have anxiolytic properties). The ratios of these various neuroactive steroids are altered in several psychiatric disorders, and allopregnanolone seems to be of particular interest with regard to depression. Allopreganolone has been reported to produce antidepressant-like actions in ovariectomized rats when given as an intra-accumbens infusion (54), and SSRI antidepressants have been shown to increase brain levels of allopregnanolone in rats at doses lower than those required to inhibit 5-HT reuptake (55). Blunted responses to allopreganaolone in women with PMDD correlate with the degree of depression in these subjects (56).

## 14.4.3  The Hypothalamic-Pituitary-Adrenal (HPA) Axis

A vast amount of literature indicates a relationship between stress and depression, and in this regard it is interesting that the HPA axis, which is involved in the

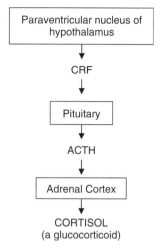

**Figure 14.7** Schematic diagram of the HPA axis.

coordination of neuroendocrine responses to stress, has been shown to be hyperactive in many depressed patients (57, 58). In their diathesis-stress hypothesis of mood disorders, Stout et al. (58) propose that the HPA axis is the principal site where genetic and environmental influences converge to cause mood disorders.

In healthy subjects, stress activates the hypothalamus, which results in the release of corticotropin releasing factor (CRF), which in turn results in release of adrenocorticotropic hormone (ACTH). The ACTH activates the secretion of the glucocorticoid cortisol from the adrenal cortex (Fig. 14.7). The HPA axis has an autoregulating mechanism mediated via negative feedback by cortisol acting at glucocorticoid receptors in the hypothalamus and the pituitary; but in many depressed patients, this regulation is impaired, which results in higher circulating levels of CRF and cortisol. This impairment probably occurs as a result of a decrease in the number or sensitivity of glucocorticoid receptors, which can be affected by gene expression, monoamines, and early childhood expression (58, 59). It has been reported that depressed patients exhibit decreased ACTH secretion in response to exogenously administered CRF and increased secretion of cortisol in response to a given ACTH level (58, 59). Antidepressants have been reported to result in return of HPA functioning to control levels (59, 60), and it has been proposed (9) that normalization of the HPA axis is a requirement for the successful attenuation of depressive symptoms.

Chronic administration of corticosterone to rats results in behaviors associated with depression (61). Indeed, inhibitors of cortisol synthesis (e.g., ketoconazole and metyrapone) and glucocorticoid receptor antagonists (e.g., mifepristone) have been tested clinically. Although preliminary studies with these drugs are promising, the former drugs have an unfavorable side effect profile, and the latter suffer from lack of specificity (61, 62). The development of CRF1 receptor antagonists has been a major thrust by several researchers and pharmaceutical companies in recent years. Increased levels of CRF in the hypothalamus and CSF and an increased number of CRF neurons in the paraventricular nucleus of the hypothalamus have been observed in major depressive disorder (10). Intracranial administration of CRF to rats or overexpression of CRF in mice leads to sleep disturbances, reduced appetite, and diminished sexual activity. These symptoms appear in many depressed patients (58, 63). At the time of writing, researchers continue to discuss whether the CRF antagonists currently under investigation are acting primarily through blockade of prefrontal and limbic CRF1 receptors rather than having an effect on the HPA axis. In addition, CRF2 receptors are observed in the brain, and although their function is less clear, drugs that act at these receptors are also of investigative interest (58).

As indicated in other parts of this review, the HPA axis interacts with most neurochemicals that are thought to be important in depression, and many researchers consider that this axis plays a central role in depression.

### 14.4.4 Substance P

The peptide Substance P acts primarily on neurokinin 1 (NK1) receptors that are coupled to the $G_q$ subunit of G proteins. Substance P has been of interest with regard to depression because of its increased expression and that of NK1 receptors in fear-and anxiety-related circuits. Substance P is released in animals in response to fear-invoking stimuli and there is a high degree of colocalization of substance P with 5-HT or its receptors in human brain (9). Kramer et al. (64) reported that chronic treatment of humans with a nonpeptide NK1 receptor antagonist resulted in clinical improvement in depressed subjects; this finding was replicated by some groups but not by others (9). It stimulated additional research on NK1 antagonists. Although clinical findings with substance antagonists have been disappointing overall, this class of drug continues to be of interest, and recent studies in laboratory animals and humans suggest that some useful antidepressant agents may develop (66–69). Recent studies indicate that NK1 antagonists act through serotoninergic and noradrenergic neurons (70), and it has been suggested that these drugs may be useful agents in combination with traditional antidepressants (71).

### 14.4.5 Cytokines

In recent years, researchers have shown interest in the role of the immune system in depression (72, 73), and the cytokine theory of depression proposes that alterations in the immune response result in behavioral, cognitive, and neuroendocrine changes in depression. Cytokines can be proinflammatory or anti-inflammatory, and it has been postulated that depression may be caused by, or at least related to, excessive amounts of proinflammatory cytokines such as interleukin-1 and tumor necrosis factor-alpha (74). Indeed, it has been observed that antiviral treatment with the cytokine interferon results in "sickness behavior" characterized by symptoms such as weight loss, anorexia, depressed mood, sleep disturbances, social withdrawal, and fatigue, which are also observed in depression (9). It is also of interest that cytokines are potent stimulators of the HPA axis via activation of CRF release, and proinflammatory cytokines can also decrease 5-HT levels by increasing 5-HT secretion and diverting metabolism of tryptophan metabolism away from formation of 5-HT by tryptophan hydroxylase as well as increasing metabolism by the indolamine-2,3-dioxygenase (IDO) pathway (75). Although the evidence for abnormal cytokine levels and/or inflammatory responses in depression is inconsistent to date, this area is very interesting for future exploration and should lead to more studies on glial cells that synthesize and release cytokines.

### 14.4.6 Intracellular Signaling Cascades and Neurotrophic Factors

As the acute increase in synaptic monoamines produced by antidepressant drugs does not correlate with timing of clinical effects, the intracellular signaling pathways that their receptors they interact with have been investigated as targets for novel antidepressants. Several such pathways are observed, and an example is given in Fig. 14.8 (76).

Several serotoninergic and DA receptors and the $\beta_1$ noradrenergic receptor activate the $G_s$ pathway, whereas others activate the $G_q$ pathway. These signaling cascades result in activation of protein kinases A and C, with a net effect of increasing phosphorylation of cyclic adenosine monophosphate (cAMP)-regulated element binding protein (CREB). CREB acts as a regulator in the expression of many genes, some of which have been implicated in neuroplasticity and neurogenesis. For example, transcription of brain-derived neurotrophic factor (BDNF), tyrosine kinase B (trkB) (which is a BDNF receptor), and the glucocorticoid receptor are activated by CREB. Inhibition of transcription of CRF and subunits of the NMDA receptor occur in the presence of CREB. Alterations in CREB levels have been observed in depressed individuals, and increases in CREB function have been associated with reduction or production of depressive symptoms in animals, which depends on the area of the brain involved (77, 78).

The most researched area of interest for antidepressant activity related to signaling pathways is the hippocampus (79–85). The volume of the hippocampus is reduced in patients with multiple episodes of major depression, as observed in imaging studies and in postmortem samples (81–84). Animal models of inescapable stress are associated with decreased hippocampal neurogenesis (84). Neurogenesis is also evident in the hippocampus of humans (85). The neurogenesis hypothesis of depression states that depression is a consequence of impaired neurogenesis in the hippocampus and that antidepressants exert their effect by stimulating neurogenesis (86). Controversy exists as to whether neurogenesis is necessary for the efficacy of antidepressant medications in animal models (87), with some evidence suggesting that when neurogenesis is inhibited, antidepressant effects are lost (84, 88).

In the hippocampus, CREB seems to mediate antidepressant effects, and a variety of antidepressants, which includes electroconvulsive therapy, increases

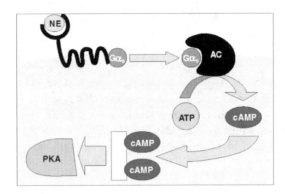

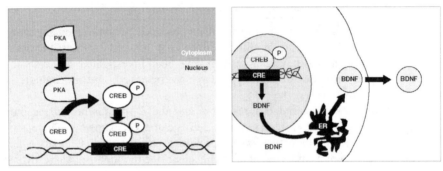

**Figure 14.8** Simplified diagram of a signaling cascade that involves NE, BDNF, and CREB after NE acts on the postsynaptic β-noradrenergic receptor. NE couples to a G protein ($G\alpha_s$), which stimulates the production of cAMP from adenosine triphosphate (ATP). This reaction is catalyzed by adenylate cyclase (AC). cAMP activates protein kinase A (PKA). Inside the cell, PKA phosphorylates (P) the CREB protein, which binds upstream from specific regions of genes and regulates their expression. BDNF is one target of cAMP signaling pathways in the brain. *CRE*, cyclic AMP regulatory element; *ER*, endoplasmic reticulum. [reprinted from Reference 76 with permission of the author and the publisher, Canadian Medical Association].

CREB expression (77). Experimentally increasing the expression of CREB in the hippocampi of depressed rats seems to have an antidepressant effect. Evidence suggests that the role of CREB in mediating antidepressant effects of medication is related to the expression of CREB-regulated neural growth factors, such as BDNF (79, 89). In humans, BDNF levels are reduced in postmortem samples of depressed individuals when compared with nondepressed controls (90). Antidepressant therapy increases serum BDNF levels in depressed humans (91). In animals, chronic stress models are associated with reduced expression of BDNF (79) and decreased cell proliferation in the hippocampus (83), which can be prevented or reversed by treatment with antidepressant drugs. However, researchers debate whether genetic disruption of the signaling pathways that involve BDNF and trkB causes depressive behavior (See Reference 92 for a review). It is also of interest that chronic administration of antidepressants has been reported to upregulate expression and activity of the neuroprotective enzyme superoxide desmutase and, depending on the dose and antidepressant, to increase immunostaining of BDNF and/or the antiapoptotic protein Bcl-2 (93, 94).

The picture with CREB is far from straightforward. Increased expression of CREB in the nucleus accumbens of rats produces depression-like effects, which include anhedonia (77, 89) and increased helplessness, in learned-helplessness models (81). Increases in BDNF expression in the mesolimbic DA system of rats are associated with induction of depressive effects in certain animal models. However, studies that use systemic administration of BDNF and activators of the cAMP-CREB cascade seem to have a net therapeutic effect in behavioral models of depression (90, 86, 89, 95).

### 14.4.7 Melatonin (N-acetyl-5-methoxytryptamine)

Depression and seasonal affective disorder, which is a form of depressive illness characterized by symptoms with onset and course related to season of the year, have been noted to be cyclic and possibly related to alterations in circadian rhythms (96). Alterations in the cyclical functioning of the HPA axis and the hypothalamic-pituitary-thyroid axis occur in some patients with major depression (96). Disruption in the sleep/wake cycle and insomnia are common in depression. Melatonin (Fig. 14.9), which is a hormone derived centrally in the pineal gland from serotonin, is instrumental in many biological functions with circadian variability. Several studies have shown that depressed individuals have a deficiency of melatonin secretion, but a body of evidence shows increased nocturnal melatonin production in the brains of depressed subjects with increased urinary excretion of its metabolite, 6-hydroxymelatonin sulfate (96, 97). This finding may reflect differences in subtypes of depression. As melatonin is secreted in a rhythmic fashion, the timing of the melatonin peak has been studied in depressed individuals. Phase shifts toward both earlier and later onset of peak secretion have been noted. Both treatment with antidepressant medications and electroconvulsive therapy have been associated with increases in melatonin excretion in depressed individuals. The pineal gland receives input from noradrenergic neurons, and it has been suggested that alterations in melatonin secretion merely reflect dysfunction in monoamine systems (96, 97). Agomelatine, which is a melatonin receptor agonist and 5HT2c receptor antagonist, has been shown to have antidepressant properties in animals and humans (98–101). Agomelatine targets the $MT_1/MT_2$ melatonin receptors and mimics melatonin in its effect on circadian rhythms. A combination of activity at melatonin receptors and monoaminergic receptors may represent a novel method of antidepressant drug action (97–100).

## 14.5 OTHER ANTIDEPRESSANT APPROACHES AND TARGETS

Electroconvulsive therapy (ECT) is considered to be the most efficacious treatment for depression, although its side-effect profile limits its use, and

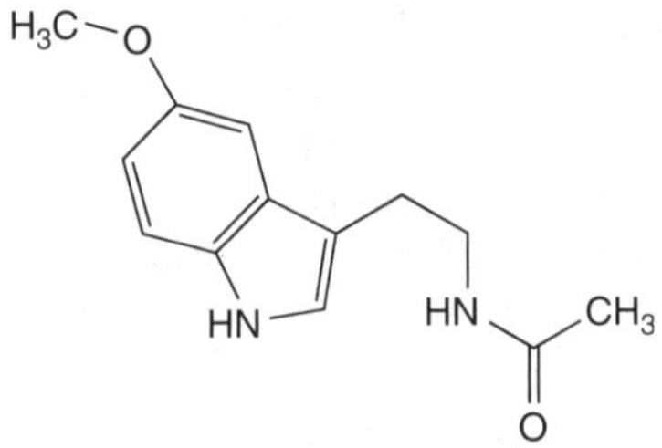

**Figure 14.9**   Structure of melatonin.

it is associated with a high relapse rate (10, 102). The exact mechanism of action of ECT is still unknown, although effects on monoaminergic systems, CRF, neurotrophic factors, and neuroendocrine systems have been suggested (102). Because DA probably plays a role in the pathophysiology of depression, triple reuptake inhibitors that inhibit reuptake of NE, 5-HT, and DA have now been developed and, at the time of writing in 2008, are undergoing preclinical and clinical testing (10). Numerous reports in the literature suggest that trace amines (so-named because their absolute levels in the brain are much lower that those of the classic neurotransmitter amines) are involved in the etiology and pharmacotherapy of several neurologic and psychiatric disorders, which include depression (103–106). Researchers have shown an increased interest in these trace amines, which include β-phenylethylamine, tyramine, tryptamine, and octopamine, in recent years with the discovery of a family of G-protein coupled receptors that bind to and are activated by these amines (107, 108). Epinephrine is present in much lower concentrations than NE and DA in the brain, but it is a major stress hormone in the periphery and could be a contributing factor in depression (109). Histamine is present in lower concentrations in brain than NE, DA, or 5-HT as well, but it has also been implicated in the etiology of depression (110).

Reports in the literature indicate that atypical (second generation) antipsychotics are useful antidepressant agents when combined with standard antidepressants (10); several researchers suggest that in some cases they and/or their metabolites may be useful antidepressants in their own right. It is interesting that some of these atypical antipsychotics have been reported to have several neuroprotective actions, which include effects on neuroprotective enzymes and factors that affect apoptosis (programmed cell death) (111). Preliminary reports suggest similar effects of antidepressants (93, 94). Some other areas related to the chemistry of depression and its treatment that contain conflicting reports in the literature but are of interest include the following: the relationship of serum cholesterol and the use of statins (cholesterol-lowering drugs) to depression, anxiety, aggression, and suicide (112, 113); omega-3 fatty acids and mood disorders (114–118); the use of herbal products such as St. John's wort extracts and S-adenosylmethionine as antidepressants (119–123); and the possible involvement of CB1 cannabinoid receptors, galanin, Neuropeptide Y, histone deacetylases, and tissue plasminogen activator in depression (See Reference 9 for a review). Other reported or putative neurobiological antidepressant treatments that we have not mentioned previously in this review include deep brain stimulation, vagus nerve stimulation, and transcranial magnetic stimulation; the reader is referred to the review of Holtzheimer and Nemeroff (10) and the references contained therein for details.

## 14.6 CONCLUDING REMARKS

Although a great deal has been learned about brain function in the search for newer antidepressants and some very interesting potential drug targets have been identified, our knowledge of the causes of depression remains inadequate. We still lack antidepressant drugs that are sufficiently rapid acting and effective in a large enough percentage of depressed patients. Problems that continue to make studies on depression difficult include the following: the heterogenous nature of depression itself, gender issues, the involvement of multiple interacting neurotransmitters and neuromodulators in the etiology of depression, the inadequacy of current animal models, the involvement of various brain regions in the symptomatology of depression, disagreements about the neurogenesis hypothesis of depression and about the relative importance of the hippocampus in depression, regional differences in the effects of transcription factors such as CREB, and the differing effects of some antidepressants on the HPA axis.

## ACKNOWLEDGMENTS

Research findings provided by the Canadian Institutes of Health Research (CIHR), the Canada Research Chairs and Canada Foundation for Innovation programs, and the Davey Endowment. The assistance of Ms. Tara Checknita in preparing this manuscript is gratefully acknowledged.

## REFERENCES

1. American Psychiatric Association. Diagnostic and Statistical Manual of Mental Disorders: DSM-IV. 4th ed. 2000. American Psychiatric Association, Washington, DC.

2. Sadock BJ, Sadock VA. Mood disorders. In: Kaplan & Sadock's Synopsis of Psychiatry: Behavioral Sciences, Clinical Psychiatry. 9th ed. Sadock BJ, Sadock, VA, eds. 2003. Lippincott Williams & Wilkins, Philadelphia, PA, pp. 534–590.

3. Papakostas GI, Petersen T, Mahal Y, Mischoulon D, Nierenberg AA, Fava M. Quality of life assessments in major depressive disorder: A review of the literature. Gen. Hosp. Psych. 2004;26:13–17.

4. Luppa M, Heinrich S, Angermeyer MC, Konig HH, Riedel-Heller SG. Cost of illness studies of major depression: a systematic review. J. Affec. Disorders. 2007;98:29–43.

5. Maletic V, Robinson M, Oakes T, Iyengar S, Ball SG, Russell J. Neurobiology of depression: an integrated view of key findings. Int. J. Clin. Pract. 2007;61:2030–2040.

6. Marchand WR, Dilda DV, Jensen, CR. Neurobiology of mood disorders. Hosp. Physician 2005;41:17–26.

7. Nestler EJ, Barrot M, Dileone RJ, Eisch A, Gold SJ, Monteggia LM. Neurobiology of depression. Neuron 2002;34:13–25.

8. Shelton RC. The molecular neurobiology of depression. Psychiatr. Clin. N. Am. 2007;1:1–11.

9. Berton O, Nestler EJ. New approaches to antidepressant drug discovery: beyond monamines. Nature Rev. Neurosci. 2006;7:137–151.

10. Holtzheimer PE III, Nemeroff CB. Advances in the treatment of depression. J. Am. Soc. Exp. Neurotherapeutics 2006;3:42–56.

11. Nutt DJ. The role of dopamine and norepinephrine in depression and antidepressant treatment. J. Clin. Psychiatry 2006;67(suppl 7):3–8.

12. Baker GB, Dewhurst WG. Biochemical theories of affective disorders. In: The Pharmacotherapy of Affective Disorders: Theory and Practice. Dewhurst WG and Baker GB, eds. 1985. Croom Helm Ltd., London, pp. 1–59.

13. Freis ED. Mental depression in hypertensive patients treated for long periods with large doses of reserpine. N. Engl. J. Med. 1954;251:1006–1008.

14. Schildkraut JJ. The catecholamine hypothesis of affective disorders: a review of supporting evidence. Am. J. Psychiatry 1965;122:509–522.

15. Lapin IP, Oxenkrug GF. Intensification of central serotonergic processes as a possible mechanism of thymoleptic effect. Lancet 1969;1:132–136.

16. Coppen AJ. The biochemistry of affective disorders. Br. J. Psychiatry 1967;113:1237–1264.

17. Firk C, Markus CR. Serotonin by stress interaction: a susceptibility factor for the development of depression? J. Psychopharmacol. 2007;21:538–544.

18. Neumeister A. Tryptophan depletion, serotonin, and depression: Where do we stand? Psychopharmacol. Bull. 2003;37:99–115.

19. Young SN, Smith SE, Pihl RO, Ervin, FR. Tryptophan depletion causes a rapid lowering of mood in normal males. Psychopharmacology 1985;87:173–177.

20. Klaasen T, Riedal WJ, van Someren A, Deutz NE, Honig A, van Praag HM. Mood effects of 24-hour tryptophan depletion in healthy first-degree relatives of patients with affective disorders. Biol. Psychiatry 1999;46:489–497.

21. Delgado PL, Price L, Miller HL, Salomon RM, Aghajanian GK, Heninger GR, Charney DS. Serotonin and the neurobiology of depression. Effects of trypophan depletion in drug-free depressed patients. Arch. Gen. Psychiatry 1994;51:865–874.

22. Delgado PL, Miller HL, Salomon RM, Licinio J, Krystal JH, Moreno FA, Heninger GR, Charney DS. Tryptophan-depletion challenge in depressed patients treated with desipramine or fluoxetine: implications for the role of serotonin in the mechanism of antidepressant action. Biol. Psychiatry 1999;46:212–220.

23. Booij L, Van der Does AJW, Riedel WJ. Monoamine depletion in psychiatric and healthy populations: review. Mol. Psychiatry 2003;8:951–973.

24. Blier P, de Montigny C. Current advances and trends in the treatment of depression. Trends Pharmacol Sci. 1994;15:220–226.

25. Bourin M, Baker GB. Do G proteins have a role in antidepressant actions? Eur Neuropsychopharmacol. 1996;6:49–53.

26. Deakin JFW, Guimaraes FS, Wang M, Hensman R. Experimental tests of the 5-HT receptor imbalance theory of affective disturbance In: 5-Hydroxytryptamine in Psychiatry – A Spectrum of Ideas. Sandler, M, Coppen A, Harnett, S. eds., 1991. Oxford University Press, Oxford, UK, pp. 143–156.

27. Vetulani J, Lebrecht U, Pilc A Enhancement of responsiveness of the central serotonergic system and serotonin-2 receptor density in rat frontal cortex by electroconvulsive treatment. Eur. J. Pharmacol. 1981;76:81–85.

28. Stockmeier CA. Involvement of serotonin in depression: evidence from postmortem imaging studies of serotonin receptors and the serotonin transporter. J. Psychiatric Res. 2003;37:357–373.

29. Neumeister A, Young T, Stastny J. Implications of genetic research on the role of serotonin in depression: emphasis on the serotonin type $1_A$ receptor and the serotonin transporter. Psychopharmacology 2004;174:512–524.

30. Vetulani J, Sulser F. Action of various antidepressant treatments reduces reactivity of noradrenergic cylic AMP-generating system in limbic forebrain. Nature 1975;257:495–496.

31. Brunello N, Mendlewicz J, Kasper S, Leonard B, Montgomery S, Nelson JC, Paykel E, Veriani M, Racagni G. The role of noradrenaline and selective noradrenaline reuptake inhibition in depression. Eur. Neuropsychopharmacol. 2002;12:461–475.

32. Leonard BE. Noradrenaline in basic models of depression. Eur. Neuropsychopharmacology. 1997;1(suppl 1): S11–16.

33. Dunlop BW, Nemeroff CB. The role of dopamine in the pathophysiology of depression. Arch. Gen. Psychiatry. 2007;64:327–337.

34. Willner P, Muscat R, Papp M. Chronic mild stress-induced anhedonia: A realistic animal model of depression. Neurosci. Biobehav. Rev. 1992;16:525–534.

35. Fibiger HC. Neurobiology of depression: focus on dopamine. Adv. Biochem. Psychopharmacol. 1995;49:1–17.

36. Krystal JH, Sanacora G, Blumberg H, Anand A, Charney DS, Marek G, Epperson CN, Goddard A, Mason GF. Glutamate and GABA systems as targets for novel antidepressant and mood-stabilizing treatments. Mol. Psychiatry. 2002;7:S71–80.

37. Paul IA, Skolnick P. Glutamate and depression: clinical and preclinical studies. Ann. NY Acad. Sci. 2003;1003:250–272.

38. Coyle JT, Duman RS. Finding the intracellular signaling pathways affected by mood disorder treatments. Neuron 2003;38:157–160.

39. Matrisciano, F, Panaccione I, Zusso M, Giusti P, Tatarelli R, Iacovelli L, Mathe AA, Gruber SH, Nicoletti F, Girardi P. Group-II metabotropic glutamate receptor ligands as adjunctive drugs in the treatment of depression: A new strategy to shorten the latency of antidepressant medication? Mol. Psychiatry 2007;12:704–706.

40. Belozertseva IV, Kos T, Popik P, Danysz W, Bespalov AY. Antidepressant-like effects of mGluR1 and mGluR5 antagonists in the rat forced swim and mouse tail suspension tests. Eur. Neuropsychopharmacol. 2007;17:172–179.

41. Zarate CA, Singh JB, Carlson PJ, Brutsche NE, Ameli R, Luckenbaugh DA, Charney DS, Manji HK. A randomized trial of an N-methyl-D-aspartate antagonist in treatment-resistant major depression. Arch. Gen. Psychiatry 2006;63:856–864.

42. Zarate CA, Sing JB, Quiroz JA, De Jesus G, Denicoff KK, Luckenbaugh DA, Manji HK, Charney DS. A double-blind, placebo-controlled study of memantine in the treatment of major depression. Am. J. Psychiatry 2006;63:153–155.

43. Mitani H, Shirayama Y, Yamada T, Maeda K, Ashby CR, Kawahara R. Correlation between plasma levels of glutamate, alanine, and serine with severity of depression. Progr. Neuro-Psychopharmacol. Biol. Psychiatry 2006;30:1155–1158.

44. Martinez-Turrillas R, Frechilla D, Del Rio J. Chronic antidepressant treatment increases the membrane expression of AMPA receptors in rat hippocampus. Neuropharmacology 2002;43:1230–1237.

45. Petty F. GABA and mood disorders: a brief review and hypothesis. J. Affec. Disorders 1995;34:275–281.

46. Sanacora G, Mason GF, Rothman DL, Behar K L, Hyder F, Petroff OA, et al. Reduced cortical gamma-aminobutyric acid levels in depressed patients determined by proton magnetic resonance spectroscopy. Arch. Gen. Psychiatry 1999;56:1043–1047.

47. Epperson CN, Haga K, Mason GF, Sellers E, Gueorguieva R, Zhang W, et al. Cortical gamma-aminobutyric acid levels across the menstrual cycle in healthy women and those with premenstrual dysphoric disorder: A proton magnetic resonance spectroscopy study. Arch. Gen. Psychiatry 2002;59:851–858.

48. Popov N, Matthies H Some effects of monoamine oxidase inhibitors on the metabolism of gamma-aminobutyric acid in rat brain. J. Neurochem. 1969;16:899–907.

49. Baker GB, Wong JT, Yeung JM, Coutts RT. Effects of the antidepressant phenelzine on brain levels of gamma-aminobutyric acid (GABA). J. Affec. Disorders 1991;21:207–211.

50. Bhagwagar Z, Wylezinska M, Taylor, M, Jezzard P, Matthews PM, Cowen PJ. Increased brain GABA concentrations following acute administration of a selective serotonin reuptake inhibitor. Am. J. Psychiatry 2004;161:368–370.

51. Dubrovsky BO. Steroids, neuroactive steroids and neurosteroids in psychopathology. Prog. Neuro-Psychopharmacol. Biol Psychiatry, 2005;29:169–192.

52. MacKenzie EM, Odontiadis J, Le Melledo JM, Prior TI, Baker GB. The relevance of neuroactive steroids in schizophrenia, depression, and anxiety disorders. Cell. Mol.Neurobiol. 2007;27:541–574.

53. Marx CE, Stevens RD, Shampine LJ, Uzunova V, Trost WT, Butterfield MI et al. Neuroactive steroids are altered in schizophrenia and bipolar disorder: relevance to pathophysiology and therapeutics. Neuropsychopharmacology 2006;31:1249–1263.

54. Molina-Hernandez M, Tellez-Alcantara NP, Garcia JP, Lopez JI, Jaramillo MT. Antidepressant-like actions of intra-accumbens infusions of allopregnanolone in ovariectomized wistar rats. Pharmacol. Biochem. Behav. 2005;80:401–409.

55. Pinna G, Costa E, Guidotti A. Fluoxetine and norfluoxetine stereospecifically and selectively increase brain neurosteroid content at doses that are inactive on 5-HT reuptake. Psychopharmacology 2006;186:362–372.

56. Girdler SS, Klatzkin R. Neurosteroids in the context of stress: Implications for depressive disorders. Pharmacol. Therapeut. 2007;116:125–139.

57. Carroll BJ, Martin FIR, Davies B. Pituitary-adrenal function in depression. Lancet 1976;I:1373–1374.

58. Stout SC, Owens MJ, Nemeroff CB. Regulation of corticotropin-releasing factor neuronal systems and hypothalamic-pituitary-adrenal axis activity by stress and chronic antidepressant treatment. J. Pharmacol. Exp. Ther. 2002;300:1085–1092.

59. Schule C. Neuroendocrinological mechanisms of actions of antidepressant drugs. J. Neuroendocrinol. 2007;19:213–226.

60. Young AH. Antiglucocorticoid treatments for depression. Aust. N.Z. Psychiatry 2006;40:402–405.

61. Johnson SA, Fournier NM, Kalynchuk LE. Effect of different doses of corticosterone on depression-like behavior and HPA axis response to a novel stressor. Brain Behav. Res. 2006;168:280–288.

62. Ravaris CL, Sateia MJ, Beroza KW, Noordky DL, Brinck-Johnsen T. Effect of ketoconazole on a hypophysectomized, hypercortisolemic, psychotically depressed woman. Arch. Gen. Psychiatry 1988;45:966–967.

63. van Gaalen MM, Marcel M, Stenzel-Poore MP, Holsboer F, Steckler T. Effects of transgenic overproduction of CRH on anxiety-like behavior. Eur. J. Neurosci. 2002;15:2007–2015.

64. Kramer MS, Cutler N, Feighner J, Shrivastava R, Carman J, Sramek JJ, et al. Distinct mechanism for antidepressant activity by blockade of central substance P receptors. Science 1998;281:1640–1645.

65. Deuschle M, Sander P, Herpfer I, Fiebich B L, Heuser I. Lieb K. Substance P in serum and cerebrospinal fluid of depressed patients: No effect of antidepressant treatment. Psychiatry Res. 2005;136:1–6.

66. Malkesman O, Braw Y, Weller A. Assessment of antidepressant and anxiolytic properties of NK1 antagonists and substance P in wistar kyoto rats. Physiol. Behav. 2007;90:619–625.

67. Dableh LJ, Yashpal K, Rochford J, Henry JL. Antidepressant-like effects of neurokinin receptor antagonists in the forced swim test in the rat. Eur. J. Pharmacol., 2005;507:99–105.

68. Chenu F, Guiard BP, Bourin M, Gardier AM. Antidepressant-like activity of selective serotonin reuptake inhibitors combined with a NK1 receptor antagonist in the mouse forced swimming test. Behav. Brain Res. 2006;172:256–263.

69. Kramer M S, Winokur A, Kelsey J, Preskorn SH, Rothschild AJ, Snavely D, Gosh K, Ball WA, Reines SA, Munjack D, Apter JT, Cunningham L, Kling M, Bari M, Getson A, Lee Y. Demonstration of the efficacy and safety of a novel substance P (NK1) receptor antagonist in major depression. Neuropsychopharmacology 2004;29:385–392.

70. Gobbi G, Blier P. Effect of neurokinin-1 receptor antagonists on serotoninergic, noradrenergic and hippocampal neurons: Comparison with antidepressant drugs. Peptides 2005;26:1383–1393.

71. Ryckmans T, Balancon L, Genicot C, Berton O., Lamberty Y Lallemand B, Pasau P, Pirlot N, Quere L, Talaga P. First dual NK(1) antagonists-serotonin reuptake inhibitors: synthesis and SAR of a new class of potential antidepressants. Bioorg. Med. Chem. Lett. 2002;12:261–264.

72. Dunn AJ, Swiergiel AH, de Beaurepaire R. Cytokines as mediators of depression: What can we learn from animal studies? Neurosci. Biobehav. Rev. 2005;29: 891–909.

73. Kim YK, Na KS, Shin KH, Jung HY, Choi SH, Kim JB. Cytokine imbalance in the pathophysiology of major depressive disorder. Prog. Neuro-Psychopharmacol. Biol. Psychiatry 2007;31:1044–1053.

74. Schiepers OJ, Wichers MC, Maes M. Cytokines and major depression. Prog. Neuro-Psychophramacol. Biol. Psychiat. 2005;29:201–217.

75. O'Brien SM, Scott LV. Dinan TG. Cytokines: Abnormalities in major depression and implications for pharmacological treatment. Hum. Psychopharmacol. 2004;19:397–403.

76. Young TL. Postreceptor pathways for signal transduction in depression and bipolar disorder. J. Psychiatr. Neurosci. 2001;26:S17–S22.

77. Carlezon WA, Duman RS, Nestler EJ. The many faces of CREB. Trends Neurosci. 2005;28:436–445.

78. Nair A, Vaidya VA. Cyclic AMP response element binding protein and brain-derived neurotrophic factor: molecules that modulate our mood? J. Biosci. 2007;31:423–434.

79. Duman RS, Monteggia LM. A neurotrophic model for stress-related mood disorders. Biol Psychiatry. 2006;59:1116–1127.

80. Schmidt HD, Duman RS. The role of neurotrophic factors in adult hippocampal neurogenesis, antidepressant treatments and animal models of depressive-like behavior. Behav. Pharmacol. 2007;18:391–418.

81. Pittenger C, Duman RS. Neuroplasticity and depression. Neuropsychopharmacol. Rev. 2008;33:88–109.

82. Duman, R. S. Depression: a case of neuronal life and death? Biol. Psychiatry. 2004;56:140–145.

83. Stockmeier CA, Mahajan GJ, Konick LC, Overholser JC, Jurjus GJ, Meltzer HY, Uylings HBM, Freidman L, Rajkowska G. Cellular changes in postmortem hippocampus in major depression. Biol. Psychiat. 2004;56:640–650.

84. Malberg JE, Duman RS. Cell proliferation in adult hippocampus is decreased by inescapable stress: reversal by fluoxetine treatment. Neuropsychopharmacology 2003;28:1562–1571.

85. van Praag H, Schinder AF, Christie BR, Toni N, Palmer TD, Gage FH. Functional neurogenesis in the adult hippocampus. Nature 2002;415:1030–1034.

86. Mackin P, Watson S, Ferrier N. The neurobiology of depression: focus on cortisol, brain-derived neurotrophic factor, and their interactions. Depression: Mind Body 2005;2:11–18.

87. Sapolsky RM. Is impaired neurogenesis relevant to the affective symptoms of depression? Biol. Psychiatry 2004;56:137–139.

88. Santarelli L, Saxe M, Gross C, Surget A, Battaglia F, Dulawa S, Weisstaub N, Lee J, Duman R, Arancio O, Belzung C, Hen R. Requirement of hippocampal neurogenesis for the behavioral effects of antidepressants. Science 2003;301: 805–809.

89. Eisch AJ, Bolanos CA, de Wit J. Simonak RD. Pudiak CM, Barrot M, Nestler EJ. Brain-derived neurotrophic factor in the ventral midbrain-nucleus accumbens pathway: a role in depression. Biol. Psychiatry. 2003;54:994–1005.

90. Castrén E, Võikar V, Rantamäki T. Role of neurotrophic factors in depression. Curr. Opinion Pharmacol. 2007;7:18–21.

91. Shimizu E, Hashimoto K, Okamura N, Koike K, Komatsu N, Kumakiri C, Nazakato M, Watanabe H, Shinoda N, Okada S, Iyo M. Alterations of serum levels of brain-derived neurotrophic factor (BDNF) in depressed patients with or without antidepressants. Biol. Psychiatry 2003;54:70–75.

92. Martinowich K, Manji H, Bai L. New insights into BDNF function in depression and anxiety. Nature Neurosci. 2007;10:1089–1093.

93. Kolla N, Wei Z, Richardson JS, Li XM. Amitriptyline and fluoxetine protect PC12 cells from cell death induced by hydrogen peroxide. J. Psychiatry Neurosci. 2005;30:196–201.

94. Xu H, Steven Richardson J, Li XM. Dose-related effects of chronic antidepressants on neuroprotective proteins BDNF, bcl-2 and Cu/Zn-SOD in rat hippocampus. Neuropsychopharmacology 2003;28:53–62.

95. Shiriyama Y, Chen ACH, Nakagawa S, Russell DS, Duman RS. Brain-derived neurotrophic factor produces antidepressant effects in behavioral models of depression. J. Neurosci. 2002;22:3251–3261.

96. Srinivasan V, Smits M, Spence W, Lowe AD, Kayumov L, Pandi-Perumal SR, Parry B, Cardinali DP. Melatonin in mood disorders. World J. Biol. Sci. 2006;7: 138–151.

97. Pacchierotti C, Iapichino P, Bossini L, Pieraccini F, Castrogiovanni C. Melatonin in psychiatric disorders: a review of melatonin in psychiatry. Frontiers Neuroendocrinol. 2001;22:18–32.

98. Olie JP, Kasper S. Efficacy of agomelatine, a MT1/MT2 receptor agonist with 5-HT2c antagonistic properties, in major depressive disorder. Int. J. Neuropsychopharmacol. 2007;10:661–673.

99. Zupancic M, Guilleminault C. Agomelatine: a preliminary review of a new antidepressant. CNS Drugs 2006;20:981–992.

100. Stahl SM. Novel mechanism of antidepressant action: norepinephrine and dopamine disinhibition (NDDI) plus melatonergic agonism. Int. J Neuropsychopharmacol 2007;10:575–578.

101. Kennedy SH, Emsley R. Placebo-controlled trial of agomelatine in the treatment of major depressive disorder. Eur. Neuropsychopharmacol. 2006;16:93–100.

102. Wahlund B, von Rosen D. ECT of major depressed patients in relation to biological and clinical variables: a brief overview. Neuropsychopharmacology 2003;28(suppl 1): S21–6.

103. Sabelli HC, Mosnaim AD. Phenylethylamine hypothesis of affective behavior. Am. J. Psychiatry 1974;131:695–699.

104. Sandler M, Ruthven CR, Goodwin BL, Coppen A. Decreased cerebrospinal fluid concentration of free phenylacetic acid in depressive illness. Clin. Chim. Acta 1979;93:169–171.

105. Davis BA, Boulton AA. The trace amines and their acidic metabolites in depression–an overview. Prog. Neuro-Psychopharmacol. Biol. Psychiatry 1994;18: 17–45.

106. Berry MD. Mammalian central nervous system trace amines. pharmacologic amphetamines, physiologic neuromodulators. J. Neurochem. 2004;90: 257–271.

107. Borowsky B, Adham N, Jones KA, Raddatz R, Artymyshyn R, Ogozalek K L, Durkin MM, Lakhlani PP, Bonini JA, Pathirana S, Boyl A, Pu A., Kouranova E, Lichtblau H, Ochoa FY, Branchek TA, Gerad C. Trace amines: Identification of a family of mammalian G protein-coupled receptors. Proc. Natl. Acad. Sci. U.S.A. 2001;98:8966–8971.

108. Lewin AH. Receptors of mammalian trace amines. AAPS J. 2006;8:E138–145.

109. Stone EA, Lin Y, Rosengarten H, Kramer HK, Quartermain D. Emerging evidence for a central epinephrine-innervated $\alpha_1$-adrenoceptor system that regulates behavioural activation and is impaired in depression. Neuropsychopharmacology 2003;28:1387–1399.

110. Kano M, Fukudo S, Tashiro A, Utsumi A, Tamura D, Itoh M, Iwata R, Tashiro M, Mochizuki K, Hongo M, Yanai K. Decreased histamine $H_1$ receptor binding in the brain of depressed patients. Eur. J. Neurosci. 2004;20:803–810.

111. Li XM, Xu H. Evidence for neuroprotective effects of antipsychotic drugs: Implications for the pathophysiology and treatment of schizophrenia. Int Rev Neurobiol. 2007;77:107–142.

112. Tanskanen A, Tuomilehto J, Viinamaki H. Cholesterol, depression and suicide. Br. J. Psychiatry, 2000;176:398–399;399–400.

113. Young-Xu, Y, Chan, KA, Liao JK, Ravid S, Blatt CM. Long-term statin use and psychological well-being. J. Am. Coll. Cardiol. 2003;42:690–697.

114. Peet M, Stokes C. Omega-3 fatty acids in the treatment of psychiatric disorders. Drugs 2005;65:1051–1059.

115. Freeman MP, Hibbeln JR, Wisner KL, Davis JM. Mischoulon D, Peet M, et al. Omega-3 fatty acids: Evidence basis for treatment and future research in psychiatry. J. Clin. Psychiatry 2006;67:1954–1967.

116. Nieminen LR, Makino KK, Mehta N, Virkkunen M, Kim HY, Hibbeln JR. Relationship between omega-3 fatty acids and plasma neuroactive steroids in alcoholism, depression and controls. Prostaglandins, Leukotrienes, Essential Fatty Acids 2006;75:309–314.

117. Lin PY, Su KP. A meta-analytic review of double-blind, placebo-controlled trials of antidepressant efficacy of omega-3 fatty acids. J. Clin. Psychiatry 2007;68:1056–1061.

118. Young SN. Fish oils for depression? J. Psychiatry Neurosci. 2008;33;(E1).

119. Hypericum Depression Trial Study Group. Effect of Hypericum perforatum (St John's wort) in major depressive disorder: a randomized controlled trial. J. Am. Med. Assoc. 2002;287:1807–1814.

120. Linde K, Mulrow CD, Berner M, Egger M. St John's wort for depression. Cochrane Database of Systematic Reviews (Online), 2005;(2):CD000448.

121. Kasper S, Anghelescu IG, Szegedi A, Dienel A, Kieser M. Superior efficacy of st. John's wort extract WS 5570 compared to placebo in patients with major depression: a randomized, double-blind, placebo-controlled, multi-center trial. BMC Med. 2006;4:14.

122. Mischoulon D, Fava, M. Role of S-adenosyl-L-methionine in the treatment of depression: a review of the evidence. Am. J. Clin. Nutrition 2002;76:1158S–1161S.

123. Shippy RA, Mendez D, Jones K, Cergnul I, Karpiak SE. S-Adenosylmethionine (SAM-e) for the treatment of depression in people living with HIV/AIDS. BMC Psychiatry 2004;4:38.

## FURTHER READING

Anisman H, Merali Z, Stead JDH. Experiential and genetic contributions to depressive and anxiety-like disorders: clinical and experimental studies. Neurosci. Biobehav. Rev. 2008; In press.

Baghai TC, Volz HP, Moller HJ. Drug treatment of depression in the 2000s: An overview of achievements in the last 10 years and future possibilities. World J. Biol. Psychiatry 2006;7:198–222.

Biol. Psychiatry 2008;63:639–722. Theme issue devoted to "The Neurobiology and Therapeutics of Antidepressant-Resistant Depression" (12 papers).

Conti B, Maier R, Barr AM, Morale MC, Lu X, Sanna PP, et al. Region-specific transcriptional changes following the three antidepressant treatments electro-convulsive therapy, sleep deprivation and fluoxetine. Mol. Psychiatry 2007;12:167–189.

De Souza E, Grigoriadis DE. Corticotrophin-releasing factor: physiology, pharmacology, and role in central nervous system disorders. In: Neuropsychopharmacology: The Fifth Generation of Progress. Davis KL, Charney D, Coyle JT, Nemeroff C, eds. 2002. Lippincott Williams & Wilkins, Philadelphia, PA, pp. 91–107.

Hasler G, Drevets WC, Manji HK, Harney DS. Discovering endophenotypes for major depression. Neuropsychopharmacology 2004;29:1765–1781.

Hauger RL, Risbrough V, Brauns O, and Dautzenberg FM. Corticotropin releasing factor (CRF) receptor signaling in the central nervous system: New molecular targets. CNS Neurol. Disorders Drug Targets 2006;5:453–479.

Maletic V, Robinson M, Oakes T. Iyengar S, Ball SG, Russell J. Neurobiology of depression: An integrated view of key findings. Int. J. Clin. Pract. 2007;61: 2030–2040.

Newport DJ, Stowe ZN, Nemeroff CB. Parental depression: animal models of an adverse life event. Am. J. Psychiatry 2002;159:1265–1283.

Pittenger C, Duman RS. Stress, depression, and neuroplasticity: a convergence of mechanisms. Neuropsychopharmacology 2008;33:88–109.

Saxe MD, Battaglia F, Wang JW, Malleret G, David DJ, Monckton JE, et al. Ablation of hippocampal neurogenesis impairs contextual fear conditioning and synaptic plasticity in the dentate gyrus. Proc. Nat. Acad. Sci. U.S.A. 2006;103:17501–17506.

Schmidt HD, Duman RS. The role of neurotrophic factors in adult hippocampal neurogenesis, antidepressant treatments and animal models of depressive-like behavior. Behav. Pharmacol. 2007;18:391–418.

Simon NM, McNamara K, Chow CW, Maser RS. Papakostas GI, Pollack MH, et al. A detailed examination of cytokine abnormalities in major depressive disorder. Eur. Neuropsychopharmacol. 2008;18:230–233.

Swaab DF, Bao AM, Lucassen PJ. The stress system in the human brain in depression and neurodegeneration. Ageing Res. Rev. 2005;4:141–194.

Tsankova N, Renthal W, Kumar A, Nestler EJ. Epigenetic regulation in psychiatric disorders. Nature Rev. Neurosci. 2007;8:355–367.

Thomson F, Craighead M. Innovative approaches for the treatment of depression: Targeting the HPA axis. Neurochem. Res. 2007. 2008;33(4): 691–707.

Young LT, Bakish D, Beaulieu S. The neurobiology of treatment response to antidepressants and mood stabilizing medications. J. Psychiat. Neurosci. 2002;27:260–265.

# 15

# OSTEOARTHRITIS

Marjolaine Gosset, Jérémie Sellam, and Claire Jacques
*Paris Universitas, Paris, France*

Francis Berenbaum
*Paris Universitas APHP Saint-Antoine Hospital, Paris, France*

Osteoarthritis (OA) is considered the most common skeletal disorder and is characterized by cartilage degradation and osteophyte formation in joints. Although it is the most common joint disease, OA is often difficult to define because it is heterogeneous in its presentation style, rate of progression, and manifestations. Recently, new insights into the molecular biological basis of chondrocyte differentiation and cartilage homeostasis have been reported. The pathology of osteoarthritis is now interpreted as the disruption of balance between anabolic and catabolic signals. This chapter focuses on the risk factors, cellular and molecular mechanisms involved in OA, tools to study OA such as animal models and biomarkers, as well as treatment. Findings with regard to the physiologic and pathologic mechanisms involved in OA have made it possible to target therapeutic approaches more accurately, which allows the development of new drugs to reduce or block the progression of the disease.

## 15.1 BIOLOGICAL AND BIOPHYSICAL BACKGROUND OF OSTEOARTHRITIS

### 15.1.1 Diagnosis

OA is the most common chronic joint disorder. The most frequently affected sites are the hands, knees, hips, and spine (1) (Fig. 15.1). It results from degeneration of

*Chemical Biology: Approaches to Drug Discovery and Development to Targeting Disease*, First Edition. Edited by Natanya Civjan.
© 2012 John Wiley & Sons, Inc. Published 2012 by John Wiley & Sons, Inc.

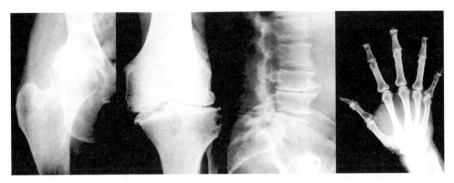

**Figure 15.1** Radiographs of OA in hip, knee, lumbar spine, and hand. Cartilage degradation shown as joint space narrowing and osteophyte formation at the edge of the joints, which are two major disorders of OA.

a synovial joint, which leads to a generally progressive loss of articular cartilage accompanied by attempted repair of articular cartilage, as well as remodeling and sclerosis of subchondral bone, subchondral cysts, and marginal osteophytes formation. This disease not only affects the articular cartilage but also involves the entire joint, which includes the subchondral bone, ligaments, capsule, and synovial membrane (Fig. 15.2).

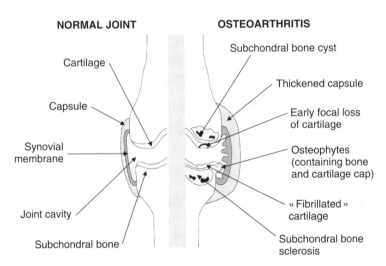

**Figure 15.2** Schematic representation of the main structures of a healthy (left side) and degenerated (right side) joint in osteoarthritis. Osteoarthritic joint is characterized by the lost or severely thinned articular cartilage ("fibrillated" cartilage), the sclerosis of subchondral bone, the formation of osteophytes, the presence of subchondral bone cyst, the thickened joint capsule, and the synovial hyperplasia. From EULAR online course, J. Sellam, G. Herrero-Beaumont and F. Berenbaum.

The diagnosis of OA requires plain radiographic examination. Usually, patients present with chronic joint pain, which is the first and predominant symptom, followed by restriction of joint motion and function, crepitus with motion, and joint effusions. The most severely affected individuals develop joint deformities and subluxations (2). The course of the disease varies but is often progressive, and the radiographic changes of OA progress inexorably (Bulletin of the World Health Organization; *http://www.who.int/whr/2007/en/index.html*).

### 15.1.2 Prevalence and Incidence

Few data are available on the incidence and prevalence of OA because of the problems of defining it and determining its onset. It is well known that OA affects people of all ethnic groups in all geographic locations in both men and women; it is the most common cause of long-term disability in most populations of elderly people (Bulletin of the World Health Organization; *http://www.who.int/whr/2007/en/index.html*). More than 80% of people over the age of 55 have radiological evidence of OA. Not all of these individuals are symptomatic, but 10–20% of those affected may report limitation of activity because of OA (3). Moreover, the World Health Organization estimates that 10% of the world's people over the age of 60 suffer from OA, and that 80% of people with OA have limitation of movement and 25% cannot perform major daily activities (Bulletin of the World Health Organization; *http://www.who.int/whr/2007/en/index.html*). Finally, the incidence of OA increases indefinitely with age (notably after age 40) and is greater in women than in men (4, 5).

### 15.1.3 Risk Factors

The precise etiology of OA in most cases is unknown, but it is generally accepted that OA is a multifactorial disorder that involves both genetic and environmental components. Among factors known to affect the progression of OA include the affected joint, increasing age, obesity, excessive physical activity, joint injury, genetic predisposition, and female gender.

*15.1.3.1 Age* Age is the most powerful risk factor for OA. The National Health and Nutrition Examination Survey found that the prevalence of knee OA increased from less than 0.1% in people 25–34 years old to 10–20% in people 35–74 years old (6). A higher prevalence of 30% between 65 to 74 years old has been found in the Framingham Study (7).

Recent works suggest that the changes observed during aging in articular cartilage are caused by deterioration of chondrocyte function. With increasing age, the ability of cells to maintain and restore cartilage matrix decreases (8). The major components of the extracellular matrix (ECM), which include type II collagen and proteoglycans, undergo changes in content, composition, and structural organization during the aging process. For example, cells synthesize smaller aggrecans and less functional link proteins, which leads to smaller more irregular proteoglycans aggregates (9). In addition, accumulation of advanced glycation end product (AGE) in the cartilage ECM could enhance collagen cross-linking (10). Finally, mitotic activity of chondrocytes declines with age, and cells become less responsive to anabolic cytokines and mechanical stimuli. This finding has been attributed to replicative senescence associated with reduction of telomere length (11). Consequently, the articular surface softens, and the tensile strength and stiffness of the matrix decreases.

**15.1.3.2 *Obesity*** Obesity increases the risk of OA, but it is unsure whether this effect is caused by excessive mechanical strains, varus malignment, or metabolic abnormalities. Increased body mass leads to abnormal strains on articular cartilage and could explain the positive association between obesity and hip and knee OA observed in many epidemiological studies (12, 13). Interestingly, the Framingham data show that being overweight as a young adult strongly predicted the appearance of knee OA in a 36-year period of follow-up (14).

The increased fat mass in obesity may therefore alter the metabolism of articular tissues such as cartilage. Recent studies have described the association between obesity and OA in non–weight-bearing joints, such as the hand (15, 16), whereas the loss of body fat seems to be more important than the loss of body weight in improving the symptoms of OA (17). Fat cells secrete a variety of proteins with the functional and structural properties of cytokines, which are called the adipokines. Adiponectin, leptin, resistin, and visfatin are the most abundant adipokines produced by adipose tissue, and their expression is regulated in obese individuals. They have been detected in synovial fluid from joints affected with OA. Recent evidence points to the involvement of leptin in OA and indicates that adiponectin and/or resistin mediates inflammation in arthritis. Visfatin can be produced not only by adipocytes but also by chondrocytes themselves, and it can induce catabolic activity of these cells (18). Thus, fat tissue is an active organ whose products contribute to inflammatory and degenerative processes underlying common joint diseases (19).

**15.1.3.3 *Excessive Repetitive Joint Loading*** Joints are highly specialized organs that allow repetitive pain-free and largely frictionless movements. This function occurs particularly because of articular cartilage and notably its ECM, which play an essential role in load transfer across the joint (20). Strains are necessary for joint physiology, as moderate exercise stimulates cartilage homeostasis and prevents the development of OA (21). Moreover, studies performed in space revealed the role of gravity in the chondrocyte physiology; the different steps of chondrogenesis are altered by long-term exposure to microgravity (22). However, repetitive loading of normal joints can exceed the tolerance of a joint and cause degeneration. Occupational overuse of the knee joint in obese subjects or in jobs that require repeated kneeling, squatting, or bending, as well as lifting of heavy loads is a risk factor for OA (23). It has been calculated that elimination of such jobs would decrease the incidence of knee OA in men by 15–30% (24). Other studies suggested that participation in sports that repetitively expose joints to high levels of impact also increase the risk of joint degeneration (25).

Cartilage matrix submitted to mechanical stress presents an alteration in pH, osmolarity, hydrostatic pressure, and reorganization of the ECM macromolecules. Chondrocytes bear receptors that respond directly to these modifications, many of which are also receptors of ECM, such as integrins or ion channels, and are located on ciliae, which was recently identified in chondrocytes (26). In response to strains, chondrocytes regulate the synthesis of matrix proteins, increase the production of inflammatory cytokines, or modulate their proliferation, differentiation, and apoptosis (27, 28). Application of static compression or excessive cyclic stress could therefore alter the ECM and stimulate an inflammatory response in cartilage, which can lead to OA.

*15.1.3.4 Joint Injury* Joint injuries frequently lead to progressive joint degeneration, which causes the clinical syndrome of posttraumatic OA. Studies of humans and animals models demonstrate that intra-articular fractures, dislocations, damage to the meniscus, and meniscal, ligament, and joint capsule tears lead to knee OA. Even if the cartilage is not involved at the time of injury, it will degenerate rapidly if the joint is unstable. The pathophysiology of posttraumatic OA has not been explained. Evidence indicates that first, the acute joint injury kills at least a few chondrocytes (29), and second, it increases the release of reactive oxygen species (ROS) in chondrocytes, which accelerate chondrocyte senescence and decreases the ability of the cells to maintain or restore the tissue (30). Furthermore, the risk of posttraumatic OA varies among individuals and among joints (31).

*15.1.3.5 Genetic Factors* Several lines of evidence indicate that genetic components affect the development of OA. Twin studies, segregation analyses, linkage analyses, and candidate gene-association studies have generated important information about inheritance patterns and the location in the genome of potentially causative mutations. Twin studies have shown that the influence of genetic factors vary from 39% to 74% in OA depending on the joints affected (32–34). Genetic defects in matrix molecules, in signaling molecules, as well as in molecules that participate in the formation of cartilage matrix and patterning of skeletal elements resulting in dysplasia may determine susceptibility to OA (35–37).

*15.1.3.6 Gender* Sex-specific differences are evident in OA. Before 50 years of age, the prevalence of OA in most joints is higher in men than in women. After about 50 years, women are more often affected with hand, foot, and knee OA than men. In most studies, hip OA is more frequent in men (Bulletin of the World Health Organization; *http://www.who.int/whr/2007/en/index.html*).

## 15.2 BIOCHEMISTRY AND MOLECULAR MECHANISM INVOLVED IN OSTEOARTHRITIS

A schematic representation of the molecular pathogenesis of osteoarthritis implicating synovial membrane, subchondral bon, and articular cartilage is shown in Figure 15.3.

### 15.2.1 Cartilage Damage in Osteoarthritis

OA is characterized by the slowly progressing destruction of the articular cartilage, which results from the failure of chondrocytes to maintain the balance between synthesis and degradation of the extracellular matrix (38, 39). Moreover, phenotypic modulations as hypertrophic differentiation of chondrocytes and chondrocytes apoptosis are known to be involved in OA development (40).

*15.2.1.1 Anabolic and Catabolic Signals of Cartilage in Osteoarthritis* A balance between anabolic and catabolic signals maintains the homeostasis of cartilage. In normal cartilage, the balance between anabolism and catabolism is

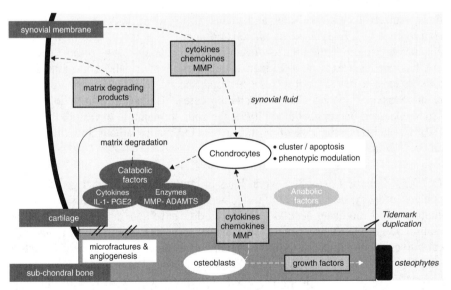

**Figure 15.3** Schematic representation of the molecular pathogenesis of osteoarthritis implicating synovial membrane, subchondral bon, and articular cartilage. During OA, chondrocytes are not still quiescent cells but undergo apoptosis or proliferation and phenotypic modulation; chondrocytes synthesize inflammatory mediators and matrix-degrading enzymes. Matrix degradation products are released in synovial fluid and trigger synovial inflammation, which leads to cytokines and proteinase release by synoviocytes. Moreover, an increased cartilage calcification and microfractures with blood vessel invasions from the subchondral bone is observed. Finally, osteoblasts release growth factors, which may participate in osteophytes formation.

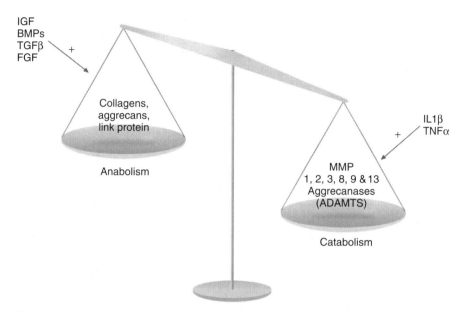

**Figure 15.4** Osteoarthritis exhibits an imbalance in cartilage matrix turnover. Anabolic and catabolic stimulatory factors are synthesized by the chondrocytes themselves in an automatic and paracrine manner. In OA cartilage, catabolism (characterized by MMPs and ADAMTS production) becomes more dominant than anabolism (characterized by collagens, aggrecans and link protein expressions), which leads to the degradation of cartilage. (*FGF*, fibroblast growth factor)

equivalent. In OA cartilage, catabolism becomes more dominant than anabolism, which leads to the degradation of cartilage (39) (Fig. 15.4).

*Anabolic Signals in Osteoarthritis*

INSULIN GROWTH FACTORS (IGFs)  In articular cartilage, IGF-1 plays a crucial role in the maintenance of homeostasis by stimulating the production of matrix proteins by chondrocytes, which counteracts their degradation (41–43). Interestingly, despite increased cartilage degradation in OA, the expression of IGF-1 messenger RNA (mRNA) is upregulated in OA cartilage, and IGF-1 protein increases in synovial fluid (44–46). The increased production of IGF-1 in OA reflects an attempt to restore cartilage homeostasis, but it is inefficacious because of the lack of bioavailability. The accessibility of IGF-1 to its receptor is regulated by extracellular IGF-binding proteins (IGFBPs), which are also key players in the failure of cartilage homeostasis during OA. IGFBP-3 is secreted by articular cartilage and chondrocytes (47), and its expression has been found to be correlated with the severity of OA (48, 49). Anabolic effects on chondrocytes by IGF-1 depend on the level of secreted IGFBPs.

Transforming Growth Factor (TGF)-β    TGF-β is an important anabolic factor in OA because it promotes type II collagen and proteoglycan synthesis and down-regulates cartilage-degrading enzymes (50–54). Moreover, TGF-β can counteract the suppression of proteoglycan synthesis by interleukin 1 (IL-1) (53). In aging and OA, the expression of TGF-β receptors and Smad2 phosphorylation are reduced in chondrocytes (55). The reduced response of chondrocytes to TGF-β may play a pivotal role in OA.

Bone Morphogenetic Proteins (BMPs)    BMPs belong to the TGF-β super-family and are considered to be one of the most important anabolic factors for articular cartilage. Recent studies showed that BMPs play a crucial role in adult articular cartilage homeostasis (56, 57). BMPs enhance proteoglycan and type II collagen synthesis *in vitro* and *in vivo* (58, 59). BMPs also counter-act catabolic cytokines such as IL-1. BMP7 has been shown to counteract matrix metalloproteinases-13 (MMP-13) induction by IL-1β. BMPs signal through Smad pathways, whereas IL-1 signals through extracellular signal regulated kinase (ERK)-p38- Jun terminal kinase(JNK)/NF-kB pathways (60). These pathways may seem to be separated but are often cross-linked in several ways. Therefore, intracellular crosstalk between the BMPs and IL-1 signaling pathways plays a part in balancing anabolism–catabolism in articular cartilage.

Fibroblast Growth Factors (FGFs)    Several studies have shown that bFGF is the most potent mitogen for chondrocytes and that it stimulates or stabilizes the synthesis of the cartilage matrix (61–63). Intra-articular administration of FGF2 enhances cartilage extracellular matrix formation in vivo (64, 65). However, oth-ers report that it was inhibitory or had no effect (66). Recent studies have shown that bFGF stored in the articular cartilage matrix is released with mechanical injury or with loading (67, 68). Moreover, bFGF has stimulated production of MMP-1, -3, and -13 instead of proteoglycan synthesis. The excessive release of bFGF from cartilage matrix may be partly responsible for cartilage matrix degradation during OA.

*Catabolic Signals in Osteoarthritis*

MMPs    The MMP family members are the major enzymes that degrade the extra-cellular matrix of cartilage. Twenty-three members of this family of neutral $Zn^{2+}$ metalloproteinases have been identified (69, 70). The MMP family is broadly classified into secreted-type MMPs and membrane-type MMPs (MT-MMPs). Because they are active at neutral pH, the MMPs can act on the cartilagi-nous matrix at some distances of the chondrocytes. They can be synthesized by chondrocytes, synoviocytes and osteoblasts after stimulation by cytokines or mechanical stress. Among MMPs, at least five are elevated in OA: collagenase-1 (MMP-1), stromelysin (MMP-3), gelatinase-92kd (MMP-9), matrisylin (MMP-7) and collagenase-3 (MMP-13). In OA, MMP-1 (also known as collagenase-1) and

MMP-13 (also known as collagenase-3) have a major role in cartilage degradation. MMP-13 is usually produced by chondrocytes in OA (71, 72). The expressions of these MMPs are low in normal cells, which contributes to normal connective tissue re-modeling. However, excessive MMP expression results in aberrant connective tissue destruction such as observed in arthritic diseases, periodontitis, and tumor invasion (73). MMPs are induced in response to cytokines and growth factors. In OA, the mechanical stress causes cytokine expression such as IL-1beta by articular chondrocytes, which subsequently express MMPs (74). MMP-1, MMP-8 (also known as neutrophil collagenase), and MMP-13 have the ability to cleave the triple helix of collagen and unwind the chains. After that, the chains become susceptible to even more degradation by other MMPs, such as MMP-9. MMP-13 plays a crucial role in collagen degradation because it is expressed by chondrocytes and it hydrolyzes type II collagen more efficiently than other collagenases (75). In an animal model, the postnatal expression of constitutively active human MMP-13 in the articular cartilage of transgenic mice leads to the degeneration of articular cartilage and joint pathology of the kind observed in OA (76).

AGGRECANASES   The ADAMTS (a disintegrin and metalloproteinase domain with thrombospondin motifs) family contains 19 members (77, 78). Family members ADAMTS-2, ADAMTS-4, ADAMTS-5, ADAMTS-9, and ADAMTS-15 are called aggrecanases. They degrade the interglobular domain separating G1 and G2 of aggrecan at a specific Glu373-Ala374 bond (78–81). Aggrecanases are active in the early phase of OA, with a later increase in MMPs activity. In surgically induced OA model mice, however, the degeneration of articular cartilage and joint pathology of OA were significantly reduced in ADAMTS-5 knockout mice with no difference in severity in ADAMTS-4 knockout mice (82, 83). In an inflammatory arthritis model, the deletion of ADAMTS-5 also prevented the progression of aggrecan degradation, whereas severe cartilage degradation was induced in ADAMTS-4 knockout mice (84). These findings suggest that ADAMTS-5 might be the key player for aggrecan degradation in arthritic disease.

PROINFLAMMATORY FACTORS   In many joint diseases, proinflammatory factors such as cytokines and prostaglandins are released at sites of inflammation, together with ROS and nitric oxide (NO) (85, 86).

*Cytokines*   The cytokines such as IL-1β and tumor necrosis factor-α (TNF-α) also play a key role in cartilage degeneration in OA. IL-1beta is one of the most prominent catabolic cytokines that plays a pivotal role in OA (87), and its level is increased in the OA cartilage (88). IL-1 not only promotes the release of degenerative enzymes including MMP-1, MMP-3, MMP-13, or ADAMTS-4 and ADAMTS-5, but also it inhibits the synthesis of extracellular matrix proteins by chondrocytes. IL-1 also induces other catabolic cytokines, such as IL-6, with synergistic effects (89). IL-6 family members have been shown to induce the JAK/STAT signaling pathways (90). The effects of IL-6 are similar to IL-1;

however, its effects are weaker than that of IL-1 in inhibiting proteoglycan synthesis. IL-6 is required for prolonged proteoglycan suppression by IL-1, and it enhances the expression and activity of MMPs and aggrecanase in articular cartilage (91). IL-1 acts through nuclear factor-kappa beta (NF-kB) and the following three classic MAPK-signaling pathways: ERK, p38, and JNK. Although IL-1beta can activate all four pathways, the ERK pathway is especially important for the induction of other cytokines such as IL-6 (89).

*Prostaglandin E2 (PGE2)* PGE2 is one of the major catabolic mediators involved in cartilage degradation and the progression of OA (92–94). The synthesis of PGE2 is the endpoint of a sequence of enzymatic reactions, which includes the release of arachidonic acid from membrane phospholipids by soluble phospholipase A2 and conversion of this substrate to prostaglandin H2 (PGH2) by cyclooxygenase (COX)-1 and COX-2, which is also known as prostaglandin H synthase 1 and 2. Then, the PGE synthase catalyzes the conversion of PGH2 to PGE2. PGE2 is a mediator of inflammation, which has numerous actions during OA. PGE2 modulates the activity of synovial cells, macrophages, and chondrocytes, and it induces bone resorption (95). The role of prostaglandins in the metabolism of articular cartilage is still a matter of debate. Some reports indicate that prostaglandins participate in the destruction of articular cartilage by degrading cartilage ECM (96, 97), whereas others show that they promote chondrogenesis and terminal differentiation (98, 99). The opposing biologic roles attributed to these compounds are a direct reflection of the molecular complexity of prostaglandins (PGs) and their unique cognate receptors (100). PGE2 is known to bind to four distinct cell surface receptors (EP1, EP2, EP3, and EP4), each of which is a "serpentine seven" G-protein–coupled receptor; the ligation of individual receptors mediates distinct intracellular signaling pathways that may contribute to the pleiotropic effects of PGs in different tissues (101, 102). Recently, two different studies report that the predominant effects of PGE2 in OA chondrocytes are catabolic and that these effects are mediated via EP4 receptor signaling (103, 104).

OXIDATIVE STRESS ROS are normal by-products of cellular metabolism. In some physiological and pathological circumstances, $O_2$ may be transformed into ROS and then may be involved in the control of various aspects of biological processes, which include cell activation, proliferation, and death. Especially low levels of ROS have been reported to act as one of the different intracellular second messenger molecules involved in the regulation of the expression of a wide variety of gene products, which include cytokines, MMPs, adhesion molecules, and matrix components (105, 106). The elevated production of ROS and/or depletion of antioxidants has been observed in a variety of pathological conditions, including inflammatory joint diseases. The intracellular and extracellular reduction-oxidation state may be causally implicated in the progression of these diseases (see review 107, 108). When the imbalance between oxidants and antioxidants is large enough to induce cellular and/or tissue structural and/or functional changes, the situation is called "oxidative stress" and is considered as an abnormal catabolic event. The overproduction of ROS in pathological cartilage has been evidenced indirectly by the accumulation of lipid peroxidation products and nitrotyrosine *in situ* (109, 110). Recently, Henrotin et al. (111) have found elevated nitrated type II collagen peptides in sera of patients with OA, which suggests the formation of ONOO- in OA cartilage. Chondrocytes produce basically cNO and $O_2c_-$ that generate derivative radicals including ONOO- and $H_2O_2$. ROS may directly oxidize nucleic acids, transcriptional factors, and membrane phospholipids, as well as intracellular and extracellular components, which leads to impaired biological activity, cell death, and breakdown of matrix components (106, 107, 112–115). ROS are highly reactive compounds with a short half-life. However, nitrotyrosine, which is formed when tyrosine is oxidized in the presence of nitrous oxide (NOO), can serve as a measure of oxidative damage *in vivo*. The presence of nitrotyrosine in cartilage was associated with older age and with osteoarthritis, which suggests a role of oxidative stress in cartilage aging and degeneration (116). Thus, oxidative stress affects chondrocyte function in joint cartilage (117, 118).

NITRIC OXIDE    NO is synthesized by way of the oxidation of L-arginine by the NO synthase; one result is constitutive NOS and the other is inducible NOS. Some studies have reported a local production of NO in the OA joint by chondrocytes and their presence in synovial fluid (119). NO synthesis is implicated as a catabolic factor in OA. The various roles of NO as a mediator of other IL-1–induced responses, which include the inhibition of aggrecan and collagen synthesis (120, 121), enhancement of MMP activity and chondrocyte apoptosis (122–124), and reduction of the production of IL-1 receptor antagonist (IL-1RA) (125), have also been suggested (126). NO may also increase chondrocyte susceptibility to injury by other oxidants such as $H_2O_2$ and contribute to resistance against the anabolic effects of IGF-I (127). Nitric oxide also has been implicated as an important mediator in chondrocyte apoptosis, and an association between NO production and apoptosis in OA cartilage has been proposed (122).

TRANSCRIPTIONAL FACTORS IN OSTEOARTHRITIS    Sox9 is a transcriptional factor with a high-mobility group DNA-binding domain and is required for cartilage formation and for expression of chondrocyte-specific genes, such as COL2A1 (128). In OA cartilage, several studies have shown that Sox9 is downregulated and that IL-1β and TNF-α inhibit the expression of cartilage-specific genes, such as COL2A1, COLL11A2, COL9A2, and aggrecan (129–131). The major mechanism by which IL-1 and TNF-α inhibit the chondrocyte phenotype is by downregulating the expression of Sox9 (132). IL-1 and TNF-α inhibit Sox9 binding to the COL2A1 enhancer, and these effects are mediated by NF-kB (133). Another study showed that Sox9 is downregulated by IL-6/sIL-6R associations at both mRNA and protein levels through the JAK/STAT signaling pathway (134). However, other studies reported that in the early phase of OA, the expressions of type II collagen and type I collagen are upregulated, which suggested that the

Sox9 type II collagen master regulation gene might be also upregulated (135). The evidence for upregulation or downregulation is mixed regarding the alteration of Sox9 gene expression during the progression of OA. It is possible that anabolic and catabolic signals might be abnormally intermingled in OA cartilage. IL-1 induction of MMP-13, MMP-1, and MMP-9 requires NF-kB (136, 137). Because inhibition of NF-kB results in a beneficial effect on OA synovial fibroblasts, such as the decreased expression of MMPs and ADAMTS-4 aggrecanase (138), the inhibition of NF-kB is anticipated as an alternative therapeutic target. Moreover, the key role of NF-kB in regulating mechanical stress in chondrocytes has been described recently in different studies (139, 140).

*15.2.1.2 Chondrocyte Proliferation and Apoptosis in Osteoarthritis*   The chondrocyte, which is the only cell type that resides in the adult cartilage matrix, has a low metabolic activity, and it survives under relatively hypoxic conditions and in the absence of a vascular supply. This cell, which is ultimately responsible for remodeling and maintaining the structural and functional integrity of the cartilage matrix, possesses poor regenerative capacity. The adult chondrocyte plays a critical role in the pathogenesis of OA in responding to adverse environmental stimuli by promoting matrix degradation and downregulating processes essential for cartilage repair. Physiologically, the chondrocytes maintain a low turnover rate of replacement of cartilage matrix proteins with a half-life for collagen of greater than 100 years (141). In contrast, the glycosaminoglycan constituents on the aggrecan core protein are more readily replaced, and the half-life of aggrecan subfractions has been estimated in the range of 3–24 years (142). The half-life of proteoglycan has been estimated in the range of 7–200 days (143). In early OA, evidence suggests the presence increased synthetic activity, which is viewed as an attempt to regenerate the matrix with cartilage-specific components, including types II, IX, and XI collagens; aggrecan; and pericellular type IV collagen (144). The aberrant behavior of OA chondrocytes is reflected in the appearance of fibrillations, matrix depletion, and cell clusters, as well as in changes in quantity, distribution, or composition of matrix proteins (145). Evidence of phenotypic modulation is reflected in the presence of collagens not normally found in adult articular cartilage, which includes the hypertrophic chondrocyte marker, type X collagen, as well as other chondrocyte differentiation genes; this finding suggests a recapitulation of a developmental program (146, 147). Several studies (148, 149) have clearly shown that (low) proliferative activity occurs in osteoarthritic chondrocytes in contrast to normal articular chondrocytes, which do not show any proliferative activity. A proliferative activity is also observed in the early stage, at a low level, which leads to the chondrocyte clustering (150–152). This increased proliferative activity, especially in the upper cartilage zone, could be explained by the exposition of chondrocytes to growth factors released by the cartilage itself, which are present in the synovial fluid. This proliferative activity is visualized as a cellular reaction to cartilage destruction to repair cartilage.

In addition to hypertrophic differentiation of chondrocytes, chondrocyte apoptosis is known to be involved in OA development (153). The intra-articular injection of a pan-caspase inhibitor has been reported to suppress cartilage degradation under OA induction in a rabbit ACLT model (154). Many studies have considered whether cell death plays a role in the pathology of OA [for review, see Aigner et al. (155)], because articular chondrocytes cannot self-renew, and therefore cell loss would be permanent. However, opinions on the prevalence the importance and the relevance of chondrocyte death for OA pathology differ widely. Early studies (156, 157) estimated the incidence of cell death as 25–50%, but the data were obtained with isolated chondrocytes and therefore may have measured a predisposition to death rather than actual incidence. Estimates of cell death based on *in vivo* staining also vary widely, from 21% (158) to less than 1% (159). Opinions also vary as to the type of cell death. Most authors classify chondrocyte death as caused by apoptosis, even though injury or mechanical damage by themselves would lead to a high level of necrosis.

### 15.2.2 Synovium Involvement in Osteoarthritis

The clinical presentation in OA joints (such as swelling, effusions, and stiffness) clearly reflects synovial inflammation lower than in rheumatoid arthritis, as a low-grade contribution to disease pathogenesis. This synovitis occurs even in early OA and can be subclinical, as arthroscopic studies suggest that localized proliferative and inflammatory changes of the synovium occur in up to 50% of OA patients (many of whom do not seem to have active inflammation) (160). Synovial histological changes include synovial hypertrophy and hyperplasia, with an increased number of lining cells, which are often accompanied by infiltration of the sublining tissue, with scattered foci of lymphocytes. In contrast to rheumatoid arthritis, synovial inflammation in OA is mostly confined to areas adjacent to pathologically damaged cartilage and bone. This activated synovium can release proteinases and cytokines that may accelerate the destruction of nearby cartilage. The synovium produces some of the chemokines and metalloproteinases that degrade cartilage, even though the cartilage itself produces most of these destructive molecules in a vicious autocrine and paracrine fashion. In turn, cartilage breakdown products, which result from mechanical or enzymatic destruction, can provoke the release of collagenase and other hydrolytic enzymes from synovial cells and lead to vascular hyperplasia in OA synovial membranes. This cascade sequentially results in the induction of synovial IL-1β and TNF-α, which extend the inflammatory outcome (161). This cytokine storm may be more likely to occur in earlier stages of the disease before end-stage damage. As shown by a recent study of 10 patients with early OA (arthroscopic specimens) and 15 patients undergoing total knee arthroplasty; synovial tissues from early OA had higher levels of IL-1β and TNF-α and increased mononuclear cell infiltration compared with late OA (162). Erosive OA likely represents a more inflammatory process, as evidenced by higher proteinase and cytokine levels One study of rapidly destructive hip OA demonstrated MMP-3 and -9 levels that were especially elevated, not only in patients' synovial cells but also in their synovial fluid, plasma, and sera (163, 164). In vitro studies from diseased human OA tissue have implicated MMP-10 expression in synovial fibroblasts, as well as in OA synovial fluid and chondrocytes stimulated with catabolic IL-1 and oncostatin M (165).

### 15.2.3 Role of Subchondral Bone in Osteoarthritis

In addition to the progressive loss of articular cartilage, OA is characterized by increased subchondral plate thickness, formation of new bone at the joint margins (osteophytes), and the development of subchondral bone cysts (166, 167). Interestingly, these abnormalities take place not only during the final stage of OA but also earlier at the onset of the disease, maybe before the cartilage degradation (168).

Subchondral bone remodeling is responsible for an alteration in mechanical load absorbance. This tissue is characterized by the accumulation of osteoid substance, which is decreased mineralization related to the production of an abnormal trimeric type I collagen that has a low affinity for calcium (169). Several studies have shown that osteoarthritic subchondral osteoblasts exhibit specific pathologic characteristics, such as overproduction of alkaline phosphate and osteocalcin, as well as transforming growth factor-beta 1 (TGF-β1), IGF-1, and urokinase (170).

Moreover, it has been shown recently that mature sclerotic osteoblasts exhibit an increased expression of osteocalcin, IL-6, IL-8, C-terminal type I procollagen propeptide, TGF-β1, osteopontin, and accumulation of osteoid substance. The expression of parathyroid hormone receptor is decreased, which explains the resistance of subchondral bone to resist to parathyroid hormone stimulation (171).

The subchondral bone is involved in the OA process (172) at the following two different levels:

1. First, osteoblasts of this area are primarily involved in the production of prodegradative agents, such as MMP-1, MMP-13, PGE2, and IL-6, after their stimulation by mechanical stress (173) and in the inhibition of cartilage matrix component synthesis. These agents can participate in the cartilage and in subchondral bone degradation (170). Moreover, cells from the OA bone modify cartilage metabolism, which affects glycosaminoglycan release significantly more than non-OA bone cell culture (174).
2. Second, increased subchondral bone expression of IGF1 and TGF-β induces new bone formation, such as osteophytes and subchondral sclerosis (170, 175).

Of note, the subchondral bone in OA is in a state of hypercoagulation, hypofibrinolysis, and thrombosis, which is related to reduced blood flow and increased production of procoagulant factors. These factors lead to venous stasis and hypertension, which are associated with thrombosis and focal ischemic bone necrosis (176).

All these modifications of structure and phenotype of osteoblasts in the subchondral bone can modify strength load on the cartilage and enhance the alteration of chondrocytes and ECM as well as the cartilage breakdown.

Angiogenesis phenomenon is observed at the junction of the articular hyaline cartilage and adjacent subchondral bone. Evidence exists to suggest vascular invasion and advancement in the zone of calcified cartilage in the region of the so-called tidemark that contributes to a decrease in articular cartilage thickness (177). A recent study showed that angiogenesis in the osteochondral junction is independent of synovial angiogenesis and synovitis but is associated with cartilage changes and clinical disease activity (178). These structural alterations in the articular cartilage and periarticular bone may lead to modification of the contours of the adjacent articulating surfaces (179, 180). The accompanying alterations in subchondral bone remodeling may contribute to the development of an adverse biomechanical environment and enhance the progression of the articular cartilage deterioration. Changes in the mineral content and thickness of the calcified cartilage and the associated tidemark advancement may be related to the localization of COL10A1, MMP-13, and Runx2 in the deep zone of OA cartilage, where the chondrocytes may attempt a defective repair response by recapitulation of the hypertrophic phenotype (181).

The link between cartilage and subchondral bone is still debated. Mechanical stress on weight-bearing joints is responsible for local lesions of cartilage, which increase stress to the local area of bone underneath this area of cartilage and enhance microfractures and sclerosis of the subchondral bone. These modifications of the bone are responsible for mechanical disturbance, which increases the lesions of the cartilage and the subchondral bone itself. Another possibility is that because of repeated microfractures, subchondral bone becomes stiffer. This structural alteration occurs before cartilage changes and can modify joint conformation and thus induce secondary cartilage alteration under load (182). Indeed, studies in animal models suggest that subchondral bone changes may precede cartilage degradation and loss. This finding has been confirmed in magnetic resonance imaging (MRI) studies in humans. Likewise, these bone changes are not a secondary manifestation in OA but may be an active part in the initiation of the OA process. Indeed, subchondral bone tissue in OA could provide growth factors, cytokines, and eicosanoids to the overlying cartilage and could promote its abnormal remodeling and metabolism, which leads to ECM degradation (183).

## 15.3 TOOLS AND METHODS FOR STUDYING OSTEOARTHRITIS

### 15.3.1 Animal Models

In an effort to clarify the pathophysiological mechanisms that lead to OA, experimental animal models have been developed. The classic models of OA have involved the induction of OA-like changes in many animal species, which include dogs, rabbits, guinea pigs, sheep, and rats using surgical intervention. Transection of ligaments combined with resection of the meniscus is used to produce instability in the knee joint. Interestingly, spontaneous forms of OA exist in various species, such as mice. But even if spontaneous models of OA are more relevant, they are more difficult and less practical (184).

Recently, transgenic and knockout mouse models, which are the most ideal animal for molecular study, have provided new tools to elucidate the mechanisms implicated in OA pathophysiology (185). The application of surgical models on these mice has been developed to characterize the *in vivo* significance of a particular gene (186–188).

Although these models are excellent for studying the time course of OA and the implication of different molecules in this event, no experimental model is totally equivalent to OA in humans.

### 15.3.2 Biomarkers

*15.3.2.1 Imaging as a "Biomarker"* Standards radiographs (X rays) are used to establish diagnosis. The radiographic hallmarks of the most common form of OA include localized joint space narrowing, subchondral bone sclerosis, formation of osteophytes, and bone cysts. However, clinical symptoms and radiographic findings are poorly correlated; many joints with radiographic evidence of OA remain asymptomatic and, conversely, the joints of many patients with severe symptoms can seem only marginally affected on X rays (189). Moreover, X rays have shortcomings with respect to the assessment of progressive disease. For example, they are insensitive to early changes within cartilage and bone and do not report synovial pathology (190).

MRI is a powerful technique that provides precocious alterations of cartilage, synovium, and bone. It represents a sensitive and specific technique for measuring disease progression. Traditional MRI is now being used for the quantitative assessment of hyaline cartilage thickness and volume, synovial hypertrophy, and bone marrow edema. Most recently, functional MRI studies (dGEMRIC, NaMRI, or T1ro), which detect biochemical changes of ECM proteins in cartilage, could be used to assess the improvement from the treatment. This study is of particular interest because cartilage thinning in OA is preceded by modification of the composition and structural organization of the collagen–proteoglycan matrix (191–193).

*15.3.2.2 Biochemical Markers* Because of the increasing knowledge of the cartilage matrix composition, molecular markers have been identified for monitoring changes in cartilage metabolism and for assessing joint damage in OA. These biochemical markers could be useful to establish state of disease, to understand the pathophysiology of OA, to predict progression, and to assess the treatment efficacy. Candidate biochemical markers of OA joint tissues present in synovial fluids, blood, or urine can reflect different actors of OA (bone, cartilage, synovitis, and systemic inflammation) as well as the different states of these tissues (i.e., degradation or synthesis). These biomarkers involve components of matrix proteins, which include several collagens as well as cross-linked derivative peptides and matrix metalloproteinases. For example, serum cartilage oligomatrix protein, urinary C-terminal cross-linking telopeptide of type II collagen, and serum hyaluronan denote the severity and extent of disease in joints (190, 194, 195). Moreover, various type II collagen epitopes have been described, including Coll 2-1, which was characterized as a disease-specific marker that is sensitive to the structural changes occurring in OA (196). Interestingly, the urinary levels of Coll 2-1 or Coll 2-1NO$_2$ over 1 year was found to be predictive of the progression joint space narrowing in OA patients (197, 198). Finally, myeloperoxidase is an innovative marker that allows oxidative stress assessment in OA: Its decrease after joint replacement indicates that neutrophil activation occurs during this disease (196). Moreover, plasma high-sensitivity C reactive protein levels reflect synovial inflammation in OA patients (199), and its levels probably reflect the disease activity of OA (200). But actually, no consensus exists to what is the optimal biomarker, and at the time of writing in 2008, there is not any indication to apply them in daily practice. Although a single marker may not be sufficient, it may be possible to use a combination of biomarkers that may discriminate between different stages of OA in different populations (201).

*15.3.2.3 Clinical Markers* Nonmodifiable risk factors, such as age, gender, body mass index, injury, or genetic background could be considered to be predictors of disease progression. The first-degree relatives of patients with generalized OA have a twofold risk of developing the disease compared with the general population (202). Moreover, monozygotic twins are significantly more concordant for hand and knee osteoarthritis than dizygotic twins (33).

Clinical endpoints can also serve as markers of disease progression. Factors that have consistently been reported as associated with radiographic OA progression are obesity, generalized OA, alignment, and synovitis (203). For example, in a 1-year follow-up study, people who suffered from OA with synovitis were found to have more rapid progression of cartilage damage (204). But, even if clinical signs can potentially be used to predict progression in OA, they are only observed in a subset of patients.

*15.3.2.4 Use of Biomarkers in Drug Development* Because the knowledge of the pathophysiological mechanisms of OA has advanced, efforts were made to identify new drugs to treat OA. Validation of clinical, chemical and imaging biomarkers should aid in this field of research. They could be used in clinical assay to track improvement, stabilization, or even progression of OA with the use of new therapeutic agents; to guide patient selection; and to reduce the number of participants needed to treat and strengthen the power of these studies (190).

## 15.4 THERAPEUTIC MANAGEMENT OF OSTEOARTHRITIS

The aims of the treatment of OA are to reduce joint pain and stiffness, to maintain and improve joint mobility, to reduce physical disability and handicap, to improve

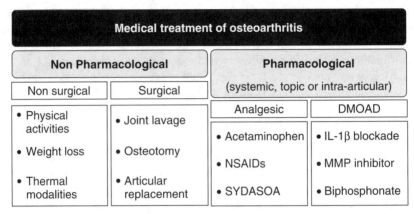

**Figure 15.5** Medical treatment modalities of osteoarthritis. Current treatments of OA include a wide range of nonpharmacological and pharmacological treatment to decrease pain, improve function, and prevent the progression of damage to joint tissues. In each category, the major therapies are listed.

health-related quality of life, to limit the progression of joint damage, and to educate patients about the nature of the disorder and its management (205).

Experts in the field of OA have developed concise, patient-focused, up to date, evidence-based recommendations for the management of hip and knee OA. These recommendations are classified in three fields: nonpharmacological therapies, surgical therapy, and drug therapy. Experts agree that optimal management of OA requires a combination of nonpharmacological and pharmacological modalities (Fig. 15.5) (205).

### 15.4.1 Nonpharmacological Therapy

*15.4.1.1 Nonchirurgical Therapy* Regular physical activities are important for patients with knee and hip OA to reduce joint pain and improve functional articular ability. Patients should be encouraged to undertake regular aerobic, muscle strengthening, and range of motion exercises, such as walking and swimming (206). Moreover, exercises in water can be effective (207). However, the effect sizes are small to moderate, with no long term benefit (208).

It is well known that obesity is a risk factor for OA (see Section 15.1). Data also indicate that weight reduction and stabilization at a lower level in patients who are overweight may result in reduction in pain and improvement in function of weight-bearing joints (14). Even a small amount of weight loss may be beneficial in patients with knee OA (209).

Some thermal modalities may be effective for relieving symptoms. Application of superficial heat, cold, or both has been widely employed for pain relief. Less-conventional treatments such as acupuncture and massage are emerging as promising options for OA treatment, but the magnitude of benefit has varied between studies (207).

Finally, patients should be given information access and education about the objectives of treatment and the importance of changes in lifestyle, exercise, pacing of activities, weight reduction, and other measures to unload the damaged joint(s).

*15.4.1.2  Surgical Therapy*  Surgical interventions in OA can be categorized in the following three groups based on the rationale and/or the main objective of the treatment (210):

1. To improve of the current symptoms using lavage of joints and remove fragments of cartilage and calcium-phosphate crystals, debridement is used, which consists of smoothing rough fibrillated articular and meniscal surfaces as well as removing inflamed synovium. The evidence of symptomatic efficacy of these techniques seems to be weak or null (211–213).
2. To prevent the risk of structural progression using osteotomy, surgical therapy could be indicated for the treatment of misalignment, which is a predisposing factor for OA progression. However, no evidence is available to suggest whether an osteotomy is more effective than conservative treatment, and it is impossible to make any formal recommendation (214).
3. To improve the symptoms related to an advanced disease by articular replacement, surgery may be required.

### 15.4.2  Drug Therapy

*15.4.2.1  Analgesic Treatment Modalities*  Acetaminophen is the treatment recommended as first-line by experts (recommendation of the European League Against Rheumatism, the American College of Rheumatology, and the Osteoarthritis Research Society International) (206, 215). Studies demonstrated the efficacy of acetaminophen (3–4 g/day) on pain relief. It could be combined with narcotic analgesics such as codeine. However, hepatic toxicity was reported in patients using acetaminophen greater than 4 g per day or with liver disease or concomitant medications. Acetaminophen may be the most appropriate initial therapy, but if pain persists, then prescription-strength nonsteroidal anti-inflammatory drugs (NSAIDs) or a more specific cyclooxygenase type 2 (COX-2) inhibitor may prove useful (216).

NSAIDs that block prostaglandin synthesis are widely used for treating OA. It includes aspirin, nonselective NSAIDS, and cyclooxygenase-2 inhibitors (coxibs), which selectively block the synthesis of the prostaglandin E2. NSAIDs provide short-term pain relief in patients with OA (217). However, adverse effects such as gastrointestinal toxicity, congestive heart failure, and renal failure were observed, and these risks increased in frequency in elderly patients. Moreover, two coxibs, rofecoxib and valdecoxib, were withdrawn from the market in 2004 because they were associated with increased incidence of cardiovascular adverse events (218).

Therefore, NSAIDs, which include both nonselective and COX-2 selective agents, should be used with caution in patients in accord with concomitant medications and comorbidities, especially in case of long-term prescription.

Compounds called symptomatic slow-acting disease osteoarthritis (SYSADOA) have been developed for the management of OA. Compared with NSAIDs and acetaminophen, they exhibit lower, delayed, and carryover effects on symptoms. Interestingly, the use of these drugs may associate with less NSAIDs requirement and could have a weak beneficial structural effect. The most frequently used SYSADOA is glucosamine sulfate, either alone or in combination with chondroïtine sulfate. Glucosamine is obtained from shrimp exoskeletons and is a component of many macromolecules, such as hyaluronic acid (an important substance in collagen formation). Chondroïtine sulfate is a macromolecule that contributes to, like hyaluronic acid, a framework for collagen formation. Solid evidence of their effectiveness and safety is available for OA treatment (219, 220). Although many studies have examined potential mechanisms of action for glucosamine and chondroïtine in OA, the exact nature of these mechanisms remains unclear. They are available commercially as nutritional supplements (221). Other examples of SYSADOA are diacerein, which may inhibit IL-1β production and reduce cartilage breakdown (222), and avocado/soybean unsaponifiables, which may counteract deleterious processes triggered by IL-1β and mechanical stress (140) and reduce pain and Lesquene index in OA patients (223). These treatments are purchased as a prescribed drug. The choice of the OA treatment depends on the benefit/risk ratio for the patient in addition to its availability, cost, and patient acceptance. Even if the effect size of acetaminophen is modest, this drug is considered as a safe treatment of OA and remains the first-line oral analgesic in the management of OA (206, 215). NSAIDs, which are more efficient to improve pain relief, are prescribed in a second time because it is not as good concerning the safety profile. Moreover, classic NSAID and coxib are similar for the symptomatic relief of OA (224). They should be used at the lowest effective dose for the shortest possible duration of treatment. However, it is not rare that the highest dose is necessary to achieve the clinical goal and sometimes on a chronic use design. The choice between these two treatments depends on comorbidities and patient history (225). Finally, SYDASOA's global effect size reveals lower efficacy, as glucosamine has shown to provide a low to medium response rate (226) and chondrotine sulfate provides a small response rate.

Topical treatments are of interest especially in patients intolerant or unwilling to take oral analgesics and especially for superficial joints. The most widely used types of topical treatment include capsaicin, topical lidocaine, and topical NSAIDs (227).

Finally, intra-articular injections of corticosteroids and hyaluronans may be useful in patients with OA. Intra-articular corticosteroids have been used for decades as adjunctive therapy, especially when local inflammation is present as indicated by the presence of a synovitis (228). Intra-articular viscosupplementations of hyaluronic acid, which is a high-molecular-weight polysaccharide that is a major component of synovial fluid and cartilage, assures relief of OA-related pain and improves joint function. Moreover, intra-articular hyaluronan formulations are well tolerated and are associated with a low incidence of adverse effects usually localized to the injected joint (229).

### 15.4.2.2 Disease-modifying Osteoarthritis Drugs (DMOADs)

DMOADs are a group of therapeutic agents that would, ideally, both control symptoms and provide structure modification. They inhibit one or more OA pathophysiological processes, such as synovial inflammation, cartilage destruction, or subchondral bone remodeling. It is a large family of molecules, the most documented of which are listed in this chapter. The following agents are under clinical investigations for the treatment of OA and are not currently used for treating OA (for review see 230).

*IL-1β Blockade* Of the cytokines thought to be involved in the pathogenesis of OA, IL-1β has attracted the most interest as a target for disease modification. Because the IL-1 receptor antagonist (IL-1Ra) anakinra has been used successfully in the treatment of the inflammation and bone destruction of rheumatoid arthritis, some have suggested that it can possibly retard the disease progression of OA. A randomized placebo-controlled study has revealed improvement in knee pain and function with intra-articular injection of anakinra treatment compared with placebo (231). The clinical benefit of a local injection of IL-1Ra in knee osteoarthritis may be limited by the antagonist's short half-life (for review, see Reference 232).

*MMP Inhibitor* Developing effective and specific inhibitors of the catabolic enzymes that suspend cartilage degradation in OA, which include MMP and ADAMTS, has been the focus of several efforts. Specific MMP inhibitors were tested in clinical trials that studied joint destruction occurring in RA, but some trials were terminated because of lack of efficacy or safety concerns. Increased understanding of the structure, regulation, and function of individual MMPs may lead to more effective strategies. Actually, research focused on the development of a novel selective MMP-13 inhibitor (233, 234).

*Biphosphonates and Strontium Ranelate* Biphosphonates (risedronate, alendronate, and zodrenate) can suppress bone turnover because of their antiosteoclastic activity. The rationale to evaluate this compound in OA is based on the potential role of subchondral bone in both the occurrence and the progression of OA. Recently, it has been shown that risedronate decreases biochemical markers of cartilage degradation but does not decrease symptoms or retard radiographic progression in patients with knee OA (235). The elucidation of risedronate's effects will require additional studies. A recent study has shown in secondary analysis that alendronate could have a structural effect on spinal OA (236). Strontium ranelate, which is another antiosteoporotic drug that specifically targets the osteoblast, could have also a beneficial effect on spinal OA with a structural and analgesic beneficial effect (237). However, all these results need additional studies to establish a potent effect in OA.

Treatment that targets NO synthesis by specific inhibitor of the inducible NO synthase and acts on bone remodeling as calcitonin represents other potential treatments in the future. For example, it has been shown that calcitonin reduces Lequesne's algofunctional index scores and decreases urinary and serum levels of biomarkers of joint metabolism, such as type II collagen neoepitopes, MMP-3, and MMP-13, in knee osteoarthritis (238).

### 15.4.3 Future Directions

Articular cartilage has limited repair and regeneration potential. The scarcity of treatment modalities for large chondral defects has motivated attempts to engineer cartilage tissue constructs with chondrocytes distributed in a three-dimensional structure, which could integrated the resident tissue. Transplantation of autologous chondrocytes has been used to repair focal cartilage defects, but this procedure could lead to variable results with the frequently formation of fibrous cartilage. In experimental studies, several differentiation factors, such as TGF-β, BMPs, or IGF-1, have been shown to stimulate chondrogenesis, which promotes the formation of functionally cartilage-like tissue repair. Therefore, the gene transfer of anabolic factors to promote chondrocytes differentiation as the production of such factors directly at the site of the lesions has been developed. However, the clinical application of such strategies is limited because of the short half-lives of the proteins *in vivo* (239, 240). Actually, various approaches are tested *in vitro* for the use of mesenchymal stem cells (MSCs) to generate tissue-engineered cartilage. MSCs are considered the cell type of choice for tissue engineering because of their multipotence and ability to differentiate into mesenchyme-derived cells, such as osteoblasts and chondrocytes. Moreover, they can be easily isolated from various adult tissues, such as bone marrow, synovium, or adipose tissue, and can be expanded. However, differentiation of MSCs into articular chondrocytes and reparation of injured cartilage are still poorly understood on the molecular level. Therefore, successful regeneration or replacement of damaged or diseased cartilage will depend on future advances in our understanding of the biology of cartilage and stem cells and technological development in engineering (241–243).

### REFERENCES

1. Brandt KD. Osteoarthritis. In: Internal Medicine 4th Edition. Stein J, ed. 1994. St. Louis, MO: Mosby Year Book, Inc.

2. Buckwalter JA, Martin JA. Osteoarthritis. Adv. Drug Deliv. Rev. 2006;58:150–167.

3. Loeser RF. Aging and the etiopathogenesis and treatment of osteoarthritis. Rheum. Dis. Clin. North Am. 2000;26:547–567.

4. Felson DT, Zhang Y, Hannan MT, Naimark A, Weissman BN, Aliabadi P, Levy D. The incidence and natural history of knee osteoarthritis in the elderly. The Framingham Osteoarthritis Study. Arthritis Rheum. 1995;38:1500–1505.

5. Buckwalter JA, Heckman JD, Petrie DP. AOA. An AOA critical issue: aging of the North American population: new challenges for orthopaedics. J Bone Joint Surg. Am. 2003;85:748–758.

6. Davis MA, Ettinger WH, Neuhaus JM, Mallon KP. Knee osteoarthritis and physical functioning: evidence from the NHANES I Epidemiologic Followup Study. J. Rheumatol. 1991;18:591–598.

7. Felson DT, Naimark A, Anderson J, Kazis L, Castelli W, Meenan RF. The prevalence of knee osteoarthritis in the elderly. The Framingham Osteoarthritis Study. Arthritis Rheum. 1987;30:914–918.

8. Aigner T, Haag J, Martin J, Buckwalter J. Osteoarthritis: aging of matrix and cells—going for a remedy. Curr. Drug Targets 2007;8:325–331.

9. Dudhia J. Aggrecan, aging and assembly in articular cartilage. Cell Mol. Life Sci. 2005;62:2241–2256.

10. Verzijl N, Bank RA, TeKoppele JM, DeGroot J. AGEing and osteoarthritis: a different perspective. Curr. Opin. Rheumatol. 2003;15:616–22.

11. Martin JA, Brown TD, Heiner AD, Buckwalter JA. Chondrocyte senescence, joint loading and osteoarthritis. Clin. Orthop. Relat. Res. 2004; S96–103.

12. Lievense AM, Bierma-Zeinstra SM, Verhagen AP, van Baar ME, Verhaar JA, Koes BW. Influence of obesity on the development of osteoarthritis of the hip: a systematic review. Rheumatology 2002;41:1155–1162.

13. Manek NJ, Hart D, Spector TD, MacGregor AJ. The association of body mass index and osteoarthritis of the knee joint: an examination of genetic and environmental influences. Arthritis Rheum. 2003;48:1024–9.

14. Felson DT, Anderson JJ, Naimark A, Walker AM, Meenan RF. Obesity and knee osteoarthritis. The Framingham Study. Ann. Intern. Med. 1988;109:18–24.

15. Cicuttini FM, Baker JR, Spector TD. The association of obesity with osteoarthritis of the hand and knee in women: a twin study. J. Rheumatol. 1996;23:1221–6.

16. Haara MM, Manninen P, Kröger H, Arokoski JP, Kärkkäinen A, Knekt P, Aromaa A, Heliövaara M. Osteoarthritis of finger joints in Finns aged 30 or over: prevalence, determinants, and association with mortality. Ann. Rheum. Dis. 2003;62:151–158.

17. Toda Y, Toda T, Takemura S, Wada T, Morimoto T, Ogawa R. Change in body fat, but not body weight or metabolic correlates of obesity, is related to symptomatic relief of obese patients with knee osteoarthritis after a weight control program. J. Rheumatol. 1998;25:2181–2186.

18. Gosset M, Berenbaum F, Salvat C, Sautet A, Pigenet A, Jacques C. Crucial role of Visfatin in matrix degradation and PGE2 synthesis in chondrocytes: possible role in osteoarthritis process. Arthritis Rheum. 2008;58:1399–1409.

19. Toussirot E, Streit G, Wendling D. The contribution of adipose tissue and adipokines to inflammation in joint diseases. Curr. Med. Chem. 2007;14:1095–1100.

20. Aigner T, Sachse A, Gebhard PM, Roach HI. Osteoarthritis: pathobiology-targets and ways for therapeutic intervention. Adv. Drug Deliv. Rev. 2006;58:128–149.

21. Roos EM, Dahlberg L. Positive effects of moderate exercise on glycosaminoglycan content in knee cartilage: a four-month, randomized, controlled trial in patients at risk of osteoarthritis. Arthritis Rheum. 2005;52:3507–3514.

22. Duke PJ, Montufar-Solis D. Exposure to altered gravity affects all stages of endochondral cartilage differentiation. Adv. Space Res. 1999;24:821–827.

23. Maetzel A, Mäkelä M, Hawker G, Bombardier C. Osteoarthritis of the hip and knee and mechanical occupational exposure--a systematic overview of the evidence. J. Rheumatol. 1997;24:1599–1607.

24. Felson DT. Preventing knee and hip osteoarthritis. Bull. Rheum. Dis. 1998;47:1–4.

25. Buckwalter JA, Martin JA. Sports and osteoarthritis. Curr. Opin. Rheumatol. 2004;16:634–639.

26. McGlashan SR, Jensen CG, Poole CA. Localization of extracellular matrix receptors on the chondrocyte primary cilium. J. Histochem. Cytochem. 2006;54:1005–1014.

27. Adams MA. The mechanical environment of chondrocytes in articular cartilage. Biorheology 2006;43:537–545.

28. Smith RL, Carter DR, Schurman DJ. Pressure and shear differentially alter human articular chondrocyte metabolism: a review. Clin. Orthop. Relat. Res. 2004; S89–95.

29. Kim HT, Lo MY, Pillarisetty R. Chondrocyte apoptosis following intraarticular fracture in humans. Osteoarth. Cartil. 2002;10:747–749.

30. Martin JA, Brown T, Heiner A, Buckwalter JA. Post-traumatic osteoarthritis: the role of accelerated chondrocyte senescence. Biorheology 2004;41:479–491.

31. Marsh JL, Buckwalter J, Gelberman R, Dirschl D, Olson S, Brown T, Llinias A. Articular fractures: does an anatomic reduction really change the result? J. Bone Joint Surg. Am. 2002;84:1259–1271.

32. MacGregor AJ, Antoniades L, Matson M, Andrew T, Spector TD. The genetic contribution to radiographic hip osteoarthritis in women: results of a classic twin study. Arthritis Rheum. 2000;43:2410–2416.

33. Spector TD, Cicuttini F, Baker J, Loughlin J, Hart D. Genetic influences on osteoarthritis in women: a twin study. BMJ 1996;312:940–943.

34. Sambrook PN, MacGregor AJ, Spector TD. Genetic influences on cervical and lumbar disc degeneration: a magnetic resonance imaging study in twins. Arthritis Rheum. 1999;42:366–372.

35. Loughlin J. Polymorphism in signal transduction is a major route through which osteoarthritis susceptibility is acting. Curr. Opin. Rheumatol. 2005;17:629–633.

36. Valdes AM, Loughlin J, Oene MV, Chapman K, Surdulescu GL, Doherty M, Spector TD. Sex and ethnic differences in the association of ASPN, CALM1, COL2A1, COMP, and FRZB with genetic susceptibility to osteoarthritis of the knee. Arthritis Rheum. 2007;56:137–146.

37. Li Y, Xu L, Olsen BR. Lessons from genetic forms of osteoarthritis for the pathogenesis of the disease. Osteoarth. Cartil. 2007;15:1101–1105.

38. Poole AR. Cartilage in health and disease. In Arthritis and Allied Conditions: A Textbook of Rheumatology. 14th Edition. Koopman W, ed. 2001. Philadelphia, PA: Lippincott Williams & Wilkins.

39. Aigner T, Soeder S, Haag J. IL-1beta and BMPs—Interactive players of cartilage matrix degradation and regeneration. Eur. Cell. Mater. 2006;12:49–56.

40. Kühn K, D'Lima DD, Hashimoto S, Lotz M. Cell death in cartilage. Osteoarth. Cartil. 2004;12:1–16.

41. McQuillan DJ, Handley CJ, Campbell MA, Bolis S, Milway VE, Herington AC. Stimulation of proteoglycan biosynthesis by serum and insulin-like growth factor-I in cultured bovine articular cartilage. Biochem. J. 1986;240:423–430.

42. Trippel SB, Corvol MT, Dumontier MF, Rappaport R, Hung HH, Mankin HJ. Effect of somatomedin-C/ insulin-like growth factor I and growth hormone on cultured growth plate and articular chondrocytes. Pediatr. Res. 1989;25:76–82.

43. Luyten FP, Hascall VC, Nissley SP, Morales TI, Reddi AH. Insulin-like growth factors maintain steady state metabolism of proteoglycans in bovine articular cartilage explants. Arch. Biochem. Biophys. 1988;267:416–425.

44. Matsumoto T, Gargosky SE, Iwasaki K, Rosenfeld RG. Identification and characterization of insulin-like growth factors (IGFs), IGF-binding proteins (IGF-BPs), and IGFBP proteases in human synovial fluid. J. Clin. Endocrinol. Metab. 1996;81:150–155.

45. Fernihough JK, Billingham ME, Cwyfan-Hughes S, Holly JM. Local disruption of the insulin-like growth factor system in the arthritic joint. Arthritis Rheum. 1996;39:1556–1565.

46. Tavera C, Abribat T, Reboul P, Dore S, Brazeau P, Pelletier JP, Martel-Pelletier J. IGF and IGF-binding protein system in the synovial fluid of osteoarthritic and rheumatoid arthritic patients. Osteoarth. Cartil. 1996;4:263–274.

47. Chevalier X, Tyler JA. Production of binding proteins and role of the insulin-like growth factor I binding protein 3 in human articular cartilage explants. Br. J. Rheumatol. 1996;35:515–522.

48. Morales TI. The insulin-like growth factor binding proteins in uncultured human cartilage: increases in insulin-like growth factor binding protein 3 during osteoarthritis. Arthritis Rheum. 2002;46:2358–2367.

49. Eviatar T, Kauffman H, Maroudas A. Synthesis of insulin-like growth factor binding protein 3 in vitro in human articular cartilage cultures. Arthritis Rheum. 2003;48:410–417.

50. Ballock RT, Heydemann A, Izumi T, Reddi AH. Regulation of the expression of the type-II collagen gene in periosteum-derived cells by three members of the transforming growth factor-beta superfamily. J. Orthop. Res. 1997;15:463–467.

51. Izumi T, Scully SP, Heydemann A, Bolander ME. Transforming growth factor beta 1 stimulates type II collagen expression in cultured periosteum-derived cells. J. Bone Miner. Res. 1992;7:115–121.

52. van Beuningen HM, van der Kraan PM, Arntz OJ, van den Berg WB. In vivo protection against interleukin-1-induced articular cartilage damage by transforming growth factor-beta 1: Age-related differences. Ann. Rheum. Dis. 1994;53:593–600.

53. van Beuningen HM, van der Kraan PM, Arntz OJ, van den Berg WB. Transforming growth factor-beta 1 stimulates articular chondrocyte proteoglycan synthesis and induces osteophyte formation in the murine knee joint. Lab. Invest. 1994;71:279–290.

54. Edwards DR, Murphy G, Reynolds JJ, Whitham SE, Docherty AJ, Angel P, Heath JK. Transforming growth factor beta modulates the expression of collagenase and metalloproteinase inhibitor. EMBO J. 1987;6:1899–1904.

55. Blancy Davidson EN, Scharstuhl A, Vitters EL, van der Kraan PM, van den Berg WB. Reduced transforming growth factor-beta signaling in cartilage of old mice: Role in impaired repair capacity. Arthritis Res. Ther. 2005;7:R1338–1347.

56. Rountree RB, Schoor M, Chen H, Marks ME, Harley V, Mishina Y, Kingsley DM. BMPreceptor signaling is required for postnatal maintenance of articular cartilage. PloS. Biol. 2004;2:e355.

57. Soder S, Hakimiyan A, Rueger DC, Kuettner KE, Aigner T, Chubinskaya S. Antisense inhibition of osteogenic protein 1 disturbs human articular cartilage integrity. Arthritis Rheum. 2005;52:468–478.

58. Goldring MB. Osteoarthritis and cartilage: the role of cytokines. Curr. Rheumatol. Rep. 2000;2:459–465.

59. Glansbeek HL, van Beuningen HM, Vitters EL, Morris EA, van der Kraan PM, van den Berg WB. Bone morphogenetic protein 2 stimulates articular cartilage proteoglycan synthesis in vivo but does not counteract interleukin-1alpha effects on proteoglycan synthesis and content. Arthritis Rheum 1997;40:1020–1028.

60. Massague J, Seoane J, Wotton D. Smad transcription factors. Genes Dev 2005; 19:2783–2810.

61. Kato Y, Gospodarowicz D. Sulfated proteoglycan synthesis by confluent cultures of rabbit costal chondrocytes grown in the presence of fibroblast growth factor. J. Cell. Biol. 1985;100:477–485.

62. Kato Y, Nomura Y, Daikuhara Y, Nasu N, Tsuji M, Asada A, Suzuki F. Cartilage-derived factor (CDF) I. Stimulation of proteoglycan synthesis in rat and rabbit costal chondrocytes in culture. Exp. Cell Res. 1980;130:73–81.

63. Kato Y, Hiraki Y, Inoue H, Kinoshita M, Yutani Y, Suzuki F. Differential and synergistic actions of somatomedin-like growth factors, fibroblast growth factor and epidermal growth factor in rabbit costal chondrocytes. Eur. J. Biochem. 1983;129:685–690.

64. Jentzsch KD, Wellmitz G, Heder G, Petzold E, Buntrock P, Oehme P. A bovine brain fraction with fibroblast growth factor activity inducing articular cartilage regeneration in vivo. Acta Biol. Med. Ger. 1980;39:967–971.

65. Cuevas P, Burgos J, Baird A. Basic fibroblast growth factor (FGF) promotes cartilage repair in vivo. Biochem. Biophys. Res. Commun. 1988;156:611–618.

66. Jones KL, Addison J. Pituitary fibroblast growth factor as a stimulator of growth in cultured rabbit articular chondrocytes. Endocrinology 1975;97:359–365.

67. Vincent T, Hermansson M, Bolton M, Wait R, Saklatvala J. Basic FGF mediates an immediate response of articular cartilage to mechanical injury. Proc. Natl. Acad. Sci. U. S. A. 2002;99:8259–8264.

68. Vincent TL, Hermansson MA, Hansen UN, Amis AA, Saklatvala J. Basic fibroblast growth factor mediates transduction of mechanical signals when articular cartilage is loaded. Arthritis Rheum 2004;50:526–533.

69. Stetler-Stevenson WG, Yu AE. Proteases in invasion: matrix metalloproteinases. Semin Cancer Biol 2001;11:143–152.

70. Murphy G, Knauper V, Atkinson S, Butler G, English W, Hutton M, Stracke J, Clark I. Matrix metalloproteinases in arthritic disease. Arthritis Res 2002;4:S39–S49.

71. Borden P, Solymar D, Sucharczuk A, Lindman B, Cannon P, Heller RA. Cytokine control of interstitial collagenase and collagenase-3 gene expression in human chondrocytes. J. Biol. Chem. 1996;271:23577–23581.

72. Mengshol JA, Vincenti MP, Coon CI, Barchowsky A, Brinckerhoff CE. Interleukin-1 induction of collagenase 3 (matrix metalloproteinase 13) gene expression in chondrocytes requires p38, c-Jun Nterminal kinase, and nuclear factor kappaB: differential regulation of collagenase 1 and collagenase 3. Arthritis Rheum. 2000;43:801–811.

73. Brinckerhoff CE, Rutter JL, Benbow U. Interstitial collagenases as markers of tumor progression. Clin. Cancer Res. 2000;6:4823–4830.

74. Shlopov BV, Lie WR, Mainardi CL, Cole AA, Chubinskaya S, Hasty KA. Osteoarthritic lesions: involvement of three different collagenases. Arthritis Rheum. 1997;40:2065–2074.

75. Mitchell PG, Magna HA, Reeves LM, Lopresti-Morrow LL, Yocum SA, Rosner PJ, Geoghegan KF, Hambor JE. Cloning, expression, and type II collagenolytic activity of matrix metalloproteinase-13 from human osteoarthritic cartilage. J. Clin. Invest. 1996;97:761–768.

76. Neuhold LA, Killar L, Zhao W, Sung ML, Warner L, Kulik J, Turner J, Wu W, Billinghurst C, Meijers T, Poole AR, Babij P, DeGennaro LJ. Postnatal expression in hyaline cartilage of constitutively active human collagenase-3 (MMP-13) induces osteoarthritis in mice. J. Clin. Invest. 2001;107:35–44.

77. Cal S, Obaya AJ, Llamazares M, Garabaya C, Quesada V, Lopez-Otin C. Cloning, expression analysis, and structural characterization of seven novel human ADAMTSs, a family of metalloproteinases with disintegrin and thrombospondin-1 domains. Gene 2002;283:49–62.

78. Somerville RP, Longpre JM, Jungers KA, Engle JM, Ross M, Evanko S, Wight TN, Leduc R, Apte SS. Characterization of ADAMTS-9 and ADAMTS-20 as a distinct ADAMTS subfamily related to Caenorhabditis elegans GON-1. J. Biol. Chem. 2003;278:9503–9513.

79. Tortorella MD, Burn TC, Pratta MA, Abbaszade I, Hollis JM, Liu R, Rosenfeld SA, Copeland RA, Decicco CP, Wynn R, et al. Purification and cloning of aggrecanase-1: a member of the ADAMTS family of proteins. Science 1999;284:1664–1666.

80. Abbaszade I, Liu RQ, Yang F, Rosenfeld SA, Ross OH, Link JR, Ellis DM, Tortorella MD, Pratta MA, Hollis JM, et al. Cloning and characterization of ADAMTS11, an aggrecanase from the ADAMTS family. J. Biol. Chem. 1999;274:23443–23450.

81. Kuno K, Okada Y, Kawashima H, Nakamura H, Miyasaka M, Ohno H, Matsushima K. ADAMTS-1 cleaves a cartilage proteoglycan, aggrecan. FEBS Lett. 2000;478:241–245.

82. Glasson SS, Askew R, Sheppard B, Carito BA, Blanchet T, Ma HL, Flannery CR, Kanki K, Wang E, Peluso D, et al. Characterization of and osteoarthritis susceptibility in ADAMTS- 4-knockout mice. Arthritis Rheum. 2004;50:2547–2558.

83. Glasson SS, Askew R, Sheppard B, Carito B, Blanchet T, Ma HL, Flannery CR, Peluso D, Kanki K, Yang Z, et al. Deletion of active ADAMTS5 prevents cartilage degradation in a murine model of osteoarthritis. Nature 2005;434:644–648.

84. Stanton H, Rogerson FM, East CJ, Golub SB, Lawlor KE, Meeker CT, Little CB, Last K, Farmer PJ, Campbell IK, et al. ADAMTS5 is the major aggrecanase in mouse cartilage in vivo and in vitro. Nature 2005;434:648–652.

85. Henrotin Y, Bruckner P, Pujol JP. The role of reactive oxygen species in homeostasis and degradation of cartilage. Osteoarth. Cartil. 2003;11:747–755.

86. Sakurai H, Kohsaka H, Liu MF, Higashiyama H, Hirata T, Kanno K Saito I, Miyasaka N. Nitric oxide production and inducible nitric oxide synthase expression in inflammatory arthritides. J. Clin. Invest. 1995;96:2357-2363.

87. Goldring MB. The role of cytokines as inflammatory mediators in osteoarthritis: lessons from animal models. Connect Tissue Res. 1999;40:1–11.

88. Tetlow LC, Adlam DJ, Woolley DE. Matrix metalloproteinase and proinflammatory cytokine production by chondrocytes of human osteoarthritic cartilage: associations with degenerative changes. Arthritis Rheum. 2001;44:585–594.

89. Fan Z, Bau B, Yang H, Aigner T. IL-1beta induction of IL-6 and LIF in normal articular human chondrocytes involves the ERK, p38 and NFkappaB signaling pathways. Cytokine 2004;28:17–24.

90. Heinrich PC, Behrmann I, Muller-Newen G, Schaper F, Graeve L. Interleukin-6-type cytokine signalling through the gp130/Jak/STAT pathway. Biochem. J. 1998;334:297–314.

91. Flannery CR, Little CB, Hughes CE, Curtis CL, Caterson B, Jones SA. IL-6 and its soluble receptor augment aggrecanase-mediated proteoglycan catabolism in articular cartilage. Matrix Biol. 2000;19:549–553.

92. Hardy MM, Seibert K, Manning PT, Currie MG, Woerner BM, Edwards D, Koki A, Tripp CS. Cyclooxygenase 2–dependent prostaglandin E2 modulates cartilage proteoglycan degradation in human osteoarthritis explants. Arthritis Rheum. 2002;46:1789–803.

93. Martel-Pelletier J, Pelletier JP, Fahmi H. Cyclooxygenase-2 and prostaglandins in articular tissues. Semin Arthritis Rheum. 2003;33:155–167.

94. Mathy-Hartert M, Burton S, Deby-Dupont G, Devel P, Reginster JY, Henrotin Y. Influence of oxygen tension on nitric oxide and prostaglandin E2 synthesis by bovine chondrocytes. Osteoarth Cartil 2005;13:74–9.

95. Goldring MB, Suen LF, Yamin R, Lai WF. Regulation of collagen gene expression by prostaglandins and interleukin-1beta in cultured chondrocytes and fibroblasts. Am. J. Ther. 1996;3:9–16.

96. Fulkerson JP, Damiano P. Effect of prostaglandin E2 on adult pig articular cartilage slices in culture. Clin. Orthop. Relat. Res. 1983: 266–269.

97. Lippiello L, Yamamoto K, Robinson D, Mankin HJ. Involvement of prostaglandins from rheumatoid synovium in inhibition of articular cartilage metabolism. Arthritis Rheum. 1978;21:909–917.

98. Clark CA, Schwarz EM, Zhang X, Ziran NM, Drissi H, O'Keefe RJ, Zuscik MJ. Differential regulation of EP receptor isoforms during chondrogenesis and chondrocyte maturation. Biochem. Biophys. Res. Commun. 2005;328:764–776.

99. Miyamoto M, Ito H, Mukai S, Kobayashi T, Yamamoto H, Kobayashi M, Maruyama T, Akiyama H, Nakamura T. Simultaneous stimulation of EP2 and EP4 is essential to the effect of prostaglandin E2 in chondrocyte differentiation. Osteoarth. Cartil. 2003;11:644–652.

100. Aoyama T, Liang B, Okamoto T, Matsusaki T, Nishijo K, Ishibe T, Yasura K, Nagayama S, Nakayama T, Nakamura T, et al. PGE2 signal through EP2 promotes the growth of articular chondrocytes. J. Bone Miner. Res. 2005;20:377–389.

101. Sugimoto Y, Narumiya S. 2007. Prostaglandin E receptors. J. Biol. Chem. 282:11613–11617.

102. Hata, A. N., and R. M. Breyer. 2004. Pharmacology and signaling of prostaglandin receptors: multiple roles in inflammation and immune modulation. Pharmacol. Ther. 103:147–166.

103. Attur M, Al-Mussawir HE, Patel J, Kitay A, Dave M, Palmer G, Pillinger MH, Abramson SB. Prostaglandin E2 exerts catabolic effects in osteoarthritis cartilage: evidence for signaling via the EP4 receptor. J. Immunol. 2008;181:5082–5088.

104. Li X, Ellman M, Muddasani P, Wang JH, Cs-Szabo G, van Wijnen AJ, Im HJ. Prostaglandin E(2) and its cognate EP receptors control human adult articular cartilage homeostasis and are linked to the pathophysiology of osteoarthritis. Arthritis Rheum. 2009; 60:513–523.

105. Clancy RM, Gomez PF, Abramson SB. Nitric oxide sustains nuclear factor kappaB activation in cytokine stimulated chondrocytes. Osteoarth. Cartil. 2004;12:552–558.

106. Lo YY, Conquer JA, Grinstein S, Cruz TF. Interleukin-1 beta induction of c-fos and collagenase expression in articular chondrocytes: involvement of reactive oxygen species. J. Cell Biochem. 1998;69:19–29.

107. Henrotin YE, Bruckner P, Pujol JP. The role of reactive oxygen species in homeostasis and degradation of cartilage. Osteoarth. Cartil. 2003;11:747–755.

108. Mates JM, Perez-Gomez C, Nunez de Castro I. Antioxidant enzymes and human diseases. Clin. Biochem. 1999;32:595–603.

109. Loeser RF, Carlson CS, Del Carlo M, Cole A. Detection of nitrotyrosine in aging and OA cartilage: correlation of oxidative damage with the presence of interleukin-1beta and with chondrocyte resistance to insulin-like growth factor 1. Arthritis Rheum. 2002;46:2349–57.

110. Bae SC, Kim SJ, Sung MK. Inadequate antioxidant nutrient intake and altered plasma antioxidant status of rheumatoid arthritis patients. J. Am. Coll. Nutr. 2003;22:311–315.

111. Henrotin Y, Deberg M, Dubuc JE, Quettier E, Christgau S, Reginster JY. Type II collagen peptides for measuring cartilage degradation. Biorheology 2004;41:543–547.

112. Tiku ML, Gupta S, Deshmukh D. Aggrecan degradation in chondrocytes is mediated by reactive oxygen species and protected by antioxidants. Free Radic. Res. 1999;30:395–405.

113. Tiku ML, Shah R, Allison GT. Evidence linking chondrocyte lipid peroxidation to cartilage matrix protein degradation. Possible role in cartilage aging and the pathogenesis of osteoarthritis. J. Biol. Chem. 2000;275:20069–20076.

114. Mendes AF, Caramona MM, Carvalho AP, Lopes MC. Hydrogen peroxide mediates interleukin-1beta-induced AP-1 activation in articular chondrocytes: implications for the regulation of iNOS expression. Cell Biol. Toxicol. 2003;19:203–214.

115. Tiku ML, Allison GT, Naik K, Karry SK. Malondialdehyde oxidation of cartilage collagen by chondrocytes. Osteoarth. Cartil. 2003;11:159–166.

116. Loeser RF, Carlson CS, Del Carlo M, Cole A. Detection of nitrotyrosine in aging and osteoarthritic cartilage: correlation of oxidative damage with the presence of interleukin-1beta and with chondrocyte resistance to insulin-like growth factor 1. Arthritis Rheum. 2002;46:2349–2357.

117. Henrotin Y, Kurz B, Aigner T. Oxygen and reactive oxygen species in cartilage degradation: friends or foes? Osteoarth. Cartil. 2005;13:643–654.

118. Afonso V, Champy R, Mitrovic D, Collin P, Lomri A. Reactive oxygen species and superoxide dismutases: role in joint diseases. Joint Bone Spine 2007;74:324–329.

119. Farrell AJ, Blake DR. Nitric oxide. Ann. Rheum. Dis. 1996;55:7–20.

120. Cao M, Westerhausen-Larson A, Niyibizi C, Hebda PA, Stefanovic-Racic M, Evans CH. Nitric oxide inhibits the synthesis of type-II collagen without altering Col2A1mRNA abundance: prolyl hydroxylase as a possible target. Biochem. J. 1997;324:305–310.

121. Taskiran D, Stefanovic-Racic M, Georgescu H, Evans C. Nitric oxide mediates suppression of cartilage proteoglycan synthesis by interleukin-1. Biochem. Biophys. Res. Commun. 1994;200:142–148.

122. Blanco FJ, Ochs RL, Schwarz H, Lotz M. Chondrocyte apoptosis induced by nitric oxide. Am. J. Pathol. 1995;146:75–85.

123. Clancy RM, Amin AR, Abramson SB. The role of nitric oxide in inflammation and immunity. Arthritis Rheum. 1998;41:1141–1151.

124. Sasaki K, Hattori T, Fujisawa T, Takahashi K, Inoue H, Takigawa M. Nitric oxide mediates interleukin-1-induced gene expression of matrix metalloproteinases and basic fibroblast growth factor in cultured rabbit articular chondrocytes. J. Biochem. 1998;123:431–439.

125. Pelletier JP, Mineau F, Ranger P, Tardif G, Martel-Pelletier J. The increased synthesis of inducible nitric oxide inhibits IL-1Ra synthesis by human articular chondrocytes: possible role in osteoarthritic cartilage degradation. Osteoarth. Cartil. 1996;4:77–84.

126. Abramson SB, Attur M, Amin AR, Clancy R. Nitric oxide and inflammatory mediators in the perpetuation of osteoarthritis. Curr. Rheumatol. Rep. 2001;3:535–541.

127. Loeser RF, Carlson CS, Del Carlo M, Cole A. Detection of nitrotyrosine in aging and osteoarthritic cartilage: Correlation of oxidative damage with the presence of interleukin-1_ and with chondrocyte resistance to insulin-like growth factor 1. Arthritis Rheum. 2002;46:2349–2357.

128. Zhao Q, Eberspaecher H, Lefebvre V, De Crombrugghe B. Parallel expression of Sox9 and Col2a1 in cells undergoing chondrogenesis. Dev. Dyn. 1997;209:377–386.

129. Goldring MB, Birkhead J, Sandell LJ, Kimura T, Krane SM. Interleukin 1 suppresses expression of cartilage-specific types II and IX collagens and increases types I and III collagens in human chondrocytes. J. Clin. Invest. 1988;82:2026–2037.

130. Lefebvre V, Peeters-Joris C, Vaes G. Modulation by interleukin 1 and tumor necrosis factor alpha of production of collagenase, tissue inhibitor of metalloproteinases and collagen types in differentiated and dedifferentiated articular chondrocytes. Biochim. Biophys. Acta 1990;1052:366–378.

131. Lum ZP, Hakala BE, Mort JS, Recklies AD. Modulation of the catabolic effects of interleukin-1 beta on human articular chondrocytes by transforming growth factor-beta. J. Cell Physiol. 1996;166:351–359.

132. Murakami S, Lefebvre V, de Crombrugghe B. Potent inhibition of the master chondrogenic factor Sox9 gene by interleukin-1 and tumor necrosis factor-alpha. J. Biol. Chem. 2000;275:3687–3692.

133. Sitcheran R, Cogswell PC, Baldwin AS, Jr. NF-kappaB mediates inhibition of mesenchymal cell differentiation through a posttranscriptional gene silencing mechanism. Genes Dev. 2003;17:2368–2373.

134. Legendre F, Dudhia J, Pujol JP, Bogdanowicz P. JAK/STAT but not ERK1/ERK2 pathway mediates interleukin (IL)-6/soluble IL-6R down-regulation of Type II collagen, aggrecan core, and link protein transcription in articular chondrocytes. Association with a down-regulation of SOX9 expression. J. Biol. Chem. 2003;278:2903–2912.

135. Lorenz H, Wenz W, Ivancic M, Steck E, Richter W. Early and stable upregulation of collagen type II, collagen type I and YKL40 expression levels in cartilage during early experimental osteoarthritis occurs independent of joint location and histological grading. Arthritis Res. Ther. 2005;7:R156–R165.

136. Vincenti MP, Brinckerhoff CE. Transcriptional regulation of collagenase (MMP-1, MMP-13) genes in arthritis: integration of complex signaling pathways for the recruitment of gene-specific transcription factors. Arthritis Res. 2002;4:157–164.

137. Lianxu C, Hongti J, Changlong Y. NF-kappaBp65-specific siRNA inhibits expression of genes of COX-2, NOS-2 and MMP-9 in rat IL-1beta-induced and TNF-alpha-induced chondrocytes. Osteoarth. Cartil. 2006;14:367–376.

138. Bondeson J, Lauder S, Wainwright S, Amos N, Evans A, Hughes CE, Feldmann M, Caterson B. Adenoviral gene transfer of the endogenous inhibitor IkappaBalpha into human osteoarthritis synovial fibroblasts demonstrates that several matrix metalloproteinases and aggrecanases are nuclear factor-kappaB-dependent. J. Rheumatol. 2007;34:523–533.

139. Chowdhury TT, Salter DM, Bader DL, Lee DA. Signal transduction pathways involving p38 MAPK, JNK, NFkappaB and AP-1 influences the response of chondrocytes cultured in agarose constructs to IL-1beta and dynamic compression. Inflamm. Res. 2008;57:306–313.

140. Gabay O, Gosset M, Levy A, Salvat C, Sanchez C, Pigenet A, Sautet A, Jacques C, Berenbaum F. Stress-induced signaling pathways in hyalin chondrocytes: inhibition by avocado-soybean unsaponifiables (ASU). Osteoarth. Cartil. 2008;16:373–384.

141. Verzijl N, DeGroot J, Thorpe SR, Bank RA, Shaw JN, Lyons TJ, Bijlsma JW, Lafeber FP, Baynes JW, TeKoppele JM. Effect of collagen turnover on the accumulation of advanced glycation end products. J. Biol. Chem. 2000;275:39027–39031.

142. Maroudas A, Bayliss MT, Uchitel-Kaushansky N, Schneiderman R, Gilav E. Aggrecan turnover in human articular cartilage: use of aspartic acid racemization as a marker of molecular age. Arch. Biochem. Biophys. 1998;350:61–71.

143. Corvol MT. the chondrocyte: from cell aging to osteoarthritis. Joint Bone Spine 2000;67:557–560.

144. Poole AR, Guilak F, Abramson SB. Etiopathogenesis of osteoarthritis. In: Osteoarthritis: Diagnosis and Medical/Surgical Management. 4th edition. Moskowitz RW, Altman RW, Hochberg MC, Buckwalter JA, Goldberg VM, eds. 2007. Philadelphia, PA: Lippincott, Williams, and Wilkins.

145. Pritzker KP, Gay S, Jimenez SA, Ostergaard K, Pelletier JP, Revell PA, Salter D, van den Berg WB. Osteoarthritis cartilage histopathology: grading and staging. Osteoarth. Cartil. 2006;14:13–29.

146. Sandell LJ, Aigner T. Articular cartilage and changes in arthritis. An introduction: cell biology of osteoarthritis. Arthritis Res. 2001;3:107–113.

147. Tchetina EV, Squires G, Poole AR. Increased type II collagen degradation and very early focal cartilage degeneration is associated with upregulation of chondrocyte differentiation related genes in early human articular cartilage lesions. J. Rheumatol. 2005;32:876–886.

148. A.G. Rothwell, G. Bentley, Chondrocyte multiplication in osteoarthritic articular cartilage. J. Bone Jt. Surg. 1973;588–594.

149. A. Hulth, L. Lindberg, H. Telhag, Mitosis in human articular osteoarthritic cartilage, Clin. Orthop. 84; 1972;197–199.

150. Aigner T, Hemmel M, Neureiter D, Gebhard PM, Zeiler G, Kirchner T, McKenna L. Apoptotic cell death is not a widespread phenomenon in normal aging and osteoarthritis human articular knee cartilage: a study of proliferation, programmed cell death (apoptosis), and viability of chondrocytes in normal and osteoarthritic human knee cartilage. Arthritis Rheum. 2001;44:1304–1312.

151. Meachim G, Collins DH. Cell counts of normal and osteoarthritic articular cartilage in relation to the uptake of sulphate (35SO4) in vitro. Ann. Rheum. Dis. 1962;21:45–50.

152. Rothwell AG, Bentley G. Chondrocyte multiplication in osteoarthritic articular cartilage. J. Bone Joint Surg. Br. 1973;55:588–894.

153. Kühn K, D'Lima DD, Hashimoto S, Lotz M. Cell death in cartilage. Osteoarth. Cartil. 2004;12:1–16.

154. D'Lima D, Hermida J, Hashimoto S, Colwell C, Lotz M. Caspase inhibitors reduce severity of cartilage lesions in experimental osteoarthritis. Arthritis Rheum. 2006;54, 1814–1821.

155. Aigner T, Kim HA, Roach H. Apoptosis in osteoarthritis, Rheum. Dis. Clin. North Am. 2004;30:639–653.

156. Blanco FJ, Guttian R, Va'zquez-Martul E, de Toro FJ, Galdo F. Osteoarthritic chondrocytes die by apoptosis. Arthritis Rheum. 1998;41:284–289.

157. Hashimoto S, Ochs RL, Komiya S, Lotz M. Linkage of chondrocyte apoptosis and cartilage degradation in human osteoarthritis. Arthritis Rheum. 1998;41:1632–1638.

158. Heraud F, Heraud A, Harmand MF. Apoptosis in normal and osteoarthritic human articular cartilage. Ann. Rheum. Dis. 2000;59:959–965.

159. Aigner T, Hemmel M, Neureiter D, Gebhard PM, Zeiler G, Kirchner T, McKenna, LA. Apoptotic cell death is not a widespread phenomenon in normal aging and osteoarthritic human articular knee cartilage: a study of proliferation, programmed cell death (apoptosis), and viability of chondrocytes in normal and osteoarthritic human knee cartilage, Arthritis Rheum. 2001;44:1304–1312.

160. Ayral X, Dougados M, Listrat V, Bonvarlet JP, Simonnet J, Amor B. Arthroscopic evaluation of chondropathy in osteoarthritis of the knee. J. Rheumatol. 1996;3:698–706.

161. Pelletier JP, Martel-Pelletier J, Abramson SB. Osteoarthritis, an inflammatory disease: potential implication for the selection of new therapeutic targets. Arthritis Rheum. 2001;44:1237–1247.

162. Benito MJ, Veale DJ, FitzGerald O, van den Berg WB, Bresnihan B. Synovial tissue inflammation in early and late osteoarthritis. Ann. Rheum. Dis. 2005;64:1263–1267.

163. Masuhara K, Nakai T, Yamaguchi K, Yamasaki S, Sasaguri Y. Significant increases in serum and plasma concentrations of matrix metalloproteinases 3 and 9 in patients with rapidly destructive osteoarthritis of the hip. Arthritis Rheum. 2002;46:2625–2631.

164. Tchetverikov I, Lohmander LS, Verzijl N, Huizinga TW, TeKoppele JM, Hanemaaijer R, DeGroot J. MMP protein and activity levels in synovial fluid from patients with joint injury, inflammatory arthritis, and osteoarthritis. Ann. Rheum. Dis. 2005;64:694–698.

165. Barksby HE, Milner JM, Patterson AM, Peake NJ, Hui W, Robson T, Lakey R, Middleton J, Cawston TE, Richards CD, et al. Matrix metalloproteinase 10 promotion of

collagenolysis via procollagenase activation: implications for cartilage degradation in arthritis. Arthritis Rheum. 2006;54:3244–3253.

166. Hill CL, Gale DG, Chaisson CE, Skinner K, Kazis L, Gale ME, Felson DT. Knee effusions, popliteal cysts, and synovial thickening: association with knee pain in osteoarthritis. J. Rheumatol. 2001;28:1330–1337.

167. Burr DB. Anatomy and physiology of the mineralized tissues: role in the pathogenesis of osteoarthrosis. Osteoarth. Cartil. 2004;12:S20–S30.

168. Lajeunesse D, Hilal G, Pelletier JP, Martel-Pelletier J. Subchondral bone morphological and biochemical alterations in osteoarthritis. Osteoarth. Cartil. 1999;7:321–322.

169. Dequeker J, Mohan S, Finkelman RD, Aerssens J, Baylink DJ. Generalized osteoarthritis associated with increased insulin-like growth factor types I and II and transforming growth factor beta in cortical bone from the iliac crest. Possible mechanism of increased bone density and protection against osteoporosis. Arthritis Rheum. 1993;36:1702–1708.

170. Mansell JP, Bailey AJ. Abnormal cancellous bone collagen metabolism in osteoarthritis. J Clin. Invest. 1998;101:1596–1603.

171. Sanchez C, Deberg MA, Bellahcene A, Castronovo V, Msika P, Delcour JP, Crielaard JM, Henrotin YE. Phenotypic characterization of osteoblasts from the sclerotic zones of osteoarthritic subchondral bone. Arthritis Rheum. 2008;58:442–455.

172. Sanchez C, Deberg MA, Piccardi N, Msika P, Reginster JY, Henrotin YE. Subchondral bone osteoblasts induce phenotypic changes in human osteoarthritic chondrocytes. Osteoarth. Cartil. 2005;13:988–997.

173. Sanchez C, Gabay O, Salvat C, Henrotin Y, Berenbaum F. Mechanical loading highly increases IL-6 production and decreases OPG expression by osteoblasts. Osteoarth. Cartil. 2009;17:473–481.

174. Westacott CI, Webb GR, Warnock MG, Sims JV, Elson CJ. Alteration of cartilage metabolism by cells from osteoarthritic bone. Arthritis Rheum. 1997;40:1282–1291.

175. van Beuningen HM, van der Kraan PM, Arntz OJ, van den Berg WB. Transforming growth factor-beta 1 stimulates articular chondrocyte proteoglycan synthesis and induces osteophyte formation in the murine knee joint. /??/Laboratory investigation; a journal of technical methods and pathology. 1994;71:279–290.

176. Ghosh P, Cheras PA. Vascular mechanisms in osteoarthritis. Best Pract. Res. Clin. Rheumatol. 2001;15:693–709.

177. Burr DB, Schaffler MB. The involvement of subchondral mineralized tissues in osteoarthrosis: quantitative microscopic evidence. Microsc. Res. Tech. 1997;37:343–357.

178. Walsh DA, Bonnet CS, Turner EL, Wilson D, Situ M, McWilliams DF. Angiogenesis in the synovium and at the osteochondral junction in osteoarthritis. Osteoarth. Cartil. 2007;15:743–751.

179. Bullough PG. The role of joint architecture in the etiology of arthritis. Osteoarth. Cartil. 2004;12:S2–S9.

180. Messent EA, Ward RJ, Tonkin CJ, Buckland-Wright C. Differences in trabecular structure between knees with and without osteoarthritis quantified by macro and standard radiography, respectively. Osteoarth. Cartil. 2006;14:1302–1305.

181. Aigner T, Bartnik E, Sohler F, Zimmer R. Functional genomics of osteoarthritis: on the way to evaluate disease hypotheses. Clin. Orthop. Relat. Res. 2004; 427:S138–143.

182. Brown AN, McKinley TO, Bay BK. Trabecular bone strain changes associated with subchondral bone defects of the tibial plateau. J. Orth. Traum. 2002;16:638–643.

183. Carlson CS, Loeser RF, Purser CB, Gardin JF, Jerome CP. Osteoarthritis in cynomolgus macaques. III: effects of age, gender, and subchondral bone thickness on the severity of disease. J. Bone Miner. Res. 1996;11:1209–1217.

184. Wollheim FA, Lohmander LS. Pathology and animal models of osteoarthritis. In: Osteoarthritis. A Companion to Rheumatology. 1st edition. Sharma L, Berenbaum F, eds. 2007. New York: Elsevier.

185. Goldring MB, Goldring SR. Osteoarthritis. J. Cell Physiol. 2007;213:626–634.

186. Glasson SS. In vivo osteoarthritis target validation utilizing genetically-modified mice. Curr. Drug Targets 2007;8:367–376.

187. Clements KM, Price JS, Chambers MG, Visco DM, Poole AR, Mason RM. Gene deletion of either interleukin-1beta, interleukin-1beta-converting enzyme, inducible nitric oxide synthase, or stromelysin 1 accelerates the development of knee osteoarthritis in mice after surgical transection of the medial collateral ligament and partial medial meniscectomy. Arthritis Rheum. 2003;48:3452–3463.

188. Kamekura S, Hoshi K, Shimoaka T, Chung U, Chikuda H, Yamada T, Uchida M, Ogata N, Seichi A, Nakamura K, et al. Osteoarthritis development in novel experimental mouse models induced by knee joint instability. Osteoarth. Cartil. 2005;13:632–641.

189. Peterfy CG. Scratching the surface: articular cartilage disorders in the knee. Magn. Reson. Imaging Clin. North Am. 2000;8:409–430.

190. Abramson S, Krasnokutsky S. Biomarkers in osteoarthritis. Bull. NYU Hosp. Jt. Dis. 2006;64:77–81.

191. Blumenkrantz G, Majumdar S. Quantitative magnetic resonance imaging of articular cartilage in osteoarthritis. Eur. Cell Mater. 2007;13:76–86.

192. Burstein D, Gray ML. Is MRI fulfilling its promise for molecular imaging of cartilage in arthritis? Osteoarth. Cartil. 2006;14:1087–1090.

193. Krasnokutsky S, Samuels J, Abramson SB. Osteoarthritis in 2007. Bull. NYU Hosp. Jt. Dis. 2007;65:222–228.

194. Charni-Ben Tabassi N, Garnero P. Monitoring cartilage turnover. Curr. Rheumatol. Rep. 2007;9:16–24.

195. Rousseau JC, Delmas PD. Biological markers in osteoarthritis. Nat. Clin. Pract. Rheumatol. 2007;3:346–356.

196. Deberg M, Dubuc JE, Labasse A, Sanchez C, Quettier E, Bosseloir A, Crielaard JM, Henrotin Y. One-year follow-up of Coll2-1, Coll2-1NO2 and myeloperoxydase serum levels in osteoarthritis patients after hip or knee replacement. Ann. Rheum. Dis. 2008;67:168–174.

197. Henrotin Y, Deberg M, Dubuc JE, Quettier E, Christgau S, Reginster JY. Type II collagen peptides for measuring cartilage degradation. Biorheology 2004;41:543–547.

198. Henrotin Y, Addison S, Kraus V, Deberg M. Type II collagen markers in osteoarthritis: what do they indicate? Curr. Opin. Rheumatol. 2007;19:444–450.

199. Pearle AD, Scanzello CR, George S, Mandl LA, DiCarlo EF, Peterson M, Sculco TP, Crow MK. Elevated high-sensitivity C-reactive protein levels are associated with local inflammatory findings in patients with osteoarthritis. Osteoarth. Cartil. 2007;15:516–523.

200. Punzi L, Ramonda R, Oliviero F, Sfriso P, Mussap M, Plebani M, Podswiadek M, Todesco S. Value of C reactive protein in the assessment of erosive osteoarthritis of the hand. Ann. Rheum. Dis. 2005;64:955–957.

201. Davis CR, Karl J, Granell R, Kirwan JR, Fasham J, Johansen J, Garnero P, Sharif M. Can biochemical markers serve as surrogates for imaging in knee osteoarthritis? Arthritis Rheum. 2007;56:4038–4047.

202. Kellgren JH, Lawrence JS, Bier F. Genetic factors in generalized osteo-arthrosis. Ann. Rheum. Dis. 1963;22:237–255.

203. Lohmander LS, Felson D. Can we identify a 'high risk' patient profile to determine who will experience rapid progression of osteoarthritis? Osteoarth. Cartil. 2004;12: S49–52.

204. Ayral X, Pickering EH, Woodworth TG, Mackillop N, Dougados M. Synovitis: a potential predictive factor of structural progression of medial tibiofemoral knee osteoarthritis—results of a 1 year longitudinal arthroscopic study in 422 patients. Osteoarth. Cartil. 2005;13:361–367.

205. Zhang W, Moskowitz RW, Nuki G, Abramson S, Altman RD, Arden N, Bierma-Zeinstra S, Brandt KD, Croft P, Doherty M, et al. OARSI recommendations for the management of hip and knee osteoarthritis, part I: critical appraisal of existing treatment guidelines and systematic review of current research evidence. Osteoarth. Cartil. 2007;15:981–1000.

206. Lee YC, Shmerling RH. The benefit of nonpharmacologic therapy to treat symptomatic osteoarthritis. Curr. Rheumatol. Rep. 2008;10:5–10.

207. Bartels EM, Lund H, Hagen KB, Dagfinrud H, Christensen R, Danneskiold-Samsøe B. Aquatic exercise for the treatment of knee and hip osteoarthritis. Cochrane Database Syst Rev. 2007: CD005523.

208. Pisters MF, Veenhof C, van Meeteren NL, Ostelo RW, de Bakker DH, Schellevis FG, Dekker J. Long-term effectiveness of exercise therapy in patients with osteoarthritis of the hip or knee: a systematic review. Arthritis Rheum. 2007;57:1245–1253.

209. Felson DT, Chaisson CE. Understanding the relationship between body weight and osteoarthritis. Baillieres Clin. Rheumatol. 1997;11:671–681.

210. Segal NA, Buckwalter JA, Amendola A. Other surgical techniques for osteoarthritis. Best Pract. Res. Clin. Rheumatol. 2006;20:155–176.

211. Ayral X. Arthroscopy and joint lavage. Best Pract. Res. Clin. Rheumatol. 2005;19:401–415.

212. Laupattarakasem W, Laopaiboon M, Laupattarakasem P, Sumananont C. Arthroscopic debridement for knee osteoarthritis. Cochrane Database Syst. Rev. 2008: CD005118.

213. Kirkley A, Birmingham TB, Litchfield RB, Giffin JR, Willits KR, Wong CJ, Feagan BG, Donner A, Griffin SH, D'Ascanio LM, et al. A randomized trial of arthroscopic surgery for osteoarthritis of the knee. N. Engl. J. Med. 2008;359:1097–1107.

214. Brouwer RW, Raaij van TM, Bierma-Zeinstra SM, Verhagen AP, Jakma TS, Verhaar JA. Osteotomy for treating knee osteoarthritis. Cochrane Database Syst. Rev. 2007;18: CD004019.

215. Zhang W, Doherty M, Arden N, Bannwarth B, Bijlsma J, Gunther KP, Hauselmann HJ, Herrero-Beaumont G, Jordan K, Kaklamanis P, et al. EULAR evidence based recommendations for the management of hip osteoarthritis: report of a task force

of the EULAR Standing Committee for International Clinical Studies Including Therapeutics (ESCISIT). Ann. Rheum. Dis. 2005;64:669–681.

216. Bannwarth B. Acetaminophen or NSAIDs for the treatment of osteoarthritis. Best Pract. Res. Clin. Rheumatol. 2006;20:117–129.

217. Bjordal JM, Ljunggren AE, Klovning A, Slørdal L. Non-steroidal anti-inflammatory drugs, including cyclo-oxygenase-2 inhibitors, in osteoarthritic knee pain: meta-analysis of randomised placebo controlled trials. BMJ 2004;329:1317.

218. Scanzello CR, Moskowitz NK, Gibofsky A. The post-NSAID era: what to use now for the pharmacologic treatment of pain and inflammation in osteoarthritis. Curr. Rheumatol. Rep. 2008;10:49–56.

219. Towheed TE, Anastassiades TP, Shea B, Houpt J, Welch V, Hochberg MC. Glucosamine therapy for treating osteoarthritis. Cochrane Database Syst. Rev. 2001: CD002946.

220. Monfort J, Martel-Pelletier J, Pelletier JP. Chondroitin sulphate for symptomatic osteoarthritis: critical appraisal of meta-analyses. Curr. Med. Res. Opin. 2008;24:1303–1308.

221. Dougados M. Management of limb osteoarthritis. In: Osteoarthritis. A Companion to Rheumatology. 1st edition. Sharma L, Berenbaum F, eds. 2007. New York: Elsevier.

222. Moldovan F, Pelletier JP, Jolicoeur FC, Cloutier JM, Martel-Pelletier J. Diacerhein and rhein reduce the ICE-induced IL-1beta and IL-18 activation in human osteoarthritic cartilage. Osteoarth. Cartil. 2000;8:186–196.

223. Christensen R, Bartels EM, Altman RD, Astrup A, Bliddal H. Does the hip powder of Rosa canina (rosehip) reduce pain in osteoarthritis patients?—a meta-analysis of randomized controlled trials. Osteoarth. Cartil. 2008;16:965–972.

224. Chen YF, Jobanputra P, Barton P, Bryan S, Fry-Smith A, Harris G, Taylor RS. Cyclooxygenase-2 selective non-steroidal anti-inflammatory drugs (etodolac, meloxicam, celecoxib, rofecoxib, etoricoxib, valdecoxib and lumiracoxib) for osteoarthritis and rheumatoid arthritis: a systematic review and economic evaluation. Health Technol. Assess. 2008;12:1–278.

225. Jones R, Rubin G, Berenbaum F, Scheiman J. Gastrointestinal and cardiovascular risks of nonsteroidal anti-inflammatory drugs. Am. J. Med. 2008;121:464–474.

226. Towheed TE, Maxwell L, Anastassiades TP, Shea B, Houpt J, Robinson V, Hochberg MC, Wells G. Glucosamine therapy for treating osteoarthritis. Cochrane Database Syst. Rev. 2005:CD002946.

227. Barron MC, Rubin BR. Managing osteoarthritic knee pain. J. Am. Osteopath. Assoc. 2007;107: ES21–7.

228. Palacios LC, Jones WY, Mayo HG, Malaty W. Clinical inquiries. Do steroid injections help with osteoarthritis of the knee? J. Fam. Pract. 2004;53:921–922.

229. Waddell DD. Viscosupplementation with hyaluronans for osteoarthritis of the knee: clinical efficacy and economic implications. Drugs Aging 2007;24:629–642.

230. Pelletier JP, Martel-Pelletier J. DMOAD developments: present and future. Bull NYU Hosp. Jt. Dis. 2007;65:242–248.

231. Chevalier X, Giraudeau B, Conrozier T, Marliere J, Kiefer P, Goupille P. Safety study of intraarticular injection of interleukin 1 receptor antagonist in patients with painful knee osteoarthritis: a multicenter study. J. Rheumatol. 2005;32:1317–1323.

232. Iqbal I, Fleischmann R. Treatment of osteoarthritis with anakinra. Curr. Rheumatol. Rep. 2007;9:31–5.

233. Milner JM, Cawston TE. Matrix metalloproteinase knockout studies and the potential use of matrix metalloproteinase inhibitors in the rheumatic diseases. Curr. Drug Targets Inflamm. Allergy 2005;4:363–375.

234. Takaishi H, Kimura T, Dalal S, Okada Y, D'Armiento J. Joint diseases and matrix metalloproteinases: a role for MMP-13. Curr. Pharm. Biotechnol. 2008;9:47–54.

235. Bingham CO 3rd, Buckland-Wright JC, Garnero P, Cohen SB, Dougados M, Adami S, Clauw DJ, Spector TD, Pelletier JP, Raynauld JP, et al. Risedronate decreases biochemical markers of cartilage degradation but does not decrease symptoms or slow radiographic progression in patients with medial compartment osteoarthritis of the knee: results of the two-year multinational knee osteoarthritis structural arthritis study. Arthritis Rheum. 2006;54:3494–3507.

236. Neogi T, Nevitt MC, Ensrud KE, Bauer D, Felson DT. The effect of alendronate on progression of spinal osteophytes and disc-space narrowing. Ann. Rheum. Dis. 2008;67:1427–1430.

237. Bruyere O, Delferriere D, Roux C, Wark JD, Spector T, Devogelaer JP, Brixen K, Adami S, Fechtenbaum J, Kolta S, Reginster JY. Effects of strontium ranelate on spinal osteoarthritis progression. Ann. Rheum. Dis. 2008;67:335–339.

238. Manicourt DH, Azria M, Mindeholm L, Thonar EJ, Devogelaer JP. Oral salmon calcitonin reduces Lequesne's algofunctional index scores and decreases urinary and serum levels of biomarkers of joint metabolism in knee osteoarthritis. Arthritis Rheum. 2006;54:3205–3211.

239. Evans CH, Gouze E, Gouze JN, Robbins PD, Ghivizzani SC. Gene therapeutic approaches-transfer in vivo. Adv. Drug Deliv. Rev. 2006;58:243–258.

240. Kuo CK, Li WJ, Mauck RL, Tuan RS. Cartilage tissue engineering: its potential and uses. Curr. Opin. Rheumatol. 2006;18:64–73.

241. Raghunath J, Salacinski HJ, Sales KM, Butler PE, Seifalian AM. Advancing cartilage tissue engineering: the application of stem cell technology. Curr. Opin. Biotechnol. 2005;16:503–509.

242. Jorgensen C, Djouad F, Bouffi C, Mrugala D, Noël D. Multipotent mesenchymal stromal cells in articular diseases. Best Pract. Res. Clin. Rheumatol. 2008;22:269–284.

243. Chen FH, Rousche KT, Tuan RS. Technology Insight: adult stem cells in cartilage regeneration and tissue engineering. Nat. Clin. Pract. Rheumatol. 2006;2:373–382.

## FURTHER READING

Bulletin of the world Health organization. *http://www.who.int/whr/2007/en/index.html*.

Sharma L, Berenbaum F. Osteoarthritis. A Companion to Rheumatology. 1st Edition. 2007. New York: Elsevier.

# 16

# HUMAN IMMUNODEFICIENCY VIRUS (HIV)

JOSEPH P. VACCA

*Merck Research Laboratories, West Point, Pensylvania*

The human immunodeficiency retrovirus (HIV), recognized widely as the causative agent of AIDS, was discovered over 25 years ago and is one of the most highly studied viruses of all time. In the 1980s through the mid 1990s, only a few treatment options were available and contracting this virus usually resulted in death. Since the introduction in the mid-1990s of HAART (Highly Active Anti-Retroviral Therapy), the disease has become manageable for many people who can tolerate the new drugs (1, 2). This review chapter will describe the way that the virus works and will highlight the many points of intervention and the small molecules that have been developed to stop the virus from replicating.

## 16.1 INTRODUCTION

HIV is now widely recognized as the causative agent of acquired immunodeficiency syndrome (AIDS), and it is now estimated that over 30 million people are infected worldwide (3). The discovery of the virus in 1983 and the ensuing epidemic spurred on the intense study of the mechanism of viral infectivity and how it might be incapacitated (4, 5). This resulted in an intense effort toward the discovery and development of drugs by academic and industrial laboratories, and now over 24 approved drugs are available as either single agents or fixed-dose combination products (Table 16.1) (6, 7). The available drugs come from

*Chemical Biology: Approaches to Drug Discovery and Development to Targeting Disease*, First Edition.
Edited by Natanya Civjan.
© 2012 John Wiley & Sons, Inc. Published 2012 by John Wiley & Sons, Inc.

four different classes: entry/fusion inhibitors, reverse transcriptase inhibitors, integrase inhibitors, and protease inhibitors. The virus has a high rate of replication and a low fidelity rate, and this leads to the development of resistance in those patients that do not adequately shut down viral replication. For this reason, all compounds must be used as part of a drug cocktail that consists of at least two different classes of drugs. Using one class of a drug (monotherapy) is never recommended because of the high risk of developing resistance against a drug class. The following chapter will describe the mechanism of viral infection, points of intervention that have been studied and the currently approved agents that have been developed against these targets.

## 16.2   THE MECHANISM OF VIRAL INFECTION

HIV is a retrovirus that works by weakening the immune system of its host such as depleting T helper CD4+ cells. This virus allows opportunistic infections by other pathogens and results in death. It is an enveloped virus that contains a conical-shaped core that contains two strands of RNA and the essential proteins needed for replication. The outside of the virus contains a surface protein gp120 that is anchored to the envelope by a gp41 protein. Both of these proteins exist as a trimeric complex, and when gp120 binds to the CD4 receptor on the surface of a CD4 cell, this unmasks other coreceptors (CCR5 or CXCR4), which also bind gp120. The gp120 unfolds to reveal the gp41 helix, which then fuses with the CD4 cell surface. The fusion inhibitor enfuvirtide acts by interacting with this gp41 protein at a specific site, and this inhibits the fusion process (8–10).

The CD4 receptors CCR5 and CXCR4 also represent a drug target (11). Cells exist with only a CCR5, or they can also have a dual CCR5 and CXCR4 receptor. A class of compounds that inhibit the binding of gp120 to the CCR5 receptor have been developed, and the first compound, maraviroc, was approved last year. Once the virus particle fuses with the T cell, then its internal core, which contains all of the essential proteins and enzymes that the virus uses for replication, is imported into the hosts' nucleus.

The core's outer proteins are dispersed, and the two strands of viral RNA are transposed into viral DNA by the essential enzyme reverse transcriptase (12). Two classes of inhibitors work at this step, and they are known as nucleoside and non-nucleoside reverse transcriptase inhibitors (NRTIs and NNRTIs). Once the viral DNA has been made, then the enzyme integrase processes it, imports it into the host cell nucleus, and incorporates it into the host's chromosome. Until recently, no drugs were available targeted for inhibiting integrase, but one compound (raltegravir) has been approved (13). Subsequent host cell transcription results in the formation of essential viral proteins and viral RNA. The viral proteins are processed by the enzyme HIV protease, and these proteins then aggregate at the host cell surface. Many HIV protease inhibitors have been developed and are available as effective drugs. There is a budding process where the nucleus of the newly formed viral particle is formed and a new particle is expelled from the cell.

This review will now describe in more detail the mechanism of each virus inhibition point and the drugs that have been developed as an intervention.

### 16.2.1  Fusion Inhibitiors

Enfuvirtide is the only available fusion inhibitor for treatment and is only approved for use in experienced patients (14, 15). It is a 36-amino-acid peptide and works by inhibiting the fusion of the gp41 protein with the CD4 cell surface. The compound is administered via injection twice per day and is usually added to a patient's current optimized therapy.

### 16.2.2  CCR5 Inhibitors

As explained, cells can be described by their tropism as either having only the CCR5 or having a mixture of CCR5 and CXCR4 receptors on their surface. The CCR5 receptor has received much more attention as a target because of the discovery that deletion of the CCR5 receptor in some people has led to a natural protection from HIV infection without any deleterious effects to the host. Many compounds have been developed and studied in people as CCR5 antagonists, but the ones that have received the most attention are maraviroc and vicriviroc. Both compounds have been studied in combination cocktails and have been shown to be effective at inhibiting the virus. Maraviroc was approved for use in experienced patients in combination with nucleosides. It is an orally administered compound and is given twice per day.

### 16.2.3  Reverse Transcriptase Inhibitors

Reverse transcriptase (RT) is an RNA-dependent DNA polymerase and is a 66-amino-acid protein. It catalyzes the production of double-stranded proviral DNA by transcription from viral-RNA. The first inhibitors of HIV were discovered to work against RT and are listed in Table 16.1. They are nucleoside analogs of the natural substrates and work via incorporation into the growing DNA strand. These drugs are prodrugs of the active species and are activated to triphosphates via intracellular kinases. Once converted to triphosphates, they act as chain terminators of the growing DNA strand. Then once incorporated, this leads to the stoppage of viral replication.

Another drug that is now used widely (tenofovir) is known as a nucleotide RT inhibitor (nRTI). This drug works via a similar mechanism as the NRTIs but exists as a mono-phosphonate prodrug. It readily undergoes uptake into the cell and is converted into a triphosphate via cellular kinases. This drug then works via a similar chain terminator mechanism and leads to inhibition of replication (16).

Another class of RT inhibitors are known as nonnucleoside inhibitors (NNRTIs) (17–19). These inhibitors are not chain terminators but work via binding to an allosteric site on RT. This binding then leads to inhibition of the enzyme

**TABLE 16.1  List of currently approved HIV therapies**

| Nucleoside inhibitor | Abbreviation | Manufacturer | Website |
| --- | --- | --- | --- |
| zidovudine | AZT | GlaxoSmith Kline | www.treathiv.com |
| zalcitabine | ddC | Hoffmann-La Roche | www.rocheusa.com/ products/hivid/ |
| didanoside | ddI | Bristol-Myers Squibb | www.bms.com |
| stavudine | d4T | Bristol-Myers Squibb | www.bms.com |
| lamivudine | 3TC | GlaxoSmith Kline | www.treathiv.com |
| emtricitabine | FTC | Gilead Sciences | www.GileadHIV.com |
| abacavir | ABC | Glaxo-Smith Kline | www.treathiv.com |
| nucleotide inhibitor | | | |
| tenofovir | TDF | Gilead Sciences | www.GileadHIV.com |
| non-nucleoside inhibitor | | | |
| nevirapine | NVP | Boehringer-Ingleheim | www.Viramune.com |
| efavirenz | EFV | Bristol-Myers Squibb | www.sustiva.com |
| delaviridine | DLV | Pfizer | |
| etravirine | | Tibotec | www.intelence-info.com/intelence/ |
| HIV Protease Inhibitors | | | |
| saquinavir | SQV | Hoffmann-La Roche | www.invirase.com |
| ritonavir | RTV | Abbott Pharmaceuticals | www.norvir.com |
| indinavir | IDV | Merck | www.crixivan.com |
| nelfinavir | NFV | Agouron Pfizer | www.viracept.com |
| amprenavir | APV | Glaxo-Vertex | www.lexiva.com |
| lopinavir/ritonavir | LPV/r | Abbott Pharmaceuticals | www.Kaletra.com |
| atazanavir | ATZ | Bristol-Myers Squibb | www.reyataz.com |
| fosamprenavir | fAPV | Glaxo-Vertex | www.lexiva.com |
| tipranavir | TPV | Boehringer-Ingleheim | |
| darunavir | DRV | Tibotec | www.prezista.com |
| Fusion Inhibitors | | | |
| enfuvirtide | ENF | Roche-Trimeris | www.fuzeon.com |
| Entry Inhibitors | | | |
| maraviroc | MRV | Pfizer | www.selzentry.com |
| Integrase Inhibitors | | | |
| raltegravir | RAL | Merck | www.isentress.com |
| Fixed dose combinations | | | |
| Truvada | TNF/FTC | Gilead Sciences | www.truvada.com |
| Combivir | AZT/3TC | GlaxoSmith Kline | www.combivir.com |
| Trizivir | ABC/AZT/3TC | GlaxoSmith Kline | www.trizivir.com |
| Atripla | EFV/TNF/FTC | Gilead Bristol-Myers Squibb | www.atripla.com |

that then leads to the stoppage of virus replication. For many years there were three approved NNRTIs, and two of them (efavirenz and nevirapine) were widely used in combination with nucleosides. These resulted in very potent regimens, but the NNRTis had a low barrier to resistance because one amino acid change within the allosteric site could lead to loss of activity. The most prevalent mutations are found with the residues K103N and Y181C. Several new compounds that work against the most prevalent mutations have been discovered and studied in patients whose virus contains the NNRTI resistant single and double mutations. One of these compounds (etravirine) has been approved for use in NNRTI treatment-experienced patients who fail their current regimens. This compound has a good antiviral effect in patients who harbor one or two documented mutations. Many other compounds are currently being studied in patients and should offer an alternative regimen for those who have exhausted other options.

### 16.2.4    HIV Integrase

HIV integrase is a 288-amino-acid enzyme that consists of three main complexes: a zinc finger region, the catalytic region, and a DNA binding domain (20, 21). The enzyme works via a metal catalyzed active site mechanism, and the site has three essential amino acids (D64;D121;E154) that are involved in metal coordination. Integrase processes the $5''$ and $3''$ ends of the proviral DNA to form a preintegration complex (22–24). This complex is then imported into the cell's nucleus and then splices the pro-viral DNA into the host cell's chromosome. Cellular enzymes then repair the gaps that were formed, and this results in a copy of the viral DNA incorporated into the host cell. Transcriptional replication of the cell then leads to production of the newly formed virus particles. Integrase was recognized for many years as a key target for inhibiting replication; however, the search for new drugs was not successful until recently. In early 2000, novel inhibitors of integrase were reported by several laboratories, but the early inhibitors were weak and were not effective in cell culture because of poor passage into the nucleus (25). Subsequent compounds were developed, and the potency was greatly improved by changing the chemical structure (26). Two compounds, raltegravir and elvitegravir, have advanced far into human clinical trials and showed good antiviral activity in patients as part of a cocktail regimen. Raltegravir was approved for use in treatment-experienced patients in combination with nucleosides and is given orally twice per day. Several other compounds are also being studied, and this class of inhibitors represents a new treatment option for many experienced patients.

### 16.2.5    HIV Protease

HIV protease is a 99-amino-acid protein that exists as a homodimer and belongs to the aspartic acid family of enzymes. The protein works by cleaving viral proteins into smaller proteins that then form the new virus particle. Inhibition of the protease leads to the production of noninfectious virus particles (27–29).

The mechanism of inhibition was based on the discovery of inhibitors of other aspartic acid proteases such as pepsin and renin. During the hydrolysis process, a water molecule is added to a peptide bond and this results in a tetrahedral intermediate that then breaks down. Inhibitors have been designed that incorporate nonhydrolyzable versions of these tetrahedral intermediates, and this leads to inhibition of the enzyme. Several compounds were first licensed in 1995 and 1996 (saquinavir, ritonavir, and indinavir), and these were combined with two nucleosides (usually AZT and 3TC or d4T) to produce very potent regimens. The use of these drug cocktails revolutionized therapy for patients, and for the first time, the death rate for AIDS patients was reduced. Although the cocktails were very potent, they also had many side effects related to them, and this limited their use. They also had complicated dosing regimens and were very expensive. This led to nonadherence by many patients, and the prevalence of resistance to the nucleosides and protease inhibitors began to emerge. This led to the discovery and development of subsequent regimens that were easier to take and had fewer side effects.

The main problem with protease inhibitors were that the compounds were complicated to synthesize and were metabolized extensively by the patients' oxidative mechanisms. This led to variable pharmacokinetics and the need to administer the compounds two or three times per day. It was subsequently found that one compound, ritonavir, was a very potent P450 inhibitor and could be administered in a low dose with other protease inhibitors to produce a regimen with high protease compound exposure. Researchers at Abbott Pharmaceuticals were the first to coadminister this agent with a protease inhibitor (lopinavir) to produce a potent regimen (30). Subsequent clinicians studied the use of ritonavir (RTV) to "boost" other protease inhibitors. This revolutionized the use of PIs and led to the use of lower doses and less frequent administration. The result was that patients experienced lower side effects and were more adherent to their regimens.

One compound approved for exclusive use in patients harboring resistant mutations is darunavir (31). This compound is potent against wild-type (WT) and resistant forms of the protease and is effective in experienced patients when coadministered with RTV. At the time of writing, this compound is currently being studied in combination with many of the new classes of drugs such as integrase and second-generation NNRTIs.

### 16.2.6 Resistance

All of the clinically effective drug combinations lose their effectiveness if the patient develops resistance against one or more of the components. HIV replicates quickly and makes many mistakes as it does so. This replication results in the probability of very low levels of resistance virus being harbored in a patient. If the drug regimen is not used properly, then this can lead to insufficient suppression of virus replication. The result is the virus replicates in the presence of suboptimal levels of a drug, and this can lead to the emergence of resistant strains. The common mutations found against the nucleoside inhibitor classes of compounds are found in the reverse transcriptase active site and are indicated with residues such as K55R and L74V. Several mutations have been identified with the use of nonnucleoside RT inhibitors and reside in the allosteric binding site where the inhibitor sits. The most common ones found are K103N, Y181C, V106A, and other various double and triple mutants. Resistance toward HIV protease inhibitors is usually a little hard to achieve, and it is thought that this class has a higher genetic barrier. Nonetheless, many mutations have been identified, usually in combination with each other, and are associated with residues found in the active site and outside of it. It is thought that the nonactive site mutations are developed to increase the enzyme's fitness. Finally, the development of HIV integrase inhibitors is just starting to show patterns of resistance toward the agents, both in clinical use and in laboratory experiments.

### 16.2.7 Fixed-Dose Combinations

One limitation to taking a three-drug combination is the number of pills that a person has to take. As the drug regimens improved, the frequency of dosing has been reduced that has led to better adherence and a lower incidence of resistance developing. To simplify the regimens even more, many drug manufacturers have started to combine two and even three drugs into one pill. Table 16.1 lists the fixed-dose combination products available at the time of writing with Atripla™ being the only regimen that contains three drugs in one pill. Several other combinations have been proposed and should be on the market within the next ten years.

## 16.3  CONCLUSION

Since the discovery of HIV over 25 years ago, many ways to attack the virus have been studied and the result has been the introduction of five classes of drugs. These compounds can be used in many different combinations and can offer the patient several different options before the same class has to be reused. This offers the patient and their families great hope that this infection can be a manageable disease. This is evident from the lowering of the death rate of patients, and now many are living full and productive lives.

## REFERENCES

1. Vermund SH. Millions of life-years saved with potent antiretroviral drugs in the United States: a celebration, with challenges. J. Infect. Dis. 2006;194:1–5.
2. Walensky RP, Paltiel AD, Losina E, Mercincavage LM, Schackan BR, Sax PE, Weinstein MC, Freedburg KA. The survival benefits of AIDS treatment in the United States. J. Infect. Dis. 2006;194:11–19.

3. Uniting the World Against AIDS. UNAIDS Epidemic report. Available: *www.unaids.org*.

4. Gallo RC, Montagnier L. The discovery of HIV as the cause of AIDS. N. Engl. J. Med. 2003;349:2283–2285.

5. Ratner L, Haseltine W, Patarca R, Livak KJ, Starcich B, Josephs SF, Doran ER, Rafalski JA, Whitehorn EA, Baumeister K, et al. Complete nucleotide sequence of the AIDS virus, HTLV-III. Nature, 1985;313:277–284.

6. De Clercq E. The design of drugs for HIV and HCV. Nature Rev. 2007;6:1001–1018.

7. De Clercq E. Antiviral drugs in current clinical use. J. Clin. Vir. 2004;30:115–133.

8. Meanwell NA, Kadow JF. Inhibitors of the entry of HIV into host cells. Curr. Opin. Drug Disc. Dev. 2003;6:451–461.

9. Matthews T, Salgo M, Greenberg M, Chung J, DeMasi R, Bolognesi D. Enfuvirtide: the first therapy to inhibit the entry of HIV-1 into host CD4 Lymphocytes. Nat. Rev. Drug Disc. 2004;3:215–25.

10. Debnath AK. Novel uses for current and future direct thrombin inhibitors: focus on ximelagatran and bivalirudin. Expert Opin. Invest. Drugs 2006;15:465–478.

11. Biswas P, Nozza S, Scartlatti G, Lazzarin A, Tambussi G. Oral CCR5 inhibitors: will they make it through? Expert Opin. Invest. Drugs 2006;15:451–464.

12. Demeter LM, Doamaoal RA. Structural and biochemical effects of human immunodeficiency virus mutants resistant to non-nucleoside reverse transcriptase inhibitors. Int. J. Biochem. Cell Biol. 2004;36:1735–1751.

13. Grinsztejn B, Nguyen BY, Katlama C, Gatell JM, Lazzarin A, Vittecoq D, Gonzalez CJ, Chen J, Harvey CM, Isaacs RD. Safety and efficacy of the HIV-1 integrase inhibitor raltegravir (MK-0518) in treatment-experienced patients with multidrug-resistant virus: a phase II randomised controlled trial. Lancet 2007;369:1261–1269.

14. Buti M, Esteban R. Adefovir Dipivoxil. Drugs of Today 2003;39:127–35.

15. LaBonte J, Lebbos J, Kirkpatrick P. Enfuvirtide. Nature Rev. Drug Disc. 2003; 345–346.

16. Srinivas RV, Robbins BL, Connelly MC, Gong Y-F, Bischofberger N, Fridland A. Metabolism and in vitro antiretroviral activities of bis(pivaloyloxymethyl) prodrugs of acyclic nucleoside phosphonates. Antimicrob. Agents Chemother. 1993;37:2247–2250.

17. Fortin C, Joly V, Yeni P. Emerging reverse transcriptase inhibitors for the treatment of HIV infection in adults. Expert Opin. Emerging Drugs 2006;11:217–230.

18. Tarby CM. Recent advances in the development of next generation non-nucleoside reverse transcriptase inhibitors. Curr. Top. Med. Chem. 2004;4:1045–1057.

19. Zhao Z, et al. Novel indole-3-sulfonamides as potent HIV nonnulceoside reverse transcriptase inhibitors (NNRT's). Bioorg. Med. Chem. Lett. 2008;18:554–559.

20. Anthony NJ. HIV-1 integrase: a target for new AIDS chemotherapeutics. Curr Top. Med Chem. 2004;4:979–990.

21. Ellison V, Brown PO. A stable complex between integrase and viral DNA ends mediates human immunodeficiency virus integration in vitro. Proc. Natl. Acad. Sci. USA. 1994;91:7316–7320.

22. Parrill AL. HIV-1 Integrase inhibition: binding sites, structure activity relationships and future perspectives. Curr. Med. Chem. 2003;10:1811–1824.

23. Pommier Y, Neamati N. Inhibitors of human immunodeficiency virus integrase. Advances in Virus Res. 1999;52:427–458.

24. Hazuda DJ, Felock PJ, Hastings JC, Pramanik B, Wolfe AL. Differential divalent cation requirements uncouple the assembly and catalytic reactions of human immunodeficiency virus type 1 integrase. J. Virol. 1997;71:7005–7011.

25. Hazuda DJ, Felock P, Witmer M, Wolfe A, Stillmock K, Grobler JA, Espeseth A, Gabryelski L, Schleif W, Blau C, Miller MD. Inhibitors of strand transfer that prevent integration and inhibit HIV-1 replication in cells. Science 2000;287:646–650.

26. Summa V, Petrocchi A, Bonelli F, Crescenzi B, Donghi M, Ferrara M, Fiore F, Gardelli C, Paz OG, Hazuda DJ, et al. Discovery of raltegravir a potent, selective orally bioavailable HIV-integrase inhibitor for the treatment of HIV AIDS infection. J. Med. Chem. 2008. In-press.

27. Toh H, Ono M, Saigo K, Miyata T. Retroviral protease-like sequence in the yeast transposon ty 1. Nature 1985;315:691.

28. Seelmeier S, Schmidt H, Turk V, von der Helm K. Human immunodeficiency virus has an aspartic-type protease that can be inhibited by pepstatin A. Proc. Natl. Acad. Sci. USA. 1988;85:6612–6616.

29. Kohl NE, Emini EA, Schleif WA, Davis LJ, Heimbach JC, Dixon RAF, Scolnick EM, Sigal IS. Active human immunodeficiency virus protease is required for viral infectivity. Proc. Natl. Acad. Sci. USA. 1988;85:4686–4690.

30. Sham HL et al. ABT-378, a highly potent inhibitor of the human immunodeficiency virus protease. Antimicrob. Agents Chemother. 1998;42:3218–3223.

31. Ghosh AK, Chapsal BD, Weber IT, Mitsuya. Design of HIV protease inhibitors targeting protein backbone: an effective strategy for combating drug resistance. H. Acc. Chem. Res. 2008;41:78–86.

# 17

# ALLERGY AND ASTHMA

Garry M. Walsh

*University of Aberdeen, School of Medicine, Aberdeen, Scotland, United Kingdom*

Current therapies for asthma and allergy are aimed at controlling disease symptoms. For most asthmatics, inhaled anti-inflammatory therapy is effective, but a subset of patients remains symptomatic despite optimal treatment creating a clear unmet medical need. Although considered a less-serious condition than asthma, allergic diseases are common and considerably impact on the quality of life of affected individuals. Innovative disease-modifying therapeutics are needed for both allergy and asthma. Biopharmaceutical approaches may identify small molecules that target key cells and mediators that drive the inflammatory responses that underlie the pathogenesis of allergy and asthma.

Asthma and allergy are important conditions whose complex pathology is influenced by both environmental and genetic factors. Although effective anti-inflammatory therapy is available for both conditions, this therapy is used when symptoms develop and is not always effective in all subjects with asthma and/or allergy. Asthma and allergy share some common pathological features including the overexpression of inflammatory mediators that result in leukocyte accumulation and changes in structural cell function. We have considerable knowledge regarding the cells mediators and factors that control the pathogenic changes in asthma and allergy. This information has informed the identification of small molecules that target aspects of the inflammatory cascade in asthma and allergy. This chapter will review the background and status of these novel compounds.

*Chemical Biology: Approaches to Drug Discovery and Development to Targeting Disease*, First Edition. Edited by Natanya Civjan.
© 2012 John Wiley & Sons, Inc. Published 2012 by John Wiley & Sons, Inc.

## 17.1  BIOLOGICAL BACKGROUND

### 17.1.1  Asthma

Asthma is now one of the most common chronic diseases in developed countries and is characterized by reversible airway obstruction, airway hyper reactivity (AHR), and airway inflammation. Key pathological features include infiltration of the airways by activated lymphocytes and eosinophils; damage to, and loss of, the bronchial epithelium; mast cell degranulation; mucous gland hyperplasia; and collagen deposition in the epithelial sub-basement membrane area. Asthma pathology is associated with the release of myriad proinflammatory substances that include lipid mediators, inflammatory peptides, chemokines, cytokines, and growth factors. In addition to infiltrating leukocytes, structural cells in the airways, which include smooth muscle cells, endothelial cells, fibroblasts, and airway epithelial cells, are all important sources of asthma relevant mediators (1). This complex scenario means that potential targets for therapeutic intervention are many and varied, and the task of successful therapy is challenging.

Anti-inflammatory therapy in asthma is largely reliant on glucocorticoids (GC)—particularly in their inhaled form—with or without the addition of short- or long-acting bronchodilators. The cysteinyl leukotriene-receptor antagonists are also an established part of the asthma armamentarium. Overall, they are less effective than inhaled GC, but some patients show a striking improvement, and a GC-sparing effect has been demonstrated. Although the symptoms of most asthmatics are satisfactorily controlled by regular use of inhaled GC, their use often raises concerns with respect to compliance, particularly in children and adolescents. Increasing evidence shows that long-term use of inhaled GC, especially in high doses, can cause systemic adverse effects, including adrenal suppression, reduced growth and reduced bone–mineral density, as well as local side effects such as dysphonia. Moreover, a significant subgroup of asthmatic patients responds poorly or not at all to high-dose inhaled or systemic GC treatment. As few alternative treatments are available, these patients can be difficult to treat and may require frequent hospitalization (1). Thus, identification of potential targets for therapeutic intervention is an important goal in asthma research.

### 17.1.2  Allergy

Allergic inflammation is characterized by an immediate immunoglobulin-E-dependent mast cell and basophil degranulation leading to the release of mediators such as histamine that is responsible for most immediate manifestations of allergic disease. Other mediators include platelet-activating factor, prostaglandins, and cytokines, which include interleukin (IL)-3, IL-4, and tumor necrosis factor-$\alpha$ (TNF-$\alpha$). Cytokines are particularly important in driving the subsequent late-phase reaction that results in the accumulation of leukocytes at the sites of inflammation. These inflammatory processes lead to various manifestations of allergic disease such as intermittent and persistent allergic rhinitis, conjunctivitis, atopic dermatitis, urticaria, and a degree of airway inflammation observed in allergic (extrinsic) asthma. The burden of allergic disease worldwide is such that it represents a serious public health problem that attracts considerable effort to identify effective and safe therapies.

Histamine has a prominent and diverse role in the pathophysiology of allergic disease; therapeutic intervention is therefore commonly focused on blocking its interaction with H1 receptors. The H1 histamine receptor is a heptahelical transmembrane molecule that transduces extracellular signals to intracellular second-messenger systems via G proteins. H1 ntihistamines act as inverse agonists that combine with the H1-receptor, which stabilizes it in the inactive form and shifts the equilibrium toward the inactive state (2). Although effective in the treatment of allergic rhinitis first-generation antihistamines, such as chlorpheniramine and promethazine, had a tendency to cross the blood–brain barrier resulting in unwanted side effects, particularly sedation and impaired psychomotor activity (3). In contrast, the second-generation histamine H1 receptor antagonists are highly effective and well-tolerated treatments for allergic disease and are among the most frequently prescribed drugs in the world (4). More recently, several novel antihistamines, which include fexofenadine, desloratadine, and levocetirizine, have been developed that are either metabolites of active drugs or enantiomers with improved potency, duration, and onset of action, together with increased predictability and safety. Anti-inflammatory effects *in vitro* and *in vivo* independent of H1-receptor blockade have been described for most antihistamines, and these may enhance their therapeutic benefit (3). Other treatments available include anticholinergics, decongestants, intranasal corticosteroids, leukotriene receptor antagonists, or desensitization with modified allergens. It is noteworthy that some of these treatments are often used in combination rather than as monotherapy (e.g. combination of an antihistamine with a nasal decongestant or with a leukotriene antagonist).

## 17.2 SMALL MOLECULES IN ASTHMA

### 17.2.1 Interleukin-5

Much inflammation in asthma is thought to be a consequence of the inappropriate accumulation of eosinophils and the subsequent release of their potent proinflammatory arsenal that includes such diverse elements as granule-derived basic proteins, mediators, cytokines, and chemokines (5). Interleukin (IL)-5 is crucial to the development and release of eosinophils from the bone marrow, their enhanced adhesion to endothelial cells that line the postcapillary venules, and their persistence, activation, and secretion in the tissues. Several animal models of asthma including the use of primates have provided good evidence that inhibiting the effects of IL-5 using specific mAb-inhibited eosinophilic inflammation and AHR (6). Given its central role in regulating eosinophil development and function, IL-5 was therefore chosen as a potentially attractive target to prevent or blunt eosinophil-mediated inflammation in patients with asthma. However, several clinical trials have reported disappointing clinical outcomes after treatment of asthmatic patients with an anti-IL-5 mAb. The first study was designed to validate the safety of the humanized anti-IL-5 mAb mepolizumab (7); it faced criticism of lack of power (8) and the validity of patient selection (9). A later placebo-controlled study (10) found that treatment of mild asthmatic patients with mepolizumab abolished circulating eosinophils and reduced airway and bone marrow eosinophils, but it reported no significant improvement of clinical measures of asthma. Critically, lung biopsy samples from the treatment group contained intact tissue eosinophils together with large quantities of eosinophil granule proteins; these findings likely explain the lack of clinical benefit after mepolizumab treatment. Similar findings were reported with the anti-IL-5 mAb SCH55700 in patients with severe asthma who had not been controlled by inhaled corticosteroid use. These authors reported profound reduction in circulating eosinophils but no significant improvement in either asthma symptoms or lung function (11). Compared with placebo, treatment of mild atopic asthmatics with mepolizumab significantly reduced the expression of the extracellular matrix proteins tenascin, lumican, and procollagen III in the bronchial mucosal reticular basement membrane. In addition, significant reductions were observed in both airway eosinophils that express mRNA for transforming growth factor-beta (TGF-$\beta$1) and the concentration of TGF-$\beta$1 in bronchoalveolar lavage (BAL) fluid (12). TGF-$\beta$1 is implicated in airway remodeling in asthma, and eosinophils are important sources of this growth factor, which thereby contributes to tissue-remodeling processes in asthma by regulating the deposition of ECM proteins. Mepolizumab may prove useful in preventing this. An alternative to the use of humanized anti-IL-5 mAb is the use of molecular modeling of the IL-5 receptor $\alpha$-chain to develop specific receptor antagonists. At the time of writing this chapter, such a compound (YM-90709) has been shown to be a relatively selective inhibitor of the IL-5 R (13), and intravenous injection of YM-90709 inhibited infiltration of eosinophils into the BAL fluid of allergen-challenged BDF1 mice (14).

## 17.2.2  Interleukin-4

Another cytokine important in eosinophil accumulation is IL-4, which together with its close relative IL-13, is important in IgE synthesis by B cells. Both cytokines signal through a shared surface receptor, IL-4 R$\alpha$, which then activates the transcription factor STAT-6 (15). Studies with soluble IL-4 R given in a nebulized form demonstrated that the fall in lung function induced by withdrawal of inhaled corticosteroids was prevented in patients with moderately severe asthma (16). However, despite these promising findings subsequent trials have not been as successful; consequently, this treatment is no longer being developed. Other approaches for blocking the IL-4 receptor include administration of antibodies against the receptor and mutant IL-4 proteins. For example, a peptide-based vaccine for blocking IL-4 was developed by antigenic prediction and structure analysis of IL-4/receptor complex. Vaccine construction involved a truncated hepatitis B core antigen as carrier with the peptide inserted using gene engineering methods. Compared with control animals, immunized allergen-challenged ovalbumin (OVA)-sensitized mice had significant reductions in Immunoglobulin-E (IgE), eosinophil accumulation in BAL, goblet-cell hyperplasia, tissue inflammation, and methacholine-induced respiratory responses (17).

### 17.2.3   Interleukin-13

IL-13 has been found in BAL after allergen provocation of asthmatic subjects, which strongly correlated with the increase in eosinophil numbers. mRNA expression for IL-13 was detected in bronchial biopsies from both allergic and nonallergic asthmatic subjects. In animal models, IL-13 mimics many proinflammatory changes associated with asthma (18). It is therefore another potential therapeutic target for the resolution of airway inflammation. Two receptors for IL-13 have been described—IL-13 $R\alpha 1$ and IL-13 $R\alpha 2$. The latter exists in soluble form, has a high affinity for IL-13, and can thus "mop up" secreted IL-13. In mice, IL-13 $R\alpha 2$ blocked the actions of IL-13, which included IgE production, pulmonary eosinophilia, and AHR (19). At the time of writing, a humanized IL-13 $R\alpha 2$ is now in clinical development as a novel therapy for asthma. Another mouse-based study reported that intratracheal administration of human IL-13 induced leukocyte infiltration in the lung, AHR, and goblet-cell metaplasia with allergic eosinophilic inflammation in the esophagus. An antihuman IL-13 IgG4 mAb (CAT-354) significantly reduced many of these parameters. In contrast, another study using mice sensitized by intranasal application of ovalbumin as a model of asthma/allergy found that the inhibition of the IL-4/IL-13 system efficiently prevented the development of the asthmatic phenotype, which includes goblet-cell metaplasia and airway responsiveness to methacholine, but it had little effect on established asthma (20). A humanized anti-IL-13 mAb called IMA638 significantly reduced eosinophils, neutrophils, eotaxin, and RANTES in BAL fluid from cynomolgus monkeys sensitized to Ascaris suum after segmental antigen challenge, compared with levels observed in control animals (21). IMA-638 also gave dose-dependent inhibition of the antigen-induced late responses and airway hyperresponsiveness in a sheep model of allergic asthma that used animals with natural airway hypersensitivity to Ascaris suum antigen (22).

As stated above, IL-4 $R\alpha$ is the signaling component of the heterodimeric receptor complex shared by both IL-4 and IL-13. Therefore, it represents an attractive target to antagonize the effects of both cytokines, as this approach may be more effective than targeting either IL-4 or IL-13 alone (23). A recombinant human IL-4 variant pitrakinra (Aerovant) was developed that competitively inhibits the IL-4 $R\alpha$ receptor complex to interfere with the actions of both IL-4 and IL-13. In two independent small-scale parallel-group Phase IIa randomized, double-blind, placebo-controlled clinical trials, patients with atopic asthma were treated with pitrakinra or placebo given either as a single subcutaneous dose or via nebulization twice daily. Active or placebo treatments were given for 4 weeks before the patients were given an inhaled inhalation challenge. Compared with placebo, allergen challenge-induced decreases in $FEV_1$ were significantly attenuated after 4 weeks of inhalation of pitrakinra. The frequency of spontaneous asthma attacks that required rescue medication use was also diminished in the study in which pitrakinra was given subcutaneously (24). These important findings support the hypothesis that dual inhibition of IL-4 and IL-13 can affect the course of the late asthmatic response after experimental allergen challenge. More large-scale clinical trials on patients with day-to-day asthma are required to establish fully whether pitrakinra is an effective and safe asthma treatment.

### 17.2.4   Interleukin-9

Another TH2 cytokine, IL-9, and its receptor are found in asthmatic airways in increased levels (25). IL-9 has several proinflammatory effects on eosinophils, which includes enhancement of eosinophil IL-5 receptor expression, differentiation in the bone marrow, and prolonged survival through inhibition of apoptosis (26). Transgene expression of IL-9 in the lungs of mice resulted in lymphocytic and eosinophilic infiltration of the lung, airway epithelial hypertrophy with mucus production, and mast cell hyperplasia, as well as production of IL-4, IL-5, and IL-13 (27). Treatment of OVA-challenged mice with an anti-IL-9 antibody significantly prevented AHR in response to a methacholine challenge together with reductions in numbers of eosinophil and levels of IL-4, IL-5, and IL-13 in BAL (28).

### 17.2.5   Tumor Necrosis Factor-α

TNF-α is expressed in asthmatic airways and may play a key role in amplifying airway inflammation through activation of transcription factors such as NF-κB and AP-1. TNF-α has proinflammatory effects on eosinophils, neutrophils, T cells, and endothelial cells. It is thought to contribute to AHR, airway remodeling, and GC resistance in asthma; therefore, it represents a potential target for therapy. Humanized anti-TNF mAb (infliximab) and soluble TNF receptor blockers (etanercept) have been developed, and preliminary clinical studies have shown significant improvements in lung function, airway hyperreactivity, and exacerbation rate, particularly in patients with severe asthma refractory to GC treatment. However, some clinical studies reported negative findings, so heterogeneity seems to exist in response to TNF-α antagonism. In addition, concerns exist regarding potential side effects in some subjects treated with anti-TNF therapy, which include higher rates of solid organ malignancies or latent TB reactivation (29). Small-molecule inhibitors of TNFα-converting enzyme have also been synthesized; and these inhibitors may be attractive therapeutic targets for asthma (30).

### 17.2.6 Immunoglobulin-E Inhibitors

IgE plays a central role in the pathogenesis of diseases associated with immediate hypersensitivity reactions, which include allergic asthma. IgE-dependent biological actions are a result of it binding to high-affinity (FcεRI) receptors on mast cells and basophils and to low-affinity (FcεRII) receptors on macrophages, dendritic cells, and B lymphocytes. Allergen molecules then crosslink adjacent Fab components of IgE on the cell surface, which thereby activates intracellular signal transduction. In mast cells, this action leads to the release of preformed mediators and the rapid synthesis and release of other mediators responsible for bronchoconstriction and airway inflammation. Therefore, blocking the action of IgE using blocking antibodies that do not result in cell activation is an attractive approach.

Omalizumab (rhuMab-E25) is a humanized monoclonal antibody directed to the FcεRI binding domain of human IgE. It inhibited early-phase and late-phase allergen-induced asthmatic reactions and reduces serum-free IgE concentrations to less than 5% of baseline; it has now progressed through clinical development (31). A large Phase II trial studied fortnightly intravenous administration of omalizumab for 20 weeks in 317 patients (32), whereas two Phase III trials, which included over 500 patients each, studied omalizumab given subcutaneously every 2–4 weeks for 12 months (33, 34). Ayres and colleagues (35) examined the effects of Omalizumab in patients with moderate-to-severe allergic asthma whose symptoms were poorly controlled by high doses of inhaled GC. Omalizumab was administered for 12 months and benefited these patients as shown by a 50% reduction in their asthma deterioration-related incidents. Another study reported that Omalizumab treatment of subjects with both persistent rhinitis and difficult-to-treat asthma resulted in significantly reduced asthma exacerbations and improved quality of life in those patients who received anti-IgE therapy over the 28-week study period (36). Omalizumab has also been shown to be beneficial as an add-on therapy in patients who have inadequately controlled, severe persistent asthma (37). Consistent findings from these trials showed that omalizumab is an effective therapy for patients with symptomatic moderate to severe allergic asthma despite treatment with GC and rescue medication. It reduced the frequency of exacerbations and improved symptom control while allowing a reduction in the use of GC and $\beta_2$-agonists. It also improved patient quality of life and produced a significant improvement in lung function as measured by PEFR and $FEV_1$. Omalizumab seems to be well tolerated with few side effects reported in these studies with no reports of circulating antibodies against omalizumab. Although more long-term studies are needed to elucidate the benefit and safety of anti-IgE therapy in asthma, its niche may be in the treatment of patients with severe asthma who are dependent on oral corticosteroids.

### 17.2.7 Eotaxin

Chemokines are a family of small, secreted proteins that control migration of monocytes, lymphocytes, neutrophils, eosinophils, and basophils. Eotaxin is an inducible, secreted chemokine that promotes selective recruitment of eosinophils from the blood into inflammatory tissues. It was first described in 1993 when intradermal injection of naive guinea pigs with BAL fluid from antigen-challenged guinea pigs resulted in the recruitment of eosinophils (38). Uniquely, the major characteristic of eotaxin is its selective ability to act on eosinophils. CCR3, which is a seven-transmembrane-spanning G protein-coupled receptor for eotaxin-1, is highly expressed on eosinophils and mediates the biological effects of other eosinophil chemokines, such as eotaxin-2, eotaxin-3, MCP-3, MCP-4, and RANTES. Furthermore, CCR3 is expressed not only on eosinophils but also on basophils (39), mast cell subpopulations (40), and activated Th2 cells (41), which might explain the coordinated recruitment of these cell types to sites of allergic inflammation. CCR3 is also expressed by airway epithelial cells (42), and although the bronchial epithelium consists of structural nonmigratory cells, expression of the CCR3 receptor may represent an autoregulatory feedback mechanism to monitor chemokine production. Furthermore, eotaxin produced by the epithelium may be sequestered by the CCR3 receptor and presented to infiltrating cells, which thereby enhances their activation; this phenomenon is observed with IL-8 and its receptor. Several clinical studies have suggested a pivotal role for CCR3 ligands/CCR3 in the eosinophilic inflammation characteristic of atopic dermatitis, asthma, and allergic rhinitis; thus, blockade of this receptor may have pronounced beneficial effects in these diseases (43). Several small-molecule CCR3 antagonists have been described. In this regard, N-(ureidoalkyl)-benzpiperidines have been identified as potent CCR3 antagonists, which inhibit eosinophil chemotaxis and calcium mobilization in the micromolar to nanomolar concentration range (44). The small-molecule selective CCR3 antagonist YM-344031 potently inhibited the binding of eotaxin-1 and RANTES to human cells transfected with CCR3, ligand-induced $Ca^{2+}$ flux, and chemotaxis, together with inhibition of eotaxin-1-induced eosinophil shape change. Furthermore, both immediate and late-phase allergic skin reactions in a mouse model were significantly inhibited by orally administered YM-344031 (45). Another small-molecule selective CCR3 antagonist YM-355179 inhibited eotaxin-induced intracellular $Ca^{2+}$ influx, chemotaxis, and eosinophil degranulation. It also inhibited eosinophil infiltration into airways of cynomolgus monkeys after segmental bronchoprovocation with eotaxin (46). The effect of a low molecular weight CCR-3 antagonist on chronic experimental bronchial asthma was examined using BALB/c mice intraperitoneally sensitized with OVA subsequently chronically challenged with OVA aerosol to induce chronic airway inflammation and airway remodeling. CCR-3 antagonist treatment resulted in a marked reduction of eosinophils in the bronchoalveolar lumen and in airway wall tissue, whereas infiltration of lymphocytes or macrophages remained unchanged. Furthermore, antagonizing CCR-3 reduced AHR, goblet-cell hyperplasia, and airway remodeling as defined by subepithelial fibrosis, and it increased accumulation of myofibrocytes in the airway wall of chronically challenged mice. Therefore, antagonizing CCR-3 represents a novel approach and promising asthma or allergy therapy

(47). Furthermore, evidence from animal models suggests that IL-5 and eotaxin may work in a synergistic fashion to promote the release of mature eosinophils from the bone marrow (48). Thus, it might be that combination therapies of CCR3 antagonist and humanized anti-IL-5 mAb might prove an effective approach to limit or prevent eosinophil toxicity in the asthmatic lung.

## 17.3 TARGETING INFLAMMATORY CELL ACCUMULATION

Asthma pathology is characterized by excessive leukocyte infiltration that leads to tissue injury. Leukocyte extravasation, migration within the interstitium, cellular activation, and tissue retention are controlled by cell-adhesion molecules (i.e. selectins, integrins, and members of the immunoglobulin superfamily). Numerous animal studies have demonstrated essential roles for these cell-adhesion molecules in lung inflammation, which include L-selectin, P-selectin, and E-selectin, ICAM-1, and VCAM-1, together with many of the $\beta 1$ and $\beta 2$ integrins. For example, compared with wild-type mice, inhaled allergen challenge of P-selectin deficient mice resulted in fewer eosinophils in BAL fluid with significant reductions in AHR. In a sheep model of allergic asthma, pretreatment with an L-selectin monoclonal antibody significantly reduced both the early and late airway response and significantly reduced AHR (49). Therefore, these therapies represent potentially important therapeutic targets and these families of adhesion molecules have been under intense investigation by the pharmaceutical industry for the development of novel therapeutics.

### 17.3.1 Selectins

Selectins are responsible for the early "rolling" adhesive events between leukocytes and the endothelial cells that line the postcapillary venules. The small-molecule selectin antagonist Bimosiamose (TBC1269) is a synthetic computer-designed selectin antagonist targeted against all three selectins *in vitro* and has proven to be efficacious in mouse, rat, rabbit, guinea pig, and sheep models of allergic asthma (50). For example, in a sheep model of allergy, inhaled TBC1269 potently inhibited allergic airway responses, histamine levels in BAL, and tissue kallikrein and neutrophilic inflammation (51). In patients with asthma, a single intravenous dose of TBC1269 had only a minor effect on sputum eosinophils or inhaled allergen-induced late asthmatic reactions (52). In contrast, inhaled TBC1269 significantly reduced late-phase asthmatic reactions by approximately 50% compared with placebo in mild asthmatic subjects (53). Thus, the inhaled route of TBC1269 may offer advantages over systemic delivery in terms of both efficacy and safety. Because selectins are also vital in early adhesive neutrophil interactions with the endothelium, TBC1269 may also prove an effective therapy for COPD.

### 17.3.2   Integrins

Good evidence from animal models indicates that α4β1 (VLA-4) is a viable drug target for asthma. Monoclonal antibodies (mAb) against the α4 subunit of VLA-4 have proven efficacious in asthma models in five different species. For example, in a mouse model of asthma, intravenously administered anti-α4 mAb eliminated eosinophilia but did not affect AHR. In contrast, when delivered intranasally, the mAb blocked both airway inflammation and AHR (54). The small-molecule VLA-4 antagonist (2S)-3-(4-Dimethylcarbamoyloxyphenyl)-2-{((4R)-5,5-dimethyl-3-(1-methyl-1H-pyrazole-4sulfonyl)thiazolidine-4-carbonyl)amino}propionic acid (WAY103) was assessed for its effects on eosinophil VLA-4-dependent functions, which include adhesion, migration, respiratory burst, and degranulation. WAY103 inhibited eosinophil adhesion to VCAM-1, transendothelial migration, and VCAM-1-stimulated eosinophil superoxide generation. However, it also enhanced cytokine-activated eosinophil-derived neurotoxin degranulation, which generated concerns that it might promote eosinophil activation in certain circumstances (55). More recently, a potent and selective, small-molecule VLA-4 inhibitor, (2S)-3-(2′,5′-dichlorobiphenyl-4-yl)-2-({(1-(2-methoxybenzoyl)piperidin-3-yl)carbonyl}amino) propanoic acid, blocked VLA-4 dependent functions on a variety of cell types and also inhibited eosinophil infiltration in a dose-dependent manner by up to 80% in an air-pouch mouse model (56). Overall, several small-molecule integrin α4β1 antagonists have been developed, but most of these have proved to be ineffective in preventing the clinical symptoms of asthma. However, a clinical trial of an oral α4β1 antagonist valategrast demonstrated significant positive effects on lung function and rates of exacerbation in asthma patients who had their inhaled GC therapy withdrawn prior to treatment with valategrast (57).

### 17.4   IMMUNOTHERAPY

Asthma and allergy are manifestations of an imbalance in cytokine and signaling pathways that mediate inflammatory and structural changes within the affected tissues. Developing treatments include strategies to alter the cytokine/chemokine balance or to skew the cytokine profile away from T helper2 (Th2) responses and toward Th1 responses. Immunotherapy may potentially attenuate symptoms by disease modification through the induction of tolerance to common environmental allergens rather than by suppressing inflammation. A major advantage is the potential for a positive effect to remain for several years after the end of the treatment period. The use of allergen-specific immunotherapy is not a new approach to asthma therapy, and until relatively recently the crude nature of the allergen extracts available meant that its use was limited by unwanted side effects, such as anaphylaxis. Strategies to overcome these problems include the use of hypoallergenic isoforms, recombinant allergens, or DNA vaccines (58).

For example, in mice with chronic airway inflammation maintained by repeated ovalbumin inhalation, mucosal administration of CpG DNA oligonucleotides significantly reversed airway hyperreactivity together with both acute and chronic markers of inflammation (59). Another approach is the use of short, synthetic, allergen-derived peptides that induce T cell tolerance but cannot cross-link IgE on mast cells or basophils and induce anaphylaxis. The main effect seems to be a shift from a Th2 to Th1 profile and induction of regulatory cytokines such as IL-10 and TGB-β. One study that used patients with asthmatic reactions to cats demonstrated that treatment with a desensitising vaccine based on many short, overlapping, HLA-binding, T-cell peptides derived from Fel d 1 inhibited both early- and late-phase reactions to a subsequent whole-allergen challenge. Changes in immunological parameters included modulation of the proliferation of blood mononuclear cells together with their production of IL-4, IL-10, and IL-13, and of interferon-γ (60). A more recent study from the same group demonstrated that treatment of cat-allergic asthmatic subjects with Fel d 1-derived T-cell peptides significantly improved clinically relevant outcome measurements, which include reductions in late asthmatic reactions to inhaled whole cat dander and significant improvements in asthma-related quality of life (61). Although the results from these studies are encouraging, we require larger, dose-ranging studies before firm conclusions on clinical efficacy of peptide allergen therapy can be made. Although it is conceivable that targeting T cells may induce Th1-type autoimmune pathology in humans, to date no evidence indicates that this occurs. In addition, nonallergic mechanisms are also likely to be important in the pathogenesis of asthma; thus, it is possible that novel treatments targeted at the allergy fraction of the phenotype may not be as efficacious as hoped.

## REFERENCES

1. Walsh GM. Novel therapies for asthma – advances and problems. Curr. Pharm. Design 2005;11:3027–3038.

2. Leurs R, Church MK, Taglialatela M. H1-antihistamines: inverse agonism, anti-inflammatory actions and cardiac effects. Clin. Exp. Allergy 2002;32:489–498.

3. Walsh GM, Annunziatto L, Frossard N, Knol K, Levander S, Nicolas JM, Taglialatela M, Tharp MD, Tillement P, Timmerman H. New insights into the second generation antihistamines. Drugs 2001;61:207–236.

4. Simons FE. Advances in H1-antihistamines. N. Engl. J. Med. 2004;351:2203–2217.

5. Walsh GM. Advances in the immunobiology of eosinophils and their role in disease. Crit. Rev. Clin. Lab. Sci. 1999;36:453–496.

6. Egan RW, Umland SP, Cuss FM, Chapman RW. Biology of interleukin-5 and its relevance to allergic disease. Allergy 1996;51:71–81.

7. Leckie MJ, ten Brinke A, Khan J, Diamant Z, O'Conner BJ, Walls CM, Mathur AK, Cowley HC, Chung KF, Djukanovic R, Hansel TT, Holgate ST, Sterk PJ, Barnes PJ. Effects of an interleukin-5 blocking monoclonal antibody on eosinophils, airway hypersonsiveness and the late asthmatic response. Lancet 2000;356:2144–2148.

8. O'Byrne PM, Inman MD, Parameswaran K. The trials and tribulations of IL-5, eosinophils and allergic asthma. J. Allergy Clin. Immunol. 2001;108:503–508.

9. Lipwoth BJ. Eosinophils and airway hyper-responsiveness. Lancet 2001;357:1446.

10. Flood-Page PT, Menzies-Gow AN, Kay AB, Robinson D. Eosinophil's role remains uncertain as anti-interleukin-5 only partially depletes numbers in asthmatic airway. Am. J. Respir. Crit. Care. Med. 2003;167:199.

11. Kips JC, O'Conner BJ, Langley SJ, Woodcock A, Kerstjens HAM, Postma DS, Danzig M, Cuss F and Pauwels RA. Effect of SCH55700, a humanised anti-human interleukin-5 antibody in severe persistent asthma – a pilot study. Am. J. Resp. Crit. Care. Med. 2003;167:1655–1659.

12. Flood-Page P, Menzies-Gow A, Phipps S, Ying S, Wangoo A, Ludwig MS, Barnes N, Robinson D, Kay AB. Anti-IL-5 treatment reduces deposition of ECM proteins in the bronchial subepithelial basement membrane of mild atopic asthmatics. J. Clin. Invest. 2003;112:1029–1036.

13. Morokata T, Ida K, Yamada T. Characterization of YM-90709 as a novel antagonist which inhibits the binding of interleukin-5 to interleukin-5 receptor. Int. Immunopharmacol. 2002;2:1693–1702.

14. Morokata T, Suzuki K, Ida K, Yamada T. Effect of a novel interleukin-5 receptor antagonist, YM-90709, on antigen-induced eosinophil infiltration into the airway of BDF1 mice. Immunol Lett. 2005;98:161–165.

15. Jiang H, Harris MB, Rothman P. IL-4/IL-13 signalling beyond JAK/STAT. J. Allergy Clin. Immunol. 2000;105:1063–1070.

16. Borish LC, Nelson HS, Lanz MJ, Claussen L, Whitmore JB, Agosti JM, et al. Interleukin-4 receptor in moderate atopic asthma. A phase I/II randomized, placebo-controlled trial. Am. J. Respir. Crit. Care. Med. 1999;160:1816–1823.

17. Ma Y, Hayglass KT, Becker AB, Halayko AJ, Basu S, Simons FE, Peng Z. Novel cytokine peptide-based vaccines: an interleukin-4 vaccine suppresses airway allergic responses in mice. Allergy 2007;62:675–682.

18. Walsh GM., McDougall CM. The resolution of airway inflammation in asthma and COPD. In: Progress in Inflammation Research. Rossi AG, Sawatzky D, eds. 2008. Birkhauser Verlag AG, Berlin. pp. 159–191.

19. Wills-Karp M, Luyimbazi J, Xu X, et al. Interleukin-13: central mediator of allergic asthma. Science 1998;282:2258–2260.

20. Hahn C, Teufel M, Herz U, Renz H, Erb KJ, Wohlleben G, Brocker EB, Duschl A, Sebald W, Grunewald SM. Inhibition of the IL-4/IL-13 receptor system prevents allergic sensitization without affecting established allergy in a mouse model for allergic asthma. J. Allergy Clin. Immunol. 2003;111:1361–1369.

21. Bree A, Schlerman FJ, Wadanoli M, Tchistiakova L, Marquette K, Tan XY, Jacobson BA, Widom A, Cook TA, Wood N, Vunnum S, Krykbaev R, Xu X, Donaldson DD, Goldman SJ, Sypek J, Kasaian MT. IL-13 blockade reduces lung inflammation after Ascaris suum challenge in cynomolgus monkeys. J. Allergy. Clin. Immunol. 2007;119:1251–1257.

22. Kasaian MT, Donaldson DD, Tchistiakova L, Marquette K, Tan XY, Ahmed A, Jacobson BA, Widom A, Cook TA, Xu X, Barry AB, Goldman SJ, Abraham WM.. Efficacy of IL-13 neutralization in a sheep model of experimental asthma. Am. J. Respir. Cell. Mol. Biol. 2007;36:368–376.

23. O'Byrne PM, Inman MD, Adelroth E. Reassessing the Th2 cytokine basis of asthma. Trends Pharmacol. Sci. 2004;25:244–248.

24. Wenzel S, Wilbraham D, Fuller R, Getz EB, Longphre M. Effect of an interleukin-4 variant on late phase asthmatic response to allergen challenge in asthmatic patients: results of two phase 2a studies. Lancet 2007;370:1422–1431.

25. Shimbara A, Christodoulopoulos P, Soussi-Gounni A, Olivenstein R, Nakamura Y, Levitt RC, et al. IL-9 and its receptor in allergic and nonallergic lung disease: increased expression in asthma. J. Allergy Clin. Immunol. 2000;105:108–115.

26. Gounni AS, Gregory B, Nutku E, Aris F, Latifa K, Minshall E, North J, Tavernier J, Levit R, Nicolaides N, Robinson D, Hamid Q. Interleukin-9 enhances interleukin-5 receptor expression, differentiation, and survival of human eosinophils. Blood 2000;96:2163.

27. Temann U-A, Ray P, Flavell RA. Pulmonary overexpression of IL-9 induces Th2cytokine expression, leading to immune pathology. J. Clin. Invest. 2002; 109:29–39.

28. Cheng G, Arima M, Honda K, Hirata H, Eda F, Yoshida N, Fukushima F, Ishii Y, Fukuda T. Anti-interleukin-9 antibody treatment inhibits airway inflammation and hyperreactivity in mouse asthma model. Am. J. Respir. Crit. Care. Med. 2002;166:409–416.

29. Brightling C, Berry M, Amrani Y. Targeting TNF-alpha: a novel therapeutic approach for asthma. J. Allergy Clin. Immunol. 2008;121:5–10.

30. Rabinowitz MH, Andrews RC, Becherer JD, Bickett, DM, Bubacz DG, Conway JG, Cowan DJ, Gaul M, Glennon K, Lambert, et al. Design of selective and soluble inhibitors of tumor necrosis factor-alpha converting enzyme (TACE). J. Med. Chem. 2001;44:425.

31. Fox H. Anti-IgE in severe persistent allergic asthma. Respirology 2007;12:22–28.

32. Milgrom H, Fick RB, Su J, Reimann JD, Bush RK, Watrous ML, Metzger WJ. Treatment of allergic asthma with monoclonal anti-IgE antibody. N. Engl. J. Med. 1999;341:1966–1973.

33. Busse W, Corren J, Lanier BQ, McAlary M, Fowler-Taylor A, Cioppa GD, van As A, Gupta N. Omalizumab, anti-IgE recombinant humanized monoclonal antibody, for the treatment of severe asthma. J. Allergy Clin. Immunol. 2001;108:184–190.

34. Soler M, Matz J, Townley R, Buhl R, O'Brien J, Fox H, et al. The anti-IgE antibody omalizumab reduces exacerbations and steroid requirement in allergic asthmatics. Eur. Respir. J. 2001;18:254–261.

35. Ayres JG, Higgins B, Chilvers ER, Ayre G, Blogg M, Fox H. Efficacy and tolerability of anti-immunoglobulin E therapy with Omalizumab in patients with poorly controlled (moderate-to-severe) allergic asthma. Allergy 2004;23:76–81.

36. Vignola AM, Humbert M, Bousquet J, Boulet LP, Hedgecock S, Blogg M, Fox H, Surrey K. Efficacy and tolerability of anti-immunoglobulin E therapy with Omalizumab in patients with concomitant allergic asthma and persistent allergic rhinitis: SOLAR. Allergy 2004;23:76–81.

37. Humbert, M; Beasley, R; Ayres, J; Slavin, R; Hebert, J; Bousquet, J; Beeh, K-M; Ramos, S; Canonica, GW; Hedgecock, S; Fox, H; Blogg, M; Surrey, K. Benefits of omalizumab as add-on therapy in patients with severe persistent asthma who are inadequately controlled despite best available therapy (GINA 2002 step 4 treatment): INNOVATE. Allergy 2005;60:309.

38. Griffiths-Johnson DA, Collins PD, Rossi AG, et al. The chemokine, eotaxin, activates guinea pig eosinophils in vitro and causes their accumulation in the lung in vivo. Biochem. Biophys. Res. Commun. 1993;197:881–887.

39. Uguccioni M, Mackay CR, Ochensberger B, Loetscher P, Rhis S, LaRosa GI, et al. High expression of the chemokine receptor CCR3 in human blood basophils. Role in activation by eotaxin, MCP-4, and other chemokines. J. Clin. Invest. 1997;100:1137–1143.

40. Romagnani P, De Paulis A, Beltrame C, Annunziato F, Dente V, Maggi E, et al. Tryptase-chymase double-positive human mast cells express the eotaxin receptor CCR3 and are attracted by CCR3-binding chemokines. Am. J. Pathol. 1999;155:1195–1204.

41. Sallusto F, Mackay CR, Lanzavecchia A. Selective expression of the eotaxin receptor CCR3 by human T helper 2 cells. Science 1997;277:2005–2007.

42. Stellato C, Brummet ME, Plitt JR, Shahabuddin S, Baroody FM, Liu MC, Ponath, Beck LA. Expression of the C-C chemokines receptor CCR3 in human airway epithelial cells. J. Immunol. 2001;166:1457–1461.

43. Erin EM, Williams TJ, Barnes PJ, Hansel TT. Eotaxin receptor (CCR3) antagonism in asthma and allergic disease. Curr. Drug Targets Inflamm. Allergy 2002;1:201–214.

44. De Lucca GV, Kim UT, Johnson C, Vargo BJ, Welch PK, Covington M, Davies P, Solomon KA, Newton RC, Trainor GL, Decicco CP, Ko SS. Discovery and structure-activity relationship of N-(ureidoalkyl)-benzyl-piperidines as potent small molecule CC chemokine receptor-3 (CCR3) antagonists. J. Med. Chem. 2002;45:3794–3804.

45. Suzuki K, Morokata T, Morihira K, Sato I, Takizawa S, Kaneko M, Takahashi K, Shimizu Y. In vitro and in vivo characterization of a novel CCR3 antagonist, YM-344031. Biochem. Biophys. Res. Commun. 2006;339, 1217–1223.

46. Morokata T, Suzuki K, Masunaga Y, Taguchi K, Morihira K, Sato I, Fujii M, Takizawa S, Torii Y, Yamamoto N, Kaneko M, Yamada T, Takahashi K, Shimizu Y. A novel, selective, and orally available antagonist for CC chemokine receptor 3. J. Pharmacol. Exp. Ther. 2006;317:244–250.

47. Wegmann M, Göggel R, Sel S, Sel S, Erb KJ, Kalkbrenner F, Renz H, Garn H. Effects of a low-molecular-weight CCR-3 antagonist on chronic experimental asthma. Am. J. Respir. Cell. Mol. Biol. 2007;36:61–67.

48. Palframan RT, Collins PD, Williams TJ, Rankin SM. Eotaxin induces a rapid release of eosinophils and their progenitors from the bone marrow. Blood 1998;91:2240.

49. Vanderslice P, Biediger RJ, Woodside DG, Berens KL, Holland GW Dixon RA. Development of cell adhesion molecule antagonists as therapeutics for asthma and COPD. Pulm. Pharmacol. Ther. 2004;17:1–10.

50. Kogan TP, Dupre B, Bui H et al. Novel synthetic inhibitors of selectin-mediated cell adhesion: synthesis of 1,6-bis(3-(3-carboxymethylphenyl)-4-(2-alpha-D- mannopyranosyloxy)phenyl) hexane (TBC1269). J. Med. Chem. 1998;41:1099–1111.

51. Abraham WM, Ahmed A, Sabater JR, et al. Selectin blockade prevents antigen-induced late bronchial responses and airway hyperresponsiveness in allergic sheep. Am. J. Respir. Crit. Care Med. 1999;159:1205–1214.

52. Avila PC, Boushey HA, Wong H, Grundland H, Liu J, Fahy JV. Effect of a single dose of the selectin inhibitor TBC1269 on early and late asthmatic responses. Clin. Exp. Allergy 2004;34:77–84.

53. Beeh KM, Beier J, Buhl R, Zahlten R, Wolff G. Influence of inhaled Bimosiamose (TBC1269), a synthetic pan-selectin antagonist, on the allergen-induced late asthmatic response (LAR) in patients with mild allergic asthma. Am. J. Respir. Crit. Care Med. 2004;169:A321

54. Henderson WR, Chi EY, Albert RK, Chu SJ, Lamm WJ, Rochon Y, Jonas M, Christie PE, Harlan JM. Blockade of CD49d (alpha4 integrin) on intrapulmonary but not circulating leukocytes inhibits airway inflammation and hyperresponsiveness in a mouse model of asthma. J. Clin. Invest. 1997;100:3083–3092.

55. Sedgwick JB, Jansen KJ, Kennedy JD, Kita H, Busse WW. Effects of the very late adhesion molecule 4 antagonist WAY103 on human peripheral blood eosinophil vascular cell adhesion molecule 1-dependent functions. J. Allergy Clin. Immunol. 2005;116:812–819.

56. Okigami H, Takeshita K, Tajimi M, Komura H, Albers M, Lehmann TE, Rölle T, Bacon KB. Inhibition of eosinophilia in vivo by a small molecule inhibitor of very late antigen (VLA)-4. Eur. J. Pharmacol. 2007;559:202-9.

57. Woodside DG, Vanderslice P. Cell adhesion antagonists: therapeutic potential in asthma and chronic obstructive pulmonary disease. BioDrugs 2008;22:85–100.

58. Kay AB. Immunomodulation in asthma: mechanisms and possible pitfalls. Curr. Opin. Pharmacol. 2003;3:220.

59. Jain VV, Businga TR, Kitagaki K, George CL, O'Shaughnessy PT, Kline JN. Mucosal immunotherapy with CpG oligodeoxynucleotides reverses a murine model of chronic asthma induced by repeated antigen exposure. Am. J. Physiol. Lung Cell. Mol. Physiol. 2003;285:L1137.

60. Oldfield WL, Larche M, Kay AB. Effect of T-cell peptides derived from Fel d 1 on allergic reactions and cytokine production in patients sensitive to cats: a randomised controlled trial. Lancet 2002;360:47.

61. Alexander C, Tarzi M, Larché M, Kay AB. The effect of Fel d 1-derived T-cell peptides on upper and lower airway outcome measurements in cat-allergic subjects Allergy 2005;60:1269.

# 18

# SCHIZOPHRENIA

Ferenc Martenyi

*Lilly Research Laboratories, Indianapolis, Indiana*

Since the discovery of chlorpromazine in 1953, the number of antipsychotic medications has increased tremendously. Most of these drugs affect the dopaminergic neurotransmission (primarily blocking dopamine $D_2$ receptors). Antipsychotics registered until 2008 could be classified into 3 groups based on receptor occupancy profile rather than using their chemical structure. Conventional (or first-generation antipsychotics, like haloperidol and fluphenazine) with a predominant $D_2$ receptor affinity have benefits in the treatment of delusions and hallucinations (positive symptoms of schizophrenia), which are parallel with a characteristic side-effect profile of extrapyramidal symptoms (EPS) and prolactin elevation. Atypical antipsychotics—which are characterized by a higher antagonist affinity to serotonin (5-hydroxytryptamine, 5HT) $5HT_2$ receptors than to dopamine $D_2$ receptors—are generally thought to be effective in alleviation of negative symptoms (such as anhedonia, and lack of motivation) beyond the positive symptoms domain of schizophrenia, with a reduced EPS burden (such as clozapine, olanzapine, risperidone). A third group of novel antipsychotics has a partial dopamine agonist profile (aripiprazole), with a lack of or minimal EPS. The nondopaminergic $N$-methyl-D-aspartate (NMDA) antagonist model of schizophrenia raised new gulatamatergic targets for treatment. NMDA antagonists ketamine and phencyclidine produce wide range of symptoms characteristic for schizophrenia, such as positive and negative symptoms as well as cognitive impairment. Compounds that blunt gulatamatergic activity (for example lamotrigine) or NMDA receptor glycine site allosteric agonists have both showed improvement of schizophrenic symptoms in combination with antipscychotics. The first compound that produced improvement in monotherapy for the treatment of

*Chemical Biology: Approaches to Drug Discovery and Development to Targeting Disease*, First Edition.
Edited by Natanya Civjan.
© 2012 John Wiley & Sons, Inc. Published 2012 by John Wiley & Sons, Inc.

schizophrenia is LY214023, which is a metabotropic Glutamate 2/3 (mGlu2/3) receptor agonist. This chapter reviews neurobiological background of treatments and clinical experiences proven in controlled clinical trials.

Schizophrenia is the most common psychotic disorder, which affects approximately 1% of the population across various cultures. It is a chronic and debilitating illness, which typically manifests during the adolescent and early adulthood stages and causes significant social and economic burden. Schizophrenia is accompanied by a mixture of symptoms such as delusions, hallucinations, and excitement, but essentially it is characterized by "weakening of the mainsprings of volition," "lowered mental efficiency," "unsteadiness of attention," and "inability to shift, arrange and correct ideas, and to accomplish mental grouping of ideas," as described for the first time by Emil Kraepelin (1). Meta-analyses of schizophrenic populations have shown the clustering of symptoms into at least three distinct major symptoms domains, which include positive symptoms (such as hallucinations, delusions thought disorder, and paranoidity), negative symptoms (anhedonia, social withdrawal, and thought poverty), and cognitive dysfunctions (inattention as well as disturbances in executive functions and working memory). Genetic, brain imaging, nosological, clinical, and pharmacological studies show the evidence that schizophrenia is a heterogeneous group of disorders.

Since 1953 when chlorpromazine was discovered, hundreds of compounds have been developed for the treatment of schizophrenia. With the development of the first generation antipsychotics, it has for the first time become possible to treat psychotic (positive) symptoms of schizophrenia, such as hallucinations and delusions. These first-generation antipsychotics are generally not considered an efficacious treatment for negative symptoms (such as anhedonia and lack of motivation) or cognitive deficits associated with schizophrenia, and they also cause significant extrapyramidal side effects (like parkinsonism and tardive dyskinesia), as well endocrine disturbances (hyperprolactinemia). After the discovery of clozapine, followed by many other second-generation antipsychotics, the treatment opportunities of schizophrenia improved by partial curability of negative symptoms and by significant decrease of extrapyramidal symptoms. However, only 10–20% of patients with schizophrenia recover fully to the premorbid level of functioning; 60–75% of patients suffer continuously from mental impairment (psychotic, affective, and cognitive impairment), and the remaining 15–20% of patients are treatment resistant (with highly troublesome outcome) (2).

## 18.1  NEUROBIOLOGY (NEUROCHEMISTRY) OF SCHIZOPHRENIA

The pathophysiology of schizophrenia is complex, with multiple factors that contribute to the disturbance of brain functions. Schizophrenia does not result in the dysfunction of one single neurotransmitter system, but in fact it originates from dysregulation of several interacting systems, which include the dopaminergic system, serotonergic (5-hydroxitryptamine, 5HT) system, and gulatamatergic system. More detailed information about neurobiology of schizophrenia see in the following handbooks: D'Haenen and den Boer, 2002, and Kenneth et al., 2002, in the Further Readings.

### 18.1.1 Dopaminergic Dysregulation

The dopaminerg model of schizophrenia has been supported by three major strains of evidence as follows: 1) all efficacious antipsychotics were effective $D_2$ receptor antagonists, 2) drugs that increase brain dopaminergic activity (DA) induce psychosis in healthy subjects or exacerbate psychotic symptoms in schizophrenics, and 3) more dopaminergic neurochemical alterations are detected in the brain of schizophrenics as compared with healthy controls.

Positive symptoms of schizophrenia have long been considered the result of increased activity of mesolimbic dopaminergic neurons, whereas negative symptoms are the consequences of the decreased activity of the mesocortical dopaminergic pathways. The mesolimbic and striatal dopaminergic system is controlled by the prefrontal cortex (PFC). The lesion of DA neurons in the PFC increased the levels of DA. Its metabolites, such as homovanillic acid (HVA) and dihydroxyphenylacetic acid (DOPAC), as well DA $D_2$ binding site in the striatum and conversely the injection of apomorphine (a DA agonist) into PFC, decreased the striatal dopaminergic activity. Results of studies that examined correlation between plasma HVA concentration and schizophrenic symptoms are ambiguous, but a positive correlation was found between plasma HVA concentration and clinical severity of symptoms in several studies that employed multiple sampling of plasma HVA (and reduced intraindividual variance). Assuming that the mesolimbic and/or the mesocortical dopamine activity is etiologically significant in schizophrenia but that most HVA in the cerebrospinal fluid (CSF) is produced by the nigrostriatal dopamine system, the predictive value of HVA measures in CSF is poor.

The dopamine overactivity model of schizophrenia was not unanimously supported by measurements of dopamine receptor binding at brain regions: the positive findings might be confounded by the effect of previous antipsychotic medication, and the increased receptor density may be a consequence of antipsychotic drug treatment maintained until death. Postmortem studies show increased $D_2$ and $D_4$ but not $D_1$ receptor affinity in various subcortical regions, such as the caudate, nucleus accumbens, and amygdala. However, most *in vivo* [positron emission tomography (PET)] studies of drug-naive patients have not supported the hypothesis of elevated $D_2$ receptor densities in the striatum. DA $D_1$ receptors did not differ significantly between subjects with schizophrenia and healthy subjects—using [(11)C]SCH 23390 binding—in any of the brain regions or for any of the binding measures studied; however, significant correlation was observed between Brief Psychiatric Rating Scale scores on the negative symptoms with the binding B(max) in the right frontal cortex.

### 18.1.2  Serotonergic Dysregutaion

The following data support the involvement of the serotonergic (5HT) system in the pathogenesis of schizophrenia: 1) $5HT_{2A}$ (partial) agonists induced psychosis and hallucinations, 2) the neuroendocrine-challenged paradigms, and 3) atypical antipsychotics show prominent $5HT_2$ antagonist profile.

The 5HT theory emerged from human observation that the lysergic acid diethylamide (LSD), or 4-iodo-2,5-dimethoxyphenylisopropylamine and mescaline, produced hallucinations predominantly through a common route of action of $5HT_{2A}$ receptor agonism. The LSD psychosis model failed to show a good face validity; visual hallucinations produced by it are highly unusual in schizophrenia. However, acoustic hallucinations, paranoid delusions, conceptual disorganization, and cognitive impairments that characterize schizophrenia are grossly absent during LSD intoxication. Most studies found no significant differences in the CSF concentration of 5HT metabolite 5-hydroxy-indol-acetilic acid (5-HIAA); however, a few papers reported decrease of 5-HIAA CSF concentration in schizophrenia (3). Neurochemical evidence suggests that $5HT_{2A}$ receptor stimulation contributes to facilitation of synthesis and release of DA, which influences either the firing rates of DA neurons and/ or the $5HT_{2A}$ heterorecptors situated on DA nerve terminals. Activation of $5HT_{2A}$ receptors in the medial PFC caused a calcium-dependent increase in the frequency of excitatory postsynaptic potentials and currents. These effects can be blocked by the application of alpha-amino-3-hydroxy-5-methyl-4-isoxazolepropionic acid receptor antagonists or the activation of group mGlu2 receptors. These results suggest that glutamate release plays a key role: It represents a common pathway for the actions of serotonergic hallucinogens that, the $5HT_{2A}$ agonists, increase 5HT release in the medial prefrontal cortex secondary to stimulation of glutamate release (4).

Neuroendocrine challenge paradigms are widely used in psychiatric research to probe the functioning of a given neurotransmitter system. Because serum prolactin and cortisol are regulated by central 5HT pathways, serotonergic agonists such as $D$-fenfluramine can be used to study the functional status of central serotonergic synapses in schizophrenia. Several studies showed a significantly enhanced prolactin response to D-fenfluramine in schizophrenic patients, and the prolactin response significantly correlated with negative symptoms. This finding suggests either sensitization or upregulation of $5HT_{2A}/5HT_{2C}$ postsynaptic receptors. In addition, substantial evidence indicates a direct role of the prefrontal $5HT_{2A}$ receptors in the cognitive process. Working memory dysfunction is a core symptom of schizophrenia (5).

Clozapine, olanzapine, and risperidone, which are atypical antipsychotics with dominant $5HT_2$ antagonist profile, blunted felnfluramine-induced prolactin response in schizophrenics. Clozapine, but not haloperidol, blocked the effect of 5HT (partially $5HT_{2A}$) agonist, m-chlorophenylpiperazine on adrenocorticotropic hormone, and prolactin release, which demonstrated that clozapine, in contrast to haloperidol, is a functional potent $5HT_{1c}$ antagonist. The postmortem literature reports increased $5HT_{1A}$ binding in the prefrontal cortex in schizophrenia;

however, a PET study using $[(11)C]WAY$ 100635 as radiotracer for $5HT_{1A}$ receptor did not detect differences in $5HT_{1A}$ binding in medication-free schizophrenics compared with normal controls (6).

### 18.1.3 Gulatamatergic Dysregulation

The anesthetics phencyclidine (PCP) and ketamine are known to produce dissociative and psychotic symptoms with considerable similarity to schizophrenic psychotic symptoms such as the following: positive and negative symptoms, as well as cognitive impairment (deficits of memory, attention, and abstract reasoning), disruptions in smooth-pursuit eye movements, and prepulse inhibition of startle. A direct comparison of healthy volunteers who received subanesthetic doses of ketamine and individuals with schizophrenia shows similar disruptions in working memory and thought disorder between the two groups. A consequence of blocking $N$-methyl $D$-aspartate (NMDA) receptors is the excessive release of glutamate in the cerebral cortex; the consequent overstimulation of postsynaptic neurons might explain the cognitive and behavioral disturbances associated with the NMDA receptor hypofunction state.

So far, clozapine and lamotrigine have been investigated in this human experimental psychosis paradigm and have been shown to be effective in suppressing the expression of psychotomimetic symptoms. Haloperidol, which is a DA $D_2$ antagonist, did not suppress ketamine-induced psychosis (7). However, drugs that facilitate $GABA_A$ neurotransmission (e.g., lorazepam) attenuate partially ketamine-induced reactions. These data support the hypothesis that the gulatamatergic neurotransmission has a key role in promoting psychotic symptoms that interact with numerous other neurotransmitter systems, like 5HT and GABA.

Brain imaging studies also provide new solid evidence of gulatamatergic dysregulation in schizophrenia. Reduced NMDA receptor binding have reported in the hippocampus of medication-free schizophrenic patients with $(^{128}I)CNS$-1261 SPECT (8). Gray matter changes, such as volume reduction in mediofrontal cortex, prior to manifestation of psychosis could be related to abnormal neuronal plasticity of excitotoxitcity caused by elevated glutamate level (9). Typical and atypical antipsychotics seem to modulate this system toward normality.

Glutamate neurons have played a key role in the orchestration of the other brain neurons that have been suggested in the pathophysiology of schizophrenia. These neurons include GABA interneurons because reduced GABA synthesis was detected in the inhibitory GABA neurons in the dorsolateral PFC in schizophrenia, dopamine neurons, and serotonergic neurons. The bursting of dopamine-containing neurons—a key response to environmental stimuli—is dependent on NMDA receptor activation.

### 18.1.4 Cholinergic Dysregulation

A shift of dynamic balance of dopaminergic/cholinergic neurotransmission (increased dopaminergic and/or reduced cholinergic activity) contributes to the development of psychotic symptoms. Neuroimaging and postmortem brain studies (using $M_1$ and $M_4$ selective antagonist [$^3$H]pirenzepine binding) show the role of muscarinic acetylcholine receptors in the pathology of schizophrenia. The numbers of both muscarinic and nicotinic receptors are decreased in schizophrenia. Functional polymorphism of 267A/C in the muscarinic 1 ($M_1$) receptors has been associated with cognitive deficits of the disorder, and the $\alpha_7$ nicotinic receptor has been linked genetically to schizophrenia. The muscarinic/acetylcholinergic system interacts in several other transmitter pathways. $M_4$ receptors are most highly expressed in brain regions rich in dopamine and dopamine receptors. The basal forebrain cholinergic complex—which projects throughout the cerebral cortex—is regulated by GABA-mediated output from the nucleus accumbens. An abnormal dopaminergic tone in the nucleus accumbens might alter the activity of cholinergic projections to the cortex, which thereby links an alteration in cortical function to subcortical dopamine dysregulation. However, $M_1$/$M_4$ agonists such as oxotremorine preferentially increase cortical DA release and inhibit amphetamine-induced DA release in the nucleus accumbens.

Additional, indirect data suggests a key role of muscarinic abnormalities in schizophrenia: Clozapine (and obviously N-desmethylclozapine), with its unique efficacy profile, shows the highest affinity to the muscarinic $M_1$ and $M_4$ receptors.

## 18.2  ANTIPSYCHOTICS

### 18.2.1  Receptor-binding Profile Antipsychotics

Small molecules in the treatment of schizophrenia are characterized by clinical and/or pharmacodynamic properties rather than their chemical structures. The first-generation antipsychotics are labeled as "conventional" compared with later developed, "atypical" antipsychotics. However, a considerable level of imprecision exists in the use of this terminology for the second-generation drugs versus atypical antipsychotics. Conventional antipsychotics vary widely in the chemical structures (see Table 18.1), but these drugs share a common route of action of blocking DA $D_2$ receptors with the same or similar affinity in the mesolimbic and nigrostriatal areas of the brain. These drugs are considered conventional based on the long-lasting experience that antipsychotic effect cannot be observed without a certain level of extrapyramidal (neuroleptic) symptoms. For establishing the neuroleptic threshold, several methods have been developed, for example the handwriting test, which can be used to detect the effect of $D_2$ receptor blockade. The appearance of the motoric symptoms such as the handwriting becomes slowed (bradykinesia) and letters decreased in size (micrigraphia), indicate the Parkinsonian (neuroleptic) threshold. Conventional antipsychotics can be classified on a potential scale of low-medium-high, which is based on the affinity to DA $D_2$ receptors and the average therapeutic dose compared with chlorpromazine (10).

**TABLE 18.1  Classification of conventional and atypical antipsychotics by chemical structure**

### Phenothiazines

#### With aliphatic side chain

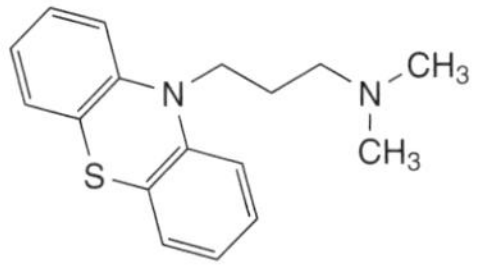

Chloropromazine

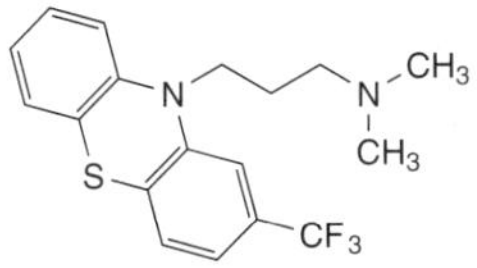

Triflupromazine

### Piperidines

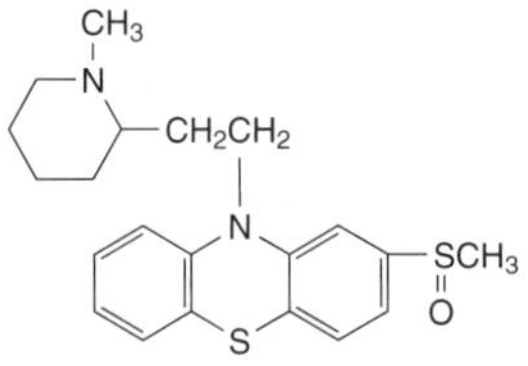

Mesoridazine

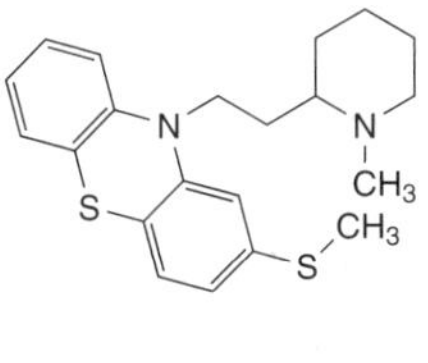

Thioridazine

### Piperazines

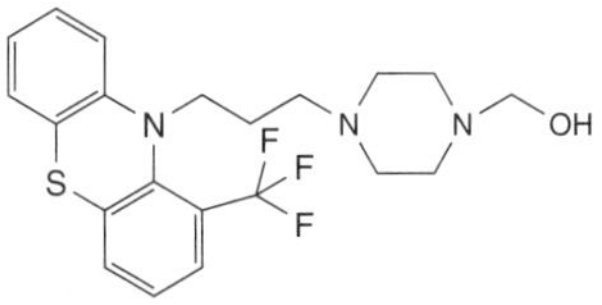

Fluphenazine

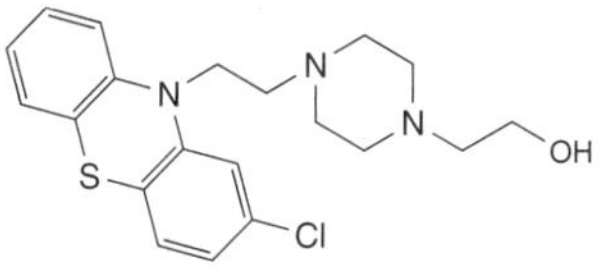

Perphenazine

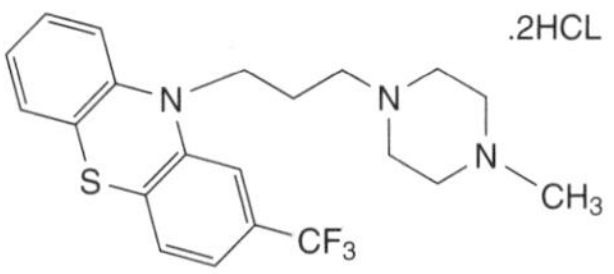

Trifluoperazine

### Thioxanthens

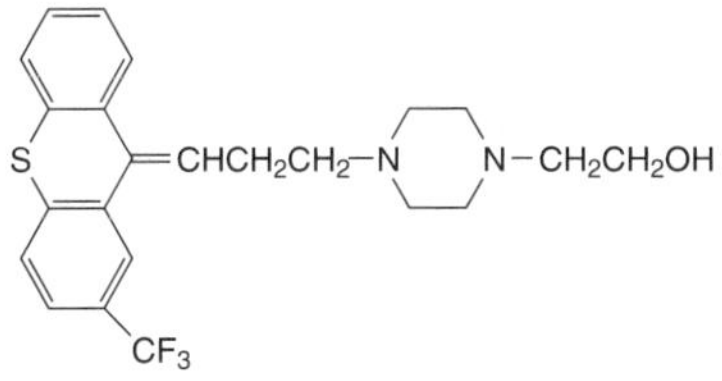

Flupentixol

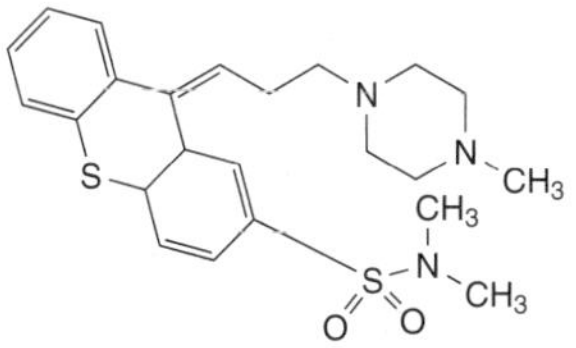

Thiothixene

(*continued*)

**TABLE 18.1**  *(Continued)*

Phenothiazines

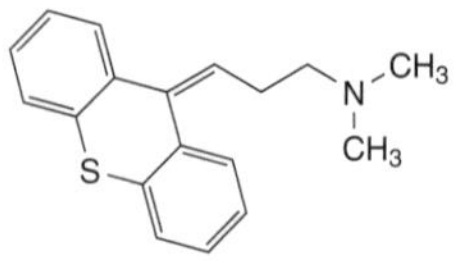

Chlorprothixene

Butyrophenones

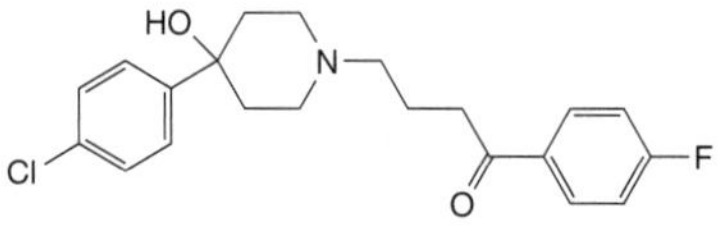

Haloperidol

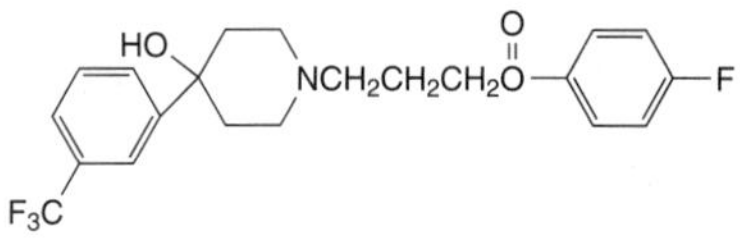

Trifluperidol

Diphenylbutylpiperidines

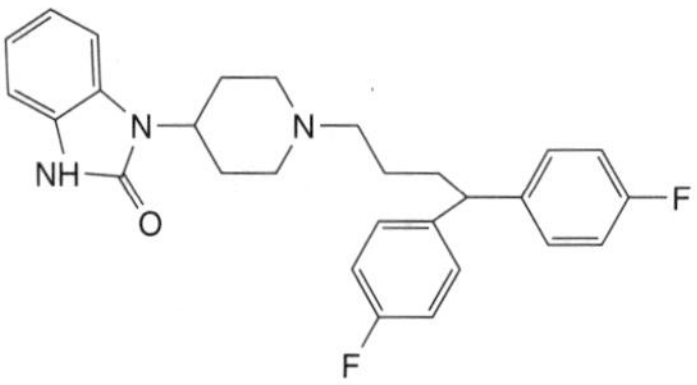

Pirrozide

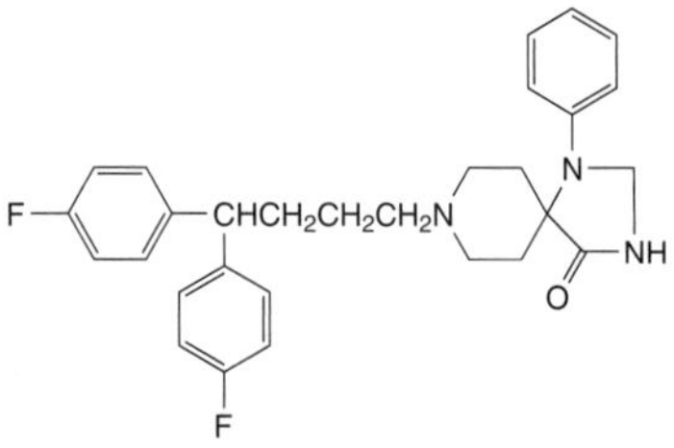

Fluspirilene

Benzamides

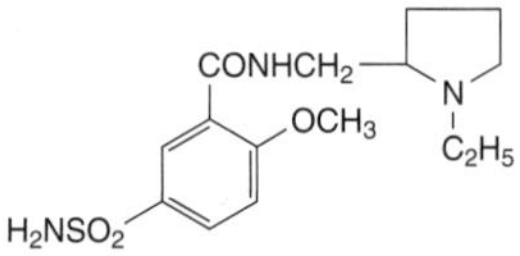

Sulpiride

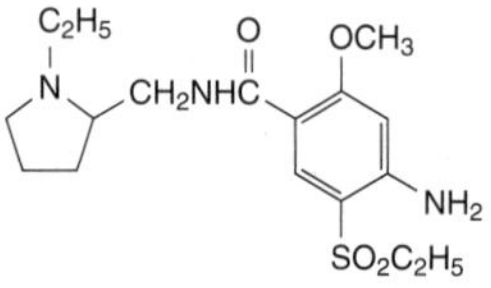

Amisulpride

Indole derivates

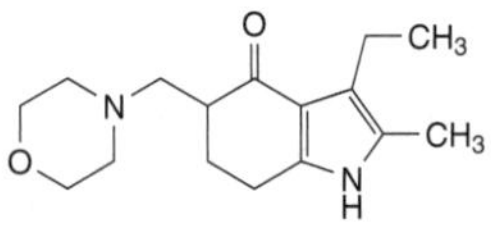

Molindone

**TABLE 18.1**   *(Continued)*

Phenothiazines

Benzoazepines

Dibenzodiazepines            Thienobenzodiazepines

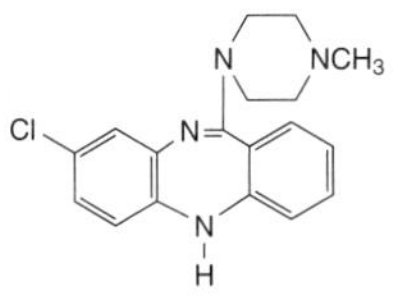

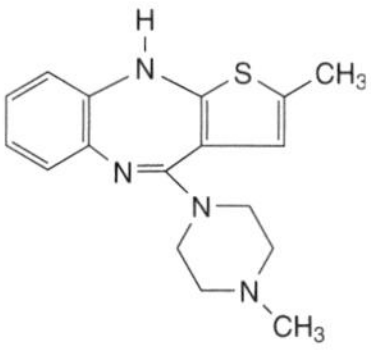

Clozapine                    Olanzapine

Dibenzothiazepine

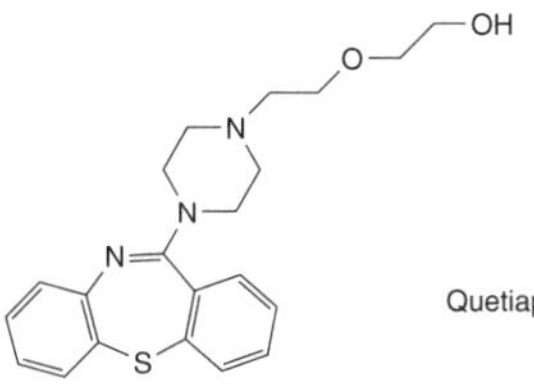

Quetiapine

Benzothiazolylpipoerazine

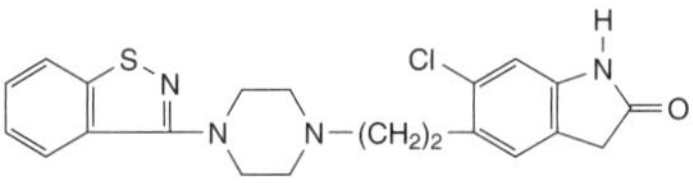

Ziprasidone

Bezisoxasoles

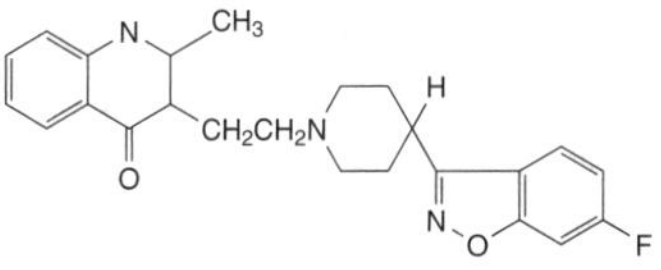

Risperidone

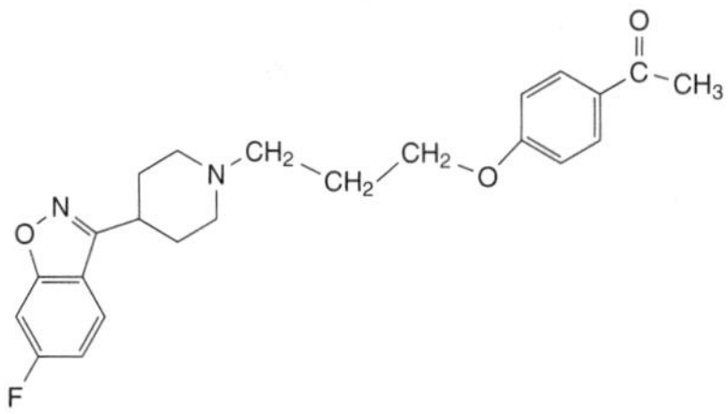

Iloperidone

*(continued)*

**TABLE 18.1** (*Continued*)

Phenothiazines

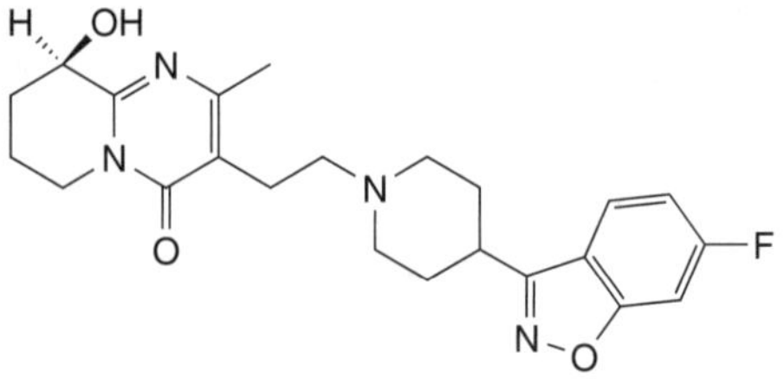

Paliperidone

Quinolinones

Imidazolidinone

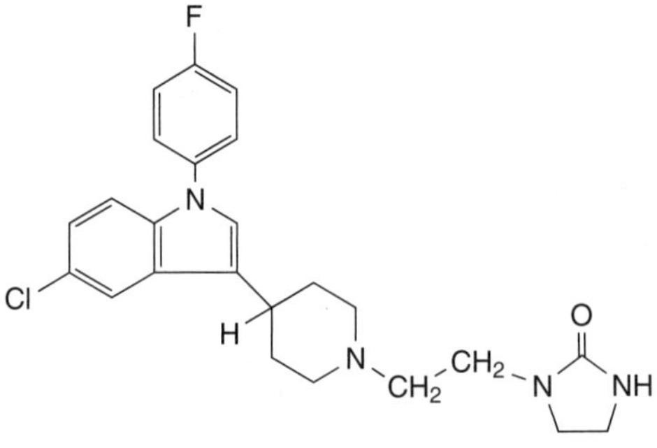

Sertindole

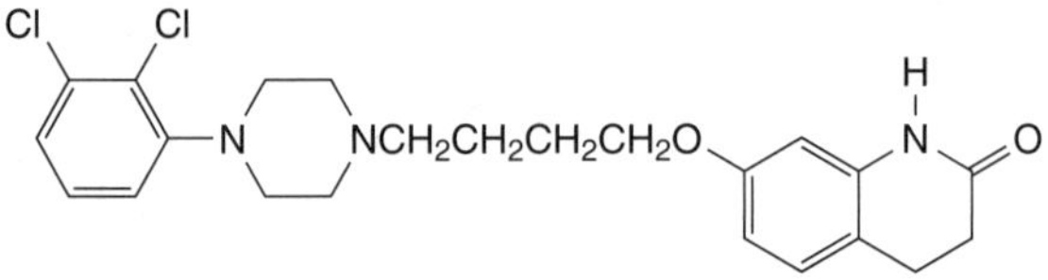

Aripiprazole

PET studies demonstrated the importance of DA $D_2$ occupancy as a predictor of antipsychotic response and adverse events. The antipsychotic effect in conventional drugs requires 65–70% $D_2$ receptor occupancy, but if it increased to more than 80% it increases significantly the risk of extrapyramidal symptoms. Atypical antipsychotics such as olanzapine do not show increased risk for development of extrapyramidal symptoms. Clozapine, at doses known to be effective in routine clinical settings, shows $D_2$ receptor occupancy of 16–68%, which is lower than risperidone (63–89%) or olanzapine (43–89%) (11).

The strength of DA $D_2$ receptor affinity (log $K_i$ $D_2$) showed a positive correlation with the clinically effective doses for conventional antipsychotics but not for atypical antipsychotics (12). The most widely accepted hypothesis postulates that a relatively weaker $D_2$ receptor affinity compared with $5HT_{2A}$ receptor activity is the characteristic for compounds with robust antipsychotic activity. With none or minimal extrapyramidal symptoms in humans and in animal models, these drugs efficaciously block d-amphetamine-induced hyperlocomotion and have weak cataleptogenic effects (13). Using the ratio of $5HT_{2C}/D_2$ rather than $5HT_{2A}/D_2$ receptor affinity [log $K_i$ ($5HT_{2A}/D_2$)] plotted against the clinically effective dose (log average dose) provides a better correlation for conventional and atypical antipsychotics.

Clozapine, which shows minimal EPS liability and robust antipsychotic potential, has partial agonist properties at $5HT_{1A}$ receptors. Several other atypical antipsychotics, such as aripiprazole and bifeprunix, have $5HT_{1A}$ receptor agonist properties. The ability of clozapine to increase DA release in the PFC in rats is partly related to its $5HT_{1A}$ antagonist properties.

Based on the receptor affinity analysis data, it is suggested that the antipsychotic potency is determined by three different factors as follows: 1) increased $D_2$ receptor binding affinity enhances antipsychotic potency, 2) increased $5HT_{2A}$ and $5HT_{2C}$ receptor affinity improves antipsychotic effect, and 3) increased $5HT_{1A}$ receptor binding reduces antipsychotic potential; however, partial or full agonism promotes an antipsychotic effect. Aripiprazole represent a somewhat unique profile: Both on $D_2$ and $5HT_{1A}$ receptors, it shows partial agonism. It was suggested that aripiprazole has "functional-selective" actions: Depending on the cellular milieu, a mixture of antagonist/partial agonist/agonist actions are likely. Agonist action might induce changes in receptor conformation, which causes changes in sensitivity.

An alternative hypothesis suggests that a faster dissociation rate ($k_{off}$) from the DA $D_2$ receptor is the key element, which results in a lower overall affinity for the $D_2$ receptor (14). However, the highly selective $D_{2/3}$ receptor antagonist, amisulpride, shares a couple of criteria of atypicality with several $5HT_2/D_2$ drugs. A simultaneous $5HT_{2A}$ and $5HT_{2C}$ receptor blockade might be more successful in mediating antipsychotic effect than inhibition of either of these receptors separately.

Several conventional and atypical antipsychotics are potent antagonists of the $\alpha_1$ and $\alpha_2$ receptors. The high $\alpha_1$ receptor affinity might explain the atypical feature of clozapine, olanzapine, risperidone, and paliperidone, which has $D_2$

affinity profile similar to numerous conventional antipsychotics. The blockade of $\alpha_1$ receptors increases the DA release in the limbic system in the shell part of the nucleus accumbens, which indicates a limbic rather than striatal effect of $\alpha_1$ antagonism. Most of the atypical antipsychotics mentioned above are also potent $\alpha_2$ receptor antagonists. The combination of idazoxan, which is an $\alpha_2$ antagonist, with the conventional antipsychotic fluphenazine, resulted in improved antipsychotic treatment efficacy similar to clozapine in a treatment-resistant schizophrenic population (15).

## 18.3  BEHAVIORAL PHENOTYPE

In rodents, locomotor activity changes are often used as a screening model for psychosis (schizophrenia). Animals treated with psychostimulants such as d-amphetamine and apomorphine (DA agonists) revealed locomotor hyperactivity as well as stereotyped or perseverative behaviors in higher doses. A blockade of d-amphetamine-induced hyperlocomotion is considered to be a predictor of antipsychotic effect via blockade of DA $D_2$-receptors in the mesocortical—mesolimbic system.

The catalepsy test (16) is a behavioral measure (time spent immobile on an inclined plane) of nigro-striatal DA $D_2$ receptor blockade.

The ratio of doses ($ED_{0.5}$) of half-minute cataleptogenic effect and ($ED_{50}$) dose of blunting hyperlocomotion followed by d-amphetamine exposure is considered a predictor of conventional versus atypical profile of an antipsychotic. The $ED_{0.5}$ for cataleptogenic effect was $>120$ μmol/kg, $>120$ μmol/kg, 13.1 μmol/kg, and 0.3 μmol/kg for clozapine, chlorpromazine, risperidone, and haloperidol, respectively. The $ED_{50}$ for cataleptogenic effect was 2.4 μmol/kg, 0.5 μmol/kg, 0.8 μmol/kg, and 0.2 μmol/kg, which provided a catalpetogenic/antihyperlocomotion ratio of 50.6, 42, 16.8, and 1.8, respectively. This model reliably differentiates clozapine from haloperidol, but chlorpromazine remained in the atypical range (17). Several other drugs, which do not share a DA and/or 5HT antagonist profile, such as xanomeline (partial agonist to $M_1/M_4$ $> M_2/M_3 > M_5$ and with modest interaction at $5HT_{1A}$, $5HT_{2A}$ receptors), have shown an atypical profile. These drugs block dopamine agonist-induced behavioral disturbances in rodents, without producing catalepsy.

Of the available animal models, subchronic treatment with PCP or ketamine seems to provide the most valid model or schizophrenia-like alterations in the animal behavior, which include hyperlocomotion, stereotyped behavior, as well marked deficits in social behavior. These effects are reversed by clozapine and olanzapine but not by haloperidol.

## 18.4  ANTIPSYCHOTICS CLINICAL EFFICACY PROFILE

Conventional versus atypical antipsychotic classification is based on numerous factors, which include behavioral phenotype in preclinical animal models, receptor-binding affinity, clinical efficacy and safety, and efficacy in "treatment-resistant" cases (see Table 18.2). The conceptualization of "atypicality" versus "conventionality" of antipsychotics from a clinical perspective was grossly determined by the fact, that clozapine—the archetype of atypical antipsychotics—provided clinically significant advantages to the established conventional antipsychotics, such as chlorpromazine or haloperidol. Controlled, double-blind studies suggest that clozapine, olanzapine, and risperidone are superior to haloperidol in controlling psychotic symptoms. Risperidone was superior to haloperidol in the treatment of positive and negative symptoms; olanzapine has demonstrated efficacy in the treatment of positive, negative, and mood symptoms of schizophrenia. In several trials, olanzapine was superior to risperidone and haloperidol in the improvement of negative symptoms; however, other studies—with different trial methodology showed a greater benefit of risperidone versus olanzapine in positive and mood symptoms (18). Quetiapine, ziprasidone, and aripirazol seem to be comparable with chlorpromazine and/or haloperidol in both positive and negative symptoms (19–22). With positive and negative symptoms, aripiprazole failed to show noninferiority in comparison with olanzapine in the positive and negative symptoms, as well in the depressive symptoms associated with schizophrenia. For further details of meta-analysis of efficacy of antipsychotics see systematic reviews in the Cochran database, and the WFSBP Treatment Guidelines for schizophrenia (Falkai et al., 2005) in the Further Readings.

### 18.4.1 Treatment-resistant Schizophrenia

Clozapine is the golden standard of antipsychotics in the treatment of schizophrenia refractory to other antipsychotic administrations. Clozapine was superior compared with chlorpromazine in patients who showed no clinical response to three previous antipsychotic treatment regimens and 6 weeks of haloperidol administration. In a sample of 268 patients, 30% of patients in the clozapine group met response criteria at 6 weeks compared with 7% of patients treated with chlorpromazine (23). Olanzapine failed to show superiority to chlorpromazine in a trial designed identically with the clozapine study mentioned above but was found more effective than haloperidol in another study (24). No difference was observed between aripiprazole and perphenazine (a conventional antipsychotic) in treatment-resistant schizophrenia, where beyond case history, treatment resistance was confirmed by 4 to 6 weeks of open-label treatment with aripiprazole (15–30 mg/day) or perphenazine (8–64 mg/day). No differences were observed in the *a priori* predefined response rates.

### 18.4.2 Cognitive Impairments Associated with Schizophrenia

Cognitive symptoms are usually present during the prodromal phase (preceding the first psychotic episode) of schizophrenia and persist during the course of the illness. The impact of cognitive impairment in schizophrenia was well

**TABLE 18.2   A pattern of clinical and preclinical criteria of "atypicality" of antipsychotics***

| Criteria | | Clozapine | Amsulpride | Aripiprazole | Olanzapine | Quetiapine | Risperidone | Ziprasidone |
|---|---|---|---|---|---|---|---|---|
| In comparison with haloperidol or chlorpromazine | more efficacious in positive symptoms | + | + | − | + | − | + | − |
| | more efficacious in negative symptoms | + | + | − | + | − | + | − |
| | more efficacious in cognitive symptoms | + | ? | − | + | − | + | − |
| | more efficacious in mood/ depressive symptoms | + | ? | − | + | +/− | + | ? |
| Efficacy in treatment-refractory schizophrenia | | + | ? | +/− | +/− | ? | +/− | ? |
| Efficacy in prevention of suicidality | | + | ? | ? | − | ? | ? | ? |
| Adverse event profile | No worsening / induction of Parkinsonism | + | +/− | + | + | + | +/− | + |
| | No induction of tardive dyskinesia | + | + | + | + | + | + | + |
| | No prolactin elevation | + | − | + | + | + | − | + |
| Combined 5HT / D2 receptor affinity profile | | + | − | + | + | + | + | + |
| Block PCP or ketamine induced behavioral abnormalities in rodents | | + | − | ? | + | + | + | ? |

+ = increasing effect; 0 = no effect; ? = inconclusive effect.

*Many characteristics are used to distinguish atypical and conventional antipsychotics. Clozapine represents the first and might be most efficacious atypical antipsychotic. Clinical evidence supports the efficacy that exceeds the effect of conventional antipsychotics (chlorpromazine and haloperidol) in the treatment of positive, negative, cognitive, and mood symptoms, as well in the treatment of treatment-refractory schizophrenia. Clozapine showed evidence in reduction of suicide in the high-risk schizophrenic population. From an adverse-event profile, the optimal atypical antipsychotic (similarly to clozapine) does not provoke parkinsonian symptoms or tardive dyskinesia and does not increase prolactin level.

From a route-of-action aspect, these drugs usually have a combined serotonergic/dopaminergic receptor affinity profile ($5HT2/D_2$). The prevention of NMDA receptor antagonists (PCP or ketamine) can induce behavioral abnormalities; this result is highly predictive for a group of atypical antipsychotics and distinguishes them from conventional antipsychotics.

Using a narrow definition of "atypicality," olanzapine and risperidone are sharing the most criteria of an atypical profile of clozapine. According to a broader definition, more antipsychotics met the typical criteria, which shows low propensity to cause extrapyramidal symptoms.

known for Kraepelin (1), but the pharmacotherapy remained ineffective in the cognitive domain until the introduction of clozapine and the newer atypical antipsychotics. Clozapine, risperidone, and olanzapine have shown superior efficacy compared with conventional antipsychotics; these drugs improve verbal fluency, attention, motor functions, and executive functions (25). In comparison with conventional antipsychotics, the low frequency or absence of drug-induced parkinsonian symptoms in case of atypical antipsychotics might contribute to the cognitive benefits. Haloperidol has a significant negative impact on cognition, sustained attention, reaction time, and speed of information processing (26). Comparing cognitive effects of various atypical antipsychotics in patients with poor response to conventional antipsychotic treatment clozapine, olanzapine and risperidone showed differences in their patterns of cognitive effects. However, in one study amisulpride was not inferior to olanzapine for any cognitive domain, comparing combined $5HT_{2A}/D_2$ receptor blockade and $D_2/D_3$ receptor inhibition. This finding suggests that $5HT_{2A}/D_2$ receptor blockade is probably not necessary for cognitive improvement by second-generation antipsychotics (27).

***18.4.2.1 Prevention of Suicide*** Suicide is one of the leading causes of death in schizophrenia. Four of six retrospective studies provide evidence for the ability of clozapine therapy to reduce suicidal behavior. So far, only one controlled clinical study has been conducted in the high-suicide-risk population, in which clozapine showed better efficacy in prevention of suicidal thoughts or suicide attempts compared with olanzapine (28).

## 18.5 ADVERSE EFFECTS ASSOCIATED WITH ANTIPSYCHOTIC TREATMENTS

### 18.5.1 Extrapyramidal (Motor) Side Effects

Acute dystonia occurs secondary to $D_2$ receptor blockade in the nigrostriatal system. Acute dystonia can appear within minutes or up to the first few days of conventional antipsychotic treatment in forms of involuntary and usually painful muscle spasms. These spasms occur particularly in the eye muscles (eyes roll upward, oculogyric crisis), spasm of the neck muscles (torticollis), spasm of laryngeal muscles (laryngospasm), or unilateral spasm of the paravertebral muscles. In addition, rotation and flexion of the upper trunk, neck, and head to one side (Pisa-syndrome) can occur.

***18.5.1.1 Medium-term Adverse Events*** Akathisia usually occurs within the first week of treatment and involves motor and/or psychic restlessness. Motor restlessness usually affects the lower limbs, with shifting from one foot to the other while standing or frequently crossing and uncrossing the legs while sitting. Akathisia is only partly the consequence of antidopaminergic effects; adrenergic, serotonergic and muscarineregic effects also play role in the development of adverse events.

Parkinsonism usually occurs after several days to weeks of antipsychotic treatment; this disorder mimics idiopathic Parkinson's disease with the typical symptoms of tremor, bradykinesia, rigidity of the muscles of extremities and face muscles ("poker face"), and loss of upper limb swing while walking. The symptomatology of parkinsonism is rather akinetic; rigid tremors (bilateral) occur less frequently.

*18.5.1.2 Chronic Extrapyramidal Adverse Events*    Tardive dyskinesia (TD) usually manifests after prolonged exposure of conventional antipsychotics. The main symptoms are the choreoathetoid involuntary movements of the tongue and mouth, as well as of the head, neck, and trunk. Movements are often suppressed by voluntary actions that involve the affected part. Based on the following facts, TD was considered as a consequence in supersensitivity of DA $D_2$ receptors. First, TD is evoked by prolonged DA antagonist exposure. Second, dyskinesias increase when the dosage is reduced or the drug is discontinued. Also, movements are reduced by readministration or increase of the dose of the conventional antipsychotic, but increasing conventional neuroleptic dosage increases the risk and severity of tardive dyskinesia in the long run. Beyond upregulation of $D_2$ receptors, degeneration in GABA-generating neurons is another significant pathway in tardive dyskinesia. Acetylcholine might play an important role in tardive dyskinesia as well. Degeneration of the striatal cholinergic neurons was observed in autopsies of patients treated with neuroleptics and who developed tardive dyskinesia. In fact anticholinergics, which are often used in the treatment of parkinsonian symptoms, worsen TD.

*18.5.1.3 Neuroleptic Malignant Syndrome (NMS)*    This syndrome is life threatening; it involves hyperthermia, muscle rigidity, severe instability of autonomic functions, and consciousness fluctuation. These symptoms are accompanied by extreme increase of creatinine phosphokinase and prolactin. NMS was rarely reported with atypical antipsychotics.

### 18.5.2 Nonmotor Side Effects of Antipsychotics

*18.5.2.1 Hematologic Side Effects*    Almost all groups of psychotropic drugs have been reported to cause dysfunctions in the hemopoietic system. In case of phenothiazines (chlorpromazine), the risk of developing agranulocytosis is approximately 0.13%; in case of clozapine, the risk is 0.85%. Reports of agranulocytosis led to the withdrawal of clozapine in the 1970s, and in most countries, clinical administration of clozapine is restricted only to treatment-resistant schizophrenia. Metabolism of clozapine, nor-clozapine, and/or clozapine-$N$-oxide to electrophilic nitrenium ions is the initial step in the events that lead to agranulocytosis. Nitrenium ions may bind to neutrophils causing either cell death or neutrophil apoptosis via oxidative stress. Antineutrophil antibody formation against the neutrophil protein + nitrenium ion hapten may also be involved in the pathophysiology of agranulocytosis induced by clozapine.

*18.5.2.2 Weight Gain and Metabolic Effects* Concerns about extrapyramidal symptoms induced by conventional antipsychotics have been replaced by weight gain, glucose deregulation, and dyslipidemia caused by atypical antipsychotics. Weight gain during atypical antipsychotic treatment varies by the length of administration and by body mass index (BMI) at baseline: Patients with low BMI (<23) tend to gain more weight compared with patients with BMI greater than 27 who saw less weight gain. Psychotropic drugs with significant histaminergic activity (such as the antidepressant mirtazapine or the low potential conventional antipsychotics) have well-known effects of weight gain. Weight gain after short-term atypical antipsychotic exposure significantly correlates with H1 histaminergic affinity of the drugs, as well with $\alpha_{1A}$, $5HT_{2C}$, and $5HT_6$ receptor bindings. Other analysis shows correlation between weight increase and affinity to $D_2$, $\alpha_2$, $5HT_{1A}$, and $5HT_{2C}$ receptors. Dopamine produces a robust decrease in feeding behavior in animal models. Blockade of $D_2$ receptors in the lateral-perifornical hypothalamus seems to increase feeding, but it decreases appetite in nucleus accumbens (29). Antipsychotics might influence the energy homeostasis indirectly or directly. Leptin and ghrelin play major roles in the control system for energy balance in humans. Elevation of adipocyte-derived hormone leptin levels has been observed in schizophrenic patients during several atypical antipsychotic treatment (olanzapine and clozapine) early after the onset of medication. Changes in the gut-derived hormone ghrelin levels were observed as well in a biphasic manner depending on the treatment duration; a decrease at was observed earlier time intervals and an increase was observed in longer ones.

The use of (atypical) antipsychotic medications is associated with an increase in adipose mass and the development of diabetes. Characteristic hormonal findings in patients with insulin-resistance syndromes include elevated levels of plasma insulin. In studies conducted on schizophrenic patients, plasma insulin levels are markedly increased compared with the control population. The changes in insulin levels associated with the use of antipsychotic medications are not acutely observed. The increase in insulin levels and insulin resistance in the absence of weight gain may be the result of an increase in the secretion of insulin counter-regulatory hormone glucagon. DA $D_2$ receptors are expressed in β cells of both rats and humans and are believed to mediate glucose-stimulated insulin secretion with insignificant effects on the basal secretion of insulin (see Table 18.3) (30).

The comparison of antipsychotic-induced weight change after 10 weeks of standard drug dose showed the rank of clozapine > olanzapine > thioridazine > sertindol > chlorpromazine > risperidone > haloperidol > fluphenazine > ziprasidone (weight neutral) > molindine (weight loss) > placebo (weight loss) (31). Genetic variation such as polymorphism of $5HT_{2C}$ receptor gene influences antipsychotic-induced weight changes (32).

Patients with schizophrenia are 2–3 times more likely to have Type 2 diabetes than adults in the general population. Large population studies found that patients treated by atypical antipsychotics, such as clozapine, quetiapine, and olanzapine are at higher risk for developing Type 2 diabetes than patients treated by conventional antipsychotics. Decreased pancreatic beta-cell responsiveness as well as

**TABLE 18.3   Metabolic effects of haloperidol and atypical antipsychotics (30)**

|  | Haloperidol | Olanzapine | Clozapine | Risperidone | Ziprasidone |
|---|---|---|---|---|---|
| Weight gain | 0 | +++++ | ++++ | +++ | ++ |
| Diabetes | 0 | +++ | +++ | ? | 0 |
| Increase of triglycerides | 0 | ++++ | +++++ | +++ | ++ |
| Decrease of high-density lipoprotein | 0 | +++ | +++ | ? | 0 |
| Leptin | 0 | +++ | ++++ | ++ | ? |
| Ghrelin | 0 | +++ | +++ | +++ | ? |
| Insulin | ? | +++ | ++++ | ++ | ? |
| Glucagon | ? | ++ | ++ | ++ | ? |

+ = increasing effect; 0 = no effect; ? = inconclusive effect.

decreased insulin sensitivity that is possibly mediated by antagonism of the H1 $5HT_{2C}$ receptor, as well $M_3$ muscarinic receptors, might lead to the development of Type 2 diabetes.

Clozapine, olanzapine, and to a lesser extent quetiapine are associated with an increase of triglyceride and serum cholesterol levels, whereas risperidone and aripirazol have minimal or no effect. The antipsychotics route of action in deregulation of lipid metabolism is unclear, but peroxisome proliferator-activated receptors might play a role (33).

## 18.6   CHOLINERGIC DRUGS (POSITIVE MODULATORS OF MUSCARINIC RECEPTORS)

The hypothesis that positive allosteric modulation of muscarinic receptors might lead to reduction of psychosis symptoms of schizophrenia was raised by a study of xanomeline in Alzheimer's Disease (AD). Xanomeline showed dose-dependent improvement in cognitive–behavioral disturbances associated with AD, which include reductions in vocal outbursts, suspiciousness, delusions, hallucinations, paranoia, and other behavioral disturbances that share similarities with the positive symptoms observed in schizophrenia (34). However, a 4-week placebo-controlled proof-of-concept trial of with xanomeline that consisted of 20 schizophrenic patients (with a dose of 225 mg/day) showed significant improvement in both positive and negative cognitive domains of schizophrenia (35).

## 18.7   GULATAMATERGIC DRUGS IN THE TREATMENT OF SCHIZOPHRENIA

From a mechanistic perspective, if NMDA antagonists (ketamine, PCP, and MK-801) provoke or worsen psychotic symptoms by increase of glutamate release, drugs that as agonists on the NMDA receptor complex might provide an antipsychotic effect. Direct NMDA agonists provoke excessive excitation and seizure should not be feasible to develop due to their toxicity. However, positive allosteric modulators of NMDA receptor glycine site, such as glycine, full agonist D-serine, and partial agonist D-cycloserine, are potential candidates. In a behavioral model, PCP-induced hyperactivity in rodents was blocked by glycine and glycine uptake inhibitor gycinedodecyclaminde (36). Studies have demonstrated the putative benefits of a D-serine and D-cycloserine in reduction of negative symptoms and the benefit of D-serine, D-alanine, and glycine treatment for improvement of cognitive symptoms. A meta-analysis of adjuvant glycine-site agonist treatments show a moderate amelioration of negative symptoms based on the data of 19 relatively heterogeneous trials (35).

Lamotrigine is a broad-spectrum anticonvulsant with a route of action that includes inhibition of glutamate release. In a placebo-controlled trial of a lamotrigine add-on to clozapine, administration in treatment-resistant schizophrenic patients showed significant improvement of positive symptoms. Lamotrigine adjuvant therapy in 400 mg was efficacious in another trial of treatment-resistant schizophrenia by adding it on to conventional and atypical antipsychotics.

Like atypical antipsychotics (e.g., clozapine and olanzapine), agonists for mGlu2/3 receptors (e.g., LY354740 and LY379268) attenuate some of the behavioral effects of PCP in animals and block increased glutamate release (36). The ability of mGlu2/3 receptor agonists to block PCP-induced behaviors seems to be linked to suppression of glutamate transmission, as it can still be observed in monoamine-depleted rats (37).

LY2140023 (the methionine prodrug of the potent mGlu2/3 receptor agonist LY404039) showed statistically significant superiority compared with placebo in monotherapy of schizophrenia, both in the positive and negative symptoms of the illness. An adverse-event profile of LY2140023 differed from any existing conventional or atypical antipsychotics. A lack of prolactin increase or worsening of extra pyramidal symptoms along with a significant reduction in akathisia, was observed in patients treated with the mGlu2/3 agonist when compared with placebo. In addition, the investigational new drug was weight neutral. This study provided the first clinical evidence about efficacy of a new, nondopaminergic monotherapy paradigm in schizophrenia (38).

After the breakthrough of the first efficacious antipsychotics during the mid-twentieth century, followed by the clinical use of atypical antipsychotics with a better efficacy/extrapyramidal adverse event profile in the 1990s, the discovery and development of new antipsychotics with novel routes of action is still in the main focus of psychopharmacology with the goal of full recovery of symptoms.

## REFERENCES

1. Kraepelin E. Dementia Praecox and Paraphrenia. 1971. Krieger Publishing, Huntington, New York.

2. Tamminga CA, Holcomb HH. Phenotype of schizophrenia: a review and formulation. Mol. Psychiatry 2005;10:27–39.

3. Bleich A, Brown SL, Kahn R, van Prag HM. The role of serotonin in schizophrenia. Schizophr. Bull. 1988;14:111–123.

4. Martin-Ruiz R, Puig MV, Celada P, Shapiro DA, Roth BL, Mengod G, Artigas F. Control of serotonergic function in medial prefrontal cortex by serotonin-2A receptors through a glutamate-dependent mechanism, J. Neurosci. 2001;21:9856–9866.

5. Cameron AM, Oram J, Geffen GM, Kavanagh DJ, McGrath JJ, Geffen LB. Working memory correlates of three symptom clusters in schizophrenia. Psychiatry Res. 2002;110:49–61

6. Frankle WG, Lombardo I, Kegeles LS, Slifstein M, Martin JH, Huang Y, Hwang DR, Reich E, Cangiano C, Gil R, et al. Serotonin 1A receptor availability in patients with schizophrenia and schizo-affective disorder: a positron emission tomography imaging study with [11C]WAY 100635. Psychopharmacology 2006;189:155–164.

7. Krystal JH, D'Souza DC, Karper LP, Bennett A, Abi-Dargham A, Abi-Saab D, Cassello K, Bowers MB Jr, Vegso S, Heninger GR, et al. Interactive effects of subanesthetic ketamine and haloperidol in healthy humans. Psychopharmacology 1999;145:193–204.

8. Pilowsky LS, Bressan RA, Stone JM, Erlandsson K, Mulligan RS, Krystal JH, Ell PJ. First in vivo evidence of an NMDA receptor deficit in medication-free schizophrenic patients. Mol. Psychiatry 2006;11:118–119

9. Deutsch, SI, Rosse, RB, Schwartz, BL, Mastropaolo J. A revised excitotoxic hypothesis of schizophrenia: therapeutic implications. Clin. Neuropharmacol. 2001;24:43–49.

10. Baldessarini RJ, Katz B, Cotton P. Dissimilar dosing with high-potency and low-potency neuroleptics. Am. J. Psychiatry 1984;141:748–752.

11. Kapur S, Zipursky RB, Remington G. Clinical and theoretical implications of 5-HT2 and $D_2$ receptor occupancy of clozapine, risperidone, and olanzapine in schizophrenia. Am. J. Psychiatry 1999;156:286–293.

12. Richtand NM, Welge JA, Logue AD, Keck PE Jr, Strakowski SM, McNamara RK. Dopamine and serotonin receptor binding and antipsychotic efficacy. Neuropsychopharmacology 2007;32:1715–1726.

13. Meltzer HY, Fang V, Young MA. Clozapine-like drugs. Psychopharmacol. Bull. 1980;16:32–34.

14. Seeman P, Tallerico T. Antipsychotic drugs which elicit little or no Parkinsonism bind more loosely than dopamine to brain $D_2$ receptors, yet occupy high levels of these receptors. Mol. Psychiatry 1998;3:123–134.

15. Litman RE, Su TP, Potter WZ, Hong, WW, Pickar D. Idazoxan and response to typical neuroleptics in treatment resistant schizophrenia. Comparison with the atypical neuroleptic clozapine. Br. J. Psychiatry 1996;168:571–579.

16. Hoffman D, Donovan H. Catalepsy as a rodent model for detecting antipsychotic drugs with extrapyramidal liability. Psychopharmacology 1995;120:128–133

17. Jackson DM, Mohell N, Bengtsson A, Malmberg A. What are atypical neuroleptics and how do they work. In Brunello N, Mendelewitz J, Rcagni G eds. New Generation of Antipsychotic Drugs: Novel Mechanism of Action. 1993. Karger, Basel, Switzerland.

18. Conley RR, Mahmoud R, Risperidone Study Group. Efficacy of risperidone vs olanzapine in the treatment of patients with schizophrenia or schizoaffective disorder. Am. J. Psychiatry 2001;158:765–774.

19. Peuskens J, Link CGG. A comparison of quetiapine and chlorpromazine in the treatment of schizophrenia. Acta Pscyh. Scand. 1997;96:4:265–273.

20. Arvanitis LA, Miller BG, Seroquel Trial 13 Study Group. Multiple fixed doses of seroquel (quetiapine) in patients with acute exacerbation of schizophrenia: a comparison with haloperidol and placebo. Biol. Psychiatry 1997;42:233–246.

21. Goff DC, Posever T, Herz L, Simmons J, Kletti N, Lapierre K, Wilner KD, Law CG, Ko GN. An exploratory haloperidol-controlled dose finding study of ziprasidone in hospitalized patients with schizophrenia or schizoaffective psychosis. J. Clin. Psychopharm. 1998;18:296–304.

22. Kane JM, Carson WH, Saha AR, McQuade RD, Ingenito GG, Zimbroff DL, Ali MW. Efficacy and safety of aripiprazol and haloperidol vs placebo in patients with schizophrenia and schizoaffective disorder. J. Clin. Psychiatry 2002;63:763–771.

23. Kane J, Honigfeld G, Singer J, Meltzer H. Clozapine for the treatment-resistant schizophrenia. A double-blind comparison with chlorpromazine. Arch. Gen. Psychiatry 1988;45:789–796.

24. Breier AF, Hamilton SH. Comparative efficacy of olanzapine and haloperidol for patients with treatment-resistant schizophrenia. Biol. Psychiatry 1999;45:403–411.

25. Keefe RS, Silva SG, Perkins DO, Lieberman JA. The effect of atypical antipsychotic drugs on neurocognitive impairment in schizophrenia: a review and meta-analysis. Schizophrenia Bull. 1999;25:201–222.

26. Saeedi H, Remington G, Christensen BK. Impact of haloperidol, a dopamine $D_2$ antagonist, on cognition and mood. Schizophr. Res. 2006;85:222–231.

27. Wagner M, Quednow BB, Westheide J, Schlaepfer TE, Maier W, Kühn KU. Cognitive improvement in schizophrenic patients does not require a serotonergic mechanism: randomized controlled trial of olanzapine vs amisulpride. Neuropsychopharmacology 2005;30:381–390.

28. Meltzer HY, Alphs L, Green AI, Altamura AC, Anand R, Bertoldi A, Bourgeois M, Chouinard G, Islam MZ, Kane J, Krishnan R, Lindenmayer JP, Potkin S, International Suicide Prevention Trial Study Group. Clozapine treatment for suicidality in schizophrenia. International Suicide Prevention Trial (InterSePT) Arch. Gen. Psychiatry 2003;60:82–91.

29. Meguid MM, Fetissov SO, Varma M, Sato T, Zhang L, Laviano A, Rossi-Fanelli F. Hypothalamic dopamine and serotonin in the regulation of food intake. Nutrition 2000;16:843–857.

30. Elias AN, Hofflich HH. Abnormalities in glucose metabolism in patients with schizophrenia treated with atypical antipsychotic medications. Am. J. Med. 2008;121:98–104.

31. Allison DB, Mentore JL, Heo M, Chandler LP, Cappelleri JC, Infante MC, Weiden PJ. Antipsychotic induced weight gain: a comprehensive research synthesis. Am. J. Psychiatry 1999;156:1686–1696.

32. Reynolds GP, Zhang ZJ, Zhang XB. Association of antipsychotic drug-induced weight gain with $5HT_{2c}$ receptor gene polymorphism. Lancet 2002;359:2086–2087.

33. Arulmozhi DK, Dwyer DS, Bodhankar SL. Antipsychotic induced metabolic abnormalities. Life Sci. 2006;79:1865–1872.

34. Boidick NC, Offen WW, Shannon HE, Satterwhite J, Lucas R, Van Lier R, Paul SM. The selective muscarinic agonist xanomeline improves both the cognitive deficits and behavioral symptoms of Alzheimer disease. Alzheimer Dis. Assoc. Disord. 1997;11:S16–22.

35. Shekhar A, Potter WZ, Lightfoot J, Lienemann J, Dubé S, Mallinckrodt C, Bymaster FP, McKinzie DL, Felder CC. Selective muscarinic receptor agonist xanomeline as a novel treatment approach for schizophrenia. Am. J. Psychiatry 2008;165:1033–1039

36. Javitt DC, Sershen H, Hashim A, Lajtha A. Reversal of phencyclidine-induced hyperactivity by glycine and the glycine uptake inhibitor gycildodeccyclamide. Neuropsychopharmacology 1997;17:202–204.

37. Tuominen HJ, Tiihonen J, Wahlbeck K. Glutaminergic drugs for schizophrenia: a systematic review and meta-analysis. Schizophrenia Res. 2005;72:225–234.

38. Cartmell J, Monn JA, Schoepp DD. Attenuation of specific PCP-evoked behaviors by the potent mGlu2/3 receptor agonist, LY379268 and comparison with the atypical antipsychotic, clozapine. Psychopharmacology 2000;148:423–429.

39. Swanson CJ, Schoepp DD. The group II metabotropic glutamate receptor agonist (-)-2-oxa-4-aminobicyclo[3.1.0.]hexane-4,6-dicarboxylate (LY379268) and clozapine reverse phencyclidine-induced behaviors in monoamine-depleted rats. J. Pharmacol. Exp. Ther. 2002;303:919–927.

40. Patil ST, Zhang L, Martenyi F, Lowe SL, Jackson KA, Andreev BV, Avedisova AS, Bardenstein LM, Gurovich IY, Morozova MA, et al. Metabotropic glutamate 2/3 receptor agonism represents a novel therapeutic approach in treating schizophrenia: results from a randomized, placebo-controlled Phase 2 clinical trial. Nature Med. 2007;13:1102–1107.

## FURTHER READING

Cochrane database of systematic reviews. Amisulprid for schizophrenia. Cochrane Database Syst. Rev. 2002;2:CD001357.

Cochrane database of systematic reviews. Olanzapine for schizophrenia. Cochrane Database Syst. Rev. 2005;2:CD001359.

Cochrane database of systematic reviews. Risperidone versus typical antipsychotic medication for schizophrenia. Cochrane Database Syst. Rev. 2003;2:CD000440.

Cochrane database of systematic reviews. Sertindol for schizophrenia. Cochrane Database Syst. Rev. 2005;3:CD001715.

D'Haenen H, den Boer JA, Willner P, eds. Biological Psychiatry. 2002. Wiley, New York.

Falkai P, Wobrock T, Lieberman J, Glenthoj B, Gattaz WF, Möller HJ, WFSBP Task Force on Treatment Guidelines for Schizophrenia. World Federation of Societies of Biological Psychiatry (WFSBP) guidelines for biological treatment of schizophrenia, Part 1: acute treatment of schizophrenia. World J. Biol. Psychiatry 2005;6:132–191.

Kenneth L, Davis KL, Charney D, Coyle JT, Nemeroff C, eds. Neuropsychopharmacology: The Fifth Generation of Progress. 2002. Lippincott Williams & Wilkins, Philadelphia, PA.

# 19

# PROTEIN MISFOLDING AND DISEASE

JOHANNA C. SCHEINOST, GRANT E. BOLDT, AND
PAUL WENTWORTH, JR.

*The Scripps–Oxford Laboratory, Department of Biochemistry, University of Oxford, UK*

*Department of Chemistry and The Skaggs Institute for Chemical Biology, The Scripps Research Institute La Jolla, CA*

The amyloidoses are a class of conformational diseases that arise from the conversion of normally unfolded or globular proteins into fibrillar aggregates that are either pathogenic or nonfunctional. At present there are more than 20 proteins that are associated with human amyloid diseases. This review focuses on three natively unfolded proteins that form fibrillar aggregates; amyloid-β, islet amyloid polypeptide ("amylin"), and α-synuclein, the diseases they contribute to and chemical and biophysical approaches that are used to investigate these proteins' aggregation.

## 19.1 INTRODUCTION

Conversion of native conformational folded proteins or peptides into highly-ordered insoluble fibrillar aggregates termed "amyloid" has been demonstrated to be clinically-relevant in several diseases. Misfolded proteins are deposited in a variety of tissues, leading to serious and even fatal human diseases. To date there are over 20 proteins and peptides that are known to induce pathological conditions associated with protein misfolding (Table 19.1). These proteins include the amyloid-β peptides, prion proteins, α-synuclein, transthyretin and polyglutamine-containing peptides (1). The hypothesis that these diseases are in fact caused by

*Chemical Biology: Approaches to Drug Discovery and Development to Targeting Disease*, First Edition. Edited by Natanya Civjan.

**TABLE 19.1** **List of protein misfolding diseases and the implicated proteins and peptides**

| Clinical Syndrome | Fibril Component |
|---|---|
| Alzheimer's Disease | Aβ peptide, 1-40, 1-42 |
| Spongiform encephalopathies | Full-length prion and fragments |
| Primary systemic amyloidosis | Intact light chain and fragments |
| Secondary systemic amyloidosis | 76-residue fragment of amyloid A protein |
| Familial amyloidotic polyneuropathy I | Transthyretin variants and fragments |
| Senile systemic amyloidosis | Wild-type transthyretin and fragments |
| Hereditary cerebral amyloid angiopathy | Fragments of cystatin-C |
| Haemodialysis-related amyloidosis | β2-macroglobulin |
| Familial amyloidotic polyneuropathy II | Fragments of apolipoprotein A-1 |
| Finnish hereditary amyloidosis | 71-residue fragment of gelsolin |
| Type II diabetes | Islet-associated polypeptide |
| Medullary carcinoma of the thyroid | Fragments of calcitonin |
| Atrial amyloidosis | Atrial natriuretic factor |
| Lysozyme amyloidosis | Full-length lysozyme variants |
| Insulin-related amyloid | Full-length insulin |
| Fibrinogen α-chain amyloidosis | Fibrinogen α-chain variants |
| Synucleinopathies | α-synuclein |

misfolded protein and the misfolded protein does not arise as a side effect of the disease is supported by familial disease, where proteins containing mutations that facilitate aggregation result in more severe and early onset disease variants.

Diseases arising as a result of protein misfolding are classified based on the location of the protein deposition. Thus, the amyloidoses are generally classified as either neuronal or systemic. The majority of conditions arise sporadically, with only 10% of the cases being linked to hereditary causes. The familial forms of the diseases are caused by mutations that make a protein more prone to aggregation, or by mutations in processing proteins like proteases. Under rare circumstances (5% of protein misfolding diseases), protein aggregation can be initiated by infection and transmission. All diseases caused by misfolded protein share the problem of definitive diagnosis as they are difficult to identify *in vivo*, especially during early stages of symptoms.

Precursors of amyloid are quite diverse, ranging from small unstructured peptides like amyloid-β peptides to large oligomeric multidomain proteins like p53. Despite the differences in structure of the monomers, the structures of the misfolded β-sheet rich forms share similar features. In recent years it has become apparent that the proteins that form amyloid share common features, and it has been suggested that many if not all proteins share the intrinsic predisposition to form amyloid under the right conditions (2, 3). This hypothesis is affirmed by the absence of a correlation between the propensity of a peptide or protein to misfold and its sequence. Precursor proteins share neither homology in sequence nor structure. However, certain factors such as hydrophobic regions and/or unstable globular protein conformations contribute to a predisposition to form amyloid.

In this review, we will discuss the aggregation of three intrinsically unstructured peptides and proteins and the diseases to which they contribute. Specifically we will concentrate on amyloid-β (Aβ), islet amyloid polypeptide (IAPP), and α-synuclein.

## 19.2  BIOLOGICAL BACKGROUND OF PROTEIN MISFOLDING DISEASES

### 19.2.1  Amyloid-β (Aβ) and Alzheimer's Disease

Alzheimer's disease (AD) is the most common neurodegenerative disease in humans. AD is a mostly sporadic, late-onset disease (90%). The risk of developing AD increases exponentially with age. In the early stage, AD is often overlooked as it manifests in mildly impaired short-term memory. The long-term memory is unaffected. With progression of the disease, the impairment of memory both short and long-term becomes apparent. Patients undergo a change in personality from having paranoid delusions to hallucinations and anxiety. At the late stages of AD, the affected person loses his/her ability to participate in daily life and fails to perform the most basic actions such as eating or walking.

A key hallmark of AD is the misfolding of two proteins, Aβ and tau, whose extra- and intracellular aggregates in the brain are coincident with and most likely causative of the disease. Therefore aggregates of these proteins are used as histological markers for diagnosis of AD *post mortem*. Tau is a protein needed for microtubule stabilization. However, upon hyperphosphorylation it forms neurofibrillar tangles (NFTs) that perturb the integrity of microtubules leading to impaired neuronal function. The presence of NFTs correlates to the severity of AD. However, NFTs are absent in 10% of AD patients and in up to 50% of mild of AD cases (4).

Aβ is the principal constituent of senile plaques found in AD patients' brains and as such has made it a key suspect in the search for causative agents of AD. Aβ is a 39-43 residue peptide (Fig. 19.1) that is derived from the proteolytic cleavage of the amyloid precursor protein APP, a type I integral membrane protein. APP is processed by three proteases α-, β-, and γ-secretases. Cleavage of APP by α- and γ-secretases releases the non-amyloidogenic fragment p3. Upon cleavage of APP by β- and γ-secretases, Aβ is released. The peptide is characterized by a hydrophilic $N$-terminus and a hydrophobic $C$-terminus, corresponding to parts of the extracellular and intramembrane domains of APP, respectively. Aβ exists mainly as a 40 or 42 residue long peptide (Aβ(1-40) and Aβ(1-42)) (Fig. 19.1). The C-terminally extended, more hydrophobic sequences like Aβ(1-42) aggregate faster *in vitro* than their shorter partners, which might explain why they are more abundant in senile plaques.

Native Aβ is unstructured and its function is still uncertain. Several possible roles have been suggested, including cholesterol transport, antioxidative

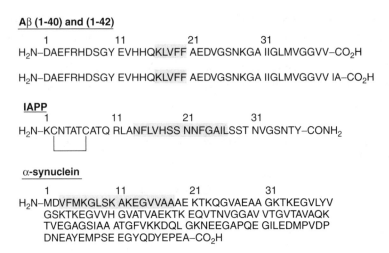

**Figure 19.1**  Sequences of Aβ, IAPP and α-synuclein.

protection and TGF-β activity (5–7). Similarly, a definite function for APP is still unidentified. APP is a putative Notch-like receptor for an unknown ligand (8) and has been suggested to play a role in intercellular adhesion (9).

The presence of extracellular plaques consisting of misfolded Aβ in the brain of AD patients led to the widely accepted hypothesis that insoluble aggregates of Aβ are the toxic conformation of the native peptide. However, there is a poor correlation between the number of mature fibrils found in brain and neuronal death (10). Studies have now shown that soluble Aβ oligomers are able to damage and kill neurons in central nervous system cultures (11), providing a direct link between Aβ and AD. More recently, dodecamers of Aβ have been identified as the causative agent in impairment of spatial memory in mice (12), although the relevance of this result in the context of other memory changes is unknown (13). These oligomeric species are most likely precursors or folding intermediates and do not emerge by a distinct pathway (14). It has already been shown that oligomers in the absence of fibrils and monomers can inhibit hippocampal long-term potentiation in rats *in vivo* (15).

Developing animal models for AD is an ongoing challenge due to the multifac-torial nature of the syndrome. Existing murine models, such as Tg2576, express mutated human APP (the Swedish mutation) and develop amyloid plaques, but lack other important characteristics of AD, such as NFTs. These mice develop memory loss but are devoid of neurodegeneration. A triple transgenic mouse model has been developed that expresses the Swedish mutant APP, a human tau mutant, as well as a human presenilin-1 (PS1) mutant (16). These triple trans-genic mice build up misfolded Aβ and tau and more accurately mimic human AD pathology.

At the time of writing, there is no generally accepted theory that accounts for the development of AD pathology. The multifactorial basis of the disease makes such a theory unlikely to be possible in the foreseeable future. In addition to the NFTs and amyloid-hypothesis of the disease, oxidative stress, systemic levels of redox active metal ions, cardiovascular disease, the apoEε4 allele and type 2 diabetes are all clear risk factors for development of AD. However, recent research supports the notion that the Aβ buildup may be a key event in AD and that other manifestations of the disease, like NFT formation, result from an imbalance between Aβ production and Aβ clearance (17).

### 19.2.2 Islet Amyloid Polypeptide (IAPP) and Type 2 Diabetes

Deposition of amyloid in pancreatic islets is a feature of type 2 diabetes in man, but the factors that contribute to this phenomenon are unknown. Progressive amyloidosis results in loss of insulin-producing cells and increased disease state. The presence of amyloid depositions in pancreatic tissue was first described over a century ago (18). However, it took much effort and time to identify the peptide component of these accumulations as islet amyloid polypeptide (IAPP, also called *amylin*) (19, 20). IAPP is a 37 residue long peptide (Fig. 19.1) which is cosecreted with insulin. It originates from a longer precursor, pro-IAPP (67 amino acids). The peptide is co-stored with insulin in β-cell secretory granules and secreted together with insulin in response to β-cell stimulation, primarily glucose intake. Its native structure both intra- and extracellular prior to aggregation *in vivo* is still unclear; under physiological concentrations *in vitro* it adopts a stable random coil structure. During maturation of β-cell secretory granules, the *N*- and *C*-termini of pro-IAPP are enzymatically cleaved by pro-hormone convertases 1/3 and 2. The IAPP peptide is naturally present in humans and other species and is usually excreted via the kidney. It is unclear if IAPP has any function *in vivo* although multiple possible roles have been suggested.

IAPP forms insoluble aggregates, confined within the islets of Langerhans, which are usually incidental with the noninsulin-dependent type 2 diabetes. In contrast to the autoimmune disease type 1 diabetes, which is characterized by the destruction of insulin secreting cells, type 2 diabetes is a late-onset disease with decreased insulin secretion and reduced sensitivity toward insulin of peripheral tissue. Misfolded IAPP is present in over 90% of type 2 diabetes patients (as diagnosed *post mortem*), but also in 15% of nondiabetics, indicating that other factors that are not diabetes related may play a role in the aggregation of the 37mer. Unlike Aβ, IAPP is present in concentrations above the critical concentration even in healthy individuals, but aggregation is prevented. It has been shown that insulin but not pro-insulin forms a complex with IAPP that stabilizes the peptide *in vitro* and possibly *in vivo*.

The role of amyloid in type 2 diabetes appears to be complex: severe islet dysfunction correlates well with extensive amyloidosis; however, the degree of amyloidosis can vary extremely in long term patients. The factors that trigger the onset of the disease are both genetic and environmental. On the other hand, a common qualification of misfolding diseases is that they occur mostly sporadic, without the necessity of mutations.

Mature fibrils formed from IAPP demonstrate β-cell toxicity *in vitro* (21). Analogous to the toxic-conformation-discussion with Aβ, it has also been proposed that smaller, soluble oligomers of IAPP have cytotoxic effects (22) and show membrane-disrupting activity.

### 19.2.3   A Link Between AD and Diabetes?

It has long been suggested that diabetes is a risk factor for AD, and a cohort study on people older that 55 years revealed that patients suffering from diabetes have a 65% increase in the risk of developing AD compared to the group that did not suffer from diabetes (23). Moreover, the two diseases share similar physiological processes, most notably degeneration and age dependence. Interestingly, insulin receptors are not only found in the peripheral system, but also in neurons in the CNS (24). Therefore, the role of these receptors is not limited to the regulation of blood sugar levels but they are also involved in neuronal differentiation and cell proliferation (25, 26), implicating an intricate connection between insulin and AD. This unexpected link of AD and diabetes has led to the postulation of a new, brain specific "type 3 diabetes" (27), which was further elucidated when it was shown that soluble oligomers of Ab disrupt insulin signaling by binding to the insulin receptors on the neuronal surface (28). Given that synaptic failure and memory dysfunction are characteristic for AD, this link is another plausible explanation for the observed symptoms of the disease.

### 19.2.4   α-Synuclein and Synucleopathies

α-Synuclein (Fig. 19.1) is the protein component associated with synucleinopathies, protein misfolding diseases that are characterized by the presence of α-synuclein deposits. Synucleinopathies include Parkinson disease, Lewy body (LB) dementia, diffuse LB dementia, and also the LB variant of AD.

α-Synuclein has been found to be the major constituent of Lewy bodies (LBs). Intriguingly, these deposits are not extracellular as with Aβ and IAPP, but intracellular inclusion bodies found in dopaminergic and nondopaminergic neurons and glia. The protein morphology within the inclusions can be distinct; at least five different conformations have been described: LBs, Lewy neurites, glial cytoplasmic inclusions, neuronal cytoplasmic inclusions, and axonal spheroids.

α-Synuclein exits in three isoforms of varying length in humans. The best characterized is the full length, 140 amino acid long protein; the other isoforms are shorter and derived by splicing. The N-terminus of the 140mer is characterized by the presence of a seven eleven amino acid long repeat containing highly conserved hexameric motifs (KTKEGV). These repeats are thought to be implicated in the interaction with lipids and adopt α-helical structure upon contact with lipid vesicles. The central region of the protein is highly prone to aggregation.

α-Synuclein is natively unstructured and a definitive function remains elusive. Possible roles include the regulation of vesicular release and other synaptic tasks as α-synuclein has been found to exist in equilibrium of free protein with vesicle-bound protein. α-Synuclein interacts *in vitro* with synthetic or purified phospholipid membranes, altering their integrity and suggesting participation in membrane lipid component organization. Moreover, it may also act as inhibitor of lipid organization. An intriguing function of α-synuclein is that it may operate as a molecular chaperone.

α-Synuclein is known to bind many proteins; over 500 proteins interacted in a proteomic analysis of a neuroblastoma cell line. Contact with some proteins or protein complexes promotes aggregation, e.g. histones and the tau protein, while others, particularly the shorter synucleins (β and γ), inhibit misfolding. Due to a high plasticity of the protein, it can adopt many different conformations. This conformational diversity of α-synuclein has had the protein labeled as a "protein chameleon." Under physiological conditions *in vitro* the protein is unstructured. In the presence of lipid vesicles, its *N*-terminus can acquire α-helical structure *in vitro*. It can form noncovalent dimers and higher order oligomers leading to insoluble aggregates. Interestingly, α-synuclein can also form covalent dimers that are cross-linked *via* a dityrosine bond, under oxidative conditions.

Evidence points toward the oligomeric state of α-synuclein as the toxic conformation causing neuronal death. It has been suggested that fibril formation is merely an escape route or protection mechanism in the brain to prevent further neuronal damage.

## 19.3 BIOPHYSICAL CHARACTERIZATION OF FIBRILS AND MOLECULAR MECHANISM OF THEIR FORMATION

Although precursor proteins of amyloid are extremely diverse in sequence, the end-product fibrils in general share several structural features such as cross-β sheet assemblies, non-branched fibrils of similar length and structural organization.

### 19.3.1 Key Intermediates in Fibril Formation

In general, there is a simple picture of the events that occur during oligomerization of natively unstructured protein and peptides: unfolded monomers undergo organized self-assembly *via* several intermediate assemblies on the pathway to fibrils. Many of these intermediates have been characterized and named based on *in vitro studies*, but there is no defined nomenclature as of yet. The following intermediates have been identified so far in the fibril polymerization process:

*Monomers* of the native peptide or protein can either be unfolded or folded. However, even natively unstructured peptides, e.g. Aβ, can adopt some secondary structure depending on the physiological environment, and some proteins exist in a native oligomerized state, like transthyretin. Monomers can arrange into *oligomers*, which are small globular, possibly spherical, micelle-like assemblies that contain some secondary structure (29). It is still unclear whether these oligomers are direct intermediates of the fibril formation pathway, or whether they occur as part of an "off-pathway." For the misfolding of Aβ, evidence points toward oligomers being precursors of fibrils (29). Moreover, these oligomers are thought to be the neurotoxic conformation in a variety of misfolding diseases, most importantly AD (11). Structural instabilities of folding intermediates make it difficult to determine the structure of these intermediates, but not impossible (24).

Two unfortunate names that can easily lead to confusion have been given to the "in-pathway" conformations *protofibrils* and *protofilaments*. Protofibrils are, similar to oligomers, mostly unstructured but linear (31). However, in most cases the term "protofibrils" is used interchangeably with the term oligomer. The name protofilaments refers to linear aggregates with β-sheet structure that have a diameter of 3 nm and are usually 50–100 nm in length. Mature, elongated protofilaments, also called filaments, intertwine to form protein *fibrils*. Fibrils are generally unbranched, approx. 10 nm in diameter and can be several μm long.

### 19.3.2 Nucleated Polymerization and Other Models of Fibril Formation

The actual mechanism of self-association of monomeric protein into amyloid is complex and three mechanisms of structure conversion have been proposed (32): In *templated assembly*, a monomeric native state peptide binds to an existing nucleus. Upon binding, there is a change in the secondary structure of the monomer as it is added to the growing chain. *Monomer directed conversion* involves the presence of a misfolded monomer that templates the structure conversion of a native monomer, followed by disassociation and chain formation. The third model is *nucleated polymerization*, which is the most widely accepted model for the fibril growth.

Nucleated polymerization is a crucial process in cells, involved in providing a rigid cytoskeleton by organization of tubulin into microtubules and actin into thin filaments. These processes are under tight control; however, nucleated polymerization of proteins can also be "accidental" and occur to proteins that, in healthy individuals, remain in their native structure. The kinetics of misfolding are characterized by the presence of a lag phase, a rapid growth or elongation phase, and a plateau phase. The overall rate of amyloid aggregation is restricted by the formation of a nucleus, a process that is thought to have either slow kinetics or is driven by an unfavorable equilibrium. A nucleus is defined as the least stable conformer in the aggregation process that is in equilibrium with monomeric protein (33). The formation of a nucleus, a "stochastic" event, is concentration dependent. Nuclei only occur above a critical protein monomer concentration (for Aβ this concentration is approx. 15 μM). However, in the presence of preformed seeds the lag phase is completely abolished. Once a nucleus or seed is present, it acts as a template for the conversion of monomers into β-sheet rich conformers that assemble rapidly into protofibrils and fibrils. Upon consumption of monomers there is a plateau phase where no more fibrils are formed. This general model of fibril formation is able to explain the spontaneous onset of most amyloidoses as well as their rapid progression. However, the kinetics of this mechanism can be influenced to a great extent by quite a few factors both *in vitro* and *in vivo*: pH, temperature of the sample, presence of preformed seeds, presence of metal ions, and oxidative damage.

Mathematical descriptions of the kinetics of protein misfolding have largely been attempted using noncomplex simulations and algorithms. It has been suggested that a complete mathematical analysis of these processes is currently beyond our capability (34).

### 19.3.3 Structure of Fibrils

At the time of writing, no high resolution structure of fibrils has been published yet, as fibrils are insoluble, noncrystalline material, and therefore the classic structure determination methods like NMR and X-ray crystallography fail.

First experiments to elucidate the structure of the proteins in the fibrillar organization were done by X-ray diffraction. Amyloid fibrils show a characteristic diffraction pattern, the so called cross-beta pattern (35), which is indicative of β-sheets parallel to the fibril axis with the protein strands perpendicular to the fibril's long axis (36, 37). The pattern of amyloid is characterized by reflections at 4.75 Å (along the fibril axis) and 10 Å (perpendicular to the fibril axis) which occur from regular repeats and stacking of monomers.

Further characterization of Aβ fibrils involved solid-state NMR to determine the structure of the peptide subunits. Depending on the experimental constraints applied, several models for full length (e.g. (38)) and fragments of Aβ (e.g. (39, 40)) have been developed, most of them proposing an antiparallel, in-register β-sheet organization. However, it has since been proposed that the monomers composing a fibril associate in parallel, not antiparallel β-sheets (41) with full length Aβ, and a model based on these constraints has been determined (42) (Fig. 19.2).

Within the IAPP polypeptide, three key regions have been identified that are responsible for its misfolding: residues 8-20 (43), 20-29 (44), and 30-37 (45). Residues 20-29 were initially found to be the amyloidogenic chore of IAPP, as peptide fragments corresponding to that sequence was found to form fibrils *in vitro*. Taken together, these three regions might participate in the formation of an intramolecular β-sheet with both parallel and antiparallel organization (43). Electron paramagnetic resonance spectroscopy of spin-labeled IAPP fibrils suggested a structure similar to that of Aβ: in-register, parallel β-sheet structure with units showing a disordered *N*-terminal segment (46).

Solid-state NMR studies of α-synuclein fibrils have shown that the β-sheet rich domain is located between at least residues 38-95, with the *N*-terminal residues being unstructured and a rigid backbone starting at residue 22 (47). These results are consistent with data obtained from electron paramagnetic resonance (EPR) of spin-labeled protein, showing an ordered core around residues 31-109, with parallel, in-register β-sheets (42, 43).

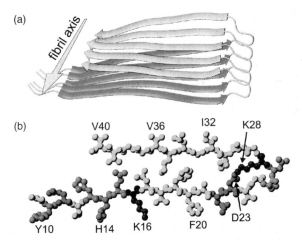

**Figure 19.2**  Structural model for Aβ 1–40 fibrils, consistent with solid state NMR constraints on the molecular conformation and intermolecular distances and incorporating the cross-β motif common to all amyloid fibrils. Residues 1–8 are considered fully disordered and are omitted. (a) Schematic representation of a single molecular layer, or cross-β unit. The yellow arrow indicates the direction of the long axis of the fibril, which coincides with the direction of intermolecular backbone hydrogen bonds. The cross-β unit is a double-layered structure, with in-register parallel β-sheets formed by residues 12–24 (light gray ribbons) and 30–40 (dark gray ribbons). (b) Central Aβ1–40 molecule from the energy-minimized, five-chain system, viewed down the long axis of the fibril. Taken from Petkova, A.T., et al. Proceedings of the National Academy of Sciences, 2002. 9926: p. 16742-16747.

### 19.3.4  Lessons for Drug Development

To date, there is no successful treatment for misfolding diseases. The development of structural models for fibrils and the elucidation of intermediates of fibril formation give insights into possible drug targets. Several approaches are possible using small molecules: the use of chemical chaperones to stabilize the native structures of the protein, possibly combined with another method, the use of β-sheet breakers to destabilize β-sheet intermediates. Other approaches include the competitive inhibition of protein-protein interactions, either for the monomer to prevent oligomerization, or for oligomers to prevent further monomer addition. An interesting, but problematic option is the use of molecules that increase the turnover or degradation of fibrils. Immunotherapy for AD was developed by injecting Aβ(1-42) into mice to immunize young animals against the disease and to reduce neuropathology in older animals (50). Initial clinical trials were performed but needed to be aborted as 6% of individuals developed meningoencephalitis (51), although beneficial effects can also be seen in these individuals (52).

### 19.3.5 Small Molecule Inhibitors of Aβ Assembly

Several small molecule inhibitors that prevent assembly of Aβ have been reported, however, the literature is far from systematic. Most of the studies listed in Table 19.2 (references for studies found in (53)) test only a few compounds and do not attempt to derive a structure-activity profile. Moreover, the chemical structures do not lend themselves to good starting points for medicinal chemistry in order to improve upon their properties. At the time of writing, the known compounds that inhibit Aβ assembly are weakly potent and are poorly bioavailable, particularly to the brain. Furthermore, progress toward development of aggregation inhibitors in the clinical setting remains elusive due to a clear disconnect between the *in vitro* and *in vivo* fibril-formation process. These points are further demonstrated by the fact that the only amyloid-aggregation inhibitors that have advanced to late-stage clinical trials, Alzhemed™, Cerebril™ and Kiacta™ do not fit the classical drug-like profile and their *in vivo* disposition and mechanism of action is not well understood.

## 19.4 METHODS USED FOR FIBRIL STRUCTURE DETERMINATION AND AGGREGATION ASSAYS

Research on amyloid fibrils was facilitated by the fact that fibrils similar to the ones found *in vivo* can be formed *in vitro* from peptides and denatured proteins. However, structure determination of amyloid fibrils has proven to be difficult due to the lack of methods and the complexity of the aggregates. Standard methods for structure determination, primarily solution-state NMR and X-ray diffraction, fail for amyloid fibrils due to the lack of soluble material or the inability to form classic protein crystals. Only recently we have acquired an in-depth understanding of the detailed conformation of fibrils.

The techniques to follow aggregation of amyloidogenic proteins and to determine their molecular conformation range from spectrometric assays (thioflavin T, Congo red) over spectroscopic assays (FTIR and CD) and visualization techniques (AFM and TEM) to methods that provide detailed insight into the atomic coordinates (solid-state NMR and X-ray diffraction).

The principal difficulty in studying amyloid formation *in vitro* is doubtlessly sample preparation and protein quality, as impurities, aggregation conditions (buffer, pH, temperature, etc.) and the chosen deseeding method can have a major impact on the fibril formation kinetics as well as the structure of the fibrils.

### 19.4.1 Spectrophotometric Assays

Congo red is a diazo dye with the name-giving characteristic red color. Amyloid stained with Congo red depicts green birefringence under polarized light and this method has been used to diagnose amyloid diseases in histological samples. To date, it is still used to classify proteins aggregates as amyloid. However, due to

**TABLE 19.2   Inhibitors of Aβ assembly**[a]

| Compound | Mode of action | Example |
|---|---|---|
| Aminoindanes | Fibrils | |
| Aminopyrazoles | Fibrils | |
| Antiparkinsonian agents | Fibrils | |
| Beta-sheet breakers peptides | Fibrils | *H*-RDLPFFPVPID-*OH* |
| Coenzyme | Fibrils | |
| Ferulic Acid | Fibrils | |

**TABLE 19.2** (*Continued*)

| Compound | Mode of action | Example |
| --- | --- | --- |
| FK506 complexes | Fibrils | 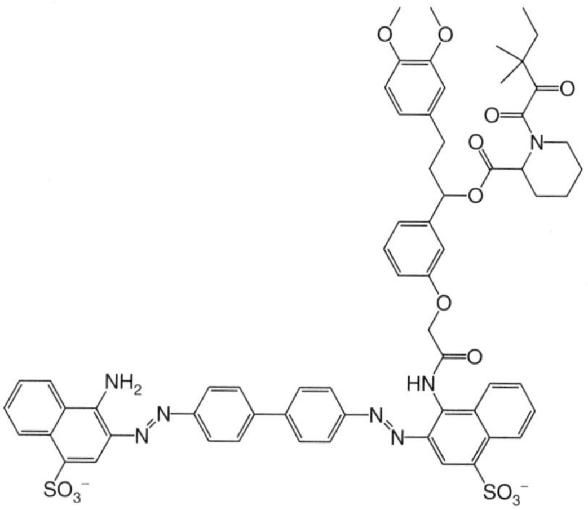 |
| Hydroxyindoles | Fibrils | 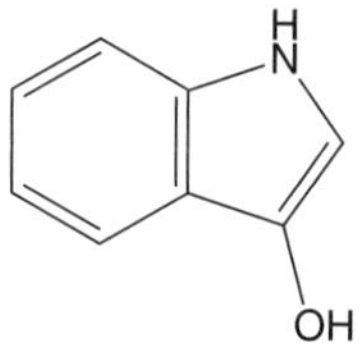 |
| Hydroxyphenyl pyridazine diamine | Fibrils | 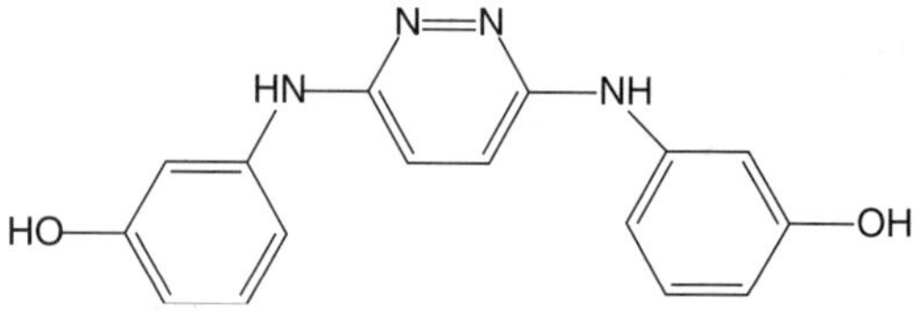 |
| Melatonin | Fibrils | 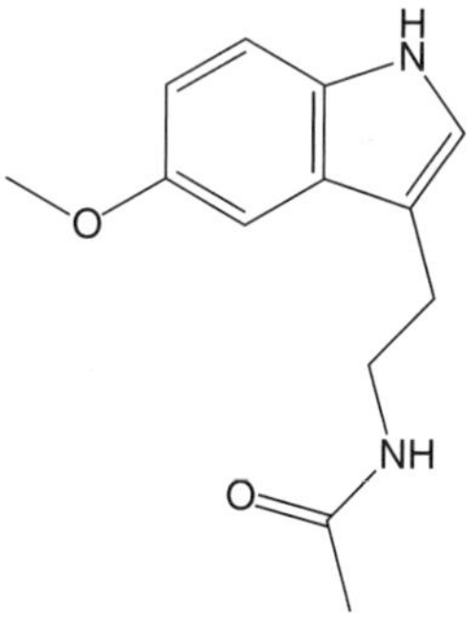 |

(*continued*)

**TABLE 19.2** *(Continued)*

| Compound | Mode of action | Example |
|---|---|---|
| Nicotine | Fibrils | |
| Nordihydro-guaiaretic acid | Fibrils | |
| NSAIDs | Fibrils | |
| Organofluorine indol-3-yl derivatives | Fibrils | racemic |
| Oxazalones | Fibrils | |

**TABLE 19.2** (*Continued*)

| Compound | Mode of action | Example |
|---|---|---|
| Phenothiazines | Fibrils | |
| Phenoxazines | Fibrils | |
| Phenylazobenzene sulfonamides | Fibrils | |
| Porphyrins | Fibrils | |

**TABLE 19.2** (*Continued*)

| Compound | Mode of action | Example |
|---|---|---|
| Rifampicin and tetracycline | Fibrils | |
| Rhodanines | Fibrils | |
| Small molecule anionic sulfonates of sulfates | Fibrils | |
| Tannic acid | Fibrils | |

**TABLE 19.2** (*Continued*)

| Compound | Mode of action | Example |
|----------|----------------|---------|
| Vitamin A | Fibrils | 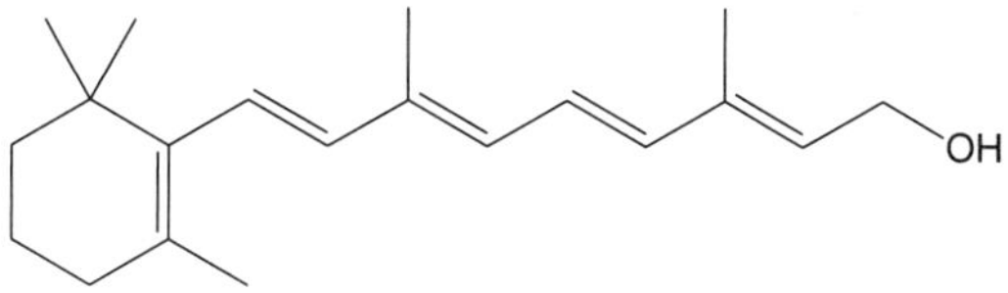 |
| Wine-related polyphenols | Fibrils | 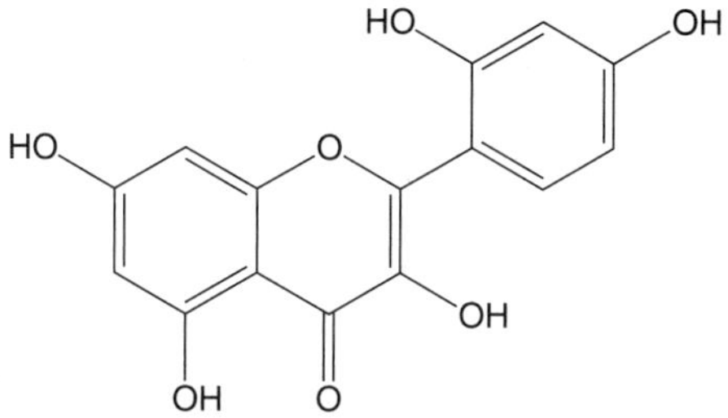 |
| Curcumin and rosmarinic acid | Fibrils, Oligomers | 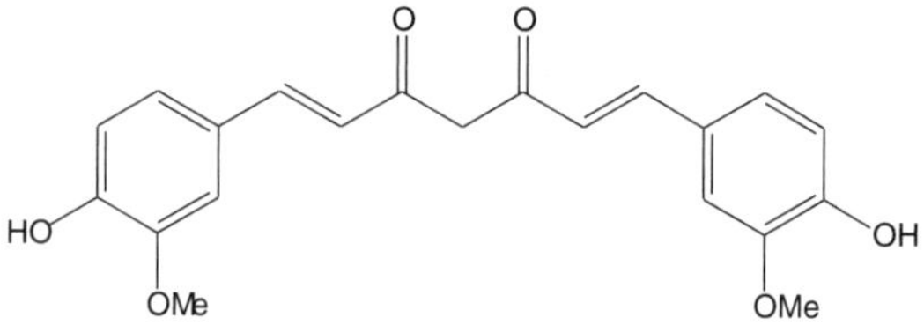 |
| Cyclohexanehexol | Fibrils, Oligomers | 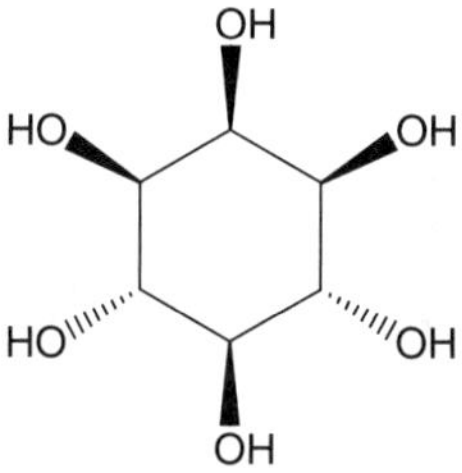 |
| Hydroxyphenyl triazine diamine | Oligomers | 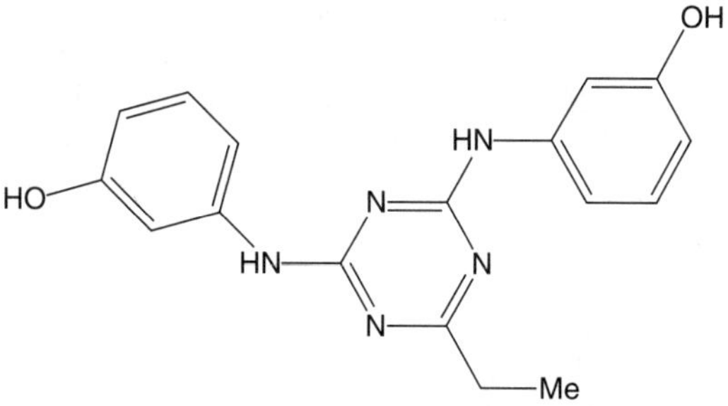 |

(*continued*)

**TABLE 19.2** (*Continued*)

| Compound | Mode of action | Example |
|---|---|---|
| Phenols | Oligomers | |
| Tricyclic pyrones | Oligomers | |

[a] References for studies found in 53.

subjective interpretation and many incidents of "false negatives," this method is to be used carefully and should be accompanied by a parallel verification method. Congo red can also be utilized in an absorbance assay to quantify the amount of misfolded material (54).

A straightforward method used frequently to follow the misfolding kinetics of previously deseeded protein is the thioflavin T assay (55). Thioflavin T is a molecule that binds rapidly to protein aggregates but not to monomeric or oligomeric intermediates. Upon binding, a red shift of the excitation and emission wavelength is monitored, allowing observing the kinetics of protein misfolding with a fluorescence microscope or by fluorescence spectroscopy. As this assay detects only β-sheet rich conformers, it is not suitable to identify unstructured intermediates of the misfolding process.

## 19.4.2 Spectroscopic Assays

To determine the presence of β-sheet in amyloid samples, spectroscopic techniques such as circular dichroism (CD) and Fourier transform infrared (FTIR) spectroscopy are routinely used. In far U.V. CD spectroscopy, the sample is examined before and after aggregation to record any change in secondary structure. Samples of soluble protein are subjected to left and right handed circular polarized light and the difference in its absorbance at far-UV wavelengths is recorded. Obtained spectra are then compared to a reference library to allow an assessment of secondary structural elements. Aβ is unstructured under physiological conditions and exhibits a classic random coil far U.V. CD spectra. However, as aggregation of Aβ progresses, and the monomers fold into protofibrils and fibrils, the CD spectra change to reflect β-sheet formation.

The presence of β-sheet in a protein solution can also be determined using FTIR by inspecting the amide I band, which occurs in the region between $1600 \text{ cm}^{-1}$ and $1700 \text{ cm}^{-1}$, and taking into account possible contributions from side chains. Amyloid fibrils typically have β-sheet peaks below $1620 \text{ cm}^{-1}$. FTIR is not suitable to determine atomic coordinates, but it has given detailed insights into protein structure, and its easy sample preparation and the applicability to most molecules have led to the development of many experimental techniques (reviewed in (56)).

### 19.4.3 Visualization Techniques (AFM and TEM)

Interesting information about the organization of fibrils has been obtained from atomic force microscopy (AFM) and transmission electron microscopy (TEM). Both methods give high resolution images of fibrils and show that fibrils are unbranched, twisted, and several μm long. These techniques are usually used to support data obtained from Thioflavin T assays, as this method gives no information about the presence of fibrils (it only confirms presence of β-sheets).

### 19.4.4 Molecular Structure Determination (solid-state NMR and x-ray diffraction)

Initial X-ray experiments on amyloid fibrils defined the classic b-sheet diffraction pattern with reflections at 4.75 Å and 10 Å, indicative of b-sheets parallel to the fibril axis, and the protein strand perpendicular to the fibril's long axis (29–31).

A more detailed view into the 3D organization of a fibril can be obtained by solid-state NMR. Solid-state NMR has the advantage that it can define the spatial arrangement of both the intrachain and intermolecular configurations. Using fibrillar, lyophilized protein the molecular packing of amyloid can be defined (57, 58). Solid-state NMR spectroscopy revealed a more detailed structure of several Aβ analogs and truncated sequences as well as the conformation of other peptides like α-synuclein, truncated peptides of transthyretin and the prion protein (Helmus J et al Molecular conformation and dynamics of the Y145Stop variant of human prion protein in amyloid fibrils. Proc Natl Acad Sci U S A. 2008 Apr 29;105(17):6284-9. in (59)). The data obtained from Aβ fibrils allowed the development of a structural model of the fibrillar form.

## REFERENCES

1. Chiti F, Dobson CM. Protein misfolding, functional amyloid, and human disease. Ann. Rev. Biochem. 2006;75:333–366.
2. Chiti F, et al. Designing conditions for in vitro formation of amyloid protofilaments and fibrils. Proc. Nat. Acad. Sci. 1999;96:3590–3594.
3. Bucciantini M, et al. Inherent toxicity of aggregates implies a common mechanism for protein misfolding diseases. Nature 2002;416:507.
4. Tiraboschi P, et al. The importance of neuritic plaques and tangles to the development and evolution of AD. Neurology 2004;62:1984–1989.
5. Yao ZX, Papadopoulos V. Function of {beta}-amyloid in cholesterol transport: a lead to neurotoxicity. FASEB J. 2002;16:1677–1679.
6. Zou K, et al. A novel function of monomeric amyloid beta -protein serving as an antioxidant molecule against metal-induced oxidative damage. J. Neurosci. 2002;22:4833–4841.
7. Huang SS, et al. Amyloid beta -peptide possesses a transforming growth factor-beta activity. J. Biol. Chem. 1998;273:27640–27644.
8. Annaert W, De Strooper B. Presenilins: molecular switches between proteolysis and signal transduction. Trends Neurosci. 1999;22:439–443.
9. Soba P, et al. Homo- and heterodimerization of APP family members promotes intercellular adhesion. Embo. J. 2005;24:3624–3634.
10. Terry RD, et al. Physical basis of cognitive alterations in Alzheimer's disease: synapse loss is the major correlate of cognitive impairment. Ann. Neurol. 1991;30:572–80.
11. Lambert MP, et al. Diffusible, nonfibrillar ligands derived from Abeta1-42 are potent central nervous system neurotoxins. Proc. Natl. Acad. Sci. USA. 1998;95:6448–6453.
12. Lesne S, et al. A specific amyloid-beta protein assembly in the brain impairs memory. Nature 2006;440:352–357.
13. Walsh DM, Selkoe DJ. Abeta Oligomers - a decade of discovery. J. Neurochem. 2007;101:1172–1184.
14. Walsh DM, et al. Amyloid beta -protein fibrillogenesis. Structure and biological activity of protofibrillar intermediates. J. Biol. Chem. 1999;274:25945–25952.
15. Walsh DM, et al. Naturally secreted oligomers of amyloid (beta) protein potently inhibit hippocampal long-term potentiation in vivo. Nature 2002;416:535.
16. Oddo S, et al. Amyloid deposition precedes tangle formation in a triple transgenic model of Alzheimer's disease. Neurobiol. Aging 2003;24:1063.
17. Hardy J, DJ Selkoe. The amyloid hypothesis of Alzheimer's disease: progress and problems on the road to therapeutics. Science 2002;297:353–356.
18. Opie E. The relation of diabetes mellitus to lesions of the pancreas. Hyaline degeneration of the islands of Langerhans. J. Exp. Med. 1901;5:527–540.
19. Westermark P, et al. Islet amyloid in type 2 human diabetes mellitus and adult diabetic cats contains a novel putative polypeptide hormone. Am. J. Pathol. 1987;127:414–417.
20. Cooper GJS, et al. Purification and characterization of a peptide from amyloid-rich pancreases of type 2 Diabetic patients. Proc. Nat. Acad. Sci. 1987;84:8628–8632.
21. Lorenzo A, et al. Pancreatic islet cell toxicity of amylin associated with type-2 diabetes mellitus. Nature 1994;368:756.

22. Demuro A, et al. Calcium dysregulation and membrane disruption as a ubiquitous neurotoxic mechanism of soluble amyloid oligomers. J. Biol. Chem. 2005;280:17294–17300.

23. Arvanitakis, Z., et al., Diabetes Mellitus and Risk of Alzheimer Disease and Decline in Cognitive Function. Arch Neurol, 2004.61(5):p.661–666.

24. Havrankova, J., J. Roth, and M. Brownstein, Insulin receptors are widely distributed in the central nervous system of the rat. Nature, 1978.272(5656):p.827.

25. Northam, E.A., D. Rankins, and F.J. Cameron, Therapy insight: the impact of type 1 diabetes on brain development and function. Nat Clin Pract Neurol, 2006.2(2):p.78–86.

26. Gispen, W.H. and G.-J. Biessels, Cognition and synaptic plasticity in diabetes mellitus. Trends in Neurosciences, 2000.23(11):p.542.

27. Steen, E., et al., Impaired insulin and insulin-like growth factor expression and signaling mechanisms in Alzheimer's disease—is this type 3 diabetes? J Alzheimers Dis, 2005.7(1):p.63–80.

28. Zhao, W.-Q., et al., Amyloid beta oligomers induce impairment of neuronal insulin receptors. FASEB J., 2008.22(1):p.246–260.

29. Yong W, et al. Structure determination of micelle-like intermediates in amyloid beta -protein fibril assembly by using small angle neutron scattering. Proc. Nat. Acad. Sci. 2002;99:150–154.

30. Huang THJ, et al. Structural studies of soluble oligomers of the alzheimer (beta)-amyloid peptide. J. Molec. Biol. 2000;297:73.

31. Murphy RM. Kinetics of amyloid formation and membrane interaction with amyloidogenic proteins. Biochim. Biophys. Acta 2007;1768:1923.

32. Kelly JW. Mechanisms of amyloidogenesis. Nat. Struct. Mol. Biol. 2000;7:824.

33. Ferrone F. Analysis of protein aggregation kinetics. Methods Enzymol. 1999; 309:256–274.

34. Wetzel R. Kinetics and thermodynamics of amyloid fibril assembly. Acc. Chem. Res. 2006;39:671–679.

35. Pauling L, Corey RB. The pleated sheet, a new layer configuration of polypeptide chains. Proc. Natl. Acad. Sci. USA. 1951;37:251–256.

36. Eanes ED, Glenner GG. X-ray diffraction studies on amyloid filaments. J. Histochem. Cytochem. 1968;16:673–677.

37. Sunde M, et al. Common core structure of amyloid fibrils by synchrotron X-ray diffraction. J. Molec. Biol. 1997;273:729.

38. Tjernberg LO, et al. A molecular model of Alzheimer amyloid beta -peptide fibril formation. J. Biol. Chem. 1999;274:12619–12625.

39. Benzinger TLS, et al. Two-dimensional structure of beta-Amyloid(10-35) fibrils. Biochem. 2000;39:3491–3499.

40. Benzinger TLS, et al. Propagating structure of Alzheimer's beta -amyloid(10-35) is parallel beta -sheet with residues in exact register. PNAS 1998; 95:13407–13412.

41. Antzutkin ON, et al. Multiple quantum solid-state NMR indicates a parallel, not antiparallel, organization of beta -sheets in Alzheimer's beta -amyloid fibrils. Proc. Nat. Acad. Sci. 2000;97:13045–13050.

42. Petkova AT, et al. A structural model for Alzheimer's beta -amyloid fibrils based on experimental constraints from solid state NMR. Proc. Nat. Acad. Sci. 2002;99:16742–16747.

43. Jaikaran ETAS, et al. Identification of a novel human islet amyloid polypeptide (beta)-sheet domain and factors influencing fibrillogenesis. J. Molec. Biol. 2001;308:515.

44. Westermark P, et al. Islet amyloid polypeptide: pinpointing amino acid residues linked to amyloid fibril formation. Proc. Nat. Acad. Sci. 1990;87:5036–5040.

45. Nilsson MR, Raleigh DP. Analysis of amylin cleavage products provides new insights into the amyloidogenic region of human amylin. J. Molec. Biol. 1999;294:1375.

46. Jayasinghe SA, Langen R. Identifying structural features of fibrillar islet amyloid polypeptide using site-directed spin labeling. J. Biol. Chem. 2004;279:48420–48425.

47. Heise H, et al. Molecular-level secondary structure, polymorphism, and dynamics of full-length alpha-synuclein fibrils studied by solid-state NMR. Proc. Natl, Acad. Sci. USA 2005;102:15871–15876.

48. Der-Sarkissian A, et al. Structural organization of {alpha}-synuclein fibrils studied by site-directed spin labeling. J. Biol. Chem. 2003;278:37530–37535.

49. Chen M, et al. Investigation of alpha-synuclein fibril structure by site-directed spin labeling. J. Biol. Chem. 2007;282:24970–24979.

50. Schenk D, et al. Immunization with amyloid-(beta) attenuates Alzheimer-disease-like pathology in the PDAPP mouse. Nature 1999;400:173.

51. Schenk D. Amyloid-(beta) immunotherapy for Alzheimer's disease: the end of the beginning. Nat. Rev. Neurosci. 2002;3:824.

52. Hock C, et al. Antibodies against (beta)-amyloid slow cognitive decline in Alzheimer's Disease. Neuron 2003;38:547.

53. Levine H. Small molecule inhibitors of A&b.beta; assembly. Amyloid 2007; 14:185–197.

54. Klunk WE, Jacob RF, Mason RP. Quantifying amyloid beta-Peptide (Abeta) aggregation using the congo red-abeta (CR-Abeta) spectrophotometric assay. Anal. Biochem. 1999;266:66.

55. Levine- III H. Thioflavine T interaction with synthetic Alzheimer's disease beta-amyloid peptides: detection of amyloid aggregation in solution. Protein Sci. 1993;2:404–410.

56. Hiramatsu H, Kitagawa T. FT-IR approaches on amyloid fibril structure. Biochim. Biophys. Acta 2005;1753:100.

57. Ritter C, et al. Correlation of structural elements and infectivity of the HET-s prion. Nature 2005;435:844.

58. Nelson R, et al. Structure of the cross-(beta) spine of amyloid-like fibrils. Nature 2005;435:773.

59. Naito A, Kawamura I. Solid-state NMR as a method to reveal structure and membrane-interaction of amyloidogenic proteins and peptides. Biochim. Biophys. Acta 2007;1768:1900.

# 20

# PROTEIN TRAFFICKING DISEASES

HEIDI M. SAMPSON AND DAVID Y. THOMAS
*McGill University, Quebec, Canada*

Many diseases are caused by defects in protein trafficking. Protein trafficking diseases occur when a mutant protein is recognized by the endoplasmic reticulum (ER) quality control system (ERQC), retained in the ER, and degraded in the cytosol by the proteasome rather than being trafficked to its correct site of action. Among these diseases are cystic fibrosis, lysosomal storage diseases (Fabry, Gaucher, and Tay-Sachs), nephrogenic diabetes insipidus, oculocutaneous albinism, protein C deficiency, and many others. A characteristic of many of these diseases is that the mutant protein remains functional, but it cannot escape the stringent ER quality-control machinery, and it is retained in the ER. This characteristic suggests that pharmacological interventions that promote the correct folding of the mutant protein would enable its escape from the ER and ameliorate the symptoms of the disease. In this review, we focus on specific examples of protein trafficking diseases in pharmacological or chemical chaperones have been shown to rescue trafficking of the mutant protein.

The etiology of several diseases can be traced to defects in protein trafficking. Studies on several different disease mutants have shown that chemicals and small molecules can correct trafficking of these mutants. As these compounds promote correct folding in a fashion analogous to the action of molecular chaperone proteins, these compounds have been termed "correctors" or "chaperones." Several different classes of chaperones can be designated based on their mechanisms of action. Chemical chaperones are the least specific and the least potent, and which often require millimolar concentrations to function. Pharmacological chaperones are small-molecule correctors that can be subdivided into two classes

*Chemical Biology: Approaches to Drug Discovery and Development to Targeting Disease*, First Edition.
Edited by Natanya Civjan.
© 2012 John Wiley & Sons, Inc. Published 2012 by John Wiley & Sons, Inc.

based on their specificities. Specialized pharmacological chaperones are protein-or mutation-specific small molecules that interact directly with the mutant protein to provide a folding template, such as enzyme active-site inhibitors. Generalized correctors are less specific and are likely to function on the endoplasmic reticulum (ER) retention machinery, ER-associated degradation (ERAD), or other signaling pathways involved in trafficking rather than through a direct interaction with mutant proteins. We will begin with a brief overview of ER quality control, followed by specific examples of protein trafficking diseases and then discuss the different classes of correctors below.

## 20.1  QUALITY CONTROL IN THE ER

Proteins that travel along the secretory pathway are subject to many quality-control checkpoints. Inside the ER, proteins must fold into their proper conformation before being sorted into vesicles destined for the Golgi apparatus. Once proteins have translocated into the ER through the translocon (Fig. 20.1), molecular chaperones, which include the heat shock chaperone family (e.g., Hsp70, BiP), the lectins calnexin and calreticulin, and the oxidoreductases (e.g., PDI), act on the nascent chain to promote the correctly folded conformation. Together with the oxidoreductase ERp57, the lectin chaperones act on the nascent protein in cycles of binding and release through the recognition of a monoglucosylated N-glycan. These cycles are controlled by N-glycan-modifying proteins including UDP-glucose:glycoprotein glucosyltransferase (UGGT) and glucosidases I and II (Fig. 20.1). Once the protein is correctly folded, the N-glycan is further modified by glucosidase II and proceeds along the secretory pathway. If several cycles of lectin binding do not result in a correctly folded protein, then the $\alpha(1-2)$-ER mannosidase I (ManI) modifies the N-glycan, which prevents it from reentering the lectin-binding cycle and targets it for degradation through ERAD in a process that involves the ER degradation-enhancing $\alpha$-mannosidase-like proteins (EDEMs), Derlin and the p97/valosin-containing protein (VCP) (1). In some cases, the misfolded protein aggregates and accumulates in the ER, which triggers the unfolded protein response.

Three different classes of ERAD-targeting components function depending on the location of the lesion: a cytosolic group, a lumenal group, and a transmembrane group. These proteins somehow sense the folding status of their substrates and triage those with defects for degradation. However, the mechanisms that regulate the decision to undergo ERAD instead of forward transport in the secretory pathway are not known. It is known that to be recruited to coat protomer II (COPII) vesicles that leave the ER, proteins must be recognized by the Sec24 machinery, either directly through an ER exit code or through another receptor that can then interact with Sec24. Several different ER exit codes have been identified, including cytosolic diacidic codes such as the one found in the cystic fibrosis transmembrane conductance regulator (CFTR) protein (2), dihydrophobic motifs like that in ER–Golgi intermediate compartment protein 53 (3),

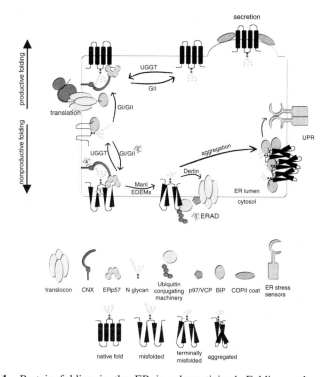

**Figure 20.1** Protein folding in the ER is schematicized. Folding pathways that lead to secretion are depicted in the upper half of the ER. Nonproductive folding and its consequences are depicted in the lower half of the ER. For simplicity, some pathways have been omitted, and some proteins are indicated by text alone. Once a protein enters the ER through the translocon, it undergoes N-glycosylation and is acted on by several chaperones including BiP. Glucosidases I and II (GI/GII) remove the terminal glucoses on the N-glycan moiety which enables recognition of the protein by calnexin (CNX) or calreticulin (not shown). The oxidoreductase ERp57 is bound to CNX and also acts on the protein. If the resulting protein has a nonnative fold, then UGGT recognizes it and adds a terminal glucose to enable rebinding with CNX. Once the protein has folded properly, the N-glycan is modified even more by GII and can be recognized by the Sec24 protein for COPII-dependent secretion. If the protein cannot be folded natively after several cycles of CNX binding, then ManI removes the terminal mannose on the N-glycan which prevents UGGT from reglycosylating it and targets it for destruction through ERAD in a process that involves the EDEMs, Derlin, and VCP. The mutant protein is retrotranslocated out of the ER, ubiquitinated, and degraded by the 26 S proteasome. In some cases, misfolded proteins have a propensity to aggregate, which overwhelms the ER folding capacity and triggers the unfolded protein response (UPR) through ER stress-sensor kinases. Correction of ER-retained mutant protein trafficking can occur through several steps of the folding pathway as indicated by numbered stars and described in the text. Pharmacological chaperones interact directly with the mutant protein (1) to stabilize its fold. Inhibitors of glucosidases (2) alter the processing of the N-glycan moiety to prevent the recognition of mutant proteins by CNX and calreticulin. Proteasomal inhibitors (3) can directly increase levels of mutant proteins and in some cases this permits some of the protein to escape ERQC. Inhibitors of SERCA calcium channels (4) correct trafficking for many different proteins by lowering the calcium levels in the ER.

and various hydrophobic signals that are found in G-protein coupled receptors (GPCRs) (4). In yeast, simply inhibiting ERAD genetically does not permit ER-retained proteins to enter COPII vesicles, which suggests that some other level of recognition by the ERQC exists. However, even when folding of substrates has improved as judged by trypsin-sensitivity assays, inhibition of ERAD still does not permit entry into COPII vesicles (5). This discrepancy could be explained by the masking of the ER exit signal in the misfolded protein. In yeast, carboxypeptidase Y mutants whose ER exit signal is not obscured can escape ERAD even though other parts of the protein are misfolded (6). For many ER-retained disease mutant proteins, the ERQC performs its function too well and retains otherwise functional proteins. Small molecules that function in any process described above could potentially correct the trafficking of these ER-retained proteins and would be useful therapeutics.

## 20.2  PROTEIN TRAFFICKING DISEASES

As many ER-retained mutant proteins are functional, rerouting them to their appropriate subcellular localization would restore the phenotype caused by their mislocalization. In some cases, it is estimated that only a small amount of functional protein (10–15%) is necessary to maintain health. Indeed, for some diseases a critical threshold seems to exist (7). Patients who express functional protein above this threshold have mild or no symptoms of the disease (7, 8), which suggests that even a modest increase in protein rerouting would improve the quality of life for many patients with these diseases. This finding provides the impetus for identifying correctors, even those with modest effects.

### 20.2.1  Properties of Correctable Protein Trafficking Disease Mutants

Mutations that destroy protein activity such as those that abolish enzyme active sites are not amenable to rescue with correctors. However, mutations that alter the stability or folding of the protein, but that do not abrogate its activity, are good candidates for rescue by correctors. Most ER-retained missense mutations identified to date fall in this class. Frequently, these mutations are temperature sensitive and show some degree of correction at permissive temperatures (usually <30°C). Another important feature of diseases that are amenable to rescue with correctors is the degradability of accumulated substrates or byproducts. For example, the lysosomal storage diseases Fabry, Gaucher, and Tay-Sachs all result in substrate accumulation in the lysosome; however, after treatment with pharmacological chaperones and rerouting of the enzymes, the corrected proteins reduce these substrate stores (9–12). In contrast, ER-retention diseases that result in the deposit of toxic or nonnative proteins, such as fibrils associated with amyloidoses, likely would not be amenable to rescue with correctors because of the cell's inability to degrade the nonnative deposits, unless treatment with the corrector begins before irreversible damage is done. Several examples of trafficking diseases and their correctors are described below.

## 20.2.2   CFTR Delta F508

One of the best-studied examples of a mutation that is amenable to rescue is the deltaF508 mutation in the chloride channel CFTR, which is the gene responsible for cystic fibrosis (13). Cystic fibrosis (CF) is the most common autosomal recessive genetic disease that affects the Caucasian population. It is a lethal disease characterized by severe dehydration of the cells lining the lung, intestine, and exocrine tissues (13). Most cases of CF can be attributed to a single mutation resulting in the deletion of a phenylalanine codon at position 508 in the protein, which is known as deltaF508. DeltaF508 CFTR can be rescued by incubation at permissive temperature (i.e., 27–30°C), by the addition of chemical chaperones such as sodium 4-phenylbutyrate, or other osmolytes such as glycerol, dimethyl sulfoxide, and trimethylamine N-oxide (Fig. 20.2a) (14–16). Modulation of ER calcium levels with curcumin and thapsigargin (Fig. 20.2b) has also been shown to correct deltaF508 CFTR trafficking; however, these results remain controversial (17, 18). Recent high-throughput screens have identified several classes of small-molecule correctors of protein trafficking, which include aminobenzothiazoles, aminoarylthiazoles, quinazolinylaminopyrimidinones, bisaminomethylbithiazoles, quinazolinones, khivorins and substituted 1-phenylsulfonylpiperazines (19–22).

Three different assays were used to screen for correctors. One assay measured directly the amount of protein that trafficked to the cell membrane using immunofluorescence against an extracellular epitope tag in deltaF508 CFTR (19). The second assay monitored deltaF508 CFTR function by iodide influx with a halide-sensitive-YFP construct as readout (20). The third assay also monitored function but through fluorescence energy transfer between a membrane-soluble voltage-sensitive dye bis-(1,2-dibutylbarbituric acid)trimethine oxonol [DiSBAC2(3)] and a plasma membrane localized fluorescent coumarin-linked phospholipid CC2-DMPE (22). The compounds identified in these screens are likely to function at different steps of the deltaF508 CFTR folding pathway. Corr4a, which is a bisaminomethylthiazole, was shown to increase the folding efficiency of deltaF508 CFTR (20). Both Corr4a and the quinazolinone VRT-325 were shown to delay ERAD of deltaF508 CFTR (20, 22). Compounds of the aminoarylthiazole class were found to act after ER folding as no increase in folding efficiency was detected, but the stability of deltaF508 was increased at the cell surface (20). VRT-325 also increased the stability of deltaF508 CFTR at the cell surface (22). This compound was shown to correct other CFTR mutants and even a mutant in the human ether-a-go-go-related gene (hERG) (see below). Derivatives of some of these compounds are now in clinical trials.

## 20.2.3   Lysosomal Storage Diseases

The lysosomal storage diseases, such as Tay-Sachs, Sandhoff, Gaucher, and Fabry disease, are autosomal recessive loss-of-function disorders. The mutant enzymes fail to degrade their respective lysosomal substrates because of their retention

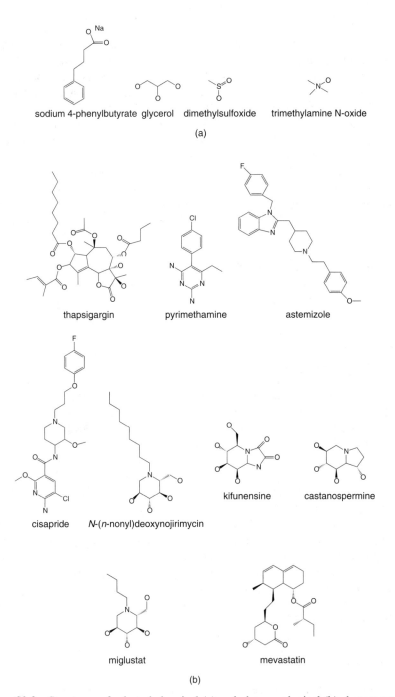

sodium 4-phenylbutyrate   glycerol   dimethylsulfoxide     trimethylamine N-oxide

(a)

thapsigargin        pyrimethamine        astemizole

cisapride    N-(n-nonyl)deoxynojirimycin       kifunensine     castanospermine

miglustat           mevastatin

(b)

**Figure 20.2**   Structures of selected chemical (a) and pharmacological (b) chaperones are shown.

in the ER and subsequent degradation. This result leads to an accumulation of substrates and a variety of phenotypes including enlargement of affected organs, skeletal lesions, neurological abnormalities, and premature death. The current therapies for these diseases include inhibition of substrate production and enzyme-replacement therapy. However, enzyme-replacement therapy is not suitable for neurological phenotypes associated with some types of these diseases because of the impermeability of the blood–brain barrier; hence, other treatments are necessary. Pharmacological chaperones for these diseases have been identified. As these proteins are enzymes, the pharmacological chaperones identified tend to be competitive active-site inhibitors. This finding may seem counterintuitive, but inhibitors are frequently trafficking correctors at subinhibitory doses (see below). The rationalization is that inhibitors stabilize the fold of the mutant proteins at the neutral pH of the ER and allow them to evade the ERQC. When the proteins reach the lysosome, the high concentration of the substrates successfully compete away inhibitor binding thereby enabling degradation of the substrates to occur. Some examples of correctors for each of these diseases are discussed below.

Several active-site specific chaperones have been identified for beta-hexosaminidase A, which is the multisubunit enzyme responsible for Tay-Sachs and Sandhoff diseases (9, 11, 23). The screening assays monitor inhibition of purified enzyme activity fluorometrically using the fluorescent artificial substrate 4-methylumbelliferyl-$\beta$-$N$-acetylglucosamine. These assays are then followed by cell-based assays to determine the amount of functional protein that traffics to the lysosome. Several distinct structural classes of pharmacological chaperones have been identified for these diseases including aza-sugars, pyrimethamine (Fig. 20.2b), and substituted bicyclic and tricyclic nitrogen-containing heterocycles (9, 11, 23).

Gaucher disease is caused by a deficiency in lysosomal beta-glucosidase activity. Deoxynojirimycins are known inhibitors of several enzymes, which include beta-glucosidase. Several alkylated deoxynojirimycins were screened for correction of N370 S Gaucher mutant lysosomal beta-glucosidase activity in an intact cell assay (10). $N$-($n$-nonyl)deoxynojirimycin (Fig. 20.2b) was found to be the most potent corrector, but consistent with its activity as a competitive inhibitor, it inhibited beta-glucosidase activity at higher concentrations. Because this deoxynojirimycin was shown to inhibit alpha-glucosidase at the effective concentrations for beta-glucosidase correction, other structural variants were tested and alpha-1-$C$-octyl-deoxynojirimycin was shown to have the most promise (24). Miglustat, which is the deoxynorijimycin derivative $N$-($n$-butyl)deoxynojirimycin (Fig. 20.2b), corrected trafficking of several mutants including N370 S in one cell type, but not in another (25). Several alkylated iminosugars, which include morpholine, piperazine, isofagomine, and 2,5-dideoxy-2,5-imino-D-glucitol, were correctors for both N370 S and G202 R mutants in cell culture (26). The addition of an adamantyl cap on the alkyl chain also increased the amount of correction achieved with the deoxynojirimycin compounds and improved the activity of the other iminosugar derivatives as well (26–28). Interestingly, the carbohydrate portion of these compounds was shown to be sufficient for chaperone activity (29, 30). Mutation-specific correctors have also been identified. The synthetic carbohydrate mimic $N$-octylvalienamine specifically corrects the trafficking of the F213I mutant, but not the N370 S mutant, which suggests that subtle differences may exist between these two mutants (31). A high-throughput screen for inhibitors of beta-glucosidase identified three classes of corrector compounds: aminoquinolines, sulfonamides, and triazines (32). These compounds were shown to be potent inhibitors and correct trafficking of the N370 S mutant. In addition to active-site inhibitors, the hydroxymethylglutaryl (HMG)-CoA reductase inhibitor mevastatin (Fig. 20.2b) was shown to correct beta-glucosidase trafficking by altering the levels of free unesterified cholesterol in severe Gaucher mutant cells (33). Work on three lysosomal storage disease proteins, including beta-glucosidase, has shown that altering the intracellular calcium levels with the voltage-gated calcium channel inhibitors diltiazem and verapamil can upregulate the ER folding capacity and correct the trafficking of these mutants (34).

Fabry disease is caused by mutations in the gene-encoding lysosomal alpha galactosidase A. Treatment of R301Q Fabry mutant cells with subinhibitory concentrations of the potent competitive inhibitor 1-deoxygalactonojirimycin results in efficient folding and stabilization of the mutant protein (35). Several derivatives of this compound were also shown to correct trafficking of Fabry mutants, but they were less effective than 1-deoxygalactonojirimycin (36). Several other mutations, including A97V, R112H, R112 C, A143 T, and L300P, were also corrected by treatment with 1-deoxygalactonojirimycin (37). No high-throughput screens for other correctors of Fabry mutants have been performed.

### 20.2.4   hERG

hERG encodes the pore-forming subunit of the rapidly activating delayed rectifier potassium channel. Mutations or drug treatments that cause retention of hERG in the ER result in cardiac arrhythmias that can lead to sudden death. Hence, drugs are now routinely screened for their effects on hERG function before being pursued in clinical trials. Several ER-retained hERG mutants can be rescued by chemical and pharmacological chaperones. The G601 S and N470D mutations can be rescued by growth at permissive temperature or incubation with known hERG channel blockers E-4031, cisapride, and astemizole (Fig. 20.2b) (38, 39). hERG G601 S has been used in a small molecule screen to identify correctors of hERG trafficking (40). This assay measured the amount of hERG G601 S that trafficked to the cell surface by chemiluminescent detection of an extracellular epitope tag. Several different hERG blockers were found to rescue the ER retained mutant, which is consistent with their action as pharmacological chaperones. Other compounds have also been shown to rescue the trafficking of certain hERG mutants selectively. For example, the sarcoplasmic/ER calcium ATPase (SERCA) inhibitor thapsigargin can rescue the trafficking of G601 S and F805 C mutants, but not N470D, whereas E-4031 rescues G601 S and N470D mutants but not F805 C mutants (41).

## 20.2.5 G Protein-Coupled Receptors

Many diseases caused by mutations in GPCRs are also the result of ER retention. Mutations in the vasopressin 2 receptor result in nephrogenic diabetes insipidus, which is a disease characterized by the kidney's inability to concentrate urine. Trafficking of several different ER-retained mutants can be corrected by small nonpeptide V2 R and V1 R antagonists (42–46). The V206D mutation can also be corrected with glycerol, DMSO, SERCA inhibitors thapsigargin and curcumin, and the calcium ionophore ionomycin (44). This mutation cannot be corrected by growth at permissive temperature or with the addition of 4-phenylbutyrate (44). In contrast, the A98P, L274P, and R113 W mutations can be corrected with the osmolytes trimethylamine N-oxide and DMSO, as well as with growth at permissive temperature (47). However, glycerol treatment does not correct the trafficking of these mutations.

Mutations in another GPCR, which is the gonadotropin-releasing hormone receptor, result in hypogonadotropic hypogonadism. Seventeen mutations in this gene have been characterized as misfolding or misrouting mutants, and most of these can be rescued by incubation with peptidomimetic antagonists (48). These pharmacological chaperones are indoles, quinolones, and erythromycin macrolides. Another example of a GPCR with ER-retained disease mutations is rhodopsin, which when mutated causes retinitis pigmentosa (49). A mild rhodopsin mutant P23H associated with night blindness can be rescued by treatment with 11-*cis*-retinal, which is its covalently bound chromophore (49).

## 20.2.6 Alpha-1-Antitrypsin Z Variant

Alpha-1-antitrypsin mutations are associated with early-onset emphysema and liver disease that results in early death (50). The Z-variant of the disease is the most common and is present in over 95% of cases. This variant has been shown to be rescued *in vitro* by treatment with proteasome inhibitors (50). Several chemical and pharmacological chaperones have also been shown to correct the secretion of the mutant protein, which include the glucosidase inhibitor castanospermine, as well as the mannosidase inhibitors kifunensine (Fig. 20.2b) and 1,4-dideoxy-1,4-imino-D-mannitol hydrochloride (51). Incubation with the chemical chaperone 4-phenylbutyrate also results in increased trafficking. However, growth at permissive temperature does not result in trafficking correction, although it does decrease the amount of mutant protein that becomes degraded (52).

## 20.3 CORRECTORS OF PROTEIN TRAFFICKING

### 20.3.1 Chemical Chaperones

Several different classes of chaperones can be designated based on their mechanisms of action. Chemical chaperones are those that act nonspecifically. Although these chaperones are nonspecific, their use as therapeutics may be limited because of the high concentrations required to achieve correction. The mechanisms of correction employed by chemical chaperones may include the induction of molecular chaperone transcription or the nonspecific stabilization of proteins through masking of hydrophobic domains. For example, 4-phenylbutyrate (4-PBA) is a histone deacetylase inhibitor that activates transcription of different genes including the heat shock proteins (53). Studies on deltaF508 CFTR suggest that 4-PBA reduces the protein levels of the constitutive Hsc70 chaperone (54). This in turn reduces the amount of deltaF508 CFTR that interacts with Hsc70. The decrease in Hsc70 protein levels induces an increase in Hsp70 levels, which has been proposed to be a more effective chaperone, thus helping deltaF508 CFTR to fold and enabling it to escape from the ERQC (55). Interestingly, other phenyl-fatty acids have also been shown to correct trafficking, and their efficacy is not correlated with their efficacy as histone deacetylase inhibitors, which suggests that they function through another mechanism (56), perhaps through their ability to bind and mask hydrophobic regions. Although 4-PBA corrects trafficking of many misfolded proteins, its use as a therapeutic may be limited. For instance, although it corrects trafficking of Fabry disease mutants, the mutant protein is not functional (57).

### 20.3.2 Specialized Pharmacological Chaperones

Pharmacological chaperones can be divided into two subclasses depending on their mechanisms of action. Specialized pharmacological chaperones are correctors that interact directly with the mutant protein to stabilize correct folding and enable it to escape ER retention mechanisms. This subclass of chaperones is the most specific and is mainly composed of inhibitors. Specialized pharmacological chaperones may also show mutation-specific profiles. For example, two common mutations in Gaucher disease N370 S and G202 R are both localized to the catalytic domain of the protein and are both corrected by active-site directed inhibitors. However, the L444P mutation, which is located in another domain, cannot be corrected with these chaperones. However, it can be corrected by incubation at permissive temperature, which suggests that simply another class of pharmacological chaperone is necessary to achieve correction (26).

Examples of specialized pharmacological chaperones include the competitive inhibitors of the lysosomal storage disease enzymes, the potent hERG channel blockers, and the cell-permeable nonpeptide antagonists of the vasopressin receptor described above. Enzyme substrates can also act as pharmacological chaperones by providing a scaffold on which the native fold can be formed. For example, the R402Q mutant of tyrosinase, which causes oculocutaneous albinism, can be rescued by the addition of its substrate (L-tyrosine) or its substrate/cofactor L-Dopa (58). Some specialized pharmacological chaperones are not inhibitors, such as the VRT-325 and Corr4a correctors of deltaF508 CFTR (20). VRT-325 is an interesting compound as it has also been shown to correct trafficking of P-glycoprotein, a protein related to CFTR, and the structurally unrelated G601 S mutant of the hERG potassium channel. This compound has not yet been demonstrated to interact directly with hERG.

### 20.3.3   Generalized Pharmacological Chaperones

Another subclass of pharmacological chaperones includes compounds that act through other mechanisms. This subclass includes generalized chaperones—those that rescue the trafficking defects of many or all ER-retained proteins through a universal mechanism. Many different mechanisms of correction are possible. Examples of this subclass of chaperones are described below.

One might expect that simply inhibiting ERAD with proteasome inhibitors would be the most generalized form of correction (Fig. 20.1, star 3); however, this strategy gives inconsistent results with different ER-retained mutants. The hERG Y611H mutant is not corrected, deltaF508 CFTR is only weakly corrected and alpha-1-antitrypsin Z variant is corrected as visualized by confocal microscopy (50, 59, 60).

As discussed above, for many ER-retained proteins, altering the intracellular calcium levels with SERCA inhibitors or with calcium ionophores corrects trafficking of the mutant proteins, presumably by altering the ER's capacity for folding (Fig. 20.1, star 4), because several ER chaperones require calcium for function. Interestingly, for lysosomal storage diseases, trafficking correction occurs through changing the levels of calcium in the cytosol and not in the ER. These two different mechanisms of correction highlight the importance of calcium signaling in protein trafficking.

Another common mechanism of correction is ER glucosidase inhibition (Fig. 20.1, star 2), which prevents the recognition of misfolded proteins by calnexin and calreticulin and presumably allows them to escape the ERQC. Inhibition of glucosidase with castanospermine and miglustat were shown to correct the trafficking of deltaF508 CFTR and alpha-1-antitrypsin (51, 61). Whether this mechanism also works on other ER-retained mutants remains to be determined.

For other small molecules, the mechanism of action is still unclear. For example, sildenafil and structural analogs have been shown to correct trafficking of deltaF508 CFTR mutants (19, 21). Sildenafil is an inhibitor of phosphodiesterase activity; however, the link between this function and CFTR trafficking correction remains to be elucidated.

## 20.4   CONCLUSIONS

The etiology of many genetic diseases can be traced to defects in protein trafficking. Several mutants are functional but are retained in the ER because of the overly stringent ERQC. Corrector compounds that permit the escape of ER-retained mutant proteins from the ERQC have great potential as therapeutics. Several types of trafficking correctors have been identified. Chemical chaperones are nonspecific and require high concentrations to be effective, which thereby limits their potential as therapeutics. Pharmacological chaperones function at much lower concentrations and show greater promise for drug development. Pharmacological chaperones can be classified into specialized chaperones that are protein or mutation-specific or generalized chaperones that can correct the trafficking of many ER-retained mutants. The identification of a generalized pharmacological chaperone that can correct all ER-retained disease mutants without grossly affecting normal proteins would enable treatment of many different protein trafficking diseases. Defining the mechanisms of action for this class of pharmacological chaperones will be the next milestone in trafficking disease research.

Additive effects of correctors have been shown for the deltaF508 CFTR mutant (62), which suggests that treatments that combine two or more chaperones may be a good option if no single potent corrector can be identified for a particular trafficking disease. Indeed combinatorial studies are likely to be the next key area of screening for correctors of protein trafficking diseases.

## ACKNOWLEDGMENTS

Research on CFTR trafficking in the Thomas lab is supported by grants from Cystic Fibrosis Foundation Therapeutics, Inc., and the Canadian Institutes of Health Research. H.M.S. is supported by a fellowship from the Canadian Cystic Fibrosis Foundation.

## REFERENCES

1. Nakatsukasa K, Brodsky JL. The recognition and retrotranslocation of misfolded proteins from the endoplasmic reticulum. Traffic. In press.

2. Wang X, Matteson J, An Y, Moyer B, Yoo JS, Bannykh S, Wilson IA, Riordan JR, Balch WE. COPII-dependent export of cystic fibrosis transmembrane conductance regulator from the ER uses a di-acidic exit code. J. Cell Biol. 2004;167:65–74.

3. Kappeler F, Klopfenstein DR, Foguet M, Paccaud JP, Hauri HP. The recycling of ERGIC-53 in the early secretory pathway. ERGIC-53 carries a cytosolic endoplasmic reticulum-exit determinant interacting with COPII. J. Biol. Chem. 1997;272:31801–31808.

4. Dong C, Filipeanu CM, Duvernay MT, Wu G. Regulation of G protein-coupled receptor export trafficking. Biochim. Biophys. Acta 2007;1768:853–870.

5. Pagant S, Kung L, Dorrington M, Lee MC, Miller EA. Inhibiting endoplasmic reticulum (ER)-associated degradation of misfolded Yor1p does not permit ER export despite the presence of a diacidic sorting signal. Mol. Biol. Cell 2007;18:3398–3413.

6. Kincaid MM, Cooper AA. Misfolded proteins traffic from the endoplasmic reticulum (ER) due to ER export signals. Mol. Biol. Cell 2007;18:455–463.

7. Amaral MD. Processing of CFTR: traversing the cellular maze--how much CFTR needs to go through to avoid cystic fibrosis? Pediatr. Pulmonol. 2005;39:479–491.

8. Mahuran DJ. Biochemical consequences of mutations causing the GM2 gangliosidoses. Biochim. Biophys. Acta 1999;1455:105–138.

9. Maegawa GH, Tropak M, Buttner J, Stockley T, Kok F, Clarke JT, Mahuran DJ. Pyrimethamine as a potential pharmacological chaperone for late-onset forms of GM2 gangliosidosis. J. Biol. Chem. 2007;282:9150–9161.

10. Sawkar AR, Cheng WC, Beutler E, Wong CH, Balch WE, Kelly JW. Chemical chaperones increase the cellular activity of N370S beta -glucosidase: a therapeutic strategy for Gaucher disease. Proc. Natl. Acad. Sci. U.S.A. 2002;99:15428–15433.

11. Tropak MB, Reid SP, Guiral M, Withers SG, Mahuran D. Pharmacological enhancement of beta-hexosaminidase activity in fibroblasts from adult Tay-Sachs and Sandhoff Patients. J. Biol. Chem. 2004;279:13478–13487.

12. Yam GH, Zuber C, Roth J. A synthetic chaperone corrects the trafficking defect and disease phenotype in a protein misfolding disorder. FASEB J. 2005;19:12–18.

13. Ratjen F, Doring G. Cystic fibrosis. Lancet 2003;361:681–689.

14. Brown CR, Hong-Brown LQ, Biwersi J, Verkman AS, Welch WJ. Chemical chaperones correct the mutant phenotype of the delta F508 cystic fibrosis transmembrane conductance regulator protein. Cell Stress Chaperones. 1996;1:117–125.

15. Denning GM, Anderson MP, Amara JF, Marshall J, Smith AE, Welsh MJ. Processing of mutant cystic fibrosis transmembrane conductance regulator is temperature-sensitive. Nature 1992;358:761–764.

16. Sato S, Ward CL, Krouse ME, Wine JJ, Kopito RR. Glycerol reverses the misfolding phenotype of the most common cystic fibrosis mutation. J. Biol. Chem. 1996;271:635–638.

17. Egan ME, Glockner-Pagel J, Ambrose C, Cahill PA, Pappoe L, Balamuth N, Cho E, Canny S, Wagner CA, Geibel J, Caplan MJ. Calcium-pump inhibitors induce functional surface expression of Delta F508-CFTR protein in cystic fibrosis epithelial cells. Nat. Med. 2002;8:485–492.

18. Song Y, Sonawane ND, Salinas D, Qian L, Pedemonte N, Galietta LJ, Verkman AS. Evidence against the rescue of defective DeltaF508-CFTR cellular processing by curcumin in cell culture and mouse models. J Biol Chem. 2004;279:40629–40633.

19. Carlile GW, Robert R, Zhang D, Teske KA, Luo Y, Hanrahan JW, Thomas DY. Correctors of protein trafficking defects identified by a novel high-throughput screening assay. ChemBioChem. 2007;8:1012–1020.

20. Pedemonte N, Lukacs GL, Du K, Caci E, Zegarra-Moran O, Galietta LJ, Verkman AS. Small-molecule correctors of defective DeltaF508-CFTR cellular processing identified by high-throughput screening. J. Clin. Invest. 2005;115:2564–2571.

21. Robert R, Carlile GW, Pavel C, Liu N, Anjos SM, Liao J, Luo Y, Zhang D, Thomas DY, Hanrahan JW. Structural analogue of sildenafil identified as a novel corrector of the F508del-CFTR trafficking defect. Mol. Pharmacol. 2007.

22. Van Goor F, Straley KS, Cao D, Gonzalez J, Hadida S, Hazlewood A, Joubran J, Knapp T, Makings LR, Miller M, Neuberger T, Olson E, Panchenko V, Rader J, Singh A, Stack JH, Tung R, Grootenhuis PD, Negulescu P. Rescue of DeltaF508-CFTR trafficking and gating in human cystic fibrosis airway primary cultures by small molecules. Am. J. Physiol. Lung Cell Mol. Physiol. 2006;290:L1117–1130.

23. Tropak MB, Blanchard JE, Withers SG, Brown ED, Mahuran D. High-throughput screening for human lysosomal beta-N-Acetyl hexosaminidase inhibitors acting as pharmacological chaperones. Chem. Biol. 2007;14:153–164.

24. Yu L, Ikeda K, Kato A, Adachi I, Godin G, Compain P, Martin O, Asano N. Alpha-1-C-octyl-1-deoxynojirimycin as a pharmacological chaperone for Gaucher disease. Bioorg. Med. Chem. 2006;14:7736–7744.

25. Alfonso P, Pampin S, Estrada J, Rodriguez-Rey JC, Giraldo P, Sancho J, Pocovi M. Miglustat (NB-DNJ) works as a chaperone for mutated acid beta-glucosidase in cells transfected with several Gaucher disease mutations. Blood Cells Mol. Dis. 2005;35:268–276.

26. Sawkar AR, Adamski-Werner SL, Cheng WC, Wong CH, Beutler E, Zimmer, KP, and Kelly, JW. Gaucher disease-associated glucocerebrosidases show mutation-dependent chemical chaperoning profiles. Chem. Biol. 2005;12:1235–1244.

27. Sawkar AR, Schmitz M, Zimmer KP, Reczek D, Edmunds T, Balch WE, Kelly JW. Chemical chaperones and permissive temperatures alter localization of Gaucher disease associated glucocerebrosidase variants. ACS Chem. Biol. 2006;1:235–251.

28. Yu Z, Sawkar AR, Whalen LJ, Wong CH, Kelly JW. Isofagomine- and 2,5-anhydro-2,5-imino-D-glucitol-based glucocerebrosidase pharmacological chaperones for Gaucher disease intervention. J. Med. Chem. 2007;50:94–100.

29. Chang HH, Asano N, Ishii S, Ichikawa Y, Fan JQ. Hydrophilic iminosugar active-site-specific chaperones increase residual glucocerebrosidase activity in fibroblasts from Gaucher patients. FEBS J. 2006;273:4082–4092.

30. Steet RA, Chung S, Wustman B, Powe A, Do H, Kornfeld SA. The iminosugar isofagomine increases the activity of N370S mutant acid beta-glucosidase in Gaucher fibroblasts by several mechanisms. Proc. Natl. Acad. Sci. U.S.A. 2006;103:13813–13818.

31. Lin H, Sugimoto Y, Ohsaki Y, Ninomiya H, Oka A, Taniguchi M, Ida H, Eto Y, Ogawa S, Matsuzaki Y, Sawa M, Inoue T, Higaki K, Nanba E, Ohno K, Suzuki Y. N-octyl-beta-valienamine up-regulates activity of F213I mutant beta-glucosidase in cultured cells: a potential chemical chaperone therapy for Gaucher disease. Biochim. Biophys. Acta 2004;1689:219–228.

32. Zheng W, Padia J, Urban DJ, Jadhav A, Goker-Alpan O, Simeonov A, Goldin E, Auld D, LaMarca ME, Inglese J, Austin CP, Sidransky E. Three classes of glucocerebrosidase inhibitors identified by quantitative high-throughput screening are chaperone leads for Gaucher disease. Proc. Natl. Acad. Sci. U.S.A. 2007;104:13192–13197.

33. Ron I, Horowitz M. Intracellular cholesterol modifies the ERAD of glucocerebrosidase in Gaucher disease patients. Mol. Genet. Metab. 2008;93:426–436.

34. Mu TW, Fowler DM, Kelly JW. Partial restoration of mutant enzyme homeostasis in three distinct lysosomal storage disease cell lines by altering calcium homeostasis. PLoS Biol. 2008;6:e26.

35. Fan JQ, Ishii S, Asano N, Suzuki Y. Accelerated transport and maturation of lysosomal alpha-galactosidase A in Fabry lymphoblasts by an enzyme inhibitor. Nat. Med. 1999;5:112–115.

36. Asano N, Ishii S, Kizu H, Ikeda K, Yasuda K, Kato A, Martin OR, Fan JQ. In vitro inhibition and intracellular enhancement of lysosomal alpha-galactosidase A activity in Fabry lymphoblasts by 1-deoxygalactonojirimycin and its derivatives. Eur. J. Biochem. 2000;267:4179–4186.

37. Shin SH, Murray GJ, Kluepfel-Stahl S, Cooney AM, Quirk JM, Schiffmann R, Brady RO, Kaneski CR. Screening for pharmacological chaperones in Fabry disease. Biochem. Biophys. Res. Commun. 2007;359:168–173.

38. Gong Q, Anderson CL, January CT, Zhou Z. Pharmacological rescue of trafficking defective HERG channels formed by coassembly of wild-type and long QT mutant N470D subunits. Am. J. Physiol. Heart Circ. Physiol. 2004;287:H652–658.

39. Zhou Z, Gong Q, January CT. Correction of defective protein trafficking of a mutant HERG potassium channel in human long QT syndrome. Pharmacological and temperature effects. J. Biol. Chem. 1999;274:31123–31126.

40. Wible BA, Hawryluk P, Ficker E, Kuryshev YA, Kirsch G, Brown AM. HERG-Lite: a novel comprehensive high-throughput screen for drug-induced hERG risk. J. Pharmacol. Toxicol. Methods 2005;52:136–145.

41. Delisle BP, Anderson CL, Balijepalli RC, Anson BD, Kamp TJ, January CT. Thapsigargin selectively rescues the trafficking defective LQT2 channels G601S and F805C. J. Biol Chem. 2003;278:35749–35754.

42. Bernier V, Lagace M, Lonergan M, Arthus MF, Bichet DG, Bouvier M. Functional rescue of the constitutively internalized V2 vasopressin receptor mutant R137H by the pharmacological chaperone action of SR49059. Mol Endocrinol. 2004;18:2074–2084.

43. Morello JP, Salahpour A, Laperriere A, Bernier V, Arthus MF, Lonergan M, Petaja-Repo U, Angers S, Morin D, Bichet DG, Bouvier M. Pharmacological chaperones rescue cell-surface expression and function of misfolded V2 vasopressin receptor mutants. J. Clin. Invest. 2000;105:887–895.

44. Robben JH, Sze M, Knoers NV, Deen PM. Rescue of vasopressin V2 receptor mutants by chemical chaperones: specificity and mechanism. Mol. Biol. Cell 2006;17:379–386.

45. Robben JH, Sze M, Knoers NV, Deen PM. Functional rescue of vasopressin V2 receptor mutants in MDCK cells by pharmacochaperones: relevance to therapy of nephrogenic diabetes insipidus. Am. J. Physiol. Renal Physiol. 2007;292:F253–260.

46. Wuller S, Wiesner B, Loffler A, Furkert J, Krause G, Hermosilla R, Schaefer M, Schulein R, Rosenthal W, Oksche A. Pharmacochaperones post-translationally enhance cell surface expression by increasing conformational stability of wild-type and mutant vasopressin V2 receptors. J. Biol. Chem. 2004;279:47254–47263.

47. Cheong HI, Cho HY, Park HW, Ha IS, Choi Y. Molecular genetic study of congenital nephrogenic diabetes insipidus and rescue of mutant vasopressin V2 receptor by chemical chaperones. Nephrology (Carlton). 2007;12:113–117.

48. Janovick JA, Brothers SP, Cornea A, Bush E, Goulet MT, Ashton WT, Sauer DR, Haviv F, Greer J, Conn PM. Refolding of misfolded mutant GPCR: post-translational pharmacoperone action in vitro. Mol. Cell Endocrinol. 2007;272:77–85.

49. Noorwez SM, Kuksa V, Imanishi Y, Zhu L, Filipek S, Palczewski K, Kaushal S. Pharmacological chaperone-mediated in vivo folding and stabilization of the P23H-opsin mutant associated with autosomal dominant retinitis pigmentosa. J. Biol. Chem. 2003;278:14442–14450.

50. Novoradovskaya N, Lee J, Yu ZX, Ferrans VJ, Brantly M. Inhibition of intracellular degradation increases secretion of a mutant form of alpha1-antitrypsin associated with profound deficiency. J. Clin. Invest. 1998;101:2693–2701.

51. Marcus NY, Perlmutter DH. Glucosidase and mannosidase inhibitors mediate increased secretion of mutant alpha1 antitrypsin Z. J. Biol. Chem. 2000; 275:1987–1992.

52. Burrows JA, Willis LK, Perlmutter DH. Chemical chaperones mediate increased secretion of mutant alpha 1-antitrypsin (alpha 1-AT) Z: A potential pharmacological strategy for prevention of liver injury and emphysema in alpha 1-AT deficiency. Proc. Natl. Acad. Sci. U.S.A. 2000;97:1796–1801.

53. Wright JM, Zeitlin PL, Cebotaru L, Guggino SE, Guggino WB. Gene expression profile analysis of 4-phenylbutyrate treatment of IB3-1 bronchial epithelial cell line demonstrates a major influence on heat-shock proteins. Physiol. Genomics. 2004;16:204–211.

54. Rubenstein RC, Zeitlin, PL. Sodium 4-phenylbutyrate downregulates Hsc70: implications for intracellular trafficking of DeltaF508-CFTR. Am. J. Physiol. Cell Physiol. 2000;278:C259–267.

55. Choo-Kang LR, Zeitlin PL. Induction of HSP70 promotes DeltaF508 CFTR trafficking. Am. J. Physiol. Lung Cell Mol. Physiol. 2001;281:L58–68.

56. Tveten K, Holla OL, Ranheim T, Berge KE, Leren TP, Kulseth MA. 4-Phenylbutyrate restores the functionality of a misfolded mutant low-density lipoprotein receptor. FEBS J. 2007;274:1881–1893.

57. Yam GH, Roth J, Zuber C. 4-Phenylbutyrate rescues trafficking incompetent mutant alpha-galactosidase A without restoring its functionality. Biochem. Biophys. Res. Commun. 2007;360:375–380.

58. Halaban R, Patton RS, Cheng E, Svedine S, Trombetta ES, Wahl ML, Ariyan S, Hebert DN. Abnormal acidification of melanoma cells induces tyrosinase retention in the early secretory pathway. J. Biol. Chem. 2002;277:14821–14828.

59. Gong Q, Keeney DR, Molinari M, Zhou Z. Degradation of trafficking-defective long QT syndrome type II mutant channels by the ubiquitin-proteasome pathway. J. Biol. Chem. 2005;280:19419–19425.

60. Vij N, Fang S, Zeitlin PL. Selective inhibition of endoplasmic reticulum-associated degradation rescues DeltaF508-cystic fibrosis transmembrane regulator and suppresses interleukin-8 levels: therapeutic implications. J. Biol. Chem. 2006;281:17369–17378.

61. Norez C, Noel S, Wilke M, Bijvelds M, Jorna H, Melin P, DeJonge H, Becq F. Rescue of functional delF508-CFTR channels in cystic fibrosis epithelial cells by the alpha-glucosidase inhibitor miglustat. FEBS Lett. 2006;580:2081–2086.

62. Wang Y, Loo TW, Bartlett MC, Clarke DM. Additive effect of multiple pharmacological chaperones on maturation of CFTR processing mutants. Biochem. J. 2007;406:257–263.

## FURTHER READING

Aridor M, Hannan LA. Traffic jam: a compendium of human diseases that affect intracellular transport processes. Traffic 2000;1:836–851.

Aridor M, Hannan LA. Traffic jams II: an update of diseases of intracellular transport. Traffic 2002;3:781–790.

Balch WE, Morimoto RI, Dillin A, Kelly JW. Adapting proteostasis for disease intervention. Science 2008;319:916–919.

Bernier V, Lagace M, Bichet DG, Bouvier M. Pharmacological chaperones: potential treatment for conformational diseases. Trends Endocrinol. Metab. 2004;15:222–228.

Conn PM, Ulloa-Aguirre A, Ito J, Janovick JA. G protein-coupled receptor trafficking in health and disease: lessons learned to prepare for therapeutic mutant rescue in vivo. Pharmacol. Rev. 2007;59:225–250.

Fan JQ, Ishii S. Active-site-specific chaperone therapy for Fabry disease. Yin and Yang of enzyme inhibitors. FEBS J. 2007;274:4962–4971.

Hebert DN, Molinari M. In and out of the ER: protein folding, quality control, degradation, and related human diseases. Physiol. Rev. 2007;87:1377–1408.

Tropak MB, Mahuran D. Lending a helping hand, screening chemical libraries for compounds that enhance beta-hexosaminidase A activity in GM2 gangliosidosis cells. FEBS J. 2007;274:4951–4961.

Yu Z, Sawkar AR, Kelly JW. Pharmacologic chaperoning as a strategy to treat Gaucher disease. FEBS J. 2007;274:4944–4950.

## SEE ALSO

Chapter 23
Chapter 19

# 21

# METABOLIC DISEASES

CYNTHIA M. ARBEENY

*Endocrine and Metabolic Diseases, Genzyme Corporation, Framingham, Massachusetts*

Obesity, type 2 diabetes (T2D), lipid disorders, and hypertension are chronic and disabling diseases that afflict hundreds of millions of individuals worldwide. In this chapter, they are collectively referred to as "cardiometabolic diseases," because they have "common ground:" They increase the risk of cardiovascular disease (CVD) morbidity and mortality. Considerable research has been performed to understand the etiology of cardiometabolic diseases and to translate this research into effective treatment paradigms. However, it has been challenging to understand the initiation and progression of cardiometabolic diseases. This difficulty is attributed to the complexities involved in metabolic regulation, a resultant "snowballing" effect, and a downward spiral as disease progresses. The first line of therapy for treating obesity, T2D, and/or dyslipidemia is lifestyle intervention, which can impact significantly on disease and decrease CV risk, but is usually not effective because of lack of patient compliance. Patients are then given a plethora of drugs to treat the individual risk factors. However, new therapies are critically needed, because most patients require multiple drugs, some drugs may treat one CVD risk factor and exacerbate another, and most patients do not meet treatment goals. A "magic bullet" that treats the underlying causes that contribute to cardiometabolic risk has been a longstanding goal but remains elusive. Ongoing research to uncover novel targets holds promise for future therapeutics that might treat multiple facets of cardiometabolic diseases. This introductory review focuses on the epidemiology and etiology of cardiometabolic diseases, current therapies, and future treatment strategies.

It is difficult to separate obesity, type 2 diabetes (T2D), dyslipidemia, and hypertension because they are a consequence of a dysregulation in metabolism, are

*Chemical Biology: Approaches to Drug Discovery and Development to Targeting Disease*, First Edition.
Edited by Natanya Civjan.
© 2012 John Wiley & Sons, Inc. Published 2012 by John Wiley & Sons, Inc.

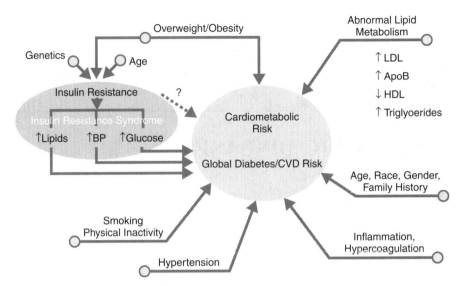

**Figure 21.1**   Risk factors that contribute to cardiometabolic risk. A model proposed by the ADA, where environmental and genetic factors lead to metabolic disorders that contribute to T2D and CVD. From Reference 1, with permission.

interrelated, and each might drive disease progression of the other. These comorbidities are usually found as a cluster within patients, but each can occur in the absence of the others. Lifestyle (high caloric intake, low physical activity, and cigarette smoking) along with a genetic predisposition result in alterations in metabolism that lead to cardiometabolic diseases and increase the risk of CVD. The multiple factors that contribute to CVD are summarized in Fig. 21.1 (1). Recent studies have provided insight into the mechanisms by which metabolic abnormalities impair insulin-receptor signaling, trigger inflammation and endothelial dysfunction, and increase CVD risk. This finding might provide new opportunities to treat the multiple risk factors that result in insulin resistance and cardiovascular disease.

Please note that numerous "landmark" studies have been published that have contributed to the understanding of cardiometabolic diseases, but the author has chosen to reference newer publications and review articles in this chapter.

## 21.1   THE GLOBAL CRISIS IN OBESITY, DIABETES, AND CV RISK

CVD is attributed to disorders of the heart and blood vessels, and it includes coronary heart disease (heart attacks), stroke, hypertension, peripheral artery disease, rheumatic heart disease, congenital heart disease, and heart failure (HF). On a global perspective, 30% of all deaths are from CVD, which ranks it the number one cause of death (2). In 2005, an estimated 17.5 million people died from CVD; 43% of the deaths are attributed to heart attacks and 32% to stroke.

Current predictions suggest that the twin epidemics of obesity and diabetes worldwide will result in an increase in CVD rates, which have sharply declined over the past 30 years, after the introduction of effective lipid-lowering and anti-hypertensive therapies. The World Health Organization (WHO) reported that in 2002, deaths from CVD outnumbered deaths from the major communicable diseases (AIDS, tuberculosis, and malaria) by 3 to 1 (2). By 2015, an estimated 20 million people will die annually from CVD. Therefore, with the advent of the new millennium, there is a sense of urgency to address the burden of chronic cardiometabolic diseases worldwide. The information on the prevalence and etiology of cardiometabolic diseases, which is cited in this section was obtained from the WHO and Centers for Disease Control (CDC) websites (2–6).

### 21.1.1 Obesity

Overweight and obesity occurs when an energy imbalance exists within the body, in which energy intake exceeds energy used, resulting in an increase in body fat. The cause of an energy imbalance for each individual may be different, and it is the result of interplay between environmental and genetic factors, which makes it a complex disease to understand and treat. Although some societies might be more genetically prone to obesity, changes in environmental factors, particularly the quantity and quality of food consumed and a sedentary lifestyle, are the key drivers for the global obesity epidemic (3).

For adults, overweight and obesity are easily defined by the body mass index (BMI). This value is calculated from a person's weight in kilograms divided by the square of the person's height in meters ($kg/m^2$); conversion tables are available for body weight in pounds and height in inches. An adult who has a BMI between 25 and 29.9 is considered overweight; an adult who has a BMI of 30 or higher is considered obese (4). Estimates of overweight and obesity for children and adolescents take age and gender into consideration.

Data obtained from the National Health and Nutrition Examination Survey for the 2003–2004 period indicated that 66% of adults in the United States were overweight (7). The prevalence of overweight and obesity had doubled for both adults and children, when compared with a previous survey for the 1976–1980 period. According to 2005 statistics from the WHO, approximately 1.6 billion adults were overweight, and at least 400 million adults were obese (3); that half of all diabetes cases would be eliminated if weight gain could be prevented. WHO projects that by 2015, approximately 2.3 billion adults will be overweight, and more than 700 million adults will be obese.

### 21.1.2 Diabetes

Epidemiologic studies have established a strong relationship between obesity and the risk for T2D (8, 9). Therefore, because of the obesity epidemic, diabetes rates are soaring. WHO data indicate that over 240 million people have diabetes worldwide; the number is expected to reach 380 million by 2025 (5). The five countries with the largest numbers of people with diabetes are India, China, the United States, Russia, and Germany. For the U.S. population, 20.6 million people or 7% of the population have diabetes. Adults with diabetes have heart disease and risk of stroke about 2 to 4 times higher than adults without diabetes (10, 11). The CDC reports that diabetes is the sixth leading cause of death among adult Americans, and that two thirds of diabetics die from CVD and stroke (6). Diabetes is the major cause of renal disease, adult blindness, and lower-limb amputation. The economic burden of diabetes in the United States totals $174 billion annually; this figure has increased by 32% since 2002 (12).

Diabetes can manifest in several forms, and this review will focus on T2D, which afflicts 90–95% of all diabetics. (For an overview of diabetes diagnosis, classification, and standard of care, see References 12–15.) T2D is a metabolic disorder that is attributed to a defect in the secretion of insulin by the β-cells of the pancreas in response to metabolic signals, combined with the inability of cells to respond to insulin (i.e., insulin resistance), which results in impaired nutrient uptake and use and increased hepatic glucose output (Fig. 21.2). T2D is considered a "silent" disease, which progresses over decades, and it is usually diagnosed after a complication is evident or a cardiovascular event occurs.

Autopsy studies in humans have indicated that a curvilinear relationship is observed between β-cell mass and fasting blood glucose concentration, and a steep increase in blood glucose concentration is associated with β-cell deficiency (16). The data have provided research incentives focused on restoring β-cell mass to reverse diabetes or preventing diabetes by protection of β-cell mass. Abnormalities in β-cell function are therefore critical in defining the risk and development of T2D, particularly in obese subjects (17). The adipose tissue of obese individuals releases high levels of non-esterified fatty acids (NEFA), glycerol, hormones, proinflammatory cytokines, and other factors that increase insulin resistance (Fig. 21.2). When insulin resistance is accompanied by impaired insulin secretion by β-cells, blood glucose levels are not controlled, which results in overt diabetes.

Diabetes is diagnosed by a Fasting Plasma Glucose Test (FPG) or an Oral Glucose Tolerance Test (OGTT) (18). A FPG level of ≥126 mg/dL indicates diabetes. In the OGTT test, a blood glucose level 2 hours after drinking a glucose solution of ≥200 mg/dL indicates diabetes. A stable measure of long-term glucose status is to determine the percent of a glycated isoform of hemoglobin in blood (termed % HbA1c), which is a slowly turning over protein within the body that is modified by high blood glucose. Current treatment guidelines are to reduce HbA1c to <7% (13), because it reduces the patient's risk of developing microvascular complications (nephropathy, retinopathy, and neuropathy) as well as macrovascular disease (cardiovascular disease and stroke) (19).

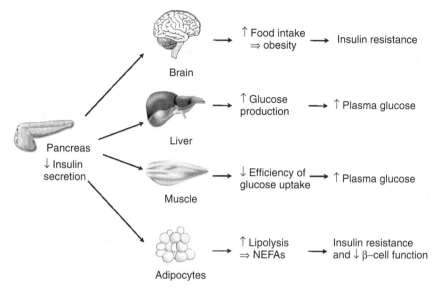

**Figure 21.2** Model for the critical role of impaired insulin release in linking obesity with insulin resistance and T2D. Impaired insulin secretion results in decreased insulin levels and decreased signaling in the hypothalamus, which leads to increased food intake and weight gain, decreased inhibition of hepatic glucose production, reduced efficiency of glucose uptake in muscle, and increased lipolysis in the adipocyte. These results lead to increased plasma NEFA levels. The increase in body weight and NEFAs contribute to insulin resistance, and the increased NEFAs suppress the β-cell's adaptive response to insulin resistance. The increased glucose levels together with the elevated NEFA levels can synergize to affect β-cell health and insulin action adversely, which is often referred to as "glucolipotoxicity." From Reference 17, with permission.

### 21.1.3 Dyslipidemia and Hypertension

Abnormal circulating lipids [i.e., high low density lipoprotein (LDL) concentration (LDLc), low HDL concentration (HDLc), hypertriglyceridemia, lipoprotein "little" a (Lp(a)), and small dense LDL], and hypertension are strongly associated with CVD risk (2, 20–22). These lipids are often found in obese and/or diabetic patients. Data from the Framingham Heart Study is used to estimate a person's 10-year risk for "hard" coronary heart disease (CHD) outcomes (myocardial infarction (MI) and coronary death), in which risk is calculated based on age, gender, total and high density lipoprotein (HDL) cholesterol, systolic blood pressure, and whether the individual is on blood-pressure–lowering medications or is a smoker. Numerous primary and secondary intervention studies have shown that improving lipid profile and lowering blood pressure significantly reduces disease morbidity and mortality (22, 23). Improved control of blood lipids in diabetics can reduce cardiovascular events by 20–50% (15, 24) In addition to lipids and blood pressure, inflammation and a procoagulant and prothrombotic state are additional CVD risk factors that need to be addressed, particularly in

obese and insulin-resistant patients (25, 26). Results from a subgroup analysis of
the Pravastatin or Atorvastatin Evaluation Trial (PROVE IT) (27) showed that
intensive lipid-lowering intervention significantly reduced acute coronary events
in diabetics, but most patients did not reach the dual goal of LDLc <70 mg/dL
and high sensitivity C-reactive protein (CRP) <2 mg/L (CRP is an inflammatory marker associated with CV risk). The data highlight the need for additional
strategies in this high-risk group, particularly those that target inflammation.

### 21.1.4  Metabolic Syndrome Versus Additive Cardiometabolic Risk Factors

Cardiometabolic diseases may present as separate diseases but more often cluster
within patients. Because of this clustering, the concept of "The Metabolic Syndrome" (also termed insulin resistance syndrome or syndrome X) has been put
forth. Several definitions of the Metabolic Syndrome are available, with the overall viewpoint that multiple interrelated risk factors increase the risk for atherosclerotic cardiovascular disease and increase the risk for T2D. Guidelines to define
the metabolic syndrome have been developed by the WHO, National Cholesterol Education Program-Adult Treatment Panel (NCEP-ATP), and International
Diabetes Federation (IDF), which are reviewed in Reference 28. Each definition includes traditional risk factors: obesity and an abdominal fat distribution,
insulin resistance, diabetes, dyslipidemia (high TG, low HDL) and hypertension.
Shown in Table 21.1 are the new IDF definition (29) and the revised National
Cholesterol Education Program-Adult Treatment Panel III (NCEP-ATPIII) definition (30), which are the most widely used. Each definition has its strengths and
weaknesses, and both groups are working toward harmonizing criteria (31).

According to CDC statistics that use the NCEP criteria (32), the metabolic
syndrome is found in 20% of the U.S. adult population. Its prevalence increases
with aging, and it is highest in Hispanic women with 35% prevalence. According
to the WHO, 25% of the world's population and at least two thirds of diabetics
have the metabolic syndrome. Considerable evidence suggests that the metabolic
syndrome is a significant predictor of CVD and T2D, and risk increases with
increasing number of risk factors (33). However, a recent assessment (34) states
that "the Metabolic Syndrome is a stronger predictor of T2D than CHD and it
does not predict CHD as well as the Framingham Risk Score, but it serves as
a simple clinical tool for identifying high-risk subjects predisposed to CVD and
T2D."

The clinical significance of the metabolic syndrome came into question in
2005, after "A Critical Appraisal of the Metabolic Syndrome" that was jointly
written by the American Diabetes Association (ADA) and the European Association for the Study of Diabetes (EASD) (35). These organizations questioned
the clinical value of diagnosing the metabolic syndrome, stating that CV risk
is not greater than the sum of the individual risk factors, and treatment of the
syndrome is no different than the treatment for each of its components. Their
recommendation was to continue efforts to understand the relationships of risk
factors that contribute to CVD. Numerous rebuttals to the ADA/EASD statement

**TABLE 21.1    The IDF and NCEP definitions of the metabolic syndrome**

| International diabetes federation | National cholesterol education program-ATPIII |
| --- | --- |
| Central obesity: waist circumference-ethnicity specific plus any two of the following: | Any three of the five criteria: Elevated waist circumference For US population: >102 cm (40 in) in men, >88 cm (>35 in) in women lower cut points for insulin resistant individuals or ethnic groups |
| High TG: >150 mg/dL (1.7 mmol/L) | High TG: >150 mg/dL (1.7 mmol/L) |
| Low HDL cholesterol: <40 mg/dL (1.03 mmol/L) in men, <50 mg/dL (1.3 mmol/L) in women | Low HDL cholesterol: <40 mg/dL (1.03 mmol/L) in men, <50 mg/dL (1.3 mmol/L) in women |
| High blood pressure: >130 mmHg systolic, >85 mmHg diastolic | High blood pressure: >130 mmHg systolic, >85 mmHg diastolic |
| Elevated plasma glucose: Fasting plasma glucose >100 mg/dL (5.6 mmol/L) or previously diagnosed T2D. If above 5.6 mmol/L an oral glucose tolerance test is strongly recommended but not necessary to define the presence of the syndrome. | Elevated fasting glucose: >100 mg/dL (5.6 mmol/L) |
| Limitations: Central obesity is required, criteria and cutoff values might need to be further defined | Limitations: Cutoff values do not consider ethnicity differences and age, criteria and cutoff values might need to be further defined |

Information was obtained from References 29 and 30.
*Central obesity not necessary if three of the other risk factors are present.

have been given, particularly by the American Heart Association (AHA) (36). In 2007, the AHA and ADA jointly issued a publication that attempted to harmonize the recommendations of both organizations where possible, and recognized areas where they differ (24). Overall, it is feasible that common ground will be reached in the near future.

## 21.2    UNDERLYING MECHANISMS THAT LEAD TO INSULIN RESISTANCE AND CVD

Chronic excessive nutrient intake leads to the deposition of fat, in not only its normal storage site, which is the adipose tissue, but also in liver and skeletal muscle. Nutrient excess also triggers an inflammatory response, with the release of inflammatory cytokines [tumor necrosis factor-alpha (TNF-$\alpha$), interleukin-6

(IL-6), and CRP]. These inflammatory mediators, along with the intracellular accumulation of lipid metabolites, lead to impaired insulin receptor signaling and defective metabolism in skeletal muscle and liver (37, 38). Nutrient excess also damages cells by generating reactive oxygen species, which results in protein modifications and in the depletion of nitric oxide (NO) that maintains vascular tone. These changes lead to cell dysfunction, alterations in blood lipids, elevated blood pressure, coagulation, fibrinolysis, and additional inflammation, which drives the downward spiral of insulin resistance, endothelial dysfunction, and atherosclerosis (39).

### 21.2.1 Metabolic Stress and Impaired Cellular Function

Nutrient excess leads to the intracellular accumulation of long chain acyl CoA and diacylglycerol, and in the activation of several serine/threonine kinases, which include protein kinase C isoforms, inhibitor of kappa B kinase (IKK), and c-jun N terminal kinase (JNK) (37). A key step in insulin receptor signaling is the tyrosine phosphorylation of IRS1 and 2, which regulates carbohydrate, lipid, and protein metabolism. The activation of several serine/threonine kinases results in serine phosphorylation of IRS1 and 2, which inhibits insulin receptor signaling. The activation of the two principal inflammatory pathways, IKKβ/NFkappa B and JNK, by nutrient excess and inflammatory cytokines, also impairs insulin receptor signaling and propagates the stress response (37). In skeletal muscle, it results in impaired insulin action on cell metabolism, which includes decreased insulin-stimulated glucose transport and decreased glycogen synthesis, leading to impaired glucose use and hyperglycemia. In the liver, glycogen synthesis is decreased, and gluconeogenesis is stimulated, which results in an increase in hepatic glucose production and hyperglycemia (38, 40).

Recently, the sphingolipid ceramide, which is a product of fatty acyl CoA, has been identified as the link between excess nutrients (i.e., saturated fatty acids) and inflammatory cytokines (i.e., TNFα), to the induction of insulin resistance. Moreover, ceramide has been shown to be toxic to pancreatic β-cells, cardiomyocytes, and endothelial cells, which contributes to diabetes, hypertension, cardiac failure, and atherosclerosis (41). However, the role of ceramide in mediating insulin-resistance humans is still unclear (42).

The endoplasmic reticulum (ER) is a network of membranes in which secreted and membrane proteins are assembled into their secondary and tertiary structures. The ER seems to be the site for sensing metabolic stress and to translate this stress into inflammatory signals (43). Under certain stress conditions, such as energy excess, lipids, and pathogens, the ER activates a complex response system known as the unfolded protein response to slow down protein synthetic pathways and to restore functional integrity to the organelle. Data from experimental models have shown that obesity leads to ER stress, which activates both JNK and IKK, and initiates pathways that trigger inflammation and insulin resistance. In the pancreatic β-cell, ER stress impairs insulin secretion and contributes to progression of T2D.

## 21.2.2  Inflammation

Obesity, and particularly the accumulation of abdominal fat, creates an inflammatory milieu that is the key driver of insulin resistance and CVD (44–46). In the normal state, adipose tissue coordinately regulates the synthesis and secretion of peptides that regulate numerous processes in the body (47, 48), which include fat mass, nutrient homeostasis and energy expenditure, the immune response, blood pressure control, hemostasis, bone mass, and reproductive function. In obesity, adipose tissue inflammation results in the secretion of proinflammatory peptides and reduction of anti-inflammatory peptides, which lead to deleterious effect on the liver, muscle, and the vasculature (Fig. 21.3). A reduction in abdominal fat improves the atherogenic lipid profile, reduces inflammation, and decreases blood pressure, which thereby decreases CV risk (44, 49). Current strategies are focused on identifying additional inflammatory markers that put the patient at cardiometabolic risk.

Chronic hyperglycemia induces numerous alterations in the vasculature that accelerate the atherosclerotic process. Several major mechanisms contribute to the pathological alterations in blood vessels in diabetes, including: 1) the nonenzymatic glycosylation of proteins and lipids, which form advanced glycation endproducts (AGEs) that can interfere with their normal function; and 2) the induction of oxidative and nitrosative stress, as well as exacerbation of proinflammatory responses (50). These abnormalities lead to impaired endogenous platelet inhibition and platelet activation, which could result in arterial thrombosis, and consequently myocardial infarction and stroke (51).

## 21.2.3  Impaired Endothelial Function

Because of metabolic stress, the endothelium loses its ability to balance vasodilating and vasoconstricting factors to maintain hemostasis. Nitric oxide (NO) is the most important mediator of vasodilation, and loss of NO bioavailibility contributes to the loss of vessel tone and damage to the endothelium. This damage is particularly evident in the insulin-resistant state, in which insulin plays a key role in maintaining endothelial function and stimulating NO production. In diabetes, defective insulin signaling in the endothelial cell results in an imbalance between the vasodilating agent NO and the vasoconstricting agent endothelin-1, which results in additional endothelial dysfunction and hypertension (52). Furthermore, markers of an activated endotheium appear prior to the presentation of overt diabetes, which suggests that the endothelium plays a primary role in the disease process.

A dysfunctional endothelium makes it susceptible to damage by adhesion molecules, inflammatory cytokines, activated platelets, and lipids, which culminates in atherosclerosis (46, 53). In the kidney, endothelial dysfunction impairs glomerular filtration and leads to the progressive loss in renal function. Endothelial dysfunction can be addressed by treating the patient with angiotensin-converting enzyme inhibitors (ACEi) and angiotensin II

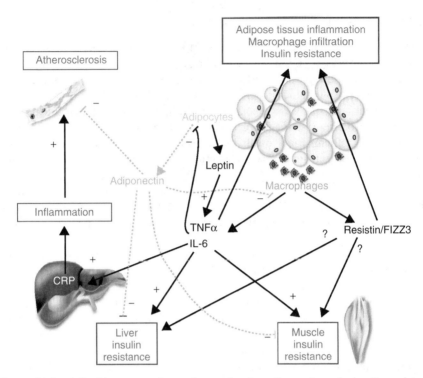

**Figure 21.3** Adipokine expression and secretion by adipose tissue in insulin-resistant, obese subjects. Obesity results in adipose tissue inflammation with macrophage infiltration. This result leads to 1) a decrease in adiponectin, which is an anti-inflammatory adipokine, that is positively correlated with insulin sensitivity and plays a protective role on the vasculature; and 2) an increase in inflammatory cytokines (TNFα, IL-6, and resistin) which causes insulin resistance, inflammation, and atherosclerosis. From Reference 47 with permission.

receptor blockers (ARBs), which block the deleterious effects of an activated renin-angiotensin-aldosterone system (RAAS) on the endothelium (54). Reviews of endothelial damage and treatment strategies can be found in References 53 and 55. Novel therapies are needed to prevent progressive endothelial damage, whereas early intervention might improve endothelial function, offering an opportunity to protect against both CVD and organ damage. The consequences of metabolic dysregulation on endothelial damage and end organ injury that culminate in CVD are summarized in Fig. 21.4.

## 21.3 CURRENT AND NEXT GENERATION THERAPEUTICS

Numerous drugs are marketed to treat metabolic diseases, which include weight reducing agents, antidiabetics, lipid-lowering treatments, and antihypertensives.

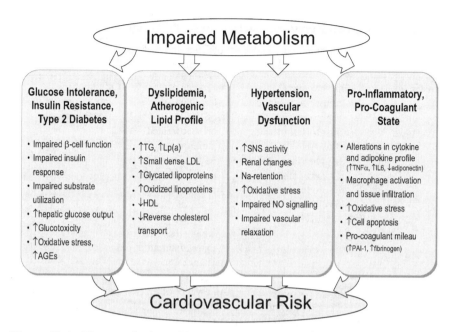

**Figure 21.4** The contribution of impaired metabolism to cardiovascular risk. A dys-regulation in nutrient metabolism contributes to cardiovascular risk by the following mechanisms: 1) impaired β-cell function, tissue damage, and insulin resistance; 2) an atherogenic lipid profile that is characterized by high TG, low HDL, and abnormal lipoproteins; 3) hypertension and vascular dysfunction; and 4) alterations in cytokines and adipokines that lead to a proinflammatory and procoagulant state.

The pharmacology, positive attributes, efficacy, and limitations for each drug class are summarized in Tables 21.2–21.5 (56, 57).

### 21.3.1 Weight-loss Agents

Three currently marketed weight-loss drugs are listed in Table 21.2. Orlistat is a pancreatic lipase inhibitor that acts at the level of the gastrointestinal (GI) tract to inhibit the absorption of dietary fat. Sibutramine and rimonabant act at different transporters or receptors to modulate central and peripheral metabolism. The paucity of weight loss agents that are available is attributed to the hurdles in developing weight-loss drugs. Large placebo-controlled trials are required; the placebo group is given a low-calorie diet and exhibits weight loss. A variation is observed in response and a significant number of nonresponders, and safety considerations often emerge. Most weight-loss treatments are associated with improvement in insulin sensitivity and lipids, but the drugs have side effects that generally limit use.

**TABLE 21.2    Currently prescribed weight-loss agents**

| Treatment | Pharmacology and positive attributes | Efficacy (clinical trials) | Limitations |
|---|---|---|---|
| Orlistat (oral) | Inhibits pancreatic lipase, blocks the digestion and absorption of dietary triglycerides (TG). Causes weight loss, improves lipid profile and insulin sensitivity. | At 1 year of treatment, 57% of the orlistat-treated patients and 31% of the placebo-treated patients lost $\geq 5\%$ of their body weight; weight loss was maintained at 2 years of treatment. | Oily spotting, abdominal pain, fecal urgency |
| Sibutramine (oral) | Inhibits the reuptake of norepinephrine, serotonin and dopamine; suppresses appetite and increase energy expenditure.<br><br>Causes weight loss, improves lipid profile and insulin sensitivity. | At 2 years of treatment, 67% of sibutramine-treated patients and 49% of placebo-treated patients lost $\geq 5\%$ of their body weight. | May increase blood pressure and heart rate |
| Rimonabant (oral) | An inverse agonist for the cannabinoid receptor CB1, reduces appetite and improves metabolism.<br><br>Causes weight loss, improves lipid profile, blood pressure and insulin sensitivity. | At 1 year of treatment, 75% of rimonabant-treated patients and 28% of placebo-treated patients lost $\geq 5\%$ of their body weight. | Nausea, dizziness, may cause severe depression. Not approved for use in the U.S. |

Information is current as of 2008 and was obtained from References 49 and 56.

**TABLE 21.3  Currently prescribed anti-diabetic agents**

| Treatment | Pharmacology and positive attributes | Expected efficacy | Limitations |
|---|---|---|---|
| Insulin (injectable or inhaled protein, multiple products) | Increases insulin levels, improves glucose tolerance, improves lipid profile. | HbA1c reductions of 1.5–3.5% (less reduction for inhaled vs injectable) | May cause hypo-glycemia, weight gain |
| Metformin (oral) | Decreases hepatic glucose output plus additional mechanisms, improves glycemic control, ↓ TG and LDLc, ↑ HDLc | HbA1c reductions of 1.0–2.0% | May cause GI problems, lactic acidosis (rare) |
| Sulfonylureas and non-SU secretagogues (oral, multiple agents) | Increases insulin secretion by pancreatic beta cells, improves glycemic control | HbA1c reductions of 1.0–2.0% for sulfonylureas 1.0–1.5 % for non-SU secretagogues | May cause hypo-glycemia, weight gain |
| α-Glucosidase inhibitors (oral, multiple agents) | Delays GI absorption of carbohydrates, improves glycemic control | HbA1c reductions of 0.5–0.8% | May cause GI problems |
| Thiazolidine-diones (oral, rosiglitazone, pioglitazone) | PPARγ agonist, increases insulin sensitivity and improves glycemic control, ↓TG (rosiglitazone may ↑TG), ↑LDLc and ↑HDLc; pioglitazone has greater beneficial effects on blood lipids) | HbA1c reductions of 0.5–1.4% | Associated with weight gain, edema; not recom-mended for patients with CHF |
| Exenatide (injectable protein) | Long acting GLP-1 analog, improves glycemic control, decreases TG and increases HDLc, causes weight loss. | HbA1c reductions of 0.5–1.0% | Associated with GI problems |
| Sitaglipin (oral) | Inhibits DPP-IV thereby potentiating the action of GLP-1; improves glycemic control, weight neutral. | HbA1c reductions of 0.5–0.8% | Recently approved, little experience |
| Pramlintide (injectable protein) | Amylin analog, improves glycemic control, causes weight loss. | HbA1c reductions of 0.6% | Associated with GI problems |

Information is current as of 2008 and was obtained from References 56 and 57.

**TABLE 21.4   Currently prescribed lipid-lowering agents**

| Treatment | Pharmacology and positive attributes | Efficacy | Limitations |
|---|---|---|---|
| HMGCoA reductase inhibitors "Statins" (oral, multiple agents) | Oral agent, inhibits HMGCoA reductase and upregulates the LDL receptor, decreases LDLc and TG, increases HDLc | LDLc ↓18–55% HDLc ↑5–15% TG ↑ 7–30% | Potential for elevating liver enzymes and causing myopathy |
| Bile acid sequestrants (oral, multiple agents) | Binds bile acids in the GI tract, resulting in the upregulation of the hepatic LDL receptor and a decrease in LDLc | LDLc ↓15–30% HDLc ↑3–5% TG no change or ↑ | May cause GI problems, decreased absorption of other drugs |
| Ezetimibe (oral) | Inhibits the intestinal absorption of cholesterol, decreases LDLc | LDLc ↓18% HDLc ↑1% TG ↓ 8% | Hypersensitivity reactions, myalgia, increase in liver enzymes |
| Fibrates (oral, multiple agents) | Decreases TG synthesis and very low density lipoprotein (VLDL) secretion and increases LPL, decreases TG, increases HDL, decreases fibrinogen and Lp(a), may cause weight loss | LDLc ↓5–20% (may increase in patients with high TG) HDLc ↑10–20% TG ↓20–50% | May cause dyspepsia, gallstones, myopathy |
| Niacin and nicotinic acid (oral, multiple agents) | Decreases TG synthesis and VLDL secretion, inhibits free fatty acid (FFA) release from adipose tissue, decreases TG, raises HDL, decreases fibrinogen and Lp(a) | LDLc ↓5–25% HDLc ↑15–35% TG ↓20–50% | May cause stomach upset, flushing, headache, may decrease glucose tolerance |

Information is current as of 2008 and was obtained from References 23 and 56.

**TABLE 21.5   Currently prescribed blood pressure-lowering agents**

| Treatment | Pharmacology and positive attributes | Conditions favoring use | Limitations |
|---|---|---|---|
| ACEi (oral, multiple agents) | Prevents conversion of Ang I to Ang II, which prevents action of Ang II at its receptor to cause vasoconstriction and cardiac stimulation | HF, post-MI, nephropathy, left ventricular (LV) hypertrophy, carotid atherosclerosis, atrial fibrillation, metabolic syndrome (MetS) | May cause cough, elevated potassium levels, low blood pressure, dizziness, headache |
| ARBs (oral multiple agents) | Blocks the action of Ang II at its receptor | HF, post MI, nephropathy, LV hypertrophy, atrial fibrillation, MetS | May cause cough, elevated potassium levels, low blood pressure, dizziness, headache |
| Aliskiren (oral) | Direct renin inhibitor, inhibits the RAAS pathway. | HF, post MI, diabetic nephropathies, hypertension, kidney disorders | Side effects include angioderma, hyperkalemia, hypotension, GI symptoms |
| Beta blockers (oral, multiple agents) | Blocks the βAR, decreases the chronotropic, inotropic and vasodilator responses to βAR stimulation | HF, post MI, angina pectoris, tachyarrhythmias, glaucoma, pregnancy | May cause weight gain, decrease insulin sensitivity and adversely affect plasma lipids |
| Thiazide diuretics (oral, multiple agents) | Inhibits $Na^+/Cl^-$ reabsorption from the distal convoluted tubules in the kidneys by blocking the thiazide-sensitive $Na+-Cl^-$ symporter | Isolated systolic hypertension (elderly), HF, hypertension in blacks | May cause hypokalemia and increased serum cholesterol; long-term use may increase homocysteine (associated with atherosclerosis) |

*(continued)*

**TABLE 21.5** (*Continued*)

| Treatment | Pharmacology and positive attributes | Conditions favoring use | Limitations |
|---|---|---|---|
| Calcium antagonists: dihyropy-ridines (oral, multiple agents) | Block L-type voltage-gated calcium channels in muscle cells of the heart and blood vessels; often used to reduce systemic vascular resistance and arterial pressure | Isolated systolic hypertension, angina pectoris, LV hypertrophy, carotid/coronary atherosclerosis, pregnancy, hypertension in blacks | Side effects include peripheral edema, dizziness, not used to treat angina (with the exception of amlodipine); contra-indicated in certain patient populations. |
| Calcium antagonists: (oral, verapamil diltiazem) | L-type calcium channel blocker, decreases impulse conduction through the AV node, protects ventricles from atrial tach-yarrhythmias; causes smooth muscle relaxation and vasodilation. | Angina pectoris, carotid atherosclerosis, supraventricular tachycardia | Side effects include headache, constipation, Side effects include dizziness, flushing, peripheral edema. (Diltiazem is contra-indicated in certain patient populations.) |

Information is current as of 2008 and was obtained from References 22 and 56.

### 21.3.2  Antidiabetic Agents

Marketed antidiabetic treatments are shown in Table 21.3. These drugs may act to: improve insulin sensitivity, decrease hepatic glucose output, decrease the absorption of glucose in the GI tract, or stimulate the secretion of insulin. Numerous studies have shown that antidiabetic and weight-loss interventions, either by lifestyle or by drug treatment, prevent the development of diabetes in patients at risk for developing that disease (58). However, antidiabetic and weight-loss drugs are not prescribed as preventative agents. Because diabetes is a complicated disease, a treatment paradigm has been developed, and multiple treatments are generally required (57). However, a patient's diabetes still progresses even when given several antidiabetic drugs in combination. This progression seems to be caused by the progressive loss of β-cell function despite multiple treatments (59).

Oral antidiabetic agents consist of several classes: biguinides, sulfonylureas (SUs), non-SU secretagogues, α-glucosidase inhibitors (AGI), thiazolidinediones (TZDs), and dipeptidyl peptidase IV (DPP-IV) inhibitors, which are used as monotherapy or in combination. Metformin is the only available biguinide in the United States, and its exact mechanisms are not clearly understood. However, it has been shown to increase insulin sensitivity by inhibiting hepatic glucose production. SU and non-SU secretagogues increase the release of insulin from pancreatic β-cells and potentiate insulin action, but hypoglycemia may result. The non-SU secretagogues differ from SUs in that they are absorbed more rapidly and are eliminated. They can minimize postprandial glucose (PPG) excursions and lower postprandial insulin levels, with reduced risk of hypoglycemia. AGIs act at the level of the GI tract to delay carbohydrate digestion and absorption, which thereby limits PPG excursions. TZDs are peroxisome-proliferator-activated receptor-γ (PPAR-γ) receptor agonists that enhance adipocyte differentiation and decrease lipolysis, increase insulin-stimulated glucose uptake in muscle, and decrease hepatic glucose production. In early 2007, a prospective review of clinical data for the thiazolidinedione (TZD) rosiglitazone has raised cardiovascular concerns. The issues are discussed in Reference 57, and a revision of the treatment paradigm for diabetic patients has resulted.

Newer targets include potentiating the effects of glucagon-like peptide-1 (GLP-1) and amylin, which are reviewed in Reference 60. GLP-1 is a key regulatory hormone that is secreted by the L-cells of the intestine. This hormone stimulates insulin secretion by the pancreatic β-cells in response to glucose, inhibits gastric emptying and glucagon secretion, and has a central effect to inhibit food intake. Two approved pharmacological treatments to promote GLP-1 action are as follows: 1) to inject a GLP-1 protein mimetic (i.e., exenatide) or 2) to potentiate endogenous GLP-1 action by inhibiting its degradation via sitaglipin, which is an oral DPP-IV inhibitor. Amylin is another regulatory hormone involved in glucose homeostasis. It is cosecreted with insulin by β-cells and decreases PPG excursions by slowing gastric emptying and decreasing glucagon secretion. Pramlintide, which is an injectable analog of amylin, is approved to treat both T1 and T2D patients.

### 21.3.3 Lipid-lowering Agents

Many lipid-lowering agents are used alone or in combination to achieve lipid goals (Table 21.4). Although numerous lipid-lowering agents are available, which are used alone and in combination, additional therapies are required to decrease other atherogenic components and to enhance reverse cholesterol transport.

The "statins" inhibit HMGCoA reductase, which results in an inhibition of cholesterol synthesis and upregulation of LDL clearance. This result leads to a marked reduction in LDLc, with additional positive effects to decrease TG modestly and to increase HDL. Bile acid sequestrants are also used alone or in combination to lower LDLc. Bile acid sequestrants also upregulate LDL clearance and decrease LDLc, but the effect is generally less than with a statin. In addition to lowering cholesterol, bile acid sequestrants may elicit an antidiabetic effect (61); a newer seqeustrant colesevelam HCl received approval to treat both T2D and high LDL cholesterol.

Ezetimibe is a newer drug that inhibits cholesterol absorption by the GI tract; it reduces LDLc modestly as monotherapy and is used in combination therapy with simvastatin (marketed as Vytorin) to achieve good cholesterol lowering and minimizing the "statin" dose to reduce side effects. However, the "Enhance" clinical trial results (62) have raised concerns about the lack of vascular benefit with ezetimide, and additional outcome trials are ongoing.

The fibrates act at the level of the liver to decrease VLDL-TG substantially and to increase HDLc secretion. Nicotinic acid also decreases TG and is considered to the most potent agent to increase HDL. The effects of fibrates and nicotinic acid on TG and HDL are generally greater than those observed with "statins," which are generally the most effective LDLc-lowering agents.

### 21.3.4   Blood Pressure Agents

Several classes of blood-pressure–lowering agents are available and are often used in combination to achieve goals (Table 21.5). Results of large clinical trials have indicated that the benefits of antihypertensive treatment is caused by the lowering in blood pressure and largely independent of the drugs employed (22). A challenge has been to understand why differential responses to a specific class or combination are observed.

Three antihypertensives (ACEi, ARB, and renin blockers) block different steps the RAAS pathway, which results in a downstream blockade of $AT_1$ receptors that resulting vasodilation, decreased secretion of vasopressin, and decreased secretion of aldosterone, contributing to a blood pressure lowering effect. The decrease in aldosterone decreases sodium and water resorption in the kidney and decreases potassium excretion, which leads to a lowering in blood pressure. These drugs are often used to treat hypertension, diabetic nephropathy, and congestive heart failure (63). ACEi also blocks the bradykinin pathway, which induces nitric oxide and vasodilation, but it is often associated with the persistent dry cough and/or angioedema that may limit ACEi therapy. This side effect is rarely observed with ARBs. Although ACEi and renin antagonism decrease circulating angiotensin II, ARBs block its activity at the AT1 receptor. ARBs increase angiotensin II levels by uncoupling the negative-feedback loop, and increase its stimulation of AT2 receptors, which is associated with beneficial and negative effects. A direct renin inhibitor has recently entered the market, which may result in more complete inhibition of the RAAS system inhibition than with ACEi or ARBs, but more clinical experience and outcome trial results are necessary to assess its potential adequately (64).

Beta adrenergic receptor (βAR) antagonists reduce cardiac output (caused by negative chronotropic and inotropic effects), decrease renin release from the kidneys, and cause smooth muscle relaxation. However, blockage may also decrease secretion of insulin from pancreatic β-cells, which limits its use in T2D. Calcium channel antagonists act on L-type voltage gated channels in the heart and blood vessels to reduce vascular resistance and arterial pressure. Diuretics are also widely used to decrease blood pressure, particularly in the elderly and hypertensive black populations.

### 21.3.5 Future Treatment Strategies

Future treatments are focused on correcting the metabolic dysregulation, which contributes to cell damage and tissue dysfunction. A list of selected novel targets for treating cardiometabolic diseases is shown in Table 21.6. The focus for new antiobesity approaches has been to target the gut–brain axis to regulate feeding behavior and energy expenditure, as well as modulating peripheral metabolism (65). These targets should also be effective in treating obese T2D. However, it has been difficult to target feeding behavior, because considerable

**TABLE 21.6 Future treatments for metabolic diseases**

| Novel pharmacological targets | Predicted positive effect | | | | |
|---|---|---|---|---|---|
| | Bwt Loss | Anti-Diab | Lipid Impr. | BP Redn | Vasc Impr. |
| ***Improve Energy Homeostasis*** | | | | | |
| Central regulation: ↓NPY[a], ↑MSH-R[b] Gut regulation: ↓Ghrelin, ↑CCK[c], ↑PYY[d] Peripheral regulation: ↓ACC[e], ↓SCD-1[f], ↑AMPK[g], ↑Sirt1[h], ↓11-βHSD, GH[i] analogs, TH[j] analogs Adipokine and cytokine regulation: ↑adiponectin, ↓inflammation | √ | √ | √ | √ | √/? |
| ***Improve Glucose Homeostasis*** | | | | | |
| Insulin signaling stimulation: ↓PTEN[k], Glycogen regulation: ↓GSK3[l], ↑GS[m] Gluconeogenesis inhibitors, β-cell restoration | √ | √/? | | | |

*(continued)*

**TABLE 21.6** *(Continued)*

| | | | |
|---|---|---|---|
| ***Improve Lipid Profile*** | | | |
| Reverse cholesterol transport stimulation: ↑HDL, ↑apo AI, ↑ABC transporter, ↓CETP[n] LDL reduction: Anti-Apo B, ↓MTP[o], ↓PCSK9[p] | √ | | √/? |
| ***Lower Blood Pressure*** | | | |
| Neutral endopeptidase inhibitors, Nitric oxide donors, Endothelin receptor antagonists | | √ | √/? |

Summary of predicted positive effects on body weight reduction, improvement in diabetes control, improvement in lipid profile, blood pressure decrease, and improvement in the vasculature.

[a]NPY, neuropeptide Y;

[b]MSH-R, melanocyte stimulating hormone receptor;

[c]CCK, cholecystokinin;

[d]PYY, peptide YY;

[e]ACC, acetyl CoA carboxylase;

[f]SCD-1, stearoyl-CoA desaturase 1;

[g]AMPK, AMP-activated protein kinase;

[h]Sirt1, silent information regulator 1;

[i]GH, growth hormone;

[j]TH, thyroid hormone;

[k]PTEN, phosphatase and tension homolog;

[l]GSK3, glycogen synthase kinase 3;

[m]GS, glycogen synthase;

[n]CETP, cholesteryl ester transfer protein;

[o]MTP, microsomal triglyceride transfer protein;

[p]PCSK9, proprotein convertase subtilisin/kexin type 9

redundancies exist in the regulation of food intake. Another focus to treat the metabolic dysregulation in insulin-resistant and diabetic patients is to augment insulin receptor signaling and also to treat β-cell impairment. Future approaches to treating vessel wall damage include activating reverse cholesterol transport and novel antihypertensive targets.

Several challenges are faced in developing novel therapeutics to treat cardiometabolic disorders. These challenges include identifying and validating novel targets, using predictive animal model of human efficacy, identifying responsive patient populations, obtaining an acceptable safety and tolerability profile for chronic treatment, and positioning a product in a huge, competitive marketplace. However, there is a high unmet medical need and room for new, differentiated products. Recent advances hold promise for novel therapies for treating multiple risk factors for T2D and CVD.

## REFERENCES

1. ADA. Factors contributing to cardiometabolic risk. Available: *http://professional. diabetes.org/Multimedia_Display.aspx?TYP=8&CID=53348*.
2. WHO website-cardiovascular disease. Available: *http://www.who.int/topics/ cardiovascular_diseases/en/*.
3. WHO website-obesity. Available: *http://www.who.int/topics/obesity/en/*.
4. CDC website-obesity. Available: *http://www.cdc.gov/nccdphp/dnpa/obesity/*.
5. WHO website-diabetes. Available: *http://www.who.int/diabetesactiononline/diabetes/ en/*.
6. CDC website-diabetes. Available: *http://www.cdc.gov/diabetes/pubs/factsheet. htm#contents*.
7. Ogden CL, Carroll MD, Curtin LR, McDowell MA, Tabak CJ, Flegal KM. Prevalence of overweight and obesity in the United States 1999-2004. JAMA 2006;295:1549–1555.
8. Must A, Spadano J, Coakley EH, Field AE, Colditz G, Dietz WH. The disease burden associated with overweight and obesity. JAMA 1999;282:1523–1529.
9. Mokdad AH, Ford ES, Bowman BA, Dietz WH, Vinicor F, Bales VS, et al. Prevalence of obesity, diabetes, and obesity-related health risk factors, 2001. JAMA. 2003;289:76–79.
10. Haffner SM, Lehto S, Ronnemaa T, Pyorala K, Laakso M. Mortality from coronary heart disease in subjects with type 2 diabetes and in nondiabetic subjects with and without prior myocardial infarction. N. Engl. J. Med. 1998;339:229–234.
11. Fox CS, Coady S, Sorlie PD, D'Agostino RB, Sr., Pencina MJ, Vasan RS, et al. Increasing cardiovascular disease burden due to diabetes mellitus: The Framingham Heart Study. Circulation 2007;115:1544–1550.
12. ADA. Economic costs of diabetes in the U.S. in 2007. Diabetes Care 2008;31:1–20.
13. ADA. Standards of medical care in diabetes 2008. Diabetes Care 2008;31:S12–54.
14. ADA. Diagnosis and classification of diabetes mellitus. Diabetes Care 2008; 31:S55–60.

15. Ryden L, Standl E, Bartnik M, Van den Berghe G, Betteridge J, de Boer MJ, et al. Guidelines on diabetes, pre-diabetes, and cardiovascular diseases: executive summary: the task force on diabetes and cardiovascular diseases of the European Society of Cardiology (ESC) and of the European Association for the Study of Diabetes (EASD). Eur. Heart J. 2007;28:88–136.

16. Ritzel RA, Butler AE, Rizza RA, Veldhuis JD, Butler PC. Relationship between beta-cell mass and fasting blood glucose concentration in humans. Diabetes Care 2006;29:717–718.

17. Kahn SE, Hull RL, Utzschneider KM. Mechanisms linking obesity to insulin resistance and type 2 diabetes. Nature 2006;444:840–846.

18. Gabir MM, Hanson RL, Dabelea D, Imperatore G, Roumain J, Bennett PH, et al. The 1997 American Diabetes Association and 1999 World Health Organization criteria for hyperglycemia in the diagnosis and prediction of diabetes. Diabetes Care 2000;23:1108–1112.

19. Intensive blood-glucose control with sulphonylureas or insulin compared with conventional treatment and risk of complications in patients with type 2 diabetes (UKPDS 33). The Lancet 1998;352:837–853.

20. Moon JY, Kwon HM, Kwon SW, Yoon SJ, Kim JS, Lee SJ, et al. Lipoprotein(a) and LDL particle size are related to the severity of coronary artery disease. Cardiology 2007;108:282–289.

21. Grundy SM, Hansen B, Smith SC, Jr., Cleeman JI, Kahn RA. Clinical management of metabolic syndrome: report of the American Heart Association/National Heart, Lung, and Blood Institute/American Diabetes Association conference on scientific issues related to management. Circulation 2004;109:551–556.

22. Mancia G, De Backer G, Dominiczak A, Cifkova R, Fagard R, Germano G, et al. 2007 Guidelines for the management of arterial hypertension: the task force for the management of arterial hypertension of the European Society of Hypertension (ESH) and of the European Society of Cardiology (ESC). Eur. Heart J. 2007;28:1462–1536.

23. NHLBI-ATPIII. Available: *http://www.nhlbi.nih.gov/guidelines/cholesterol/index.htm*.

24. Buse JB, Ginsberg HN, Bakris GL, Clark NG, Costa F, Eckel R, et al. Primary prevention of cardiovascular diseases in people with diabetes mellitus: a scientific statement from the American Heart Association and the American Diabetes Association. Circulation 2007;115:114–126.

25. Festa A, D'Agostino R, Jr., Howard G, Mykkanen L, Tracy RP, Haffner SM. Chronic subclinical inflammation as part of the insulin resistance syndrome: the Insulin Resistance Atherosclerosis Study (IRAS). Circulation 2000;102:42–47.

26. Stratton IM, Adler AI, Neil HAW, Matthews DR, Manley SE, Cull CA, et al. Association of glycaemia with macrovascular and microvascular complications of type 2 diabetes (UKPDS 35): prospective observational study. BMJ. 2000;321:405–412.

27. Ahmed S, Cannon CP, Murphy SA, Braunwald E. Acute coronary syndromes and diabetes: is intensive lipid lowering beneficial? Results of the PROVE IT-TIMI 22 trial. Eur. Heart J. 2006;27:2323–2329.

28. Rana JS, Nieuwdorp M, Jukema JW, Kastelein JJP. Cardiovascular metabolic syndrome - an interplay of, obesity, inflammation, diabetes and coronary heart disease. Diabetes Obes. Metab. 2007;9:218–232.

29. Alberti KGMM, Zimmet P, Shaw J. The metabolic syndrome–a new worldwide definition. The Lancet 2005;366:1059–1062.

30. Grundy SM, Cleeman JI, Daniels SR, Donato KA, Eckel RH, Franklin BA, et al. Diagnosis and management of the metabolic syndrome: an American Heart Association/National Heart, Lung, and Blood Institute Scientific Statement. Circulation 2005;112:2735–2752.

31. Assmann G GR, Fox G, Cullen P, Schulte H, Willett D, Grundy SM. Harmonizing the definition of the metabolic syndrome: comparison of the criteria of the Adult Treatment Panel III and the International Diabetes Federation in United States American and European populations. Am. J. Cardiol. 2007;99:541–548.

32. Ford ES, Giles WH, Dietz WH. Prevalence of the metabolic syndrome among US Adults: Findings from the Third National Health and Nutrition Examination Survey. JAMA. 2002;287:356–359.

33. Sattar N, Gaw A, Scherbakova O, Ford I, O'Reilly DSJ, Haffner SM, et al. Metabolic syndrome with and without C-reactive protein as a predictor of coronary heart disease and diabetes in the West of Scotland Coronary Prevention Study. Circulation 2003;108:414–419.

34. Wannamethee SG, Shaper AG, Lennon L, Morris RW. Metabolic syndrome vs. Framingham risk score for prediction of coronary heart disease, stroke, and type 2 diabetes mellitus. Arch. Int. Med. 2005;165:2644–2650.

35. Kahn R, Buse J, Ferrannini E, Stern M. The metabolic syndrome: time for a critical appraisal: joint statement from the American Diabetes Association and the European Association for the Study of Diabetes. Diabetes Care 2005;28:2289–2304.

36. Grundy SM. Metabolic syndrome: a multiplex cardiovascular risk factor. Clin. Endocrinol. Metab. 2007;92:399–404.

37. Hotamisligil GS. Inflammation and metabolic disorders. Nature 2006;444:860–867.

38. Savage DB, Petersen KF, Shulman GI. Disordered lipid metabolism and the pathogenesis of insulin resistance. Physiological Reviews 2007;87:507–520.

39. Van Gaal LF, Mertens IL, De Block CE. Mechanisms linking obesity with cardiovascular disease. Nature 2006;444:875–880.

40. Yu C, Chen Y, Cline GW, Zhang D, Zong H, Wang Y, et al. Mechanism by which fatty acids inhibit insulin activation of insulin receptor substrate-1 (IRS-1)-associated phosphatidylinositol 3-kinase activity in muscle. J. Biological Chemistry 2002;277:50230–50236.

41. Summers SA. Ceramides in insulin resistance and lipotoxicity. Progress in Lipid Research 2006;45:42–72.

42. Boden G. Ceramide: a contributor to insulin resistance or an innocent bystander? Diabetologia. 2008. *http://www.springerlink.com/content/5718433253x46717/*.

43. Marciniak SJ, Ron D. Endoplasmic reticulum stress signaling in disease. Physiological Reviews 2006;86:1133–1149.

44. Kopp HP, Kopp CW, Festa A, Krzyzanowska K, Kriwanek S, Minar E, et al. Impact of weight loss on inflammatory proteins and their association with the insulin resistance syndrome in morbidly obese patients. Arterioscler Thromb Vasc. Biol. 2003;23:1042–1047.

45. Bloomgarden ZT. Inflammation, atherosclerosis, and aspects of insulin action. Diabetes Care 2005;28:2312–2319.

46. Libby P, Plutzky J. Inflammation in diabetes mellitus: Role of peroxisome proliferator-activated receptor-[alpha] and peroxisome proliferator-activated receptor-[gamma] agonists. The Am. J. Cardiol. 2007;99:27–40.

47. Bastard J, Maaci M, Lagathu C, Kim MJ, Caron M, et al. Recent advances in the relationship between obesity, inflammation, and insulin resistance. Eur. Cytokine Netw. 2006;17:4–12.

48. Tilg H, Moschen AR. Adipocytokines: mediators linking adipose tissue, inflammation and immunity. Nat. Rev. Immunol. 2006;6:772–783.

49. Scheen AJ, Finer N, Hollander P, Jensen M, Van Gaal L, Group ftR-DS. Efficacy and tolerability of rimonabant in overweight or obese patients with type 2 diabetes: a randomised controlled study. Lancet 2006;368:1660–1672.

50. Aronson D. Hyperglycemia and the pathobiology of diabetic complications. Adv. Cardiol. 2008;45:1–16.

51. Schäfer A, Bauersachs J. Endothelial dysfunction, impaired endogenous platelet inhibition and platelet activation in diabetes and atherosclerosis. Curr. Vasc. Pharmacol. 2008;6:52–60.

52. Jansson PA. Endothelial dysfunction in insulin resistance and type 2 diabetes. J. Int. Med. 2007;262:173–183.

53. John S, Schmieder RE. Potential mechanisms of impaired endothelial function in arterial hypertension and hypercholesterolemia. Curr. Hypertens. Rep. 2003;5:199–207.

54. Dzau V. The cardiovascular continuum and renin-angiotensin-aldosterone system blockade. J. Hypertens. Suppl. 2005;23:S952–1178.

55. Wilkinson IB, McEniery CM. Arterial stiffness, endothelial function and novel pharmacological approaches. Clin. Exp. Pharm. Physiol. 2004;31:795–799.

56. Physician's Desk Reference, 61st ed. 2007. Montvale, NJ, Thomson PDR.

57. Nathan DM, Buse JB, Davidson MB, Ferrannini E, Holman RR, Sherwin R, et al. Management of hyperglycemia in type 2 diabetes: a consensus algorithm for the initiation and adjustment of therapy: update regarding thiazolidinediones: a consensus statement from the American Diabetes Association and the European Association for the Study of Diabetes. Diabetes Care 2008;31:173–175.

58. Gillies CL, Abrams KR, Lambert PC, Cooper NJ, Sutton AJ, Hsu RT, et al. Pharmacological and lifestyle interventions to prevent or delay type 2 diabetes in people with impaired glucose tolerance: systematic review and meta-analysis. BMJ. 2007;334:299.

59. U.K. prospective diabetes study 16. Overview of 6 years' therapy of type II diabetes: a progressive disease. U.K. Prospective Diabetes Study Group. Diabetes. 1995;44:1249–1258.

60. Triplitt C. New technologies and therapies in the management of diabetes. Am. J. Manag. Care. 2007;2:S47–54.

61. Staels B, Kuipers F. Bile acid sequestrants and the treatment of type 2 diabetes mellitus. Drugs 2007;67:1383–1392.

62. Merck/Schering-Plough Pharmaceuticals Provides Results of the ENHANCE Trial. Press Release. January 14, 2008. *http://www.merck.com/newsroom/press—releases/product/2008–0125.html*

63. Werner C, Baumhäkel M, Teo K, Schmieder R, Mann J, Unger T, et al. RAS blockade with ARB and ACE inhibitors: current perspective on rationale and patient selection. Clinical Research in Cardiology. 2008. *http://www.ncbi.nlm.nih.gov/pubmed/18220940*

64. Shafiq MM, Menon DV, Victor RG. Oral direct renin inhibition: premise, promise, and potential limitations of a new antihypertensive drug. Am. J. Med. 2008;121:265–271.

65. Aronne LJ, Thornton-Jones ZD. New targets for obesity pharmacotherapy. Clin. Pharmacol. Ther. 2007;81:748–752.

## FURTHER READING

Bamba V, Rader DJ. Obesity and atherogenic dyslipidemia. Gastroenterology 2007; 132:2181–2190.

Lebovitz, HE. Insulin resistance - a common link between type 2 diabetes and cardiovascular disease. Diabetes Obes. Metab. 2006;8:237–249.

Woods SC, Benoit S and Clegg DJ. The brain-gut-islet connection. Diabetes 2006;55:S114–S121.

Yach D, Stuckler D and Brownwell KD. Commentary: epidemiological and economic consequences of the global epidemic of obesity and diabetes. Nat. Med. 2006;12:62–66.

# 22

# MITOCHONDRIAL MEDICINE

Richard J. T. Rodenburg and Jan A. M. Smeitink
*Nijmegen Center for Mitochondrial Disorders, Radboud University Nijmegen Medical Centre, The Netherlands*

Mitochondrial medicine is an emerging field in medicine that focuses on diseases in which the mitochondrial energy generating system plays a central role. These diseases include the "classical" mitochondrial disorders, in which a (genetic) defect in the mitochondrial energy generating system is the primary cause of the disease, but also more common disorders such as Parkinson and cancer, in which mitochondrial energy metabolism plays an important role in the pathogenesis. In this review, we present an overview of the mitochondrial energy generating system, starting at the conversion of pyruvate into acetyl-CoA by pyruvate dehydrogenase, via the TCA cycle in which reduction equivalents are formed, to the oxidative phosphorylation system, where the reduction equivalents are used to convert ADP into ATP. The mitochondrial energy generating system can be examined by global assays that measure the rate of pyruvate conversion, the rate of oxygen consumption, or the rate of ATP production. In addition, an overview is given of the spectrophotometric assays that are available to measure PDHc, several TCA cycle enzymes, and the enzymes of the oxidative phosphorylation. In the diagnostic analysis of patients suspected to suffer from a mitochondrial disorder, these assays are applied to evaluate the functioning of the mitochondrial energy generating system in muscle biopsies and other types of patient samples. In addition, the results provide clues for further investigations at the molecular genetic level. Thus, the biochemical analysis of patient material is an important step in establishing the diagnosis of a mitochondrial disorder.

The importance of mitochondria in health and disease has given rise to a new area in medicine, called mitochondrial medicine. The foundation for this medical

*Chemical Biology: Approaches to Drug Discovery and Development to Targeting Disease*, First Edition.
Edited by Natanya Civjan.
© 2012 John Wiley & Sons, Inc. Published 2012 by John Wiley & Sons, Inc.

discipline lies in the recognition of diseases in which disturbances in one of the many steps of mitochondrial energy production are present. The role of mitochondria in energy metabolism disorders is well recognized and has been the subject of study for many decades. The increasing awareness of the relationship between mitochondria and several more common disorders, like Parkinson disease and cancer, makes a thorough understanding of the chemistry of the mitochondrial energy generating system, and the analytical methods to examine the functioning of this system, necessary for an increasing variety of medical and biochemical specialists. Here, we review a) how the cell's energy currency, ATP, is produced, and b) assays to determine the overall capacity of the system as well as single enzymes in relation to genetic disorders of energy production.

## 22.1 BIOLOGICAL BACKGROUND

### 22.1.1 Mitochondrial Disorders

Mitochondrial disorders can be defined as disorders that are caused by a defect in the mitochondrial energy generating system (MEGS). The clinical spectrum is very broad, but in almost all cases involve one or more tissues that have a high energy demand, in particular skeletal muscle and neuronal tissue. In addition, heart, liver, kidney, and other tissues can be involved as well. The severity and the course of the disease are extremely diverse, ranging from severe neonatal systemic disorders to mild exercise intolerance presenting at a high age. In addition to these "classical" mitochondrial disorders, it has become apparent that the mitochondrion and its energy generating system plays a role in other, more frequently occurring diseases as well, such as Parkinson (1), diabetes (2), and cancer (3). This review will be restricted to "classical" mitochondrial disorders, although the assays and principles described here could be of value for the investigation of other disorders as well.

### 22.1.2 The Mitochondrial Energy Generating System

The production of ATP by the MEGS is a complex process involving many different transporters and enzymes (4). Mitochondrial ATP is the end product of the oxidation of pyruvate (alpha-keto propionic acid). Pyruvate is the final product of the glycolysis, the anaerobic catabolism of glucose. Other substrates for the MEGS are fatty acids and several amino acids, in particular glutamine. Pyruvate is transported into the mitochondria where it is metabolized into acetyl CoA. The acetyl CoA is oxidized in the tricarboxylic acid (TCA) cycle, during which both NADH and $FADH_2$ are produced. These reducing equivalents are oxidized by the respiratory chain, which leads to the translocation of protons out of the mitochondrial matrix. The mitochondria use the resulting proton-motive force to generate ATP from ADP and phosphate. The ATP is released in the mitochondrial matrix, and can be exported to the cytosol. The MEGS is described in more detail in the next section.

## 22.2 THE CHEMISTRY OF THE OXPHOS SYSTEM

### 22.2.1 The Conversion of Pyruvate to Acetyl-CoA

The conversion of pyruvate into acetyl-CoA involves two key players. Pyruvate is imported into the mitochondrion by a specific transporter. At the time of writing this chapter only one case has been described in which a pyruvate carrier deficiency could be shown by functional assays (5). However, no genetic defect responsible for pyruvate carrier deficiency has been identified yet. After pyruvate has entered the mitochondrial matrix, it is converted into acetyl-CoA by the pyruvate dehydrogenase complex, PDHc. This large enzyme complex has a molecular mass of approximately 9 MDa and contains multiple copies of three enzymatic entities: 20-30 copies of alpha-ketoacid dehydrogenase (E1; EC 1.2.4.1), 60 copies of dihydrolipoamide acyltransferase (E2; EC 2.3.1.12), and 6 copies of dihydrolipoamide dehydrogenase (E3; EC 1.8.1.4), as well as 12 copies of the structural building block E3 binding protein (6). The E1 subcomplex consists of a tetramer of two E1α and two E1β subunits. PDHc decarboxylates pyruvate and esterifies the resulting acetyl-group to CoA. During this reaction, $NAD^+$ is reduced to NADH. PDHc requires 5 different cofactors, namely $NAD^+$, CoA, thiamine pyrophosphate, FAD and lipoic acid. The activity of PDHc is tightly regulated. First of all, the activity is controlled in an allosteric manner by the reaction products acetyl-CoA and NADH. In addition, two specific enzymes regulate PDHc activity via a phosphorylation site in the E1 subunits. PDH kinase (EC 2.7.11.2) is ATP dependent and inactivates PDHc by phosphorylating the E1 component when the ATP/ADP ratio is high (7). This enzyme is also activated by high $NAD^+$ and acetyl-CoA levels, and inactivated by high pyruvate levels. Four isoforms of PDH kinase have been identified, each having a different tissue distribution. By contrast, when the ATP/ADP ratio is low and the pyruvate levels are high, PDH kinase is not active and PDHc E1 is dephosphorylated by PDH phosphatase (EC 3.1.3.43), of which two isoforms are known (8). Due to this tight regulation, the oxidation rate of pyruvate by PDHc is directly coupled to the mitochondrial ATP production rate.

### 22.2.2 The Generation of Reduction Equivalents in the TCA Cycle

The acetyl-CoA generated by PDHc fuels the TCA cycle. Three dehydrogenases of the TCA cycle are responsible for the reduction of $NAD^+$ into NADH. These are isocitrate dehydrogenase (EC 1.1.1.41), α-ketoglutarate dehydrogenase (consisting of 3 enzymatic subunits E1 (EC 1.2.4.2), E2 (2.3.1.61), and E3 (EC 1.8.1.4)), and malate dehydrogenase (EC 1.1.1.37). In addition, the TCA cycle enzyme succinate dehydrogenase (EC 1.3.5.1) converts FAD into $FADH_2$. Furthermore, succinate-CoA ligase generates GTP or ATP, depending on the isotype of this enzyme complex (EC 6.2.1.4 and EC 6.2.1.5, respectively). The formation of these high-energy molecules is accompanied by the release of $CO_2$ in two of the reactions of the TCA cycle. In addition to acetyl-CoA, other metabolites can fuel the TCA cycle as well, This includes glutamine, which enters the TCA cycle at the site of 2-oxoglutarate. In tissues that catabolize fatty acids, the end product of the beta-oxidation is acetyl-CoA, which is converted into the ketone bodies 3-hydroxybutyrate and acetoacetate (in the liver) or enters the TCA cycle (in most other tissues) and in this way contributes to the synthesis of ATP.

### 22.2.3   The Oxidative Phosphorylation System

NADH and $FADH_2$ are oxidized by the OXPHOS system to generate ATP. This is a coordinated multistep process that involves 5 large enzyme complexes: the respiratory chain complexes I, II, III, and IV, and ATP synthase (complex V). Complex I (NADH:ubiquinone oxidoreductase; EC 1.6.5.3) is by far the largest respiratory chain enzyme complex. It consists of 45 different subunits and has a molecular weight of approximately 1 MDa. Complex I oxidizes NADH and the electrons that are released from NADH are transferred to a flavin mononucleotide present in complex I and subsequently via a channel of 8 iron sulfur clusters within the peripheral arm of the complex toward $CoQ_{10}$ that is present in the inner mitochondrial membrane (9). The redox reaction of complex I is directly coupled to the pumping of protons by complex I from the mitochondrial matrix across the mitochondrial inner membrane to the mitochondrial intermembrane space, which is in direct connection with the cytosol for small ions. Complex II (succinate dehydrogenase; EC 1.3.5.1) oxidizes $FADH_2$ and translocates electrons toward $CoQ_{10}$. Complex II is the only respiratory chain complex that does not contribute to the mitochondrial proton gradient. Complex III (ubiquinol:cytochrome c oxidoreductase; EC 1.10.2.2) translocates the electrons from $CoQ_{10}$ to cytochrome c, a small, heme-containing protein that acts as an electron carrier between complexes III and IV of the respiratory chain. The last step of the respiratory chain is complex IV (cytochrome c oxidase; EC 1.9.3.1), that oxidizes cytochrome c and transfers the electrons to molecular oxygen, which leads to the generation of water. The pumping of protons by the respiratory chain complexes I, III, and IV across the inner mitochondrial membrane increases the pH of the mitochondrial matrix and generates a potential difference between matrix and inter membrane space. Complex V (adenosine triphosphatase; EC 3.6.1.3) utilizes the potential energy of the proton gradient to convert ADP and phosphate into ATP.

### 22.2.4   Defects in the Mitochondrial Energy Generating System

In theory, a defect in any of the transport proteins and enzymes mentioned above could result in a reduced mitochondrial energy generating capacity. Primary defects at the protein and DNA level have been found in PDHc, fumarase (EC 4.2.1.2) (10), α-ketoglutarate dehydrogenase (2-oxoglutarate dehydrogenase) (11), succinate-CoA ligase (12, 13), complexes I (9), II (14), III (15), IV (16), and V (17), the phosphate carrier (18), and ANT. A functional defect in the pyruvate carrier has been identified as well (5). In addition to these structural components of the mitochondrial energy generating system, there are many additional proteins involved in the production of these structural components. These include the chaperones required to assemble the enzyme complexes of the respiratory chain. The structural building blocks of the OXPHOS system are encoded by multiple genes. Except for complex II, these genes are located in both the nuclear and the mitochondrial genomes. Most nuclear genetic defects in these structural genes result in an isolated enzyme deficiency, although several examples have been described of patients with a mutation in a complex I gene that also result in reduced enzyme activities of other enzyme complexes, e.g. complex III and PDHc, indicating that these complexes have a higher order of organization that can be disturbed by defects in one of the complexes. The existence of the so-called supercomplexes has become more apparent by functional and structural studies. Many different pathogenic point mutations and rearrangements in the mtDNA have been found in the last two decades. Depending on the type of mutation, these can cause either isolated or combined deficiencies of complexes I, III, IV, and V (19). The mitochondrial genome (or mtDNA) is replicated by a dedicated polymerase, POLG. Defects in the POLG gene leads to depletion of, and/or deletions in, the mtDNA. This causes isolated or combined deficiencies of the mtDNA encoded OXPHOS complexes I, III, IV, and/or V. Furthermore, defects in proteins involved in nucleotide metabolism can lead to mtDNA depletion and OXPHOS deficiencies as well. Depletion has also been observed in patients suffering from a defect in the TCA cycle enzyme succinate-CoA ligase, although the underlying mechanism is not yet fully understood (12, 20). The transcription and translation of mtDNA encoded proteins involves several mitochondria-specific proteins, in which defects also lead to combined OXPHOS enzyme deficiencies (21, 22). A special class of defects are those leading to a $CoQ_{10}$ deficiency. The biosynthesis of $CoQ_{10}$ is a multistep process that has been completely elucidated in yeast. In humans, several steps of this process have been shown to exist as well, and three different genetic defects in this pathway have been identified (23).

## 22.3 TOOLS AND TECHNIQUES TO STUDY THE OXPHOS SYSTEM

There are several approaches to perform biochemical analyses of the OXPHOS system. This section will focus on assays to perform structural analyses, enzyme activity assays, and ATP production, oxygen consumption and substrate oxidation assays.

### 22.3.1 Structural Analysis of the OXPHOS System

Blue-Native PAGE is a technique that is very suitable to study the relative amounts and the assembly status of OXPHOS complexes (24, 25). It can be performed as either a 1D or a 2D assay. In the 1-dimensional approach, the complexes are separated on a non-denaturing acrylamide gel containing the Coomassie dye Serva Blue G. All 5 complexes can be visualized in this way. In the 2-dimensional approach, the second dimension is a denaturing SDS-PAGE, resulting in the separation of the OXPHOS complexes into their individual protein building blocks. After blotting, more specific staining methods can be performed, e.g. using anti-OXPHOS complex antibodies. This can provide detailed information on the assembly status of the OXPHOS complexes. Another powerful tool to perform structural analysis of OXPHOS complexes is by immunopurification followed by mass spectrometric analysis of the isolated complexes. Using this approach, it has been shown that bovine complex I (and presumably also human complex I) consists of 45 different subunits (26). This technique has lead to the identification of Ecsit as a protein involved in complex I assembly, as it was found to be associated with complex I (27). This is quite an unexpected finding, as Ecsit was previously known as a cytosolic protein involved in a pro-inflammatory signal-transduction pathway from a Toll-like receptor and in embryonic development (28). By immunopurification and mass spectrometry, it could be shown that Ecsit associates with complex I, and that the N-terminal part of Ecsit appears to be required for mitochondrial import (27).

### 22.3.2   Enzyme Activity Assays

The traditional way to determine OXPHOS enzyme activities is by spectrophotometry. Several assays have been described for all 5 OXPHOS complexes. The assays are performed in homogenates of tissue samples or cultured cells, in crude mitochondria-enriched 600 g supernatants of tissue/cell homogenates, or in mitochondrial preparations from 14000 g pellets derived from 600 g supernatants. Obviously, the higher the purity of the mitochondrial preparation, the higher the specific activity of the enzymes measured. Therefore, very high specific activities can be achieved by using immunopurified complexes (29), although this approach is not widely applied yet. In addition to enzyme activity assays in solution, BN-PAGE can be used to estimate OXPHOS enzyme activities by in-gel activity assays (30). In-gel activities are particularly suitable to monitor relative activities, e.g. to evaluate changes in enzyme activities under different experimental conditions. For quantitative enzyme activity measurements, spectrophotometric analysis is the method of choice. Table 22.1 contains a summary of the most commonly used spectrophotometric enzyme assays for measurement of the OXPHOS enzymes. Below, a brief description of these assays is given.

#### Complex I

Spectrophotometric assays for measuring the activity of complex I (or NADH:ubiquinone oxidoreductase) are usually based on measurements of NADH, which is oxidized by complex I to $NAD^+$. In addition, the assay requires CoQ as a co-substrate for complex I. Usually $CoQ_1$ or decylubiquinone are used because these have better solubility than $CoQ_{10}$. Assays usually contain bovine serum albumin, which probably is required to stabilize the protein sample and to aid the solubilization of CoQ analogues. The specific activity of complex I can be determined by measuring the rate of NADH conversion at 340 nm in the absence and presence of the specific complex I inhibitor rotenone (31). Recently, our lab described a new assay that measures complex I by including

**TABLE 22.1  Summary of respiratory chain, complex V and PDHc activity assays. All assays are spectrophotometric assays except for the PDHc assay with $CO_2$ detection, which is a radiochemical assay. The specific inhibitor indicated is used for blank measurements. Non-standard abbreviations: $UQ_1$: ubiquinone-$Q_1$; DQ: decylubiquinone; DCIP: 2,6-dichlorophenolindophenol, PK: pyruvate kinase; LDH: lactate dehydrogenase; AABS: *p*-[*p*-(aminophenyl)azo]benzene sulfonic acid; ArAt: arylamine acetyltransferase**

| Enzyme | Activity measured | Substrates | Specific inhibitor | Detection | References |
|---|---|---|---|---|---|
| complex I | NADH:ubiquinone oxidoreductase | NADH, $UQ_1$ | rotenone | NADH (340 nm) | 31, 33, 35 |
| complex I + III | NADH:cytochrome c oxidoreductase | NADH, DQ<br>NADH, cytochrome c | rotenone<br>rotenone | DCIP (600 nm)<br>cytochrome c (550 nm) | 32<br>33 |
| complex II | succinate:ubiquinone oxidoreductase | succinate, DQ | malonate | DCIP (600 nm) | 33, 35 |
| complex II + III | succinate:cytochrome c oxidoreductase | succinate, cytochrome c | malonate | cytochrome c (550 nm) | 33, 35 |
| complex III | ubiquinol:cytochrome c oxidoreductase | DQ (red), cytochrome c | antimycin A | cytochrome c (550 nm) | 33, 35 |
| complex IV | cytochrome c oxidase | cytochrome c (red) | none | cytochrome c (550 nm) | 33, 35 |
| complex V | $F_1$-ATPase | ATP | oligomycin | NADH (via PK/LDH at 340 nm) | 33, 35 |
| PDHc | pyruvate dehydrogenase | pyruvate, NADH, CoA | none | $CO_2$ (radio chemically)<br>NADH (340 nm)<br>Acetyl-CoA (via AABS/ArAt at 460 nm) | 35, 37 |

451

2,6-dichlorophenolindophenol (DCIP) as a terminal acceptor of electrons that are derived from the oxidation of NADH and the subsequent reduction of the CoQ-analogue decylubiquinone (32). DCIP reduction can be followed spectrophotometrically at 600 nm. As the molar absorption coefficient of DCIP at 600 nm is more than 3 times higher than that of NADH at 340 nm, this new assay has a much higher sensitivity than the assay that measures NADH, with similar specificity. Complex I can also be measured as NADH:cytochrome c oxidoreductase, in which the combined activity of complex I + $CoQ_{10}$ + III is measured by addition of NADH and oxidized cytochrome c as substrates. The assay measures the rotenone-sensitive reduction of cytochrome c, which can be followed spectrophotometrically at 550 nm.

### Complex II

Complex II is usually measured in two ways: either as succinate:ubiquinone oxidoreductase or as succinate:cytochrome c oxidoreductase. The most commonly used assay for succinate:ubiquinone oxidoreductase (or isolated complex II) uses DCIP in the same way as described above for the new complex I assay, only in this case succinate is added as a substrate (instead of NADH). The specificity of DCIP reduction can be determined by measuring in the presence or absence of malonate, a specific inhibitor of complex II. The assay for succinate:cytochrome c oxidoreductase (or complex II + III) uses succinate and oxidized cytochrome c as substrates and measures the reduction of cytochrome c, which can be followed spectrophotometrically at 550 nm. The assay is also suitable to screen for coenzyme Q deficiencies, as it is dependent on the endogenously present $CoQ_{10}$. In case of a CoQ deficiency, a reduced succinate:cytochrome c oxidoreductase activity will be observed that can be normalized by addition of exogenous CoQ to the reaction mixture (33).

### Complex III

Complex III (ubiquinol:cytochrome c oxidoreductase) reduces cytochrome c and oxidizes reduced CoQ. In addition to these substrates, the assay contains a strong complex IV inhibitor (e.g. potassium cyanide) to prevent re-oxidation of reduced cytochrome c. The complex III activity can be derived from the rate of cytochrome c reduction, which can be followed at 550 nm. The specificity of the assay is determined by measuring in the absence or presence of antimycin A, a specific inhibitor of complex III activity.

### Complex IV

Complex IV (cytochrome c oxidase) is measured by addition of reduced cytochrome c to the reaction mixture. The oxidation of cytochrome c can be followed at 550 nm. This assay has a low background activity, and measurement in the absence and presence of a selective inhibitor is not necessary.

### Complex V

The activity of the ATP-forming enzyme complex V is usually assessed by determining the reverse reaction ATP $\rightarrow$ ADP + Pi. The reaction is coupled to reactions catalyzed by pyruvate kinase (ADP + phosphoenolpyruvate $\rightarrow$ pyruvate + ATP) and lactate dehydrogenase (pyruvate + NADH $\rightarrow$ lactate + $NAD^+$). This final reaction can be followed spectrophotometrically by measuring NADH at

340 nm. The activity of complex V (ATPase) can be derived from the rate of NADH conversion in the presence and absence of the specific complex V inhibitor oligomycin.

**Other enzymes**

In addition to assays for the OXPHOS enzymes, assays have been described for many other enzymes involved in the MEGS. For PDHc, various types of diagnostic assays have been described and are widely used (33). Also for the TCA cycle enzymes various assays have been described, although to date, pathogenic defects have only been found in α-ketoglutarate dehydrogenase (38), succinate-CoA ligase (12), and fumarase (39).

### 22.3.3   The Mitochondrial Energy Generating System

As outlined above, the generation of ATP from pyruvate by the mitochondrion involves many different transporters and enzymes. A subset of the individual components can be assayed individually (e.g. the respiratory chain enzymes). In addition, the process of ATP generation can be studied by using assays that provide information of the mitochondrial energy generating system (MEGS) *in toto*, and moreover, on the functioning of individual enzymes in the context of the intact mitochondrion. Several types of assays have been developed and are described below. For all these assays, a crucial factor is the integrity of the mitochondria, since the generation of ATP is dependent on an inner mitochondrial membrane potential. Therefore, it is important to include control experiments that test the coupling of the ATP synthesis to the respiratory chain, as this will provide information on the integrity of the inner mitochondrial membrane.

*22.3.3.1   ATP Production Assays*   Two types of ATP production assays can be discriminated. The first type is performed in cultured cells permeabilized by digitonin (40), or in mitochondria-enriched fractions from tissue homogenates (41). Different substrates can be added to monitor different metabolic routes that lead to mitochondrial ATP synthesis. For example, by addition of pyruvate and malate, pyruvate enters the mitochondria via a pyruvate carrier and is subsequently metabolized by pyruvate dehydrogenase to acetyl-CoA that is further metabolized in the citric acid cycle where it is coupled to oxalate that is the product of malate dehydrogenase. As the ATP production rate is under the control of the ADP/ATP ratio, an excess of ADP should be present in the assay. In this way, the pathway from the pyruvate carrier to ATP can be monitored. Any primary defect in this pathway will result in a decreased ATP production rate. Therefore, this type of assay is very suitable as a diagnostic tool to screen for a defect in the MEGS. The second type of ATP production assay is restricted to cultured cells. In this assay, luciferase is used as an ATP sensor *in vivo*. Luciferase, an enzyme derived from the firefly, converts its substrate luciferin under the emission of light. This reaction requires ATP, and thus, the amount of light produced is a measure for the amount of ATP. Cells can be stably transfected with an expression vector encoding luciferase (42), infected with a virus that encodes luciferase (43), or microinjected with plasmid DNA encoding luciferase (44). By addition of an appropriate targeting sequence, luciferase can be expressed specifically in the mitochondria in order to monitor intramitochondrial ATP. The ATP levels can be monitored by luminometry. A more sophisticated approach is to determine sub-cellular (e.g. intramitochondrial) ATP levels by a microscope coupled to a CCD camera. When cells are placed in a flow cell under the microscope/CCD camera system, they can be exposed to different stimuli that regulate ATP production, in particular hormones that lead to intracellular $Ca^{2+}$ fluxes, which stimulate mito-chondrial ATP production (43). This can be monitored real-time by means of the camera. In contrast to the first type of ATP assay, this real-time ATP assay is not only dependent on the integrity of the mitochondrial ATP generating machin-ery, but also on the intracellular mechanisms that regulate ATP production. This set-up is very suitable to perform functional studies of the mitochondrial energy generating system *in vivo* and the factors that affect the activity of this system, but is less suitable as a diagnostic tool.

*22.3.3.2 Substrate Oxidation Rate Measurements* Substrate oxidation rate measurements provide detailed information on the functioning of the MEGS (41). The MEGS performs three decarboxylation reactions, at the level of PDHc, isocitrate dehydrogenase, and 2-oxoglutarate dehydrogenase. By using radiolabeled substrates in which the radiolabel is present at a carboxyl residue, the activity of the MEGS can be followed by measuring the amount of released radiolabeled $CO_2$. Usually, combinations of radiolabeled substrates and unlabeled co-substrates are used, either in the presence or absence of specific inhibitors. An example is the use of radiolabeled pyruvate, which is decarboxylated by PDHc. To force the reaction to proceed at maximum rate, the reaction contains an excess of ADP, in order to maintain a high ADP/ATP ratio. In addition, the acetyl-CoA formed by PDHc has to be removed by addition of an appropriate co-substrate. When carnitine is used as co-substrate, it will be coupled to the acetyl group of acetyl-CoA by carnitine-acetyl transferase. When malate is used as a co-substrate, this will be converted to oxalate in the TCA cycle which is subsequently coupled to the acetyl group of acetyl-CoA by citrate synthase. When these assays are performed in a control sample, e.g. a muscle sample from a healthy individual, the ratio of the pyruvate oxidation in the presence of carnitine or malate will be approximately 1. Also in a sample from a PDHc deficient patient, this ratio will be near 1, however, in that case both reaction rates will be equally reduced due to the PDHc defect. Interestingly, in case of a respiratory chain defect, the ratio between these two reactions will be around 2 in favor of the reaction with carnitine (41). As this latter reaction does not directly involve a TCA cycle enzyme, an unfavorable $NADH/NAD^+$ ratio will have much less effect on the PDHc activity when carnitine is used as a co-substrate compared to the reaction with malate as a co-substrate. As a final example, in case of a complex V defect, the reaction of pyruvate + malate will have a lower rate than in a control sample. Addition of the uncoupler CCCP will result in a normalization of the reaction rate, a phenomenon that is also observed in case of a defect in the phosphate carrier or the ATP:ADP antiporter ANT.

*22.3.3.3 Oxygen Consumption Assays* In principle, oxygen consumption rate assays can be used for the same purposes as substrate oxidation rate assays. Also in this case, the pathway from the substrate of choice down to molecular oxygen conversion by complex IV can be evaluated, and even the steps beyond complex IV that are of influence on the activity of complex IV, such as complex V (45). The assay can be performed either in cellular or tissue extracts or in whole cells *in vivo* or *ex vivo*. The classical way to detect molecular oxygen is electrochemically by using a Clark-type oxygen electrode (45). More recently, molecular probes have been developed that made it possible to perform oxygen consumption rate measurements by fluorimetry (46, 47). This latter type of assay has the advantage that small volumes can be tested in 96-well plates using relatively simple laboratory equipment. By using appropriate combinations of substrates, the oxygen consumption rate measurements are suitable to locate primary defects in the MEGS, in a similar manner as with radiochemical substrate oxidation rate assays.

## 22.4 DIAGNOSTIC BIOCHEMICAL ANALYSIS OF THE MITOCHONDRIAL ENERGY GENERATING SYSTEM

The mitochondrial energy generating system requires efficient interplay between a large number of different proteins and protein complexes, which by themselves can be made up of large numbers of individual subunits that have to be assembled in an ordered manner. This complexity of the mitochondrial energy generating system is probably one of the main reasons for the heterogeneity of mitochondrial disorders. Due to this clinical diversity, establishing the diagnosis "mitochondrial disorder" usually requires a combination of clinical, chemical, biochemical, and genetic examination of the patient suspected for a mitochondrial disorder (48–50). The biochemical analysis of a muscle biopsy is the corner stone of the diagnostic examination for mitochondrial disorders. The results will show whether or not the MEGS functions properly in this tissue. Unfortunately, the number of (potentially) mitochondrial disease causing candidate genes is very large, and molecular genetic techniques to rapidly sequence hundreds of candidate genes in a diagnostic setting are not yet available. The biochemical results are not only diagnostic in their own right, but, in combination with the clinical features of the patient, also provide important clues that are used to select candidate genes for molecular genetic analysis. Nevertheless, the diagnosis mitochondrial disorder cannot always be made at the molecular genetic level, in particular in those cases in which a comprehensive biochemical analysis has not been performed.

The biochemical diagnostic analysis of a patient suspected for a mitochondrial disorder is usually performed on a muscle biopsy, as this tissue has a very high energy demand and is more likely to exhibit signs of mitochondrial dysfunctioning than tissues with a relatively low energy conversion rate. Depending on the clinical features, it could be considered to examine other types of tissue, such as liver or heart. It has been shown that in mitochondrial depletion syndromes with liver involvement, e.g. due to mutations in DGUOK or MPV17, muscle tissue may not always show biochemical abberrations while liver shows clear signs of enzyme deficiencies (51, 52). Obvious drawbacks of organ or muscle biopsies are that an invasive procedure is required to obtain the tissue sample. Technically, it is possible to measure in fibroblasts or even blood samples, but these do not always express the mitochondrial defect. Nevertheless, these types of samples do have a very important added value to the biochemical diagnosis. First of all, a positive or a negative fibroblast result in combination with a positive muscle result has consequences for the selection of candidate genes for subsequent molecular genetic analysis. For example, in case of a mitochondrial depletion syndrome due to mutations in the POLG gene, very severe enzyme deficiencies can be observed in muscle and/or liver, whereas fibroblasts often show normal enzyme activities (53). By contrast, in case of a mitochondrial translation defect, respiratory chain enzyme deficiencies are observed both in muscle and in fibroblasts. A second important aspect of fibroblast (or lymphocyte) analysis is that a systemic expression of a biochemical defect may allow for prenatal diagnosis on the basis of biochemical analysis of chorionic tissue or amniocytes in families of mitochondrial patients in which the underlying molecular genetic defect has not (yet) been identified (54).

## REFERENCES

1. Bogaerts V, Theuns J, and van Broeckhoven C. Genetic findings in Parkinson's disease and translation into treatment: a leading role for mitochondria? Genes Brain Behav. 2007;(Epub).
2. Rabol R, Bushel R, Dela, F. Mitochondrial oxidative function and type 2 diabetes. Appl. Physiol. Nutr. Metab. 2006;31:675–683.
3. Modica-Napolitano JS, Kulawiec M, Singh KK. Mitochondria and human cancer. Curr. Mol. Med. 2007;7:121–131.
4. Smeitink J, van den Heuvel L, Dimauro S. The genetics and pathology of oxidative phosphorylation. Nat. Rev. Genet. 2001;2:342–352.
5. Brivet M, Garcia-Cazorla A, Lyonnet S, Dumez Y, Nassogne MC, Slama A, Boutron A, Touati G, Legrand A, Saudubray JM. Impaired mitochondrial pyruvate importation in a patient and a fetus at risk. Mol. Genet. Metab. 2003;78:186–192.
6. Smolle M, Prior AE, Brown AE, Cooper A, Byron O, Lindsay JG. A new level of architectural complexity in the human pyruvate dehydrogenase complex. J. Biol. Chem. 2006;281:19772–19780.
7. Ling M, McEachern G, Seyda A, MacKay N, Scherer SW, Bratinova S, Beatty B, Giovannucci-Uzielli ML, Robinson BH. Detection of a homozygous four base pair deletion in the protein X gene in a case of pyruvate dehydrogenase complex deficiency. Hum. Mol. Genet. 1998;7:501–505.

8. Huang B, Gudi R, Wu P, Harris RA, Hamilton J, Popov KM. Isoenzymes of pyruvate dehydrogenase phosphatase. DNA-derived amino acid sequences, expression, and regulation. J. Biol. Chem. 1998;273:17680–17688.

9. Janssen RJ, Nijtmans LG, van den Heuvel LP, Smeitink JA. Mitochondrial complex I: structure, function and pathology. J. Inherit. Metab Dis. 2006;29:499–515.

10. Gellera C, Uziel G, Rimoldi M, Zeviani M, Laverda A, Carrara F, DiDonato S. Fumarase deficiency is an autosomal recessive encephalopathy affecting both the mitochondrial and the cytosolic enzymes. Neurology 1990;40:495–499.

11. Bonnefont JP, Chretien D, Rustin P, Robinson B, Vassault A, Aupetit J, Charpentier C, Rabier D, Saudubray JM, Munnich A. Alpha-ketoglutarate dehydrogenase deficiency presenting as congenital lactic acidosis. J. Pediatr. 1992;121:255–258.

12. Elpeleg O, Miller C, Hershkovitz E, Bitner-Glindzicz M, Bondi-Rubinstein G, Rahman S, Pagnamenta A, Eshhar S, Saada A. Deficiency of the ADP-forming succinyl-CoA synthase activity is associated with encephalomyopathy and mitochondrial DNA depletion. Am. J. Hum. Genet. 2005;76:1081–1086.

13. Ostergaard E, Christensen E, Kristensen E, Mogensen B, Duno M, Shoubridge EA, Wibrand F. Deficiency of the alpha Subunit of Succinate-Coenzyme A Ligase Causes Fatal Infantile Lactic Acidosis with Mitochondrial DNA Depletion. Am. J. Hum. Genet. 2007;81:383–387.

14. Briere JJ, Favier J, El Ghouzzi V, Djouadi F, Benit P, Gimenez AP, Rustin P. Succinate dehydrogenase deficiency in human. Cell. Mol. Life Sci. 2005;62:2317–2324.

15. *www.mitomap.org*.

16. Pecina P, Houstkova H, Hansikova H, Zeman J, Houstek J. Genetic defects of cytochrome c oxidase assembly. Physiol. Res. 2004;53:213–223.

17. Houstek J, Pickova A, Vojtiskova A, Mracek T, Pecina P, Jesina P. Mitochondrial diseases and genetic defects of ATP synthase. Biochim. Biophys. Acta 2006;1757:1400–1405.

18. Mayr JA, Merkel O, Kohlwein SD, Gebhardt BR, Bohles H, Fotschl U, Koch J, Jaksch M, Lochmuller H, Horvath R, Freisinger P, Sperl W. Mitochondrial phosphate-carrier deficiency: a novel disorder of oxidative phosphorylation. Am. J. Hum. Genet. 2007;80:478–484.

19. Taylor RW, Turnbull DM. Mitochondrial DNA mutations in human disease. Nat. Rev. Genet. 2005;6:389–402.

20. Carrozzo R, Dionisi-Vici C, Steuerwald U, Lucioli S, Deodato F, Di Giandomenico S, Bertini E, Franke B, Kluijtmans LA, Meschini MC, Rizzo C, Piemonte F, Rodenburg R, Santer R, Santorelli FM, van Rooij A, Vermunt-de Koning D, Morava E, Wevers RA. SUCLA2 mutations are associated with mild methylmalonic aciduria, Leigh-like encephalomyopathy, dystonia and deafness. Brain 2007;130:862–874.

21. Jacobs HT, Turnbull DM. Nuclear genes and mitochondrial translation: a new class of genetic disease. Trends Genet. 2005;21:312–314.

22. Spinazzola A, Zeviani M. Disorders of nuclear-mitochondrial intergenomic communication. Biosci. Rep. 2007;27:39–51.

23. Dimauro S, Quinzii CM, Hirano M. Mutations in coenzyme Q10 biosynthetic genes. J. Clin. Invest. 2007;117:587–589.

24. Schagger H, von Jagow G. Blue native electrophoresis for isolation of membrane protein complexes in enzymatically active form. Anal. Biochem. 1991;199:223–231.

25. Nijtmans LG, Henderson NS, Holt IJ. Blue Native electrophoresis to study mitochondrial and other protein complexes. Methods 2002;26:327–334.

26. Carroll J, Fearnley IM, Skehel JM, Shannon RJ, Hirst J, Walker JE. Bovine complex I is a complex of 45 different subunits. J. Biol. Chem. 2006;281:32724–32727.

27. Vogel RO, Janssen RJ, van den Brand MA, Dieteren CE, Verkaart S, Koopman WJ, Willems PH, Pluk W, van den Heuvel LP, Smeitink JA, Nijtmans LG. Cytosolic signaling protein Ecsit also localizes to mitochondria where it interacts with chaperone NDUFAF1 and functions in complex I assembly. Genes Dev. 2007;21:615–624.

28. Kopp E, Medzhitov R, Carothers J, Xiao C, Douglas I, Janeway CA, Ghosh S. ECSIT is an evolutionarily conserved intermediate in the Toll/IL-1 signal transduction pathway. Genes Dev. 1999;13:2059–2071.

29. Murray J, Schilling B, Row RH, Yoo CB, Gibson BW, Marusich MF, Capaldi RA. Small scale immunopurification of cytochrome c oxidase for a high throughput, multiplexing analysis of enzyme activity, and amount. Biotechnol. Appl. Biochem. 2007;48:167–178.

30. Zerbetto E, Vergani L, Dabbeni-Sala F. Quantification of muscle mitochondrial oxidative phosphorylation enzymes via histochemical staining of blue native polyacrylamide gels. Electrophoresis 1997;18:2059–2064.

31. Fischer JC, Ruitenbeek W, Trijbels JM, Veerkamp JH, Stadhouders AM, Sengers RC, Janssen AJ. Estimation of NADH oxidation in human skeletal muscle mitochondria. Clin. Chim. Acta 1986;155:263–273.

32. Janssen AJ, Trijbels FJ, Sengers RC, Smeitink JA, van den Heuvel LP, Wintjes LT, Stoltenborg-Hogenkamp BJ, Rodenburg RJ. Spectrophotometric assay for complex I of the respiratory chain in tissue samples and cultured fibroblasts. Clin. Chem. 2007;53:729–734.

33. Kirby DM, Thorburn DR, Turnbull DM, Taylor RW. Biochemical assays of respiratory chain complex activity. Methods Cell Biol. 2007;80:93–119.

34. Reisch AS, Elpeleg O. Biochemical assays for mitochondrial activity: assays of TCA cycle enzymes and PDHc. Methods Cell Biol. 2007;80:199–222.

35. Rustin P, Chretien D, Bourgeron T, Gerard B, Rotig A, Saudubray JM, Munnich A. Biochemical and molecular investigations in respiratory chain deficiencies. Clin. Chim. Acta 1994;228:35–51.

36. Lopez LC, Schuelke M, Quinzii CM, Kanki T, Rodenburg RJ, Naini A, Dimauro S, Hirano M. Leigh syndrome with nephropathy and CoQ10 deficiency due to decaprenyl diphosphate synthase subunit 2 (PDSS2) mutations. Am. J. Hum. Genet. 2006;79:1125–1129.

37. Sterk JP, Stanley WC, Hoppel CL, Kerner J. A radiochemical pyruvate dehydrogenase assay: activity in heart. Anal. Biochem. 2003;313:179–182.

38. Bonnefont JP, Chretien D, Rustin P, Robinson B, Vassault A, Aupetit J, Charpentier C, Rabier D, Saudubray JM, Munnich A. Alpha-ketoglutarate dehydrogenase deficiency presenting as congenital lactic acidosis. J. Pediatr. 1992;121:255–258.

39. Zinn AB, Kerr DS, Hoppel CL. Fumarase deficiency: a new cause of mitochondrial encephalomyopathy. N. Engl. J. Med. 1986;315:469–475.

40. Wanders RJ, Ruiter JP, Wijburg FA, Zeman J, Klement P, Houstek J. Prenatal diagnosis of systemic disorders of the respiratory chain in cultured chorionic villus fibroblasts by study of ATP-synthesis in digitonin-permeabilized cells. J. Inherit. Metab Dis. 1996;19:133–136.

41. Janssen AJ, Trijbels FJ, Sengers RC, Wintjes LT, Ruitenbeek W, Smeitink JA, Morava E, van Engelen BG, van den Heuvel LP, Rodenburg RJ. Measurement of the energy-generating capacity of human muscle mitochondria: diagnostic procedure and application to human pathology. Clin. Chem. 2006;52:860–871.

42. Gajewski CD, Yang L, Schon EA, Manfredi G. New insights into the bioenergetics of mitochondrial disorders using intracellular ATP reporters. Mol. Biol. Cell. 2003;14:3628–3635.

43. Visch HJ, Rutter GA, Koopman WJ, Koenderink JB, Verkaart S, de Groot T, Varadi A, Mitchell KJ, van den Heuvel LP, Smeitink JA, Willems PH. Inhibition of mitochondrial Na + -Ca$^{2+}$ exchange restores agonist-induced ATP production and Ca$^{2+}$ handling in human complex I deficiency. J. Biol. Chem. 2004;279:40328–40336.

44. Kennedy HJ, Pouli AE, Ainscow EK, Jouaville LS, Rizzuto R, Rutter GA. Glucose generates sub-plasma membrane ATP microdomains in single islet beta-cells. Potential role for strategically located mitochondria. J. Biol. Chem. 1999;274:13281–13291.

45. Hofhaus G, Shakeley RM, Attardi G. Use of polarography to detect respiration defects in cell cultures. Methods Enzymol. 1996;264:476–483.

46. Hynes J, Floyd S, Soini AE, O'Connor R, Papkovsky DB. Fluorescence-based cell viability screening assays using water-soluble oxygen probes. J. Biomol. Screen. 2003;8:264–272.

47. Will Y, Hynes J, Ogurtsov VI, Papkovsky DB. Analysis of mitochondrial function using phosphorescent oxygen-sensitive probes. Nat. Protoc. 2006;1:2563–2572.

48. Bernier FP, Boneh A, Dennett X, Chow CW, Cleary MA, Thorburn DR. Diagnostic criteria for respiratory chain disorders in adults and children. Neurology 2002;59:1406–1411.

49. Morava E, van den Heuvel L, Hol F, de Vries MC, Hogeveen M, Rodenburg RJ, Smeitink JA. Mitochondrial disease criteria: diagnostic applications in children. Neurology 2006;67:1823–1826.

50. Wolf NI, Smeitink JA. Mitochondrial disorders: a proposal for consensus diagnostic criteria in infants and children. Neurology 2002;59:1402–1405.

51. Mandel H, Szargel R, Labay V, Elpeleg O, Saada A, Shalata A, Anbinder Y, Berkowitz D, Hartman C, Barak M, Eriksson S, Cohen N. The deoxyguanosine kinase gene is mutated in individuals with depleted hepatocerebral mitochondrial DNA. Nat. Genet. 2001;29:337–341.

52. Spinazzola A, Viscomi C, Fernandez-Vizarra E, Carrara F, D'Adamo P, Calvo S, Marsano RM, Donnini C, Weiher H, Strisciuglio P, Parini R, Sarzi E, Chan A, Dimauro S, Rotig A, Gasparini P, Ferrero I, Mootha VK, Tiranti V, Zeviani M. MPV17 encodes an inner mitochondrial membrane protein and is mutated in infantile hepatic mitochondrial DNA depletion. Nat. Genet. 2006;38:570–575.

53. de Vries MC, Rodenburg RJ, Morava E, van Kaauwen EP, Ter Laak H, Mullaart RA, Snoeck IN, van Hasselt PM, Harding P, van den Heuvel LP, Smeitink JA. Multiple oxidative phosphorylation deficiencies in severe childhood multi-system disorders due to polymerase gamma (POLG1) mutations. Eur. J. Pediatr. 2007;166:229–234.

54. Niers L, van den Heuvel L, Trijbels F, Sengers R, Smeitink J. Prerequisites and strategies for prenatal diagnosis of respiratory chain deficiency in chorionic villi. J. Inherit. Metab Dis. 2003;26:647–658.

## FURTHER READING

Oxidative Phosphorylation in Health and Disease. Edited by: Jan Smeitink, Rob C.A. Sengers and J.M. Frans Trijbels. ISBN: 0-306-48232-0.

Mitochondria, 2nd edition. Methods in cell biology, vol 80. Edited by: Liza A. Pon and Eric A. Schon. ISBN: 0–12-544173–8.

Neuromuscular Disease Center website: *http://www.neuro.wustl.edu/neuromuscular/*

GeneReview of mitochondrial disorders website: *http://www.geneclinics.org/profiles/mt-overview*

Uppsala University Human Mitochondrial Genome Database: *http://www.genpat.uu.se/mtDB/*

# 23

# LYSOSOMAL DISORDERS

DOUG A. BROOKS

*Molecular Medicine Sector, Sansom Institute, University of South Australia, Adelaide, Australia*

MARIA FULLER

*Lysosomal Diseases Research Unit, Department of Genetic Medicine, Women's and Children's Hospital, Adelaide, Australia*

Lysosomal storage disorders are a group of over 50 different genetic diseases that result from lysosomal dysfunction. This disruption of lysosomal function can involve either a specific lysosomal hydrolase deficiency, a defect in lysosomal protein processing, or impaired lysosomal biogenesis. To appreciate the pathogenesis of lysosomal disorders fully, it is important to understand the dynamics of endosome–lysosome organelles, their capacity for the uptake and degradation of complex macromolecules, and how lysosomal biogenesis and hydrolysis are altered by substrate storage. The focus of this chapter will be on recognized lysosomal disorders and what is known about the composition and the function of endosome–lysosome organelles in these diseases. The lysosomal network will be discussed with a view to correlating the main site of endosome–lysosome degradation and the site of substrate accumulation in lysosomal disorders. A major unanswered question for lysosomal storage disorders is how an enzyme deficiency and the resulting storage of undegraded macromolecules impact on cells to cause organ dysfunction and disease.

A patient with possible Hurler syndrome was first described by Berkhan in 1907 (1), and it may be the first report of a lysosomal storage disorder. Nonetheless, the first detailed clinical description of a patient with a lysosomal storage

*Chemical Biology: Approaches to Drug Discovery and Development to Targeting Disease*, First Edition.
Edited by Natanya Civjan.
© 2012 John Wiley & Sons, Inc. Published 2012 by John Wiley & Sons, Inc.

disorder was by Charles Hunter, who described two brothers who are now recognized as having Hunter syndrome (2). Soon after this report, Meinhard von Pfaundler and Gertrud Hurler described Hurler-Pfaundler syndrome, which is now known as Hurler syndrome (3). A more detailed history of syndrome identification and the subsequent recognition of other lysosomal storage disorders have been documented by Whitley in 1993 (4).

In the 1960s, Hers and colleagues (5, 6) recognized that Pompe disease was caused by a deficiency of α-glucosidase and, using electron microscopy, evidence of storage vacuoles was reported. The identification of this enzyme deficiency, together with De Duve and colleagues' description of lysosomes and their contents (7, 8), led to the concept of "lysosomal storage disorders." This finding resulted in two seminal publications by Hers and Van Hoof that described "Lysosomes and Storage Diseases" (9, 10). It should be noted that much of the basic knowledge about the cell biology of lysosomes was contributed to strongly by these initial investigations on lysosomal storage disorders.

In 1999, Meikle and colleagues (11) reported the prevalence of lysosomal storage disorders as 1 in 7,700 live births for an Australian study that involved 27 different diseases. Since then, several new disorders have been recognized, and it is now accepted that, as a group, more than 50 different lysosomal storage disorders exist. Moreover, in some populations the prevalence of certain lysosomal storage disorders has been reported to be high, including 1 in 18,500 for aspartylglucosaminuria in the Finnish population (12) and 1 in 3,900 for Tay-Sachs in the Ashkenazi Jewish population (13). This finding led to past speculation that the combined incidence of lysosomal storage disorders may be as high as 1 in 1500 births (14), and more recent estimates suggest the prevalence is approximately 1 in 1000 births (*http://www.science.org.au/sats2007/hopwood.htm*). Lysosomal storage disorders are now recognized as a substantial group of genetic diseases that result in lysosomal dysfunction, leading to a failure to degrade specific substrates, which then accumulate in endosome–lysosome organelles.

## 23.1  ENDOSOMES AND LYSOSOMES

From early observations, De Duve defined lysosomes as cytoplasmic particles that were associated with a range of acid hydrolases (7). At the electron microscope level, membrane-bound vacuoles or compartments were recognized and shown to contain acid hydrolases that were detectable by cytochemical staining (15). The definitive description of a lysosome (16) includes a membrane-bound organelle compartment that is acidic and contains a range of mature acid hydrolases (e.g., proteases and glycosidases). This organelle represents the most distal compartment in the endocytic pathway (Fig. 23.1) and is distinct from the prelysosomal compartment and endosomal compartments based on the absence of mannose-6-phosphate receptors (the receptors that are responsible for the targeting and trafficking of soluble lysosomal enzymes; see below). Lysosomes are heterogeneous in size, shape, and composition, and they exhibit high density in organelle

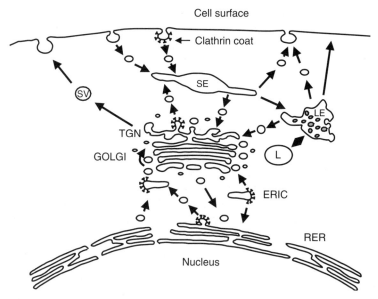

**Figure 23.1** Intracellular organelles/compartments and some basic pathways of vesicular traffic involved in lysosomal biogenesis and function. Lysosomal enzymes are synthesised in the rough endoplasmic reticulum (RER) and then traffic via the endoplasmic reticulum intermediate compartment (ERIC, involved in the retrieval of processing proteins back to the RER), to the Golgi. After glycoprocessing in the Golgi, soluble lysosomal proteins may then either exit the cell by the default secretory pathway in secretory vesicles (SV) or be targeted from the trans-Golgi network (TGN) by a mannose-6-phosphate receptor system to endosome compartments (sorting endosome, SE; late endosome, LE) for delivery to lysosomes (L). Soluble lysosomal enzymes may also traffic from the cell surface to the lysosome via a sorting and late endosomal pathway. Lysosomal membrane proteins can traffic by both of the latter intracellular routes, from either the trans-Golgi or the cell-surface to lysosomes, but cytoplasmic-based, tail sequence-targeting mechanisms are used to control this traffic.

fractionation experiments. The lysosome is the most distal compartment for lysosomal enzymes trafficking from the biosynthetic pathway. As defined by Storrie in 1988 (16), the lysosome must be the principal domicile of a "lysosomal protein", but the presence alone of a lysosomal protein in an organelle structure does not necessarily establish that organelle as a lysosome (for example lysosomal proteins can also be detected in endosomes and phagosomes).

Lysosomes are now known to be essential organelles involved in the turnover and the reuse of cellular macromolecules such as proteins, lipids, glycoproteins, and glycosaminoglycans. Lysosomes have other diverse functional roles in immune function, pigmentation, signaling, cell adhesion/motility, and membrane repair, and essentially they are a dynamic interface with the extracellular environment. Numerous subcellular events are required for the synthesis and the delivery of functional degradative enzymes to lysosomes and for the assembly of functional lysosomal organelles (17). This process is referred to as lysosomal biogenesis, and the transport of newly synthesized lysosomal proteins proceeds through another set of organelle compartments called endosomes (Fig. 23.1).

Endosomes mediate the delivery of degradative enzymes from the biosynthetic compartments to both lysosomes and the extracellular milieu and function in the processing and transport of secretory products (Fig. 23.1). Endocytic organelles are also involved in the internalization and the delivery of material from the extracellular milieu and cell surface to compartments inside the cell (18, 19). These two main intracellular pathways from the biosynthetic compartment and cell surface are convergent (Fig. 23.1) (20). Other specialist compartments exist within the network of endocytic organelles, which include phagosomes that sequester cytoplasmic material, and organelles for turnover and recycling. This complexity of endosome and lysosome organelles necessitates strict control of targeting and trafficking events. Intracellular traffic between these compartments can occur through specific vesicle formation that involves budding and fusion of membrane to and from different organelles, but other types of transient interaction are possible (see below).

## 23.2   COMPONENTS OF ENDOSOMES AND LYSOSOMES

One crucial role of the membrane that encloses the endosomal and lysosomal compartments is the isolation of the potent acid hydrolases that are key constituents of these compartments (21). The limiting membrane of lysosomes has proteins involved in membrane structure, compartment acidification (ATPase), ion transport, and vesicular traffic. Approximately 20–30 major polypeptides of molecular mass 15–200 kDa exist in lysosomal extracts, and most of these are highly glycosylated (16, 22). Typically, the mannose-6-phosphate receptors that target luminal lysosomal proteins are absent from the end-stage lysosomal compartment but are present in endosomes. The lysosome is composed of at least 20 known membrane proteins and over 50 luminal lysosomal proteins (22). The 50 or more known lysosomal storage disorders mostly relate to a dysfunction of one or more of the soluble acid hydrolases that are normally involved in macromolecular break-down, which includes proteases, glycosidases, sulphatases, phosphatases, and lipases. However, several lysosomal storage disorders involving membrane proteins, transporters/channels, and altered vesicular trafficking machinery have been recognized (23).

The major lysosomal associated membrane proteins—LAMP-I and LAMP-II are type-1 integral membrane proteins; they have a single membrane-spanning sequence, a highly glycosylated luminal domain, and a short cytoplasmic tail sequence involved in targeting/trafficking (24). Two other major lysosomal integral membrane proteins—LIMP-I/CD63 and LIMP-II—are both type II integral membrane proteins, with four and two membrane-spanning domains, respectively. Newly synthesized LAMP and LIMP molecules traffic via the trans-Golgi network (from the biosynthetic compartment) to endosomal/lysosomal compartments based on either tyrosine (LAMP-I, LAMP-II, LIMP-I/CD63) or di-leucine (LIMP-II)-sorting signals in the cytoplasmic domains of these molecules (25, 26). At steady-state, most LAMP-I and LAMP-II molecules are localized to the limiting membrane of endosomes and lysosomes. However, LAMP-I and LAMP-II are also detected in autophagic vacuoles. A major proportion of LIMP-II molecules are localized to endosomes and seem to have a role in the biogenesis of these organelles (27). LIMP-I/CD63 is primarily localized to the internal mem-branes of endosomal vesicles of multivesicular bodies and, to a lesser extent, is found at the limiting endosome–lysosome membrane and cell surface. Function-ally, LAMP-II has been shown to have an important role in autophagy and seems to have the capacity to compensate for a LAMP-I deficiency (28). The exact func-tion(s) of LIMP-I/CD63 is unclear, but it seems to have a role in immune cell activation, where it is subsequently expressed at the cell surface. LIMP-I/CD63 is a member of the tetraspannin family of proteins (29), which have also been implicated in the control of membrane and cell volume, as well as cell adhesion, cell motility, and antigen presentation.

Specific membrane and membrane-associated proteins also control the organelle traffic and fusion events that are associated with the movement of proteins between different intracellular compartments. Membrane proteins of the secretory and endocytic pathways depend on sorting signals that reside in their cytoplasmic domains. Many proteins exhibit multiple signals that determine their passage along these diverse pathways. Therefore, the steady-state distribution of any given membrane protein is dictated by the specific combination of sorting signals in the protein and the interaction of these signals with specific recognition molecules (26). This intracellular network of organelles and vesicles (Fig. 23.1) is in constant and dynamic flux.

Glycerophospholipids, cholesterol, and sphingolipids are the essential build-ing blocks for all eukaryotic cell membranes (30). The lysosomal membrane, like other eukaryotic membranes, is composed not only of highly glycosylated proteins but also is enriched in amphiphilic lipids (30, 31). The lipid and pro-tein composition of the lysosomal membrane is believed to be very complex, and this complexity ensures selective degradation such that the lysosomal membranes remain intact. This finding has led to the assumption that at least two distinct pools of lipid exist in the lysosomal membrane (32). Glycosphingolipids are important components of lysosomal membranes, which are involved directly in lipid rafts, and these are involved in both membrane transport and signaling (31). The early studies on the glycosphingolipidoses (a subgroup of lysosomal storage disorders) suggested that aggregates of lipids accumulated as multivesicular stor-age bodies in lysosomal compartments (33, 34). Notably, cholesterol is normally enriched in the membranes of early endocytic organelles but not in lysosomes.

## 23.3 TARGETING OF PROTEIN CONSTITUENTS TO LYSOSOMES

Most lysosomal proteins have either high mannose or complex oligosaccharide side chains that are attached to asparagine residues in the polypeptide at consensus

(NXS, NXT sites: where N is asparagine, S is serine, T is tyrosine, and X is any amino acid) glycosylation sites (16, 35, 36). In the case of soluble lysosomal proteins, N-linked glycosylation is attached to the growing polypeptide chain in the endoplasmic reticulum, and this forms the basic structure that is modified to generate the mannose-6-phosphate targeting signals that are involved in targeting these proteins to the lysosome (37). This latter processing event occurs in the cis-Golgi apparatus and involves the enzyme N-acetylglucosamine-1-phosphotransferase, which adds an N-acetylglucosamine-1-phosphate to certain mannose oligosaccharides. The removal of the N-acetylglucosamine residue in the trans-Golgi exposes the mannose-6-phosphate targeting signal that then allows soluble lysosomal hydrolases to interact with mannose-6-phosphate receptors for trafficking to endosomes and delivery to lysosomes (reviewed in References 37–39).

Thus, within the lumen of organelle compartments, some integral membrane proteins act as cargo receptors that recruit soluble molecules for traffic within the network of organelles comprising the secretory, endocytic, and lysosomal pathways. For example, the cation-dependent mannose-6-phosphate receptor [CD-MPR; 46-kDa dimer, which requires calcium (40)] and cation-independent mannose-6-phosphate receptor [CI-MPR; 300-kDa, also called the IGF II receptor (38)] are involved in binding soluble lysosomal proteins in the trans-Golgi network for targeted delivery of these enzymes to the endosome–lysosome system. Mannose-6-phosphate receptors are type I glycoproteins (have a single transmembrane sequence) and recycle between the trans-Golgi network, endosomal compartments, and the cell surface. Surprisingly, this recycling process does not seem to depend on whether the receptor is loaded with ligand (41). Mannose-6-phosphate receptors normally release their lysosomal enzyme cargo molecules in the prelysosomal compartment as a result of the low pH environment (pH 5.0–6.0). The free mannose-6-phosphate receptor is then returned to the trans-Golgi for subsequent cargo delivery. Therefore, lysosomes are defined as acid hydrolase-rich organelles that lack both the CD-MPR and CI-MPR, which distinguishes them from endosomes (42).

Sorting signals that are either tyrosine-based (NPXY, YXX: where N is asparagine, P is proline, and X is any amino acid) or leucine-based [(DE)XXXL(LI), DXXLL: (where D is aspartic acid, E is glutamic acid, L is leucine, I is isoleucine, and X is any amino acid] are found within the cytosolic domains of transmembrane proteins (26). Sorting occurs through coated areas of membranes comprised of proteins such as the adaptor protein (AP) complexes and Golgi gamma ear adaptin (GGA 1-3) proteins, which act to bind targeting motifs (26). The NPXY motif, which is found within a subset of type I membrane proteins, mediates internalization from the plasma membrane alone. The YXX motif is found in the CI-MPR and the CD-MPR, LAMP-1, LAMP-2, and CD63, and it is involved in a wide variety of sorting processes. Positioning of the motif relative to the membrane is also critically important for the recognition of these targeting signals. The YXX-AP-2 interaction is facilitated by a conformational change induced by phosphorylation, which is part of targeting regulation. The dileucine signal (DE)XXXL(LI) found in many type I, type II, and multispanning proteins also binds to AP complexes. The LL and LI motifs exhibit a distinct preference for the complexes that are critical for the fine specificity of targeting. The DXXLL motif is found within several transmembrane receptors, CI-MPRs and CD-MPRs, which cycle between the trans-Golgi network and the endosomes. Mannose-6-phosphate receptors bind the VHS domain of GGA proteins only via this signal, which is regulated by phosphorylation of serine residues. GGA1 and GGA3 also have the DXXLL motif within their hinge regions. Binding the hinge region with the VHS domain invokes an auto-inhibitory effect. Phosphorylation of the GGA serine results in auto-inhibition and transfer of the mannose-6-phosphate receptors from GGA1 to AP-1. Ubiquitin has been shown to be involved in sorting at the cell surface, endosomes, and trans-Golgi network. Endocytic proteins, which include epsin, Hrs, and STAM, have a ubiquitin-interacting motif that binds directly to ubiquitin, which suggests that they could act as adaptors for sorting ubiquitinated cargo at discrete intracellular sites (26).

## 23.4   INTERACTION OF ENDOSOMES AND LYSOSOMES AND THE SITE OF SUBSTRATE HYDROLYSIS

Many ligands that enter the endocytic pathway via receptor interaction are either sorted for traffic along specific organelle pathways (Fig. 23.1) or transit to the late endosome where ligands are dissociated from the receptors by the acid pH environment. For lysosomal enzymes, this dissociation event from mannose-6-phosphate receptors allows additional traffic to the most distal element of the endocytic machinery, the lysosomal compartment. A constant flux of membrane proteins seems to occur between late endosomes and lysosomes (43). It has been postulated that lysosomes fuse with endosomes to form a transient compartment that seems to be the major organelle involved in macromolecule degradation [i.e., "the cell stomach" (44)]. This finding implies that lysosomes are effectively a reservoir for acid hydrolases that are then tapped when required, whereas the late endosome is the degradative compartment. Moreover, lysosomal constituents can be recovered from the prelysosomal compartment and reformed as a lysosomal organelle (43). This theory helps to explain the heterogeneity of endocytic organelles, but it is consistent with the initiation of proteolysis/hydrolysis of macromolecules in endosome and prelysosomal compartments. The molecular machinery that mediates these vesicular traffic, fusion, and budding processes is yet to be defined fully. However, studies on a FYVE finger protein localized to early endosomes called Hrs and the endosomal-sorting complex required for transport (ESCRT) have indicated that a group of proteins that recognize ubiquitin motifs are involved directly in endosomal sorting and recruitment of proteins into multivesicular endosomes (45–48).

The mechanism for the transfer of endocytosed material between endosomes and lysosomes has generated many theories, which include maturation of endosomes into lysosomes, vesicular transport between endosomal and lysosomal compartments, transient interaction via channel formation ("kiss and run"), and direct fusion (44). Depending on the cargo involved, all of these mechanisms seem to be used (perhaps in different degrees) to regulate the level of hydrolytic capacity in any given compartment. Thus, lysosomes can interact with not only endosomal compartments but also phagosomes, autophagosomes, and the plasma membrane to effect different functional roles. For a lysosomal storage disorder, this model of lysosomal function would predict that endocytic cargo, such as glycosaminoglycan that is destined for degradation after internalization from the cell surface, would be delivered to a late endosome for degradation. In an attempt to degrade the substrate, lysosomes would either infuse lysosomal hydrolases into the late endosome and/or fuse directly to the late endosome to generate a hybrid organelle. The latter would provide maximum delivery of hydrolases, and a mechanism by which the cell could try to compensate for the reduced catalytic capacity that develops from a hydrolase deficiency. However, a failure to degrade the substrate contents may result in the maintenance of these hybrid endosome–lysosome organelles.

Evidence that the latter model of endosome–lysosome fusion and lysosomal recovery is valid comes from the lysosomal storage disorder mucolipidosis type IV (MLIV), which involves a defect in a $Ca^{++}$ channel that seems to prevent the retrieval of lysosomes from the hybrid organelle (49). Thus, intra-organelle $Ca^{++}$ is presumed to be required for the condensation process that leads to lysosomal organelle reformation. The trafficking of the glycosphingolipid lactosylceramide is inhibited in MLIV and, in common with other lipids, it may then accumulate inappropriately because of this blockage. In other lysosomal storage disorders that involve a hydrolase(s) deficiency, the failure to degrade intraorganelle substrate(s) may signal an incomplete process of degradation and, in the same way, potentiate the hybrid organelle and lead to the same adverse effects on cell function. Similarly, a defect in the organelle trafficking machinery can also impact on endosome–lysosome degradation to generate similar compartments, which also then accumulate undegraded storage material. Thus, a report of a defect in the endosomal sorting complex required for transport (ESCRT-III) machinery (50) impacts directly on late endosomes (multivesicular bodies), which generates compartments with a morphology remarkably similar to that observed in other lysosomal storage disorders that are caused by hydrolase deficiencies (e.g., Sanfilippo syndrome). This finding raises questions about the similarity between the vesicular pathology in different lysosomal storage disorders and whether common points for pathogenesis lead to similar clinical outcome.

## 23.5  COMMONALITIES FOR LYSOSOMAL STORAGE DISORDERS IN TERMS OF CLINICAL PHENOTYPE AND VESICULAR PATHOLOGY

Most patients with lysosomal storage disorders are born with no clinically obvious signs of disease, but in the severe forms of the disease, onset and progression of symptoms is rapid. In most cases, the severe form of each disorder is devastating (see Table 23.1) (51, 52) and results in an early death. Lysosomal storage disorders have been classified based either on clinical presentation, substrates stored, or similarities in the defect (53). For the purposes of this discussion, we have grouped some representative examples of lysosomal storage disorders according to the type of defect: mucopolysaccharidosis (defects in glycosaminoglycan degradation), oligosaccharidoses (defects in glycoprotein and glycogen degradation), sphingolipidoses and lipidoses (defects in glycolipid degradation), and finally protein processing and transport defects (Table 23.1). Short stature, skeletal dysplasia, coarse facial features, joint problems, visceromegaly, cardiac disease, CNS (central nervous system) pathology, and early death are all common disease manifestations, but these features are not present in all disorders (Table 23.1). This finding implies that a common mechanism for pathology may occur in some disorders, but some unique features presumably relate to the type of substrate being stored and/or the relative organ distribution and turnover rate of the substrate.

Potentially compartment specific differences exist between each lysosomal storage disorder. For example, in Pompe disease, the storage of glycogen in endosome–lysosome organelles results from sequestration of cytoplasmic material and involves a phagasomal compartment. In contrast, other lysosomal storage disorders involve material that has undergone traffic from different endosome/phagosome compartments. Thus, compartment-specific storage effects may occur, although it may still result in a common molecular mechanism by impact at a certain point in the endosome–lysosome pathway. Some commonalities observed in lysosomal storage disorder pathology may result from substrate being turned over in similar organs. Thus, the rate of substrate turnover will impact directly on the level of storage and thus cell and organ dysfunction.

## 23.6  PRIMARY AND SECONDARY STORAGE MATERIALS

Most known lysosomal storage disorders result from a single gene defect that results in the reduction of a single catabolic event, and hence the accumulation of a specific substrate. Although this concept is apparently simple, the reality is that the storage of a primary compound can result in a complex cascade of dysfunction (51). Glycosphingolipids such as the gangliosides GM2 and GM3, and unesterified cholesterol can, for example, accumulate after the deposition of undegraded glycosaminoglycans, and this accumulation is evident in the lysosomal storage disorders mucopolysaccharidosis types I, II, and III. Thus, progressive disease manifestations such as skeletal dysplasia, heart disease, and CNS dysfunction may be more the result of this secondary storage than from the primary storage material. In turn, these manifestations may reflect a general disruption to lysosomal dysfunction and account for some commonalities in clinical presentation for lysosomal storage disorders.

**TABLE 23.1  A representative sample of 37 known (50 or more) lysosomal storage disorders and the common clinical symptoms observed in these patients at the severe end of the clinical spectrum (adapted from References 51 and 52)**

| Lysosomal storage disorder (syndrome) | Enzyme deficiency | Substrate Stored | Clinical symptoms (common in severe form) |
|---|---|---|---|
| *Mucopolysaccharidoses: Defects in glycosaminoglycan degradation* | | | |
| Mucopolysaccharidosis I (Hurler, Hurler-Scheie and Scheie syndromes) | α-L-Iduronidase | HS, DS | Short stature, skeletal dysplasia, coarse facial features, joint stiffness, visceromegaly, cardiac disease, corneal clouding, CNS involvement |
| Mucopolysaccharidosis II (Hunter Syndrome) | Iduronate 2-sulphatase | HS, DS | Short stature, skeletal dysplasia, coarse facial features, joint stiffness, visceromegaly, cardiac disease, corneal clouding, CNS involvement |
| Mucopolysaccharidosis IIIA (Sanfilippo A syndrome) | Heparan *N*-sulphatase | HS | Coarse hair, CNS involvement, aggressive behavior, dysmorphic features (+/−), skeletal dysplasia (+/−) |
| Mucopolysaccharidosis IIIB (Sanfilippo B syndrome) | α-N-Acetylglucosaminidase | HS | Coarse hair, CNS involvement, aggressive behavior, dysmorphic features (+/−), skeletal dysplasia (+/−) |

| | | | |
|---|---|---|---|
| Mucopolysaccharidosis IIIC (Sanfilippo C syndrome) | Acetyl-CoA α-glucosamine N-acetyltransferase | HS | Coarse hair, CNS involvement, aggressive behavior, dysmorphic features (+/−), skeletal dysplasia (+/−) |
| Mucopolysaccharidosis IIID (Sanfilippo D syndrome) | N-Acetylglucosamine 6-sulphatase | HS | Coarse hair, CNS involvement, aggressive behavior, dysmorphic features (+/−), skeletal dysplasia (+/−) |
| Mucopolysaccharidosis IVA (Morquio A syndrome) | N-Acetylgalactosamine 6-sulphatase | KS | Short stature, skeletal dysplasia, cardiac disease, corneal clouding |
| Mucopolysaccharidosis IVB (Morquio B syndrome) | β-Galactosidase | KS | Short stature, skeletal dysplasia, cardiac disease, corneal clouding |
| Mucopolysaccharidosis VI (Maroteaux-Lamy syndrome) | N-Acetylgalactosamine 4-sulphatase | DS, CS | Short stature, skeletal dysplasia, coarse facial features, joint stiffness, visceromegaly, cardiac disease, corneal clouding |
| Mucopolysaccharidosis VII (Sly syndrome) | β-D-Glucuronidase | HS, DS | Short stature, skeletal dysplasia, coarse facial features, joint stiffness, visceromegaly, CNS involvement, cardiac disease |

*(continued)*

**TABLE 23.1** (*Continued*)

| Lysosomal storage disorder (syndrome) | Enzyme deficiency | Substrate Stored | Clinical symptoms (common in severe form) |
|---|---|---|---|
| Mucopolysaccharidosis IX | Hyaluronidase | HA | Short stature, joint stiffness |
| *Oligosaccharidoses: Defects in glycoprotein and glycogen degradation* | | | |
| α-Mannosidosis | α-Mannosidase | α-Mannosides | Short stature, skeletal dysplasia, coarse facial features, joint stiffness, visceromegaly, cardiac disease, CNS involvement |
| β-Mannosidosis | β-Mannosidase | β-Mannosides | Short stature, skeletal dysplasia, coarse facial features, joint stiffness, visceromegaly, cardiac and renal disease, angiokeratoma, CNS involvement |
| α-Fucosidosis | α-Fucosidase | α-Fucosides Glycolipids | Skeletal dysplasia, joint stiffness, visceromegaly, cardiac disease, angiokeratoma, CNS involvement |
| Sialidosis | α-Sialidase (Neuraminidase) | Sialyloligosaccharides | Short stature, skeletal dysplasia, CNS involvement |
| Galactosialidosis | α-Sialidase, α-galactosidase, protective protein. | Oligosaccharides | Short stature, skeletal dysplasia, coarse facial features, joint stiffness, visceromegaly, cardiac and renal disease, CNS involvement |

| Disease | Enzyme | Storage Material | Clinical Features |
|---|---|---|---|
| Aspartylglucosaminuria | Aspartylglucosaminidase | Aspartylglucosamine | Skeletal dysplasia, visceromegaly, CNS involvement |
| Schindler, Kanzaki | α-Galactosidase B | Galactosaminides Glycolipids | Cardiac and renal disease, angiokeratoma, CNS involvement |
| Pompe disease | α-D-Glucosidase | Glycogen | Muscle weakness, cardiac disease |

*Sphingolipidoses and lipidoses: Defects in glycolipid degradation*

| Disease | Enzyme | Storage Material | Clinical Features |
|---|---|---|---|
| GM1 gangliosidosis | β-Galactosidase | GM1-gangliosides, oligosaccharides, KS, glycolipids | Cardiac disease, CNS involvement |
| GM2 gangliosidosis (Sandhoff and Tay-Sachs) | Hexosaminidase A and B | GM2-gangliosides, oligosaccharides, glycolipids | CNS involvement, visual impairment |
| Gaucher | Glucocerebrosidase | Glucoceramide globoside | Bone disease, visceromegaly, CNS involvement (severe form) |
| Fabry | α-Galactosidase | α-Galactosyl sphingolipids, oligosaccharides | Heart and kidney disease, angiokeratoma, CNS involvement (recurrent stroke) |
| Wolman | Acid lipase | Cholesterol esters | Visceromegaly, CNS involvement |
| Nieman-Pick types A and B | Sphingomyelinase | Sphingomyelin | Visceromegaly, CNS involvement |

(continued)

**TABLE 23.1** (*Continued*)

| Lysosomal storage disorder (syndrome) | Enzyme deficiency | Substrate Stored | Clinical symptoms (common in severe form) |
|---|---|---|---|
| Farber | Acid ceramidase | Ceramide | Nodular swelling around joints, hypotonia, visceromegaly, CNS involvement, angiokeratoma |
| Krabbe | Galactosylceramidase | Galactosylceramides | Spasticity, ataxia, weakness, hypotonia, CNS involvement |
| Metachromatic leukodystrophy | Arylsulphatase | Sulphatides | Weakness, hypotonia, psychoses, ataxia, CNS involvement, behavior changes |
| *Protein processing and transport defects* | | | |
| Multiple sulphatase deficiency | Endoplasmic reticulum sulphatase modifying factors → sulphatases | MPS, sulphatides, glycolipids | Short stature, skeletal dysplasia, coarse facial features, joint stiffness, visceromegaly, CNS involvement |
| Mucolipidosis II/III I-cell disease | Golgi N-acetylgalactosamine 1-phosphtransferase → lysosomal hydrolases | Oligosaccharides and glycolipids | Short stature, skeletal dysplasia, coarse facial features, joint stiffness, visceromegaly, cardiac disease, corneal clouding, CNS involvement |
| Mucolipidosis IV | Mucolipin 1, Ca$^{++}$/cation ion channel | PC, PL, MPS, SL, gangliosides | Corneal clouding, CNS involvement |

| Disease | Transporter/protein | Storage material | Clinical features |
| --- | --- | --- | --- |
| Sialic acid storage disorder Salla disease | Sialic acid transporter | Sialic acid | Visceromegaly, renal disease, CNS involvement |
| Cystinosis | Cysteine transporter | Cysteine | Renal disease, CNS involvement |
| Cobalamine deficiency type F | Cobalamine transporter | Cobalamine | CNS involvement |
| Hermansky-Pudlak syndrome (disorders of lysosomal biogenesis) | AP3 adaptor protein, or BLOC-1-3 (biogenesis of lysosome-related organelle complex), or HOPS (homotypic vacuolar protein sorting or VPS class C complex). | Ceroid lipofuscin | Lung and kidney disease, oculocutaneous albinism, CNS involvement |
| Fronto-temporal dementia | Endosomal sorting complex required for transport (ESCRT) including CHMP2B | | CNS involvement |

*HS*, heparin sulphate; *DS*, dermatan sulphate; *KS*, keratan sulphate; *HA*, hyaluronic acid; *PC*, phosphatidylcholine; *PL*, phospholipid; *SL*, sphingolipid.

## 23.7 IMPACT OF STORAGE ON VESICULAR STRUCTURE AND TRAFFIC

Secondary storage in response to a primary defect and its associated stored substrate suggests a common process of lysosomal dysfunction in some storage disorders. Thus, GM2 and GM3 gangliosides have been observed to not only accumulate in GM1 gangliosidosis, Tay Sachs, and Sandhoff patients, but also in a range of other storage disorders including mucopolysaccharidosis types I, II, and III, and Niemann-Pick type C (54). In these disorders, vesicular structures with characteristic multilamellar inclusions, membrane swirls, and internal vesicles have been observed, which resemble either autophagosome or multivesicular endosome structures (52). Moreover, in mucopolysaccharidosis type III, the storage compartments have been reported to contain different amounts of the primary storage material heparin sulphate, GM2, and GM3 gangliosides (54). In addition, spheroid structures that contain ubiquitin have been reported (54). Chloroquine toxicity also results in vesicular structures that have similar morphology, which include whorled inclusions of lipid, multivesicular endosome-like structures and zebra bodies. In turn, this toxicity has been reported to be almost identical to the vesicular pathology in Fabry disease (55). This commonality of vesicular structures and pathology could be explained by the general impact on lysosomal function created by the primary storage material. Although these examples involve an enzyme deficiency and could therefore tend to support this concept, other lysosomal storage disorders involve dysfunction in either membrane transporters or vesicular machinery, and they generate similar vesicular pathology. For example, a defect in the lysosomal transporter mucolipin-1 (the cation channel that is involved in calcium export and necessary for the recovery of lysosomes from late endosome–lysosome hybrid organelles) causes the storage of lipids and gangliosides in enlarged multiconcentric lamellar structures. Moreover, a dysfunction of the ESCRT-III vesicular machinery, which results in neuropathology, has also been reported to cause the formation of similar multivesicular structures in a mouse model and these vesicles are almost identical to that reported in a mouse model of mucopolysaccharidosis IIIA (56). Possible explanations for these commonalities in vesicular pathology might therefore include the impact of storage on a critical event in lysosomal function, such as vesicle formation/recovery or vesicular traffic.

## 23.8 TREATMENT STRATEGIES AND ABILITY TO CORRECT RESIDUAL PATHOLOGY

Treatment strategies are available for lysosomal storage disorders and include hematopoietic stem cell transplantation, enzyme replacement therapy, substrate reduction therapy, chemical chaperones, and gene therapy (reviewed in References (57–61)). In practice, however, the clinical spectrum observed in lysosomal storage disorders means that—at the present time—a single therapeutic strategy is unlikely to treat all sites of pathology effectively. Thus, hematopoietic stem cell marrow transplantation is used currently for many patients with neuropathology, but it is not optimal (not effective for some disorders) and has significant risks. At the time of writing, hematopoietic stem cell transplantation is recommended for patients at risk of cognitive impairment and has been evaluated in combination with enzyme replacement therapy (62). Enzyme replacement therapy by intravenous infusion is being used for patients at the attenuated end of the clinical spectrum, in some lysosomal disorders [e.g., MPS I (mucopolysaccharide) (58, 62)], but it is of limited use for the treatment of those disorders with neuropathology. Small molecule therapeutic strategies have been employed as potentially alternative or adjunct treatment strategies (reviewed in Reference 61). For example, substrate deprivation therapy uses small molecule inhibitors to reduce substrate synthesis, whereas enzyme enhancement therapy has been investigated using chemical chaperones (63) to improve the folding of mutant protein and thus enhance the level of residual enzyme activity in patient cells. Both new therapeutic strategies have been investigated in Gaucher disease (64), and substrate reduction therapy is in clinical practice.

The complex pathology that can result from lysosomal storage makes it potentially difficult to treat all sites of pathology and particularly to clear residual pathology. Thus, in general, the earlier the treatment is implemented the more likely an effective therapeutic outcome will be achieved. The optimum treatment strategy to cure lysosomal storage disorders may be gene replacement therapy, but it is still in the developmental stages and will likely be a long way from widespread clinical practice. Nonetheless, therapeutic strategies that result in significant improvement in the quality of life for lysosomal storage disorder patients are in clinical practice.

## REFERENCES

1. Berkhan O. Zwei falle von skaphokephalie. Arch. Anthrop. 1907;34:8.

2. Hunter C. A rare disease in two brothers: Evaluation of scapala, limitation of movement of joints and other abnormalities. Proc. R. Soc. Med. 1917;10:104–116.

3. Hurler G. Ueber einen typ multiper abartungen, vor-weigend am skeletsystem. Z. Kinderheilk. 1919;24:220–234.

4. Whitley CB. The mucopolysaccharidoses. In: McKusick's Heritable Disorders of Connective Tissue, 5th edition. Beighton P, ed. 1993. Mosby Year Book Inc., St Louis, MO.

5. Hers HG. Alpha-glucosidase deficiency in generalized glycogen-storage disease (Pompe's Disease). Biochem. J. 1963;86:11.

6. Van Hoof F, Hers HG. L'ultrastructure des cellules hepatiques dans la maladie de Hurler (gargoylism). CR. Acad. Sci. 1964;259:1281–1283.

7. DeDuve C, Pressman BC, Gianetto R, Wattiaux R, Appelmans F. Tissue fractionation studies: 6. Intracellular distribution patterns of enzymes in rat-liver tissue. Biochem. J. 1955;60:604.

8. DeDuve C, Wattiaux R. Functions of lysosomes. Annu. Rev. Physiol. 1966; 28:435–492.

9. Hers HG, van Hoof F. The genetic pathology of lysosomes. Prog. Liver Dis. 1970; 3:185–205.

10. Hers HG, van Hoof F. Lysosomes and Storage Diseases. 1973. Academic Press, New York.

11. Meikle PJ, Hopwood JJ, Clague AE, Carey WC. Prevalence of lysosomal storage disorders. JAMA 1999;281:249–254.

12. Arvio M, Autio S, Louhiala P. Early clinical symptoms and incidence of aspartylglucosaminuria in Finland. Acta Paediatr. 1993;82:587–589.

13. Petersen GM, Rotter JI, Cantor RM, Field LL, Greenwald S, Lim JS. The Tay-Sachs gene in north American Jewish populations: geographic variations and origin. Am. J. Hum. Genet. 1983;35:1258–1269.

14. Scriver CR, Beaudet AL, Sly WS, Valle D, eds. Lysosomal enzymes. In: The Metabolic and Molecular Bases of Inherited Disease. 1995. McGraw-Hill, New York.

15. Novikoff AB, Beaufay H, DeDuve C. Electron microscopy of lysosome-rich fractions from rat liver. J. Biophys. Biochem. Cytol. 1956;179–184.

16. Storrie B. Assembly of lysosomes: perspectives from comparative molecular cell biology. Int. Rev. Cytol. 1988;111:53–105.

17. Kornfeld S, Mellman I. The biogenesis of lysosomes. Annu. Rev. Cell Biol. 1989;5:483–525.

18. Helenius A, Mellman I, Wall D, Hubbard A. Endosomes. Trends Biochem. Sci. 1983;245–250.

19. Griffiths G, Gruenberg J. The arguments for pre-existing early and late endosomes. Trends Cell Biol. 1991;1:5–9.

20. Geuze HJ, Slot JW, Strous GJ, Hasilik A, von Figura K. Possible pathways for lysosomal enzyme delivery. J. Cell Biol. 1985;101:2253–2262.

21. Saftig P. Lysosomal membrane proteins. In: Lysosomes. Saftig P, ed. 2005. Springer Science and Business Media, New York.

22. Landgrebe J, Lubke T. Lysosomal proteome and transcriptome. In: Lysosomes. Saftig P, ed. 2005. Springer Science and Business Media, New York.

23. Verheijen FW, Mancini GMS. Lysosomal transporters and associated diseases. In: Lysosomes. Saftig P, ed. 2005. Springer Science and Business Media, New York.

24. Fukuda M. Lysosomal membrane glycoproteins. Structure, biosynthesis, and intracellular trafficking. J. Biol. Chem. 1991;266:21327–21330.

25. Peters C, von Figura, K. Biogenesis of lysosomal membranes. FEBS Lett. 1994;346:108–114.

26. Bonifacino JS, Traub LM. Signals for sorting of transmembrane proteins to endosomes and lysosomes. Annu. Rev. Biochem. 2003;72:395–447.

27. Kuronita T, Eskelinen EL, Fujita H, Saftig P, Himeno M, Tanaka Y. A role for the lysosomal membrane protein LGP85 in the biogenesis and maintenance of endosomal and lysosomal morphology. J. Cell. Sci. 2002;115:4117–4131.

28. Eskelinen EL, Illert AL, Tanaka Y, Schwarzmann G, Blanz J, Von Figura K, Saftig P. Role of LAMP-2 in lysosome biogenesis and autophagy. Mol. Biol. Cell 2002;13:3355–3368.

29. Hannah MJ, Williams R, Kaur J, Hewlett LJ, Cutler DF. Biogenesis of Weibel-Palade bodies. Semin. Cell Dev. Biol. 2002;13:313–324.

30. Kolter T, Sandhoff, K. Sphingolipid metabolism diseases. Biochim. Biophys. Acta 2006;1758:2057–2079.

31. Sillence D. New insights into glycosphingolipid functions--storage, lipid rafts, and translocators. Int. Rev. Cytol. 2007;262:151–189.

32. Furst W, Sandhoff K. Activator proteins and topology of lysosomal sphingolipid catabolism. Biochim. Biophys. Acta 1992;1126:1–16.

33. Suzuki K. Biochemical pathogenesis of genetic leukodystrophies: comparison of metachromatic leukodystrophy and globoid cell leukodystrophy (Krabbe's disease). Neuropediatrics 1984;15:32–36.

34. Harzer K, Massenkeil G, Frohlich E. Concurrent increase of cholesterol, sphingomyelin and glucosylceramide in the spleen from non-neurologic Niemann–Pick type C patients but also patients possibly affected with other lipid trafficking disorders. FEBS Lett. 2003;537:177–181.

35. Goldberg D, Gabel C, Kornfeld S. In: Lysosomes in Biology and Pathology 7. Dingle JT, Dean RT, Sly WS, eds. 1984. North Holland Publishing, Amsterdam, The Netherlands.

36. von Figura K, Hasilik A. Lysosomal enzymes and their receptors. Annu. Rev. Biochem. 1986;55:167–193.

37. von Figura K. Molecular recognition and targeting of lysosomal proteins. Curr. Opin. Cell Biol. 1991;3:642–646.

38. Kornfeld S. Structure and function of the mannose 6-phosphate/insulinlike growth factor II receptors. Annu. Rev. Biochem. 1992;61:307–330.

39. Hille-Rehfeld A. Mannose 6-phosphate receptors in sorting and transport of lysosomal enzymes Biochim. Biophys. Acta 1995;1241:177–194.

40. Stein M, Meyer HE, Hasilik A, von Figura K. 46-kDa mannose 6-phosphate-specific receptor: purification, subunit composition, chemical modification. Biol. Chem. Hoppe-Seyler 1987;368:927–936.

41. Sandholzer U, von Figura K, Pohlmann R. Function and properties of chimeric MPR 46-MPR 300 mannose 6-phosphate receptors. J. Biol. Chem. 2000;275:14132–14138.

42. Sahagian GG, Neufeld EF. Biosynthesis and turnover of the mannose 6-phosphate receptor in cultured Chinese hamster ovary cells. J. Biol. Chem. 1983;258:7121–7128.

43. Bright NA, Reaves BJ, Mullock BM, Luzio JP. Dense core lysosomes can fuse with late endosomes and are re-formed from the resultant hybrid organelles. J. Cell Sci. 1997;110:2027–2040.

44. Luzio JP, Pryor PR, Bright NA. Lysosomes: fusion and function. Nat. Rev. Mol. Cell Biol. 2007;8:622–632.

45. Bache KG, Brech A, Mehlum A, Stenmark H. Hrs regulates multivesicular body formation via ESCRT recruitment to endosomes. J. Cell Biol. 2003;162:435–442.

46. Bonifacino JS, Glick BS. The mechanisms of vesicle budding and fusion. Cell 2004;116:153–166.

47. Bonifacino JS. The GGA proteins: adaptors on the move. Nat. Rev. Mol. Cell Biol. 2004;5:23–32.

48. Bonifacino JS, Lippincott-Schwartz J. Coat proteins: shaping membrane transport. Nat. Rev. Mol. Cell Biol. 2003;4:409–414.

49. LaPlante JM, Sun M, Falardeau J, Dai D, Brown EM, Slaugenhaupt SA, Vassilev PM. Lysosomal exocytosis is impaired in mucolipidosis type IV. Mol. Genet. Metab. 2006;89:339–348.

50. Lee JA, Beigneux A, Ahmad ST, Young SG, Gao FB. ESCRT-III dysfunction causes autophagosome accumulation and neurodegeneration. Curr. Biol. 2007;17:1561–1567.

51. Wraith JE. Clinical aspects and diagnosis. In: Lysosomal Disorders of the Brain. Platt FM, Walkley SU, eds. 2004. Oxford University Press, New York.

52. Applegarth DA, Dimmick JE, Hall JG. Organelle Diseases; Clinical Features, Diagnosis, Pathogenesis and Management. 1997. Chapman and Hall, London.

53. Platt FM, Walkley SU. Lysosomal defects and storage. In: Lysosomal Storage Disorders of the Brain. Platt FM, Walkley SU, eds. 2004. Oxford University Press, New York.

54. Walkley SU. Pathogenic cascades and brain dysfunction. In: Lysosomal Storage Disorders of the Brain. Platt FM, Walkley SU, eds. 2004. Oxford University Press, New York.

55. Fischer EG, Moore MJ, Lager DJ. Fabry disease: a morphologic study of 11 cases. Mod. Pathol. 2006;19:1295–1301.

56. Crawley AC, Gliddon BL, Auclair D, Brodie SL, Hirte C, King BM, Fuller M, Hemsley KM, Hopwood JJ. Characterization of a C57BL/6 congenic mouse strain of mucopolysaccharidosis type IIIA. Brain Res. 2006;1104:1–17.

57. Matzner U. Therapy of lysosomal storage disorders. In: Lysosomes. Saftig P, ed. 2005. Springer Science and Business Media, New York.

58. Wraith JE. Limitations of enzyme replacement therapy: current and future. J. Inherit. Metab. Dis. 2006;29:442–447.

59. Bruni S, Loschi L, Incerti C, Gabrielli O, Coppa GV. Update on treatment of lysosomal storage diseases. Acta Myol. 2007;26:87–92.

60. Vellodi A. Lysosomal storage disorders. Br. J. Haematol. 2005;128:413–431.

61. Beck M. New therapeutic options for lysosomal storage disorders: enzyme replacement, small molecules and gene therapy. Hum. Genet. 2007;121:1–22.

62. Pastores GM, Arn P, Beck M, Clarke LA, Guffon N, Kaplan P, Meunzer J, Norato DY, Shapiro E, Thomas J, Viskochi D, Wraithe JE. The MPS I registry: design, methodology, and early findings of a global disease registry for monitoring patients with mucopolysaccharidosis type I. Mol. Genet. Metab. 2007;91:37–47.

63. Brooks DA. Getting into the fold. Nat. Chem. Biol. 2007;3:84–85.

64. Yu Z, Sawkar AR, Kelly JW. Pharmacologic chaperoning as a strategy to treat Gaucher disease. FEBS J. 2007;274:4944–4950.

## FURTHER READING

Fuller M, Meikle PJ, Hopwood JJ. Epidemiology of lysosomal storage diseases: an overview. In: Beck M, Mehta A, Sunder-Plassmann G, Widmer U, eds. 2006. Oxford PharmaGenesis Ltd. Oxford, UK.

Swiedler SJ, Beck M, Bajbouj M, Giugliani R, Schwartz I, Harmatz P, Wraith JE, Roberts J, Ketteridge D, Hopwood JJ, Guffon N, Sa Miranda MC, Teles EL, Berger KI, Piscia-Nichols C. Threshold effect of urinary glycosaminoglycans and the walk test as indicators of disease progression in a survey of subjects with Mucopolysaccharidosis VI (Maroteaux-Lamy syndrome). Am. J. Med. Genet. 2005;134:144–150.

Tollersrud OK, Berg T. Lysosomal storage disorders. In: Lysosomes. Saftig P, ed. 2005. Springer Science and Business Media, New York.

Wraith JE, Hopwood JJ, Fuller M, Meikle PJ, Brooks DA. Laronidase treatment of mucopolysaccharidosis I. BioDrugs 2005;19:1–7.

Wraith JE. Lysosomal disorders. Semin. Neonatol. 2002;7:75–83.

# INDEX

Abbreviations, 438–439

Aβ aggregation, 207. *See also* Amyloid-β (Aβ)

Aβ assembly
inhibitors of, 390–396
small molecule inhibitors of, 389

Aβ fibrils, structure of, 387, 388

Aβ oligomers, 382

Absorption. *See also* ADME (absorption, distribution, metabolism, excretion) properties
carrier-mediated, 151
transport mechanisms for, 105

Absorption and transit (ACAT) model, 32

Absorption phase, 85

ABT594, 176–177

Academic screening efforts, 7

ACAT model, 32

"Access" protomer, 123, 125

Accumulation Index ($R$), 87. *See also* $R$-values

Acetaminophen, in treating osteoarthritis, 311

Acetylcholine, 372

Acetyl-CoA, pyruvate conversion to, 447

Acquired immunodeficiency syndrome (AIDS), 331. *See also* Human immunodeficiency retrovirus (HIV)

drug cocktails for, 336

AcrAB efflux system, 127

AcrB–drug complex structure, 125

AcrB transporter, 120, 121, 123–125
transport mechanism of, 124

Actinomycin family, 180

Active compounds, docking, 50

Active molecules, *de novo* design and, 54

Active-site specific chaperones, 407

Acute dystonia, 371

Acyclovir, 175

ADAMTS family, 301
in treating osteoarthritis, 313

Additive cardiometabolic risk factors, Metabolic Syndrome *vs.*, 424–425

Adhesion molecules, in asthma pathology, 349

Adipokines, 296

Adiponectin, 296

ADME (absorption, distribution, metabolism, excretion) properties, of drugs, 101–114

ADME pharmacokinetic processes, 152

ADME process, 148–151
role of transporters in, 151

ADMET analysis, 30

ADME/T optimization, 31–35

*Chemical Biology: Approaches to Drug Discovery and Development to Targeting Disease*, First Edition.
Edited by Natanya Civjan.
© 2012 John Wiley & Sons, Inc. Published 2012 by John Wiley & Sons, Inc.